ALLOYING

Edited by
John L. Walter
Melvin R. Jackson
Chester T. Sims

Metals Park, Ohio 44073

Library of Congress Catalog Card Number: 88-70119
ISBN: 0-87170-326-2
SAN: 204-7586

Editorial and production coordination by
Carnes Publication Services, Inc.

PRINTED IN THE UNITED STATES OF AMERICA

Foreword

Development and design of alloys are driven by the need for materials whose properties allow better performance or lower cost, or both. The pace of such development is generally quite gradual, except for occasional perturbations. The present frenzied effort on oxide superconductors, for example, has resulted in such a perturbation in activity. Concerns of a few years back regarding availability of critical or strategic materials, particularly related to the element cobalt, were also responsible for a perturbation in activity driven by a sudden increase in cost of commercial materials. These concerns had an impact on the development of nickel-base superalloys, as well as on materials for cutting tools.

As of this writing, another perturbation has been under way for the past two or three years, as major efforts have been initiated on developing materials for an advanced aircraft engine and for the aerospace plane concepts. Researchers of these systems have considered aluminum, titanium, nickel, niobium, and copper-based alloys and metal-matrix composites; intermetallic compounds such as Ni_3Al, NiAl, Ti_3Al and TiAl; carbon-carbon materials and schemes to protect those materials from oxidation; and oxide-, nitride- or carbide-based ceramics and ceramic-matrix composites.

These advanced propulsion systems have generated substantial new efforts not only in the practice of alloy development, but also in the theory of alloy design. In early 1986, Chester Sims (now of Rensselaer Polytechnic Institute) and John Walter discussed the value of a lecture series on alloy theory and practice to serve as a refresher and update. Out of those discussions grew a program of seminars held at roughly one-month intervals at General Electric Company's Corporate Research and Development. The program was open to the public, with attendees coming from neighboring universities and government facilities, as well as from GE. The program was enthusiastically sponsored by Harvey Schadler, Manager of the Materials Research Laboratory at GE CRD.

The list of speakers focusing on the theory of alloy design included Leo Brewer and Bill Pearson, who have worked in developing foundations to understanding materials since the 1940's. A number of other researchers reviewed their own excellent work in improving earlier theories or in establishing new and provocative approaches, including Phil Clapp, Sam Faulkner, Dick Messmer, Juan Sanchez and Ulrich Gösele.

Presentations in the lecture series dealing with the practice of alloy design and development covered a broad array of materials, extending beyond aircraft-engine alloys. Ed Starke covered aluminum, Harry Paxton treated steels, and Bill Reed talked on stainless steels. Nickel alloys were discussed by Norm Stoloff, refractory metals were described by Bill Buckman, and titanium-based systems were detailed by Ted Collings. The field of alloying in ceramics was considered to be too diverse for in-depth coverage. However, Ken Jack illustrated the development of a specific family of ceramics, the sialons, as representative of the practice of alloying in ceramics.

The series could have been extended further, covering other topics related to both theory and practice; there have been subsequent speakers whose subjects would easily fit in the series. However, when ASM International expressed an interest in publishing the collection of presentations in a single volume, the scope of the collection had to be limited for timely publication. Every speaker in our original program was willing to prepare a manuscript for the book.

We hope the format of this unique collection proves to be of use to researchers, both present and future, in the theory and practice of alloy design and development. Meeting the authors and discussing material interests was a very enjoyable exercise for the editors. We thank the authors for their contributions to both the lecture series and this publication.

John L. Walker
Melvin R. Jackson
Chester T. Sims

Contents

1

Chemical Bonding Theory Applied to Metals

LEO BREWER
Lawrence Berkeley Laboratory and Department of Chemistry
University of California

Characterizing the phases and phase boundaries of a typical binary phase diagram can require preparation and study over a range of temperatures of several dozen compositions. A study of the combinations of just the 30 most common metals would require 435 binary diagrams. In a ternary diagram each of the tie-lines would require as much work as a binary diagram, one would have to evaluate a number of tie-lines, and there would be 12,180 ternary diagrams. As one increased the number of components, the number of combinations would surpass two billion. It is essential to have predictive models to predict which of the astronomical number of diagrams might produce the materials with a given set of desired properties.

The variation of properties of elements or compounds across the Periodic Table has been commonly used for interpolation or extrapolation of properties on the assumption that properties will vary smoothly and consistently with change of atomic number, but there are many exceptions. For example, the enthalpies of formation of the solid alkali iodides become more negative as one goes from lithium to cesium. However, just the opposite trend is found for the fluorides. This can be resolved when it is recognized that there can be several factors that contribute to the property that may vary in different ways with atomic number. The difficulty with alkali halides can be resolved by the Born-Haber model that expresses the enthalpy of formation in terms of two separate steps. The first step is the enthalpy of formation of the gaseous ions, and the second is the lattice energy of the solids relative to the gaseous ions. These two contributions vary in opposite directions as one goes from lithium to cesium, with the variation of lattice energies predominating for the fluorides and the variation of enthalpies of formation of the gaseous cations predominating for the iodides. A similar valence-bond model will be presented here to separate the important factors that contribute to bonding of metals.

Electrons are the universal glue for all materials. An understanding of the way they operate allows one to understand the variations of the properties of materials and to predict the properties of materials that have not been studied. The chemical bonding description of the interactions of electrons with one another and with the nuclei of atoms is a general model that is equally applicable to metallic and nonmetallic materials. The key feature is the recognition that electrons do not have free choice of their distribution about an atomic nucleus. Quantum mechanics describes the restriction of electrons in space in terms of orbitals. In the mode that will be used, the important characterizations of these orbitals are the values of the total quantum number, n, and the values of the angular momentum quantum number, l, that distinguishes the subshell orbitals of each main shell. For values of l ranging from zero to three, the electrons will be designated as s, p, d, and f electrons, respectively. The maximum value of l in any main shell is $n - 1$. For each value of l, there are $2l + 1$ subshells. Thus there is one s orbital, three p orbitals, five d orbitals, and seven f orbitals in the $n = 4$ main shell.

VALENCE-BOND MODEL

In the valence-bond model, the cohesion of atoms is ascribed to overlapping of atomic orbitals to allow the electrons to occupy bonding orbitals in which they can interact with two or more nuclei and hold them together. The Pauli exclusion principle establishes that filled orbitals will repel one another and will not form chemical bonds. Compilations[(1)] of the energy levels of the atomic elements and their electronic configurations are available, and with them one can predict the bonding capabilities of different electronic configurations. For most atoms with two or more valence electrons, the electrons beyond the atomic core of filled electron shells, the electronic configuration of the ground state or lowest-energy state of the atom has the outer s orbital filled with a nonbonding pair of electrons. For effective bonding in the solid, it is necessary to promote to various excited electronic configurations to be able to make effective use of the valence electrons in bonding. This difference between the electronic configuration of the ground state of the gaseous atom and the electronic configurations in the solid metal is the origin of the irregular variation across the Periodic Table of boiling points and enthalpies of sublimation of the metals. In a manner similar to that of the Born-Haber model, the enthalpy of sublimation is separated into (1) the enthalpy of promotion of the gaseous atom from the ground electronic configuration to the electronic configuration corresponding to that in the metal and (2) the bonding enthalpy of the promoted electronic state upon formation of the solid.

BOND ORDER

The bonding effectiveness of a given electronic configuration depends on the number of unpaired electrons available, the types of orbitals in which the electrons reside, and the bond order, or the number of electron pairs used in each bond, when the atoms are brought together. The bonding capability of the electrons in inner-shell d and f orbitals will be considered when the transition metals and lanthanides and actinides are discussed. The effect of varying the number of outer-shell s and p electrons and the order of bonding is illustrated in Table 1, where bonding enthalpies at 0 K are given for the elements from sodium to argon, first brought together to form diatomic molecules(2) and then condensed to the solid.(3) In the diatomic molecules, all of the bonding electrons are concentrated in a single bond and the bond order can vary from one to three. Pauling(4) demonstrated many years ago that the effectiveness of the bonding electrons depends on how crowded they are in the bond. In contrast to the high bond orders in the diatomic molecules, the bond order in solid sodium is only 1/8 because the single electron pair is spread over bonds to eight nearest neighbors. The bond order increases to 1/6 for magnesium metal and to 1/4 for aluminum metal. These low orders for the metals greatly increase the bonding strength compared with the diatomic molecules. As Table 1 indicates, the bonding of the diatomic molecules is due only to the unpaired electrons in the ground-state configuration of the gaseous atom. If an excited electronic configuration involving the promotion of an s electron to a p orbital were used, more electrons would be available for bonding. However, the promotion energy required would be larger than the additional bonding in the high-order bonds. For the solids with much lower bond orders, the extra bonding achieved by promotion of an s electron to a p orbital provides two additional bonding electrons that provide more than enough additional bonding energy to offset the promotion energy required. Even silicon, which has a bond order of one in the diamond structure, gains sufficient bonding energy to promote from the s^2p^2 to the sp^3 configuration that allows use of all four valence electrons in bonding. In diatomic Si_2, the additional bonding would not offset the promotion energy. Similarly for aluminum, promotion from s^2p to sp^2 does not occur for Al_2 because the triple bond does not have a high enough bonding energy. In the solid, with a bond order of 1/4, the use of all three valence electrons is achieved.

Table 1. Bonding in diatomics compared with solids

Gaseous atom	Na	Mg	Al	Si	P	S	Cl	Ar
Ground-state configuration	3s	$3s^2$	$3s^23p$	$3s^23p^2$	$3s^23p^3$	$3s^23p^4$	$3s^23p^5$	$3s^23p^6$
Bonding electrons per atom (e/a)	1	0	1	2	3	2	1	0
$\frac{1}{2}M_2(g) = M(g)$ $\Delta H_0^\circ/R$, kK	4.2	0.3	10	18.7	29.2	25.4	14.4	0.06
Electronic configuration of solid	3s	3s3p	$3s3p^2$	$3s3p^3$	$3s^23p^3$	$3s^23p^4$	$3s^23p^5$	$3s^23p^6$
Bonding e/a	1	2	3	4	3	2	1	0
$M(s) = M(g)$ $\Delta H_0^\circ/R$, kK	12.9	17.6	39.3	53.7	39.7	33.1	16.2	0.9
Structure	bcc	hcp	ccp	diamond	↖	$<$	—	

For the diatomic molecules, the increase in the number of electrons from sodium to magnesium greatly reduces the bonding in the diatomic because the electrons in the filled s subshell repel one another. The 3s electrons are used for bonding only in Na_2 and are nonbonding for Mg_2 to Ar_2. However, all of the p electrons contribute to bonding as long as the three p orbitals in the atoms do not contain more than one electron per orbital. Thus there is a maximum in stability for P_2 with a reduction in bond strength as each additional electron is added, resulting in a nonbonding pair. The order of the bonds or the number of electron pairs formed in the bond is high for diatomic molecules, varying from one in Na_2 or Cl_2 to a maximum of three with the triple bond of P_2. For the solids, the maximum stability is at silicon with four bonding electrons after promotion. For either the diatomics or the solid, the bonding enthalpies on either side of the maximum are higher on the right-hand side because of the increase of nuclear charge that contracts the atomic core of filled electron shells. The bonding electrons are thus closer to the nuclear charges and tie the atoms together more strongly.

Figure 1 illustrates the role of promotion for magnesium. The spectroscopic data[1] for magnesium show that the first excited state is the $^3P_0$ state of the 3s3p configuration, and the right-hand side of Fig. 1 shows its enthalpy relative to the ground $^1S_0$ state. The abscissa of Fig. 1 represents the average internuclear distance between the magnesium atoms in a mole of magnesium gas at various pressures. As the gas is compressed, the atoms in the s^2 $^1S_0$ ground state repel other magnesium atoms, increasing the energy. For the sp excited state, both orbitals may overlap and form bonding orbitals with neighboring atoms with the sp configuration, and their energy drops. When compression has reached the density of the metal phase, the bonding between the sp atoms has reduced their enthalpy far below that of atoms in the s^2 configuration, and all of the atoms in the solid have been promoted to the sp configuration. Because the bond order is only 1/6, the

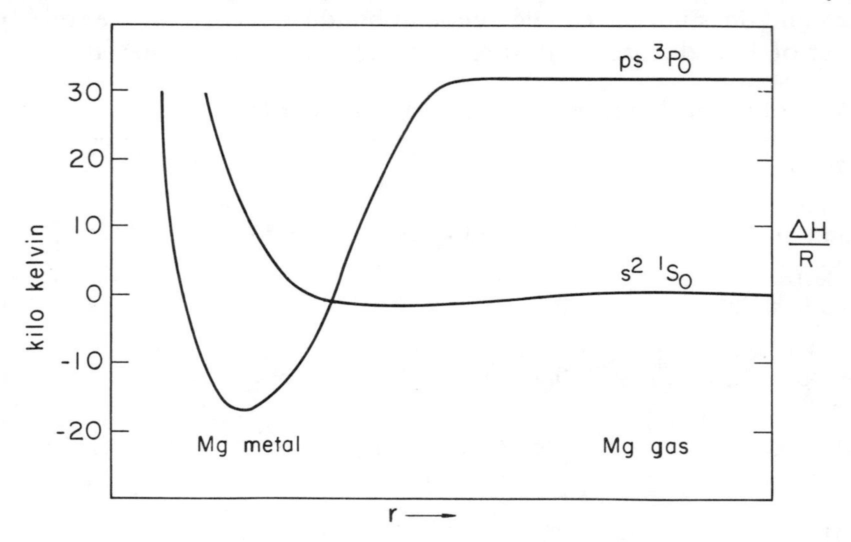

Fig. 1. Energies of s^2 and ps configurations of magnesium as functions of internuclear distance

atoms are strongly bound. In Mg_2, where promotion would result in a bond order of two, the bonding is not sufficient to offset the promotion energy. If magnesium gas is not compressed sufficiently to allow for promotion, the magnesium atoms with s^2 configuration will behave like a noble gas. It has been demonstrated[5] that magnesium gas is a permanent gas at room temperature in a container coated with a hydrocarbon. A nucleus of a substantial number of magnesium atoms is required for sufficient bonding to offset the promotion energy.

Table 1 and Fig. 1 illustrate the following five important features of chemical bonding.

1. Electrons are the glue that binds nuclei together.
2. Electrons are restricted to orbitals in physical space.
3. An orbital filled with a pair of electrons repels a similar filled orbital of another atom. Such electrons are not available for bonding.
4. Bonding requires overlapping of orbitals to allow concentration of the electrons between the nuclei.
5. Elements with two or more valence electrons usually have a ground electronic state with a pair of electrons in an s orbital, and promotion of an s electron to another orbital is required for use of those electrons in bonding between the elemental atoms in a solid.

PROMOTION TO THE VALENCE STATE

To determine which valence states predominate in the solid, one has to examine the spectroscopic data for the gaseous atoms.[1] The right-hand side of Fig. 2 illustrates the energies of the lowest states of the three lowest electronic configurations of tungsten gas. Bonding energies per electron are available[6] for the 6s, 6p, and 5d electrons to provide the bonding energies of each configuration. The ground-state configuration, d^4s^2, has only four bonding electrons as the filled s orbital is

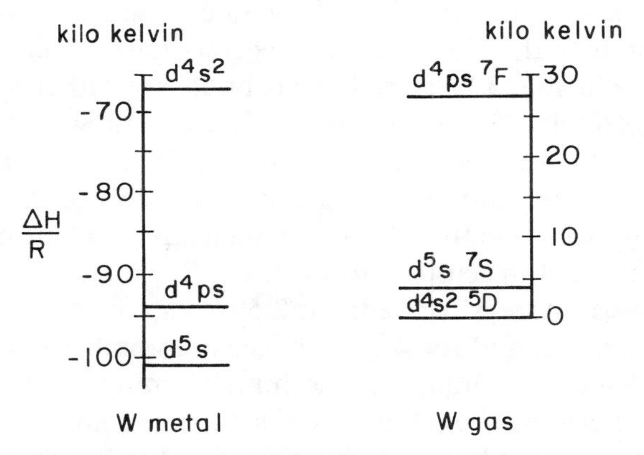

Fig. 2. Energies of d^4s^2, d^5s, and d^4sp configurations of tungsten for metallic internuclear distance and for the isolated atom

nonbonding. If the gaseous atoms were compressed as in the magnesium example of Fig. 1, the $5d^4 6s^2$ configuration would not drop in energy as much as the $5d^5 6s$ and $5d^4 6s6p$ configurations, which have six bonding electrons per atom. The left side of Fig. 2 shows the resulting energies relative to zero energy for the gaseous ground state. The outer-shell 6s and 6p orbitals are more effective in overlapping with the corresponding orbitals of approaching atoms than are the somewhat localized 5d orbitals of the inner shell. Tungsten in the d^4sp valence state will bond more strongly and decrease in energy more than in the d^5s valence state, but the difference is smaller than the larger promotion energy and the predominant configuration in solid tungsten is the d^5s configuration. In Fig. 1 and 2, the enthalpy of sublimation of the solid metal is given by the difference between the enthalpy level of the ground state of the gas and the level corresponding to the condensed solid, because the measured enthalpy of sublimation corresponds to the gas vaporizing in the ground state and not in the valence state.

RELATION BETWEEN ELECTRONIC CONFIGURATION AND CRYSTAL STRUCTURE

The combination of spectroscopic data with bonding enthalpies allows one to calculate properties such as the enthalpy of sublimation. It also provides prediction of the crystal structure of the solid. Table 1 shows that each element from sodium to chlorine has a distinctive electronic configuration ranging from s and sp on the left to s^2p^5 on the right, and each has a distinctive crystal structure, fixed by the electronic configuration. Hume-Rothery,[(7)] over 50 years ago, was able to extend this correlation between electronic configuration and crystal structure of the pure elements to metallic alloys. He noted that many of the nontransition intermetallic compounds with the same average number of valence electrons per atom had the same crystal structure. He was not able to extend this correlation to the transition metals because the use of the total number of valence electrons including the inner-shell d electrons did not fit. Engel,[(8)] who combined Pauling's[(4)] description of metallic bonding with the Hume-Rothery correlation, recognized that the d electrons should not be included because their orbitals are rather localized compared to the outer-shell orbitals. The outer s and p electrons range far out into the metal and can control the long-range order. The d electrons bond primarily to nearest neighbors and add to the stability of the solid, but do not directly influence the structure. For alloys with less than 1.5 outer-shell s,p electrons per atom (e/a), the body-centered cubic (bcc) structure is formed. For 1.7 to 2.1 s,p e/a, the hexagonal close-packed (hcp) structure is formed; for 2.5 to 3 s,p e/a, the cubic close-packed (ccp) structure is formed; and for 4 s,p e/a, the diamond structure is formed. For the sequence of elements sodium, magnesium, aluminum, and silicon with only outer-shell s,p valence electrons, the crystal structure changes from bcc to hcp to ccp to diamond with each addition of an electron. The sequence of structures for the solid transition metals near the melting point is given below. The two regions that are enclosed in boxes have other structures at lower temperatures that will be discussed shortly. Including the alkali metals, all 18 metals of the first six groups show the bcc structure.

II	III	IV	V	VI	VII	VIII	IX	X	XI	
bcc	bcc	bcc	bcc	bcc	bcc	bcc	ccp	ccp	ccp	3d
bcc	bcc	bcc	bcc	bcc	hcp	hcp	ccp	ccp	ccp	4d
bcc	bcc	bcc	bcc	bcc	hcp	hcp	ccp	ccp	ccp	5d

In contrast to the abrupt change in structure of the nontransition elements with each addition of an electron, the transition metals show wide ranges of stability for each structure. Engel recognized that the availability of vacant d orbitals drains off electrons from the outer-shell orbitals and delays the buildup of electrons in the p orbitals. For the 4d and 5d transition series, there is an abrupt transition from bcc to hcp in going from molybdenum and tungsten with six electrons per atom to technetium and rhenium with seven electrons per atom. If the seventh electron were added to the d orbitals, a pair of nonbonding electrons would form, leaving only five bonding electrons. The configuration d^5sp with seven bonding electrons yields an hcp solid that is more stable than the bcc d^6s solid even though a promotion-energy penalty must be paid. As additional electrons are added for the elements to the right of technetium and rhenium, they will be distributed between the d orbitals with loss of bonding ability and the p orbitals for which a high promotion energy is required. Some of the loss of bonding ability with more than five d electrons is offset by promotion to the range $d^{n-2.5}sp^{1.5}$ to $d^{n-3}sp^2$, where n is the total number of valence electrons. This results in a wide range of ccp structures from manganese to copper, rhodium to silver, and iridium to gold.

If the spectroscopic data[(1)] are examined, one finds for many metals that only one configuration is important, as for the example of tungsten in Fig. 2. For some metals, there are two configurations of comparable stability in the solid state, and rarely there are more than two. Some examples are given here.

Single Configuration	Two Configurations
$d^{n-1}s$ V, Nb, Ta, Cr, Mo, W $d^{n-2}sp$ Tc, Re	$d^{n-1}s$ and $d^{n-2}sp$ Sc, Y, Ti, Zr, Hf $d^{n-2}sp$ and $d^{n-3}sp^2$ Co
Three Configurations	**Six Configurations**
$d^{n-1}s$, $d^{n-2}sp$, $d^{n-3}sp^2$ Mn, Fe	f^5d^2s, f^6sp, f^5dsp, f^6ds, f^5sp^2, f^4d^3s Pu

PROMOTION ENERGIES TO VALENCE STATES

Figure 3 compares the relative energies of the electronic states of the $d^{n-1}s$ and $d^{n-2}sp$ configurations of the gaseous atoms for the 4d elements from strontium to molybdenum. At the far left of the Periodic Table, the energies of the d orbitals are high. As the nuclear charge is increased upon going to the right, the inner-

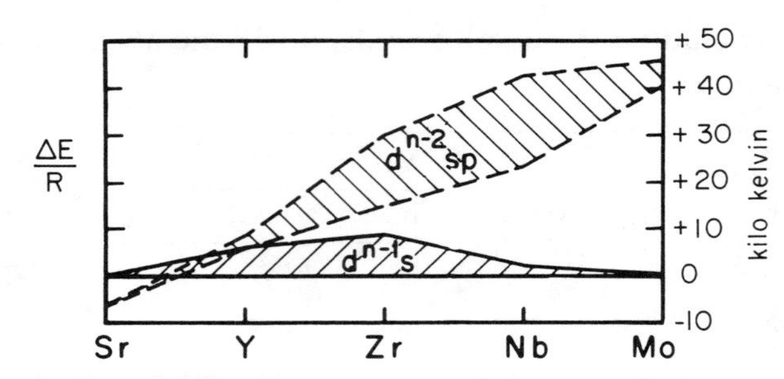

Fig. 3. Relative energies of $d^{n-1}s$ and $d^{n-2}sp$ configurations for gaseous atoms of the 4d transition series

shell d orbitals are contracted and stabilized relative to the outer-shell p orbital. In Fig. 3, with the energy of the lowest level of the $d^{n-1}s$ configuration at zero, the energies of the $d^{n-2}sp$ states increase with increasing nuclear charge. Although the bonding ability of the configuration with more outer-shell electrons is higher, its promotion energy for niobium and molybdenum is too large for it to compete with the $d^{n-1}s$ configuration, as shown in Fig. 2 for tungsten. Similar behavior is found for all the 3d and 5d elements of Groups V and VI. For the metals of Group IV, the promotion energy of d^2sp compared with d^3s has been reduced so that both configurations are of comparable stability in the solid. The d^2sp configuration, which yields the hcp structure, is lower in energy, but the bcc structure with a coordination number of only eight has lower vibrational frequencies and a higher heat capacity and entropy. This results in a transformation from hcp to bcc on heating. The same is true for yttrium. Table 2 shows the complete list of structures given in order of temperature stability when there is more than one structure.

When the orbitals are partially empty, there are several possible distributions of the electrons among the orbitals and there will be, in general, several energy states corresponding to a given configuration. This is illustrated in Fig. 3 by representing the high multiplicity levels of each configuration as a band. The spectra of the elements give us the energy differences between states of the various electronic configurations. Many of the states can be involved in bonding, and the energy of each configuration is often described in terms of an average over the states of a configuration. It has been shown[(6)] that a simpler and equally accurate procedure is to use the energy of the lowest state of a configuration to characterize the bonding energy of the configuration. This procedure will be followed in the discussion of the bonding abilities of different configurations.

A striking illustration of the role of promotion energies in understanding the properties of metals is given by the lanthanide metals. The melting point drops from lanthanum to cerium and then rises with increasing nuclear charge from cerium to samarium and from gadolinium to thulium with drops in melting point at europium and ytterbium and finally the highest melting point at lutetium. The boiling points show a steady drop from lanthanum to europium with a jump at gadolinium followed by another drop with increasing nuclear charge to ytterbium and finally a large jump to lutetium. The compression of the filled 5s5p shells with increasing nuclear charge would be expected to increase the effectiveness of bond-

Table 2. Crystal structures of the elements(a)

Li I												Be I II	B * *	C IV =(f)	N ≡(b)	O =(b)	F —(c)
Na I												Mg II	Al III	Si IV	P ↖ (e)	S < (d)	Cl —
K I	Ca I III	Sc I II	Ti I II	V I	Cr I	Mn I III β α	Fe I III I	Co III II	Ni III	Cu III	Zn II	Ga AII	Ge IV	As ↖	Se <	Br —
Rb I	Sr I III	Y I II	Zr I II	Nb I	Mo I	Tc II	Ru II	Rh III	Pd III	Ag III	Cd II	In (III)(g)	Sn tI4 IV	Sb ↖	Te <	I —
Cs I	Ba I	La I II III	Hf I II	Ta I	W I	Re II	Os II	Ir III	Pt III	Au III	Hg hRl	Tl I II	Pb III	Bi ↖	Po	At
Fr	Ra I	Ac III														

Ce I II III	Pr I II	Nd I II	Pm	Sm I (II)(g)	Eu I	Gd I II	Tb I II	Dy I II	Ho II	Er II	Tm II	Yb I III	Lu II
Th I III	Pa III tI2	U I σ oC4	Np I tP4 oP8	Pu I * III ***	Am III? II	Cm ? II	Bk ? II	Cf ? II	Es III?	Fm	Md	No	Lr

(a) I: Body-centered cubic, cI2. II: Hexagonal close-packed, hP2 and hP4. III: Cubic close-packed, cF4. IV: Diamond, cF8. Asterisk denotes complex structure. The structures are listed in order of temperature stability from room temperature up to the MP except for metastable diamond. (b) Diatomic molecules with double or triple bonds. (c) Diatomic molecule with a single electron-pair bond. (d) Atoms which form two single bonds per atom to form rings or infinite chains. (e) Three single bonds per atom, corresponding to a puckered planar structure. (f) The graphite structure where one resonance form consists of two single bonds and one double bond per atom. (g) Parentheses indicate slight distortions.

ing. This results in an increase in melting point with increase of nuclear charge as is observed except for europium and ytterbium. If one examines the various electronic configurations that are available for the lanthanides,[9,10,11] one finds that the $4f^{n-3}5d^26s$ and $4f^{n-3}5d6s6p$ configurations are the most important configurations for most of the lanthanides. However, as the nuclear charge increases from lanthanum to europium, the configuration $f^{n-2}s^2$ drops in energy and becomes the ground electronic state of gaseous atoms. The promotion of two electrons, one from the f orbital and one from the s orbital, is needed to make three electrons available for bonding. The additional bonding offsets the promotion energy for all of the lanthanides except for europium with a half-filled f shell and ytterbium with a filled f shell. For those two lanthanides, promotion of only one electron from the filled s orbital to a d or p orbital is possible, and these two metals have melting and boiling points comparable to those of the alkaline earth metals. The other lanthanides use three electrons in bonding, but most vaporize to a gaseous atom in a $f^{n-2}s^2$ ground state. Even though the bonding of the three electrons increases with increasing nuclear charge, resulting in increasing melting points, the vaporization is to a divalent ground state that drops in energy rapidly with increasing nuclear charge from lanthanum to europium and then again from gadolinium to ytterbium so that the boiling point drops. If, on the other hand, one calculates the bonding energy of the trivalent valence states in formation of the solid, one finds that the bonding does increase with increasing nuclear charge in agreement with the melting-point behavior. There are many other unusual properties of the lanthanides and actinides that are readily understood if one thinks of the bonding from the valence state and not from the divalent gaseous ground state.[9,10,12]

BONDING CONTRIBUTION PER ELECTRON

It is necessary now to deal quantitatively with the difference of bonding abilities of inner-shell d electrons and outer-shell s,p electrons. For all of the transition metals except for rhodium and palladium, atomic spectroscopy[1] has yielded complete data on the relative energies of the $d^{n-2}s^2$, $d^{n-2}sp$, and $d^{n-1}s$ configurations. From these data, the enthalpies of sublimation[3] can be combined with the promotion energy to the lowest level of the $d^{n-1}s$ configuration for bcc solids or to the $d^{n-2}sp$ configuration for the hcp solids to obtain the total bonding enthalpy of the valence state. From the variation of bonding energy of nontransition elements, it is clear that the s,p bonding is a function of internuclear distance, with bonding ability decreasing as one goes down in the Periodic Table because of the increasing size of the filled shells and, thus, increasing internuclear distance. The bonding contributions of the s,p electrons vary smoothly with atomic size, as plotted as a function of nuclear charge in Fig. 4 for the 5s,p and 4s,p electrons of the elements listed on top of each figure. Subtraction of the bonding contribution per s,p electron times the number of bonding s,p electrons yields the bonding contributions of the d electrons. This value divided by the number of bonding d electrons is plotted in Fig. 4. As noted above, the d orbitals are expanded at the

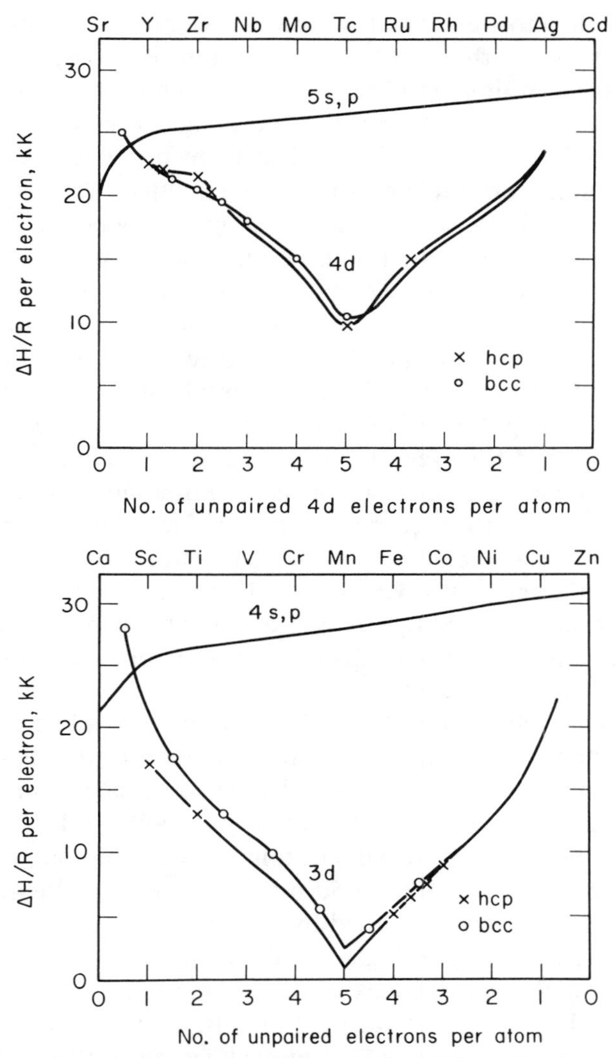

(Above) In upper curve, bonding enthalpy of 5s or 5p electrons is plotted for each element. Bottom curve plots bonding enthalpy of 4d electrons against number of unpaired d electrons in the valence state. (Below) Same information for 4s,p electrons and for 3d electrons.

Fig. 4. Valence-state bonding enthalpy per unpaired electron

left-hand side of the Periodic Table and can bond effectively. However, as one moves from left to right, the increasing nuclear charge contracts the d orbitals and reduces their ability to overlap with neighboring orbitals, which is necessary to provide a sufficient electron density between the atoms to attract both nuclei. Also, in addition to poorer bonding due to higher bond order, there is a crystal field effect in the solid that distorts d orbitals so that some are more extended than the average and some are more contracted. When there are only a few d electrons being used, they occupy the most extended d orbitals to achieve the best overlap and to reduce the Gibbs energy of the system to a minimum. However, as one approaches five d electrons per atom, it is necessary to use the poorly bonding orbitals as well as the more extended ones, and the average bonding energy per electron drops off rapidly. As additional electrons are added to form nonbonding pairs, these will occupy most contracted orbitals and the more extended orbitals will be used for bonding. As shown in Fig. 4, the average bonding energy per electron increases as the number of d orbitals used for bonding decreases from five to one.

Another very important property of the d orbitals is illustrated in Fig. 4, which shows how poorly bonding the 3d orbitals are in comparison with the 4d orbitals. As one goes from chromium to molybdenum to tungsten, the d orbitals become much more extended relative to the filled s,p orbitals of the same shell. The s electrons have no orbital angular momentum and can be described as moving right through the nucleus, and thus feel the increased nuclear charge strongly. In contrast, the d electrons could be described as moving in almost circular orbits. As the nuclear charge is increased from chromium to tungsten, the s and p filled subshells do not expand as much as the 5d orbital, which can thus overlap more effectively with d orbitals of adjacent atoms. One can understand why the melting and boiling points and other properties fixed by strength of bonding increase for transition metals as one goes down in the Periodic Table, while the opposite trend exists for nontransition metals.

As one moves from left to right through the 3d metals, some of the 3d orbitals become so contracted that there is no interaction with adjoining atoms and electrons can remain unpaired as in the gas and yield ferromagnetism. The 4d and 5d orbitals are so much more extended that all orbitals will overlap with neighboring orbitals to form bonding pairs or will contain nonbonding pairs, and ferromagnetism is not observed for the 4d and 5d metals. However, the 4f orbitals of the lanthanides, which are largely contained by the filled 4d, 4s, 4p, 5s, and 5p shells, are contracted enough to contain single unpaired electrons. In moving down to the actinides with 5f orbitals, they have expanded more than the closed 6s,p subshells, and the early actinides, as for the early 3d metals, do not show ferromagnetism. With increasing nuclear charge, the 5f orbitals become contracted enough to display magnetism.

EFFECT OF PRESSURE ON TRANSITION-METAL STRUCTURES

The poor overlapping of d orbitals compared with the outer-shell electrons allows one to make clear predictions of the effect of pressure on the stability of competing structures and electronic configurations. As Table 3 illustrates, compression of the solid by pressure improves the d bonding and favors the structure with the most bonding d electrons. For example, if one has titanium, zirconium, or hafnium at the temperature where the bcc (d^3s) structure is in equilibrium with the hcp (d^2sp) structure, one predicts that the bcc structure with the greater number of bonding

Table 3. Effect of pressure on phase stability

6 e/a	Bonding d electrons	8 e/a	Bonding d electrons
bcc . . . d^5s	5	bcc . . . d^7s	3
hcp . . . d^4sp	4	hcp . . . d^6sp	4
ccp . . . d^3sp^2	3	ccp . . . d^5sp^2	5
Pressure stabilizes bcc over hcp; hcp over fcc.		Pressure stabilizes fcc over hcp; hcp over bcc.	

d electrons will be stabilized over a wider temperature range relative to the close-packed structure. It would be true also for Group V and Group VI metals that pressure would destabilize the close-packed structures relative to the bcc structure. On the other hand, for elements with more than five d electrons, such as iron, the bcc structure (d^7s) with the most d electrons would have fewer bonding d electrons than the hcp structure (d^6sp), and the prediction is that pressure destabilizes the bcc structure compared with the close-packed structure. The predictions would not apply to the far left of the Periodic Table where the d orbitals are almost as extended as the outer-shell orbitals; but for Groups IV through XI, the predictions have been 100% confirmed.[(13)]

EFFECTS OF SOLUTES ON TRANSITION-METAL STRUCTURES

The partially localized d orbitals do not overlap with the extended outer-shell orbitals as well as with other d orbitals. The consequences are illustrated by the example of the effects of various substitutional transition-metal solutes on the zirconium transition from bcc (d^3s) to hcp (d^2sp). Any transition metal with more than 2.5 bonding d electrons would find better accommodation for d electron bonding in the bcc phase with more bonding d electrons than in the hcp phase. This is illustrated in Fig. 5, where the number of d electrons in molybdenum, technetium, and rhodium valence states are compared with those of the two forms of zirconium. Although the three solutes have different structures, all stabilize the bcc structure over the hcp structure. Transition metals to the left with less than 2.5 bonding d electrons, or nontransition metals, would have just the opposite effect.

Similarly, one can predict that interstitial elements that do not interfere with the d or f bonding, and merely add to the s,p concentration, would favor hcp over bcc, and ccp over hcp. The recognition of the role of partial delocalization of d and f orbitals makes possible the prediction of many other properties. The relationships that have just been discussed can be summarized as follows:

1. Promotion that is required to increase the number of bonding electrons changes the proportion of inner-shell and outer-shell electrons.
2. Pressure favors inner-shell bonding over outer-shell bonding.
3. Substitutional nontransition-metal neighbors destabilize inner-shell bonding.

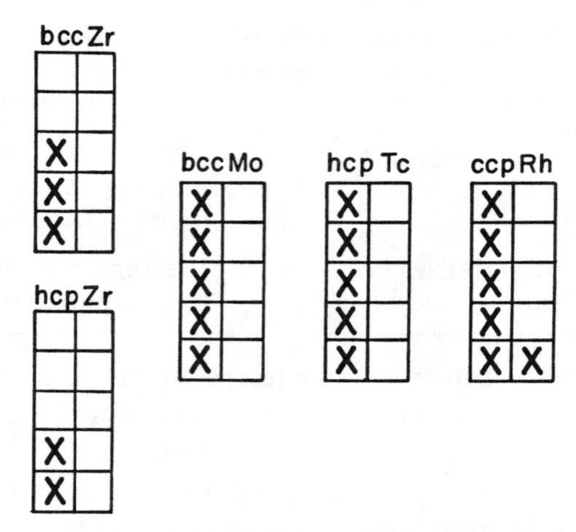

Fig. 5. Comparison of number of bonding d electrons of molybdenum, technetium, and rhodium with those of bcc and hcp zirconium

4. Substitutional transition-metal neighbors stabilize inner-shell bonding.
5. Interstitial elements such as boron, carbon, nitrogen, and oxygen merely add to the number of electrons per atom.
6. Magnetism arises from nonpaired localized electrons not used in bonding.
7. Inner-shell orbitals split with nonbonding electron pairs in the most localized orbitals.

TO WHAT EXTENT DO RIGHT-HAND TRANSITION METALS REQUIRE PROMOTION?

It was noted earlier that the Group V and Group VI transition metals have only one electronic configuration of importance in the solid state. The Group IV transition metals have two electronic configurations of comparable stability in the solid state. Other than the lanthanides and actinides that have been briefly discussed, all of the left-hand transition metals use all of their valence electrons in bonding. This continues with technetium and rhenium, but elements farther to the right cannot use all of their electrons in bonding. With six or more d electrons, the valence state will have one or more d orbitals filled with electron pairs. The electron pairs of filled orbitals repel one another as the atoms approach, and the electrons cannot concentrate between the nuclei to pull the nuclei together. The number of electron pairs in d orbitals can be reduced by promoting to p orbitals. For every d electron promoted, two bonding electrons become available. However, the bonding due to these electrons in d and p orbitals must be sufficient to offset the promotion energy required.

One can employ the promotion energies from spectroscopic measurements[1,9,10,11, 12] to determine possible valence states, as was done with the lanthanides. Spectroscopic data are lacking for the $d^{n-3}sp^2$ configuration for some of the metals at the right-hand side of the transition-metal group, and one must use other procedures to determine the degree of promotion for valence states corresponding to $d^{n-3}sp^2$ or mixtures of this configuration with $d^{n-2}sp$ yielding fractional average electron populations in a given orbital. The $d^{n-3}sp^2$ configuration is important because the ccp structure of manganese to copper, rhodium to silver, and iridium to gold will be assigned to configurations in the range $d^{n-2}sp^{1.5}$ to $d^{n-3}sp^2$ in subsequent discussion. To achieve these configurations, at least 0.5 d electrons must be promoted for manganese to nickel and iridium, and at least 1.5 d electrons must be promoted for rhodium, platinum, and copper to gold. Palladium, with a $4d^{10}$ ground atomic configuration, would have only weak van der Waals bonding (as for the noble gases) if it did not promote d electrons to s and p orbitals. It must promote at least 2.5 d electrons to achieve the $d^{7.5}sp^{1.5}$ to d^7sp^2 range that is correlated to the ccp structure. Let us now consider the justification of this degree of promotion, recognizing that all of the properties of the metals that are enhanced by increasing cohesion between the atoms will depend strongly on the degree of promotion to increase the number of bonding electrons.

What limits are there on the net bonding contributions per electron? This question can be answered by reviewing the data for the almost 80 elements for which the number of bonding electrons is indisputable. In Table 4, the atomization enthalpies of the solid phases have been divided by *n*, the number of bonding electrons, to yield $\Delta H_0^\circ/nR$ per gram-atom of bonding electrons. The first column, headed by hydrogen, lists in order of decreasing bond strength the values for elements with one valence electron per atom. The second column, headed by beryllium, lists atomization enthalpies divided by two for elements with two valence electrons per atom. The division by families continues to the seventh column, where the atomization enthalpies of rhenium, technetium, and neptunium divided by seven are listed. In each column, the values are tabulated in order of decreasing bond

Table 4. $\Delta H_0^\circ/nR$ (kK per bonding electron) for atomization of solid at 0 K with *n* bonding electrons per atom

$n = 1$	$n = 2$	$n = 3$	$n = 4$	$n = 5$	$n = 6$	$n = 7$
H 26	Be 19	B 22	C 21	Ta 19	W 17	Re 13
Li 19	S 17	N 19	Hf 19	Nb 18	Mo 13	Tc 11
Cl 16	O 15	Ce, La, Lu, Y 17	Zr, Th 18	Pa 14	U 11	Np 8
Be 14	Se 14	Ac, Gd, Tb 16	Ti 14	V 12		
Na 13	Te 13	Cm, Sc 15	Si 13			
	Ba 11					

strength. With the restriction to those elements with indisputable numbers of bonding electrons, no elements with more cohesion per electron than the last member of each column have been omitted. It is of interest to note that metals and nonmetals are interspersed in Table 4. With the exception of hydrogen, for which the electrons acting on bare nuclei can achieve an unusually small internuclear distance and unusually large cohesion, there is no element for which the net bonding enthalpy exceeds 22 kK per mole of bonding electrons. After the very tightly bound boron and carbon, no element achieves a net bonding enthalpy above 19 kK per gram-atom of electrons (i.e., 1.65 eV per electron). This upper limit will be important in dealing with the degree of promotion of elements for which not all of the valence electrons are promoted to bonding status. For the nontransition elements of the fifth to seventh groups, which have in the ground state at least one electron in each of the three p orbitals, promotion from the s to the p orbital does not increase the number of bonding electrons, and the bonding configuration will be close to the gaseous atomic ground-state configuration. Almost all other elements benefit by promotion to increase the number of bonding electrons.

There are a number of properties that can be used as measures of strength of cohesion in addition to the enthalpy of atomization. Melting points and internuclear distances are not always direct measures of cohesion as substantial changes in coordination number in the melting process or substantial changes in electronic

Table 5A. Cohesion trends

	Ca	K	
MP, °C	840	63	
R, Å	3.9	4.6	
$\Delta H_S/R$, kK	21.4	10.7	
	Zn	Cu	Ni
MP, °C	420	1085	1455
R, Å	2.7	2.6	2.5
$\Delta H_S/R$, kK	15.7	40.6	51.8
	Sr	Rb	
MP, °C	774	40	
R, Å	4.3	5.0	
$\Delta H_S/R$, kK	19.9	9.7	
	Cd	Ag	Pd
MP, °C	321	962	1554
R, Å	3.0	2.9	2.75
$\Delta H_S/R$, kK	13.5	34.3	45.3
	Ba	Cs	
MP, °C	729	29	
R, Å	4.35	5.25	
$\Delta H_S/R$, kK	21.9 ± 1	9.2	
	Hg	Au	Pt
MP, °C	−39	1065	1772
R, Å	3.0	2.9	2.8
$\Delta H_S/R$, kK	7.4	44.3	67.9

configuration can produce abnormal behavior, but the comparisons made in Tables 5A and 5B are for metals similar enough so that variation of melting point and contraction of the lattice can be taken as measures of the degree of cohesion.

In Table 5A, the divalent elements of the first column all have the ground electronic state s^2, and the alkali metals of the second column have the configuration s. Nickel, palladium, and platinum have the ground-state configurations d^8s^2, d^{10}, and d^9s, respectively. Copper, silver, and gold have the ground-state configuration $d^{10}s$. Clearly the two electrons per atom of calcium achieved a much greater cohesion than the one electron per atom of potassium with respect to higher melting point, smaller internuclear distance, and larger atomization enthalpy. The same behavior is seen for strontium compared with rubidium and for barium compared with cesium. The trends in melting point, internuclear distance, and atomization enthalpy go hand-in-hand with the number of bonding electrons. Under each comparison of the divalent alkaline earth metal with the univalent alkali metal to its left in the Periodic Table is the similar comparison in the same period of divalent zinc, cadmium, and mercury with the elements to the left. For example, divalent cadmium is compared with "univalent" silver and "zerovalent" palladium. The data clearly indicate that silver does not remain univalent and palladium does not remain zerovalent, and that promotion must have occurred. The same behavior is found as one moves to the left in the Periodic Table from zinc, cadmium, or mercury. The melting points and cohesive enthalpies rise sharply and, most significantly, the internuclear distance decreases even though the nuclear charge is decreasing. With a decrease of nuclear charge, the core of filled electronic orbitals will expand if not offset by enhanced bonding.

Table 5B compares the atomization enthalpies as in Table 5A, but with exten-

Table 5B. Bonding electrons of copper, silver, and gold

	Sc	Ca	K	
n	3	2	1	
$\Delta H_S/R$, kK	45	21	11	
	Ga	Zn	Cu	Ni
n	3	2	4	5
$\Delta H_S/R$, kK	33	16	41	52
	Y	Sr	Rb	
n	3	2	1	
$\Delta H_S/R$, kK	51	20	10	
	In	Cd	Ag	Pd
n	3	2	4	5
$\Delta H_S/R$, kK	29	13.5	34	45
	La	Ba	Cs	
n	3	2	1	
$\Delta H_S/R$, kK	52	22	9	
	Tl	Hg	Au	Pt
n	≤ 3	2	4	5
$\Delta H_S/R$, kK	22	7	44	68

sion to the trivalent metal to the left to give a better idea of the number of bonding electrons in copper and nickel, for example. The trend from scandium to calcium to potassium is what is expected for three to two to one bonding electrons per atom. Comparison with Table 4 shows that the atomization enthalpies per electron of 15, 10.5, and 11 kK for scandium, calcium, and potassium are in line with values found generally throughout the Periodic Table. From gallium to zinc, the atomization enthalpy varies as expected for three to two bonding electrons, but the atomization enthalpy of 40.5 kK for copper clearly cannot be due to only one bonding electron. Comparison with Table 4 shows that even hydrogen with a bare nucleus cannot obtain more than 26 kK cohesion from one electron. Below hydrogen, even the most strongly bound boron and carbon obtain only 22 and 21 kK per electron, and all other elements fall below these values. In every instance, the "univalent" metals copper, silver, and gold achieve higher atomization enthalpies than the trivalent metals gallium, indium, and thallium. The value of n, the number of bonding electrons, is given as 4 for copper, silver, and gold, indicating a promotion from $d^{10}s$ for 1.5 d electrons to achieve a valence state of $d^{8.5}sp^{1.5}$ with four unpaired electrons available for bonding. To achieve five bonding electrons, palladium must promote 2.5 d electrons to achieve a valence state of $d^{7.5}sp^{1.5}$. For the same valence state, platinum must promote 1.5 d electrons and nickel must promote 0.5 d and 1.0 s electrons to p orbitals.

A more precise characterization of the number of bonding electrons and of the valence-state electronic configuration for copper to gold and for nickel to platinum requires consideration of promotion energies and valence-state bonding energies as for the lanthanides, as briefly described above and in more detail by Brewer.[9,10] However, it is simpler to examine the net cohesive enthalpies for which there is a clear maximum limit of cohesion per electron, as shown in Table 4. There can be no question upon examining Tables 4 and 5 that copper, silver, and gold could not attain their measured properties using only one bonding electron per atom. The comparison was made in Table 5A using melting point, internuclear distance, and atomization enthalpy. One could also use bulk modulus, coefficient of expansion, compressibility, and a wide range of properties that depend on strength of cohesion. All of these properties indicate that copper, silver, and gold are using at least four bonding electrons per atom.

PREDICTIONS OF MULTICOMPONENT PHASE DIAGRAMS

The Engel correlation and the properties of inner-shell orbitals compared with outer-shell orbitals can be used to predict[14] multicomponent phase behavior. Figure 6 is a projection along the temperature axis of 57 phase diagrams including one six-component, six five-component, 15 four-component, 20 three-component, and 15 two-component diagrams. A horizontal line between molybdenum and osmium (8 e/a) would show the maximum phase boundaries as seen by projecting down the temperature axis. The ordinate represents the average number of valence elec-

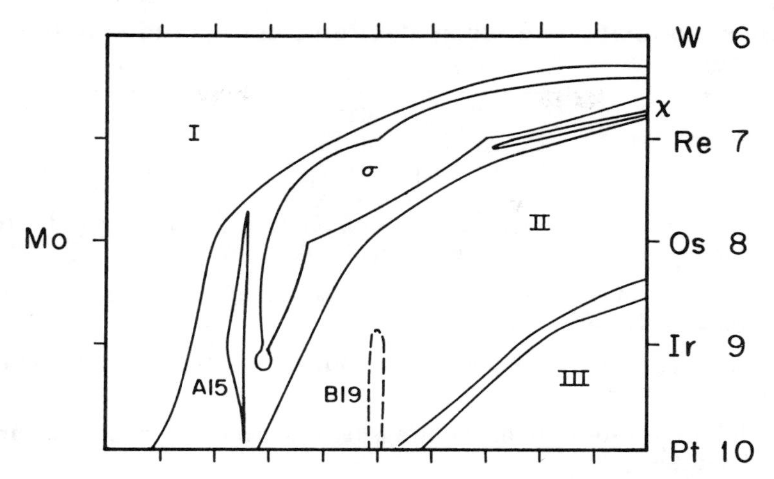

Fig. 6. Multicomponent phase diagram of molybdenum with 5d transition metals, tungsten to platinum, projected along the temperature axis. Abscissa gives molefraction of molybdenum in alloys.

trons per atom for any mixture of the five elements on the right. For a mixture with an average e/a of 7.5, a horizontal line at the 7.5 level would intersect the maximum phase boundaries at the optimum temperature. The bcc phase region with less than 1.5 outer electrons per atom is labeled I, the hcp phase region is labeled II, and the ccp phase region is labeled III. The sloping boundaries represent constant e/a lines corresponding to the outer-electron e/a limits for each phase. The stabilities of the other phases require consideration of size differences and other factors that are discussed in detail in other papers.[14,15] The B19 phase indicated by the dashed line is an ordered phase that precipitates out at low temperature and cannot be seen when sighting along the temperature axis from high temperature.

Table 6 presents a summary[14,16,17] of the general predictions that can be applied with high reliability to the phase diagrams of transition metals. Reviews[16,18,19] of the available data show agreement with the predictions of the Engel model of better than 98% for (a) and (b) and better than 99% for (b′) and (c). The most striking aspect of the Engel model is the prediction of intermetallic compounds of extremely high stability for mixtures of transition metals on the left-hand side of the Periodic Table having vacant d orbitals with platinum-group metals that have nonbonding electron pairs. This prediction has been confirmed most strikingly.[20-22] Many other properties can be predicted using the same chemical bonding model. The improved overlapping of d orbitals upon compression allows one to predict abnormally high coefficients of compression compared with those of normal metals of similar cohesion. The large impairment of bonding upon expansion provides the prediction that coefficients of expansion will be abnormally large compared with those of normal metals of similar cohesion. As expansion upon heating reduces cohesion, the heat capacities must be abnormally high. Similarly, the expansion upon fusion must result in substantial bond breaking and in very large enthalpies of fusion and abnormally high melting points. Data are frag-

Table 6. Engel correlation between electronic configuration and crystal structure

I	bcc	$d^{n-1}s$
II	hcp	$d^{n-2}sp$
III	ccp	$d^{n-3}sp^2$

The effects of alloy additions in small amounts on the relative stabilities of the three structures can be summarized for the transition metals as follows:

(a) Small additions of interstitial electron donors (B,C,N,O,H) favor structures with the most p electrons: III > II > I.

Interstitial electron acceptors (F, higher H conc.) favor structures with the fewest p electrons: I > II > III.

(b) Small additions of substitutional nontransition elements favor the structure with the fewest d bonds.
For Group VI and to the left: III > II > I.
For Group VIII and to the right: I > II > III.
For Group VII: I,III > II or β-Mn.

(b′) Since pressure favors the structure with the most d bonds, the following corollary results: The effect of pressure will be just the reverse of the summary under (b).

(c) For transition metals of Group V and to the left, the addition of any transition metal to the right of the solvent metal will stabilize the structure with the most d bonds (most d electrons for these groups): I > II > III.

For the same metals, the addition of any transition metal to the left of the solvent metal will stabilize the structure with the fewest d electrons: III > II > I.

These results also apply to the metals of Group VI except that the order of stabilization upon adding transition metals to the right reverses for addition of metals of Group VIII and beyond.

For transition metals of Group VII and to the right, the addition of any transition metal to the right of the solvent metal will stabilize the structure with the most p electrons: III > II > I.

For transition metals of Group VII and to the right, the addition of any transition metal (4d and 5d) to the left of the solvent metal will stabilize the structure with the fewest p and the most d electrons: I > II > III.

mentary in the high-temperature region where the abnormalities will be most prominent, but recent measurements[23,24] indicate that they exist. Another example of the power of using the above principles for prediction of the properties of metallic systems has been the presentation of thermodynamic values and complete phase diagrams for 100 of the binary systems of molybdenum metal.[18]

GENERALIZED LEWIS ACID-BASE REACTIONS OF METALS

One of the surprises of the Manhattan Project was the observation that platinum interacts very strongly with cerium and uranium.[22] This was not understood until

Engel's theory of bonding became available. It was clear from Engel's model why platinum interacted so strongly with cerium and uranium. Platinum atoms have the ground-state electronic configuration $5d^9 6s$, which has four nonbonding electron pairs and only one electron in d orbitals that can form bonds with neighboring atoms. By promotion of two d electrons to the 6p orbitals of the $5d^7 6s 6p^2$ configuration of the ccp structure, four additional electrons, or a total of six electrons, can be used in bonding. The additional bonding energy offsets the high promotion energy that must be paid. Although Pt in the ccp structure has six bonding electrons, four valence electrons are paired in nonbonding d orbitals. If one starts with lutetium with only three valence electrons and moves to the right to hafnium, tantalum, tungsten, and rhenium, the melting and boiling points rise markedly as more electrons are available for bonding until the $d^5 s$ and $d^5 ps$ configurations of tungsten and rhenium that utilize all of the d orbitals in bonding are reached. If one moves farther to the right to osmium ($d^6 sp$), iridium ($d^6 sp^2$), platinum ($d^7 sp^2$), and gold ($d^8 sp^2$), the melting and boiling points drop since additional electrons going into the d orbitals will produce nonbonding pairs. However, if the platinum-group metals are mixed with transition metals on the left-hand side of the Periodic Table that are Lewis acids because they have vacant d and p orbitals, the platinum-group metals can serve as Lewis bases and utilize their d electron pairs that are non-bonding in the pure platinum-group metal to bond as a Lewis base. Figure 7 illustrates the interaction of zirconium with platinum, showing just the d electrons and orbitals. Four of the d electrons are nonbonding in pure platinum. When zirconium and platinum are combined to allow the electron pairs of platinum to utilize the vacant d orbitals of zirconium, all ten d electrons of zirconium and platinum can be used in bonding the nuclei together, resulting in an increase of four bonding electrons. It has not been recognized how strong these acid-base interactions can be. It is generally expected that intermetallic phases would not have very negative enthalpies of formation. The availability of empty d and p orbitals

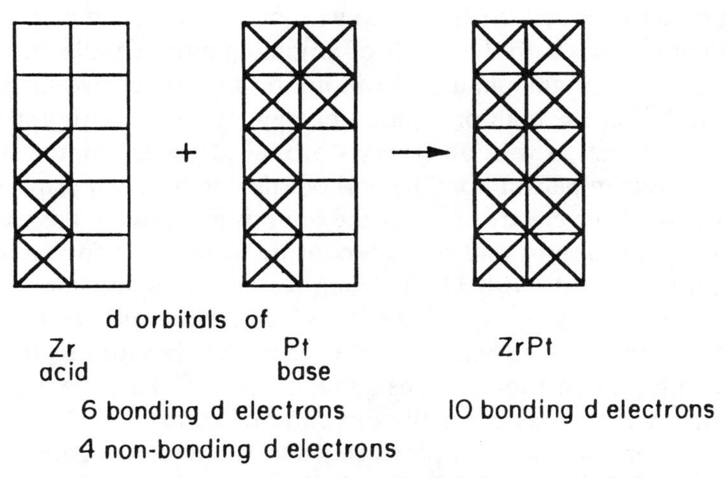

Fig. 7. Utilization of nonbonding d electrons of platinum in bonding to zirconium

of hafnium can allow each platinum atom to use at least four additional electrons in bonding than can be used in pure platinum. The enthalpy of formation of $HfPt_3$ has been determined[25] to be $\Delta H^0/R = -66 \pm 5$ kilokelvins. This intermetallic phase is much more stable than the carbide, nitride, boride, silicide, or selenide of hafnium. Only the oxide and fluoride are significantly more stable. The platinum-group metals are generally considered to be rather noble metals that do not react strongly. At high temperatures where diffusion rates become large enough to provide significant contact between the platinum-group metals and compounds of left-hand-side transition metals, many unexpected reactions take place. For example, the oxides of the lanthanides are very difficult to reduce. However, if platinum is present to reduce the thermodynamic activity of the lanthanide metal through formation of an extremely stable intermetallic phase, the oxide can be readily reduced by hydrogen.[26] Mixtures of the pure metals can be very dangerous. Once the temperature of a mixture of hafnium and platinum metal powders gets to around 1000 °C, where diffusion rates become significant, the formation of the intermetallic phase will increase the temperature by more than 5000 °C. The reaction can become so rapid that the sample can detonate and destroy the apparatus.[22,25] Many unexpected reactions will take place because of the strong Lewis acid-base interactions of platinum-group metals with transition metals with vacant d orbitals.

The platinum-group metals are amphoteric and can also serve as Lewis acids when combined with metals, such as aluminum or gallium, for which the gaseous atoms have the ground-state configuration s^2p and a large promotion energy to sp^2 is required to make all three valence electrons available for bonding in the pure metals. Gaseous palladium has the ground-state configuration d^{10} and would be a noble gas if it did not promote d electrons to s and p orbitals to achieve the ccp configuration f^7sp^2. For each electron promoted, two bonding electrons are obtained, with one d electron remaining in the d orbital and one electron in the outer-shell s or p orbital. The promotion energy is very high, and significant bonding by the d electron is required to offset the promotion energy. Because the d orbitals are somewhat localized, their overlap with outer-shell s and p orbitals is much poorer than their overlap with d orbitals of adjoining atoms. Palladium atoms surrounded by aluminum or gallium atoms will not be able to achieve sufficient d electron bonding to offset the high promotion energy, and the configuration remains d^{10}. The d^{10} configuration with an empty s orbital offers aluminum the opportunity to use its s electron pair in bonding without having to pay a promotion-energy penalty. Thus, the three electrons from the aluminum provide an average of 1.5 bonding electrons per atom, and the ccp coordination of 12 for the pure metals changes to eight in AlPd, with CsCl structure, as one would predict from the Hume-Rothery correlation of s,p electron concentration and crystal structure.

The fact that electrons are being transferred from platinum to the more electropositive hafnium or cerium disturbs some people.[22] They do not realize that the primary factor that determines the direction of transfer is the availability of orbitals and nonbonding electron pairs and that the electronegativity difference determines the distribution of electrons in the bond after the electron pair has been transferred. It is well established that the more electronegative arsenic (s^2p^3) transfers a nonbonding electron pair to the empty orbital of gallium metal (sp^2) so that each has the electronic configuration (sp^3) with tetrahedral coordination.[4,22] However, the distribution of electrons in the gallium-arsenic bond is shifted toward the arsenic to reduce the negative charge on the gallium. Another example is the formation of triple-bonded CO from carbon with four valence electrons and oxygen with six valence electrons. The oxygen contributes more electrons to the bond than does carbon, and the carbon formal charge is -1. Dipole measurements on CO indicate that the carbon is negative relative to the oxygen, but the negative charge is greatly reduced due to distortion of the electrons in the bond toward the oxygen because of its greater electronegativity. Yet another example is $Cr(CO)_6$, where the CO molecules contribute six electron pairs to the very electropositive chromium to give it a formal charge of -6. However, it is understood that backbonding from the chromium occurs to reduce the negative charge. Thus it is not surprising that in the examples given earlier, the more electronegative palladium can accept electrons from aluminum but can also transfer electrons to niobium or zirconium.

If we add to the complication of the crystal field effect on the bonding capability of d orbitals the effect of electron transfer, one can see that the accurate prediction of the acid-base interactions becomes difficult. One factor to consider is that the bonds formed in the acid-base reaction through sharing of a pair of electrons from one atom with the vacant orbital of another atom are not as strong as bonds formed by contribution of one electron from each of the interacting atoms. If the bonding enthalpies found for d orbitals of the pure metals were used to calculate the enthalpy of formation of an intermetallic such as $NbRh_3$, for example, one would calculate $\Delta H^0/R = -42$ kK, compared with the experimental[27] $\Delta H^0/R = -25$ kK. If the electrons were evenly distributed as in the pure metal, the niobium would have a formal charge of -3. This charge must be reduced by backbonding with movement of other bonding electrons away from the niobium. To be able to predict accurately the acid-base interactions, one must have a more quantitative model for the acidity of vacant orbitals and the basicity of filled orbitals.

It has been possible to characterize the interactions of zirconium, at low concentrations, with transition metals on the right-hand side of the Periodic Table.[28,29,30] In the previous discussion, the electronic configurations have been discussed in terms of simple integral electron assignments—e.g., $5d^6 6s 6p^2$ for iridium. The Hume-Rothery correlation assigns the ccp structure to s,p electron concentrations between 2.5 and 3. Thus iridium could have a configuration between $d^{6.5}sp^{1.5}$ and d^6sp^2. Because of the high promotion energy, the actual configuration will be closer to $d^{6.5}sp^{1.5}$; $sp^{1.5}$ will be used for the ccp configurations of iridium to gold in the following assignment of nonbonding electron pairs. Consider zirconium added to the 5d metals from rhenium to gold, as in Fig. 8. There is no acid-base interaction with rhenium (d^5sp) because it uses all seven valence electrons in bonding. Osmium ($d^{5.9}sp^{1.1}$) with approximately one nonbonding electron pair reduces the excess partial molal Gibbs energy of zirconium by less than 9 kK. Iridium ($d^{6.5}sp^{1.5}$) with 1.5 nonbonding electron pairs and platinum ($d^{7.5}sp^{1.5}$) with 2.5 nonbonding electron pairs reduce the excess partial molal Gibbs energy of zirconium by more than 50 kK, whereas gold ($d^{8.5}sp^{1.5}$) with 3.5 nonbonding

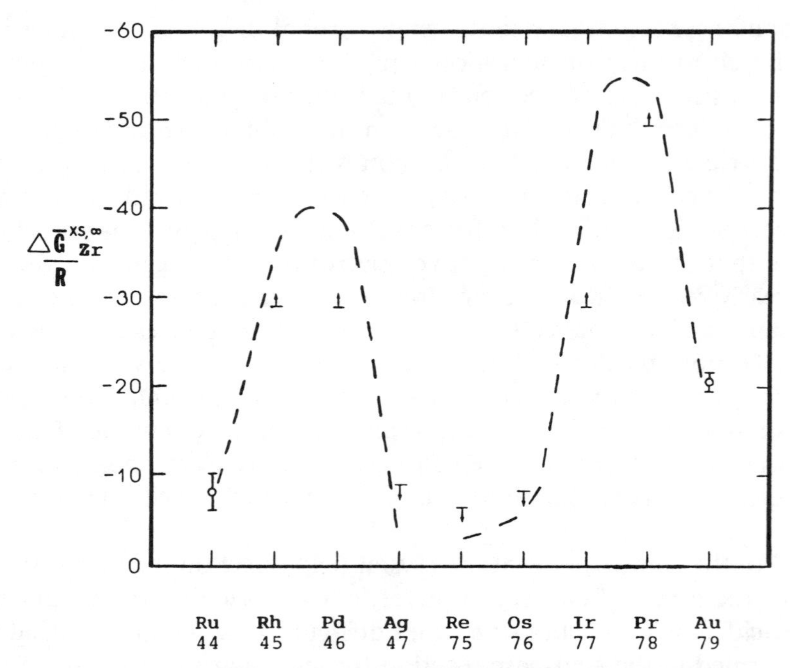

Fig. 8. Limiting excess partial molal Gibbs energies of solution of zirconium, in kK, in group VII to XI transition metals of the 4d and 5d periods

pairs reduces the excess partial molal Gibbs energy of zirconium by 20 kK. Because these data were obtained[28] by extrapolation to infinitely dilute zirconium, these values are a measure of the base strength of the electron pairs in the most extended orbit containing nonbonding pairs. There is a competition between two trends. As the nuclear charge is increased, the average extension of the d orbitals beyond the filled 5s5p shell is reduced. On the other hand, as the number of nonbonding electron pairs increases, higher orbitals with greater extension and better bonding relative to the average will be used. The maximum base strength is reached at two or slightly more nonbonding electron pairs between platinum and iridium. Although gold has a substantial acid-base interaction with zirconium, it is greatly reduced due to the greater degree of localization of the 5d orbitals. The 3d metals from iron to copper show no indications of an acid-base interaction because not only the nonbonding electron pairs but also some of the unpaired electrons are in orbitals so localized that bonding overlap with orbitals of nearest neighbors is negligible.

The above discussion deals with very dilute solutions of zirconium and requires consideration of the base strength only in the most extended orbitals containing nonbonding pairs. Characterization of the variation of base strength as a function of concentration requires a quantitative measure of the bonding capabilities of all of the orbitals containing nonbonding pairs. This requires a titration of the base with the acid with measurements of the excess partial molal Gibbs energy or activity coefficient as a function of concentration. This has been done for the Nb-Pd system using equilibration of pairs of niobium oxides with palladium metal with analysis of the palladium to determine the equilibrium niobium concentration and by high-temperature solid-electrolyte galvanic cell measurements.[30] At low concentrations, niobium will be a very effective acid using not only its one vacant d orbital but also its vacant p orbitals to receive electron pairs. Palladium will be a very effective base because it will be providing electron pairs from its most extended d orbitals containing nonbonding pairs. There will be the reduction in palladium basicity as electron pairs in more localized orbitals are used. At 1000 °C, the niobium activity coefficient at low concentrations is $10^{-9.5}$. As the niobium concentration in palladium is increased, the activity coefficient rises rapidly, going above 10^{-7} before a mole fraction of $x_{Nb} = 0.1$ is reached. At $x_{Nb} = 0.2$, the activity coefficient has risen above 10^{-4} and is up to $10^{-1.5}$ around $x_{Nb} = 0.3$. There is clearly a large crystal field effect, and the basicity of the more localized d orbitals of palladium is greatly reduced. This large crystal field effect can be illustrated with additional data. The measurements of niobium activity coefficients in rhodium by Kleykamp[27] have been extended to x_{Nb} greater than 0.3 using oxide equilibration.[30] At low niobium concentration, the activity coefficient of niobium at 1000 °C is above 10^{-8} compared with the activity coefficient in palladium of $10^{-9.5}$. As the niobium content in rhodium is increased, the activity coefficient does not rise as rapidly as in palladium. At $x_{Nb} = 0.2$, the activity coefficient in rhodium of $10^{-5.5}$ is much below the value of $10^{-3.8}$ in palladium. A crystal field model is being developed to characterize the acid-base titration curves. The model indicates considerable complexity for acid-base bonding. Although in a mole of gas, there are not many electronic levels corresponding to a given electronic configuration because of the high degeneracy of many of the levels and the equivalence of levels of all the atoms in the gas. When the atoms are condensed to the solid metal, there are the same total number of levels in the mole of solid as in the mole of gas, but the degeneracies have been removed, including the equivalence of the different atoms, so that there are several times Avogadro's number of levels that form the metallic bands. The variation of the bonding contributions of the electrons in these levels must be characterized. This extension of the Engel model will greatly extend the ability to predict the thermodynamic properties and phase diagrams for a large variety of systems.

APPENDIX

The presentation at a symposium of the above valence-bond model in Ref 6 was followed by a discussion by the participants of the symposium. One of the editors (J.L. Walter) has suggested that it would be helpful to readers if objections to the model were reviewed and discussed. The following material covers the discussion of pages 241-243, 246-249, 344-346, and 560-568 in Ref 6.

Two objections are raised by Lomer. One is that promotion is taken only to the lowest level of the valence state, which would be a high-multiplicity level, when the distinction between high and low multiplicity is wiped out in the solid. This objection can be broadened to ask why the lowest level is taken instead of an average of the various levels that would be occupied in the solid. The answer is that both approaches were tried and it was found that the energy difference between the lowest level and the average of the band of levels of the gaseous valence state varied in a smooth way with nuclear charge for almost all metals and certainly for the transition metals. If one took the higher promotion energy to the average of the levels of the valence state, then the procedure described for fixing the bonding contributions of the d electrons would increase the bonding contributions by the difference. Since both approaches yield smoothly varying bonding curves, either approach would yield the same numerical predictions for stabilities of solid phases. In view of this, the simpler procedure was used and the curves of bonding energies per d electron were reduced by the difference between the lowest level of the configuration and the average energy of all the utilized levels.

The second point raised by Lomer is also repeated by Hume-Rothery. How can it be assumed that copper, silver, and gold promote at least 1.5 d electrons from the gaseous ground electronic configuration $d^{10}s$ to provide at least four bonding electrons when everyone knows that these metals have only one bonding electron? Hume-Rothery acknowledges that it is not possible to reconcile the enthalpies of sublimation and other properties with only one bonding electron per atom and proposes that the d electrons can account for the high cohesion through van der Waals interaction of the filled shells. This proposal has been made by a number of people, but it has been impractical because van der Waals bonding of closed shells is very weak and could not possibly account for the properties of copper, silver, and gold. It is clear that the previous interpretations that assign only one bonding electron to copper, silver, and gold must be abandoned. A wide variety of properties is inconsistent with such an assumption.

Hume-Rothery states that the assumption of promotion of d electrons to p orbitals for copper, silver, and gold would be inconsistent with the Hume-Rothery rules that require that these metals have a valence of one when alloyed with zinc or cadmium. Chemists have long recognized that an element can have a variety of valences depending on the environment. When copper, silver, gold, nickel, etc. are surrounded by atoms with bonding d orbitals so that the energy of promoting d electrons to p orbitals can be offset by the additional bonding, then these metals will have s,p valences much higher than for the gaseous ground state. However, when they are surrounded by nontransition metals that do not have orbitals that would overlap well with the somewhat localized d orbitals, then it is not possible to offset the promotion energy. Thus, in agreement with the Hume-Rothery rules, copper, silver, and gold will have valences of one, and nickel, palladium, and platinum will have valences of zero when alloyed with nontransition metals. But they will have high valences when alloyed with one another or with other transition metals. This has been demonstrated in a very striking manner by M. Ellner[31] who determined the atomic volumes of copper, silver, gold, nickel, palladium, and platinum when mixed with one another and when mixed with zinc, cadmium, gallium, indium, silicon, germanium, tin, etc. When the transition metals are alloyed with one another, Vegard's Law holds and it is clear that the character of bonding does not change as one changes the composition. However, if the nontransition metals are added to any of the transition metals, one has an initial linear variation of atomic volume until enough nontransition metal has been added to start interfering with the d bonding. Then there is a rapid expansion as the Group XI metals collapse to a $d^{10}s$ configuration with only one bonding electron and the Group X metals collapse to a d^{10} configuration. This is a general behavior. To be able to predict accurately the properties of the many combinations of the metals, it is most important to recognize this difference in valence depending on the environment. Massalski raises the objection that for some of the intermediate compositions of the Cu-Ga system, it is possible for a fixed composition to have hcp structure at low temperature, ccp at intermediate temperature, and bcc at high temperature. This is quite consistent with the fact that, even for some pure metals, one can have two or more structures if the Gibbs energies of the several structures happen to be close. As noted in the main text, the $d^{n-1}s$ and $d^{n-2}sp$ configurations are close in energy for the third and fourth group transition metals. The hcp structure with the lower enthalpy is stable at low temperature, but the bcc structure with higher entropy becomes stable at high temperature. As one alloys copper with gallium, the d^9sp and $d^{10}s$ configurations will approach the d^8sp^2 configuration in stability, and it is not surprising that all three structures can be found for one composition at the boundary before the complete collapse to the $d^{10}s$ configuration that will be found for higher gallium content.

Pearson objects to the designation of d^6sp^2 for the valence state of iridium because the high-spin configuration would be higher in energy than a low-spin configuration. This is correct, but the high-spin configuration does exist for the atom at high energies. The promotion energy is high, but the additional bonding contributions are large enough to offset the high promotion energy required to get the high-spin valence state. His example of not all d orbitals mixing is correct. Some of the 3d orbitals of the right-hand transition metals are so localized that they will not overlap with orbitals of neighboring atoms. That is why the bonding energies of the fourth and fifth d orbitals of the 3d metals are so low. One has to recognize this variation in bonding ability of different d orbitals caused by the crystal field effect that extends some and improves their bonding ability and contracts others. This has been discussed in detail.[30]

Friedel objected to a model in which the d electrons contribute strongly to bonding but do not fix the structure. This objection is not valid. The d electrons do contribute to the fixing of the most stable configuration—for example, $d^{7.5}sp^{1.5}$ for palladium compared with d^8sp, d^9s, or d^{10}. When the most stable configuration has been fixed, the number of s and p electrons determine the long-range order. The d electrons that bond primarily with nearest neighbors do not fix the long-range order. The theoretical paper by Altmann, Coulson, and Hume-Rothery[32] demonstrated this by showing that the combination of one s electron with varying numbers of d electrons all yielded the bcc structure.

A number of other issues were raised, but most of these were adequately discussed or have been covered in the main section of this chapter.

REFERENCES

1. L. Brewer and J.S. Winn, *Faraday Disc., Chem. Soc.*, No. 14, 1980, pp 126-135
2. R. Hultgren, P.D. Desai, D.T. Hawkins, M. Gleiser, K.K. Kelly, and D.D. Wagman, *Selected Values of Thermodynamic Properties of the Elements*, American Society for Metals, Metals Park, OH, 1973

3. C.E. Moore, *Atomic Energy Levels*, Vol I (1949), Vol II (1952), Vol III (1958), U.S. Govt. Printing Office, Washington, DC
4. L. Pauling, *Proc. Roy. Soc. London Ser. A.*, Vol 196, 1949, p 343; *The Nature of the Chemical Bond*, 3rd Ed., Cornell Univ. Press, Ithaca, NY, 1960
5. L.B. Knight, Jr., R.D. Brittain, M. Duncan, and C.H. Joyner, *J. Phys. Chem.*, Vol 79, 1975, p 1183
6. L. Brewer, *Phase Stability in Metals and Alloys*, edited by P. Rudman, J. Stringer, and R.I. Jaffee, McGraw-Hill, New York, 1967, pp 39-61, 241-245, 344-345, 560-568
7. W. Hume-Rothery, *The Metallic State*, Oxford Press, Oxford, 1931; *Structures of Metals and Alloys*, Institute of Metals, London, 1936
8. N. Engel, Ingenioeren N101 (1939) and a series of papers cited in his later papers, such as Amer. Soc. Metals, *Trans. Quart.*, Vol 57, 1964, p 610; and *Acta Met.*, Vol 15, 1967, p 557
9. L. Brewer, *J. Opt. Soc. Am.*, Vol 61, 1971, pp 1101-11, 1666-86
10. L. Brewer, "Systematics of the Properties of the Lanthanides," in *Systematics and the Properties of the Lanthanides*, NATO ASI Series C: Mathematical and Physical Sciences No. 109, edited by S.P. Sinha, D. Reidel, Boston, 1963, pp 17-69
11. W.C. Martin, R. Zalubas, and L. Hagan, "Atomic Energy Levels—The Rare Earth Elements," NSRDS-NBS 60, U.S. Govt. Print. Off., Washington, DC, 1978
12. L. Brewer, *High Temp. Sci.* Vol 17, 1984, pp 1-30
13. L. Brewer, AIP Conference Proceedings No. 10, *Magnetism and Magnetic Materials-*1972, Sect. 1, 1973, pp 1-16
14. L. Brewer, *High Strength Materials*, edited by V.F. Zackay, Wiley, New York, 1965, pp 12-103
15. L. Brewer, *Calculation of Phase Diagrams and Thermochemistry of Alloy Phases*, TMS-AIME Conference Proceedings, edited by Y.A. Chang and J.F. Smith, 1980, pp 197-206
16. L. Brewer, *J. Nucl. Mat.* Vol 51, 1974, pp 2-11
17. L. Brewer, *Structure and Bonding in Crystals I*, edited by M. O'Keefe and A. Navrotski, Academic Press, London, 1981, pp 155-174
18. L. Brewer and R. Lamoreaux, *Molybdenum, Physiochemical Properties of its Compounds and Alloys, Atomic Energy Review*, Special Issue No. 7, International Atomic Energy Agency, Vienna, 1980, pp 1-356
19. L. Brewer, *Amer. Inst. Phys. Conf. Proc.* 1972, edited by H.C. Wolfe, No. 10 (Pt. 1), 1973, pp 1-16
20. L. Brewer, *Acta Met.* Vol 15, 1967, pp 553-556
21. L. Brewer and P.R. Wengert, *Met. Trans.* Vol 4, 1973, pp 83-104
22. L. Brewer, *J. Chem. Educ.* Vol 61, 1984, pp 101-104
23. A. Cezairliyan, M.S. Morse, H.A. Berman, and C.W. Beckett, *J. Res. Nat. Bur. Stand.*, Sect. A., Vol 74, 1970, p 65
24. J.W. Shaner, G.R. Gathers, and C. Minichino, *High Temp. High Press* Vol 8, 1976, p 425; and Vol 9, 1977, p 331
25. V. Srikrishnan and P.J. Ficalora, *Metall. Trans.* Vol 5, 1974, pp 1471-1475
26. W. Bronger and W. Klemm, *Z. Anorg. Allg. Chem.*, Vol 319, 1962, pp 58-91
27. H. Kleykamp, *J. Less-Common Met.* Vol 83, 1982, pp 105-113
28. L. Brewer, *Pure and Applied Chem.* (IUPAC) Vol 60, 1988, pp 281-286
29. J.K. Gibson, L. Brewer, and K.A. Gingerich, *Metall. Trans. A*, Vol 15A, 1984, pp 1984-2075
30. M.J. Cima, Lawrence Berkeley Lab Report LBL-21951, July, 1986
31. M. Ellner, *J. Less-Common Met.*, Vol 60, 1978, pp 15-39; Vol 75, 1980, pp 5-16; Vol 78, 1981, pp 21-32
32. S.L. Altmann, C.A. Coulson, and W. Hume-Rothery, *Proc. Roy. Soc. London Ser. A*, Vol 240, 1957, p 145

2

Bonding in Metals and Alloys From a Valence-Bond Viewpoint

R. P. MESSMER
Department of Physics, University of Pennsylvania
General Electric Corporate Research and Development
R. C. TATAR
Department of Physics, University of Pennsylvania
C. L. BRIANT
General Electric Corporate Research and Development

For decades metallurgists have sought answers to basic questions such as: Why does a given element, alloy, or intermetallic compound exhibit a particular structure? What factors determine the solid solubility of one element in another? Why do some alloying additions often have such profound effects on the properties of a material? Concise and insightful answers to these questions would represent not only a significant step forward in the science of materials but also an advance in the art of materials design.

These questions and others of a similar nature have gone unanswered to a great extent because of many factors. We touch upon only a few here. An intrinsic obstacle to obtaining a simple microscopic description of metals is that a very large number of particles (electrons and nuclei) are involved and a quantum mechanical description is necessary. Thus one must solve the Schrödinger equation, $H\Psi = E\Psi$, for the many-electron wave function, Ψ, in the presence of a set of nuclear positions, where the electronic Hamiltonian is given by

$$H = \sum_{i=1}^{N} T_i - \sum_{i=1}^{N}\sum_{k=1}^{M} \frac{1}{|\mathbf{r}_i - \mathbf{R}_k|} + \sum_{i=1}^{N}\sum_{j>i}^{N} \frac{1}{|\mathbf{r}_i - \mathbf{r}_j|} \qquad \text{(Eq 1)}$$

In this equation the first sum is over the kinetic energy operators for the N electrons of the system, the second set of terms represents the Coulombic nuclear attraction of the N electrons to the M nuclei, and the third set of terms describes the Coulombic repulsions among the N electrons.

Although it has been recognized that the answers to many basic questions will undoubtedly come from an understanding of materials at an electronic level,[1,2] the nature of chemical bonding among the atoms in metals has not been described in any simple manner. Most quantum mechanical approaches to the solid state have sought to address these problems from a band-theory viewpoint.[3] However, many band-theory concepts are based on reciprocal space, making it difficult to develop an intuitive picture of the bonding that occurs in solids. The notions of an electron gas or Bloch functions do not allow the development of a direct physical picture of the chemical bonds that are responsible for the cohesion of the solid and affect so many of its properties. This situation has made it difficult to extract concepts from band theory which might be applied to more complex metallurgical systems. Furthermore, band calculations usually can be performed only on rather simple systems containing less than a dozen atoms per unit cell.

Quite recently, however, considerable progress has been made in describing the bonding of metals in terms of real-space concepts. Some simple concepts have evolved from fully quantum mechanical computations which can be applied to metallurgical systems. It is the purpose of this paper to describe these ideas and to begin the task of applying them to various problems of interest in metallurgy. Since the concepts used are derived from the valence-bond theory of molecules, we begin with a brief description of this approach.

Valence-Bond Theory

The concepts of valence-bond (VB) theory have their roots in the empirical observations of rules for combining atoms to form compounds. Historically, the empirical rules together with atom and electron theories led to both the concept of *valence* as an intrinsic property of an atom, and the concept of the *covalent bond*.

The notion of covalent bond has two important components which are central to the ensuing discussion: (1) electron pairing and (2) the sharing of electron pairs between or among atoms. While the concepts of covalent bond and valence preceded the development of quantum mechanics, the quantum mechanical valence-bond treatment of the hydrogen molecule by Heitler and London[4] in 1927 and the subsequent development of their approach has provided a very satisfactory picture not only for valence but also for the shared electron pair in a wide variety of molecules and covalent solids.[5] This is particularly true of the valence-bond approach to many molecules, where the wave function has a rather direct correspondence with the classical ideas of bonding and provides a concrete microscopic electronic picture of them. In addition, such valence-bond wave functions have the advantage of providing insight into the geometric structure of molecules and solids, and the relationships among them.[5]

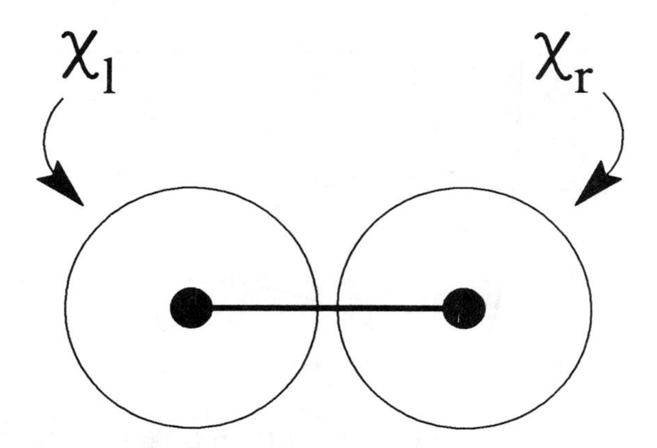

The large circles represent one-electron orbitals in the many-body wave function and are labeled to correspond to Eq 4. The black dots represent individual electrons and indicate that each orbital contains one electron. The solid line connecting the two black dots indicates that the two electrons are singlet-coupled (see text).

Fig. 1A. Schematic representation of the ground-state valence-bond wave function for H_2

In order to establish the terminology to be used later, it is convenient to review a few simple cases in some detail.* The simplest example is that of the hydrogen molecule, H_2. As it is well known, the ground state of the hydrogen atom has a single electron in a *1s* orbital, $\chi(1)$, where the argument of the function represents the spatial coordinates of electron "1." For the H_2 molecule, we might assume an approximate wave function of the form $\chi_\ell(1)\,\chi_r(2)$ (see Fig. 1A), where electron 1 is in the *1s* orbital at the left (χ_ℓ) and electron 2 is in the *1s* orbital on the right (χ_r). Thus we might write the many-electron (two in this case) wave function as

$$\Phi(1,2) \cong \chi_\ell(1)\,\chi_r(2) \qquad \text{(Eq 2)}$$

But electrons are fermions and the wave function must be antisymmetric (i.e., must change sign) upon interchanging any two electrons. Thus we must have

$$P_{12}\Phi(1,2) = -\Phi(2,1) \qquad \text{(Eq 3)}$$

where P_{12} interchanges the two electrons in $\Phi(1,2)$. For this to be true, Eq 2 can-

*The reader should not be discouraged by any of the mathematical equations, however, since our description of metallurgical systems will be based entirely on the physical picture that develops. The mathematical detail will not be required to understand the rest of the chapter, but is included for completeness.

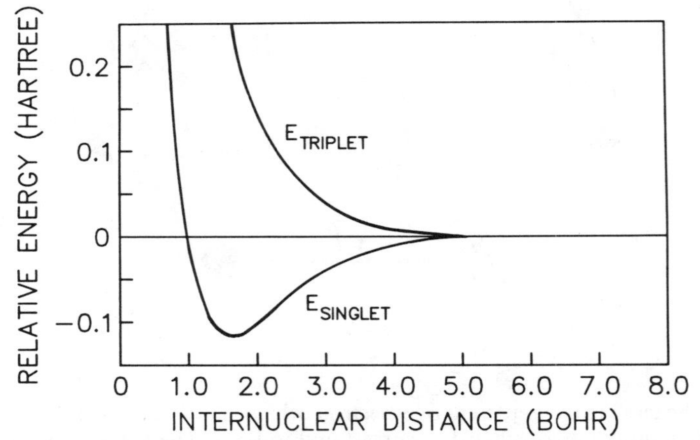

The zero on the energy scale is chosen to be the energy at infinite internuclear separation. Note that the triplet state is not stable at any finite separation and, thus, does not form a bound state.

Fig. 1B. Energy vs. internuclear separation for singlet and triplet wave functions of H_2

not be correct. Besides, we have omitted something else: the electron also has a spin which has two possible values, α or β. Let us try

$$\begin{aligned}\Phi_S^{VB}(1,2) &\cong [\chi_\ell(1)\chi_r(2) + \chi_\ell(2)\chi_r(1)][\alpha(1)\beta(2) - \alpha(2)\beta(1)]\\ &= [\chi_\ell(1)\chi_r(2) + \chi_r(1)\chi_\ell(2)][\alpha(1)\beta(2) - \beta(1)\alpha(2)]\\ &= [\chi_\ell\chi_r + \chi_r\chi_\ell][\alpha\beta - \beta\alpha]\\ &= \det[\chi_\ell\chi_r(\alpha\beta - \beta\alpha)] = \det[\chi_\ell\chi_r\theta_S] \qquad \text{(Eq 4)}\end{aligned}$$

which is the Heitler-London valence-bond wave function. This wave function satisfies Eq 3 and is a proper wave function containing both spin and space parts. The space part is symmetric under interchange of electrons and the spin part is antisymmetric ($\theta_S = \alpha\beta - \beta\alpha$); together they form a proper antisymmetric wave function. A convenient way to achieve the antisymmetric character is to use a determinant, as in the last line of Eq 4. This represents the ground state of H_2, a singlet state (i.e., the spins are paired—*up* and *down*), responsible for the bond in H_2 (see Fig. 1B). The spin function is said to represent "singlet spin coupling" or "spin pairing" and is schematically depicted in Fig. 1A by a line connecting the two orbitals which are singlet-coupled or spin-paired. The simple diagram of Fig. 1A is a shorthand way of representing the wave function of Eq 4.

Another possible wave function which is a proper eigenstate of the system is

$$\begin{aligned}\Phi_T^{VB}(1,2) &\cong [\chi_\ell\chi_r - \chi_r\chi_\ell][\alpha\alpha]\\ &= \det[\chi_\ell\chi_r(\alpha\alpha)] = \det[\chi_\ell\chi_r\theta_T] \qquad \text{(Eq 5)}\end{aligned}$$

which corresponds to the *triplet state* (i.e., both electrons have the same spin, *up* or *down*) of H_2 and does not form a bond (see Fig. 1B). Thus, triplet coupling does not result in two-electron covalent bond formation.

If the orbitals of Eq 4 are allowed to change during formation of the bond—i.e., they are not restricted to be *1s* atomic orbitals (χ)—then this results in a better wave function (in the sense that it yields a lower energy and is a more accurate description of the molecule). In particular, when the orbitals are obtained from a numerical procedure that is based on the *variational principle* (variationally optimized), a *generalized valence-bond* (GVB) description of the molecule is obtained:

$$\Phi_S^{GVB} = \det[\varphi_\ell \varphi_r \theta_S] \qquad \text{(Eq 6)}$$

The same schematic picture, Fig. 1A, can be used to represent Eq 6 also, if we remember that the orbitals of Fig. 1A are not exactly atomic hydrogen *1s* orbitals. Near the equilibrium internuclear separation, the orbitals of Eq 6 exhibit a high overlap (i.e., $\langle\varphi_\ell|\varphi_r\rangle \approx 0.8$). This is a general characteristic of a strong two-electron covalent bond.

The wave functions given by Eq 4 and 6 and shown schematically in Fig. 1A illustrate "electron correlation." In Fig. 1A, the two electrons of the bond (which repel one another because of the Coulombic repulsion in the Hamiltonian) can avoid each other (i.e., correlate their motions) by one electron staying on the left atom while the other electron stays on the right atom. If we simplify the wave function in Eq 6 by restricting both electrons in the pair to occupy the same spatial orbital (i.e., $\varphi_\ell = \varphi_r = \phi = \chi_\ell + \chi_r$), thereby having an orbital overlap of unity, we obtain the molecular orbital (MO) approximation:

$$\begin{aligned}\Phi_S^{MO} &= \det[\phi(1)\,\phi(2)\theta_S] \\ &= \det[\{\chi_\ell(1)\,\chi_\ell(2) + \chi_r(1)\,\chi_r(2) + \chi_\ell(1)\,\chi_r(2) + \chi_r(1)\,\chi_\ell(2)\}\,\theta_S] \qquad \text{(Eq 7)}\end{aligned}$$

Here it can be seen that all terms enter with a coefficient of unity and hence there is the same probability for the two electrons to be on the same atom (first two terms) as there is for the electrons to be on different atoms (last two terms). This results in a poorer approximation to the electronic structure due to the neglect of electron correlation effects. When the MO approximation is applied to solids, the method is known as *band theory* and the corresponding molecular orbitals, ϕ_i, are called Bloch functions.

We now consider a somewhat more complicated case to illustrate the generalization of the valence-bond approach to more than one bond. For this purpose CH_4 is a useful example. On the basis of simple physical ideas dating back to 1931,[6,7] we might expect the orbitals involved in the four bonds of methane to resemble those shown schematically in Fig. 2(a). The carbon atom has four "sp^3" hybrids pointing in tetrahedral directions toward the four hydrogen atoms. Each hydrogen orbital is singlet-coupled to a carbon orbital to form a two-electron cova-

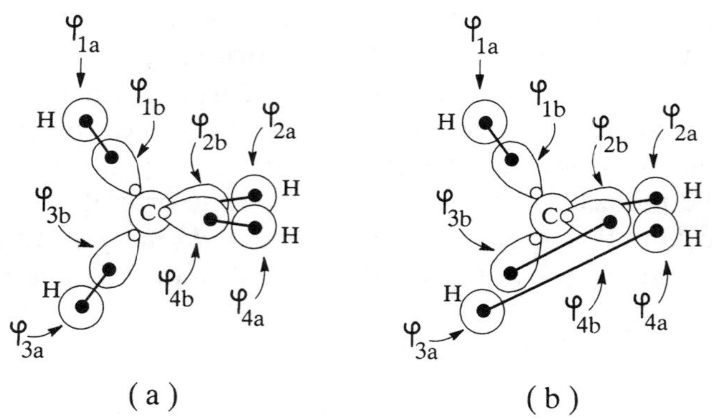

As in Fig. 1A, the black dots denote that each orbital contains one electron and the connecting lines indicate singlet coupling of the particular electron pairs. (a) In the perfect-pairing approximation (see text), this valence-bond structure (the shapes of the orbitals are variationally optimized) is taken as the complete ground-state wave function of methane. The labels on the orbitals correspond to the wave function in Eq 8. (b) Alternative spin pairing of the ground-state orbitals. If this configuration is included in a more accurate wave function that also includes the terms in Eq 8, as in Eq 9, the coefficient of this valence-bond structure would be much smaller than the coefficient of the valence-bond structure in (a).

Fig. 2. Schematic representation of a valence-bond structure for methane

lent bond. In analogy to Eq 6, we might expect the many-electron wave function to take the form

$$\begin{aligned}\Phi^{PP} &= \det[\{\text{core}\}\varphi_{1a}\varphi_{1b}(\alpha\beta-\beta\alpha)\varphi_{2a}\varphi_{2b}(\alpha\beta-\beta\alpha)\varphi_{3a}\varphi_{3b}(\alpha\beta-\beta\alpha)\varphi_{4a}\varphi_{4b}(\alpha\beta-\beta\alpha)]\\ &= \det[\{\text{core}\}\varphi_{1a}\varphi_{1b}\varphi_{2a}\varphi_{2b}\varphi_{3a}\varphi_{3b}\varphi_{4a}\varphi_{4b}\theta_S(8)] \qquad \text{(for methane)}\\ &= \det\left[\{\text{core}\}\prod_{i=1}^{N/2}\varphi_{ia}\varphi_{ib}\theta_S(N)\right] \qquad \text{(in general)} \qquad \text{(Eq 8)}\end{aligned}$$

where we see that the wave function is just the determinant of the product of four valence bond pairs and a set of "core" orbitals that are spatially localized near the nucleus and do not contribute to the bonding. In general, all spatial orbitals can mutually overlap. In the second line of Eq 8, we rearrange the factors to have the spatial functions collected together and the spin factors collected into $\Theta_S(8)$, which represents the singlet spin function for the eight electrons shown schematically in Fig. 2(a). In principle, a number of different singlet spin couplings for eight valence electrons are possible. For example, a valence bond structure with an alternative spin pairing for CH_4 is shown in Fig. 2(b). In the case of CH_4, simple physical arguments would lead one to expect that the spin coupling in Fig. 2(b) is much less significant than that of Fig. 2(a). However, in both more complicated molecules and in metals such alternative spin couplings can be quite important. When only one spin coupling (one valence bond structure) is considered, the wave function is referred to as *perfect pairing* (PP)—i.e., a unique pairing can be assumed without significant error. Equation 8 and Fig. 2(a) illustrate a VB-PP wave function (if the orbitals are not strictly atomic orbitals but are variationally optimized, it is referred to as a GVB-PP wave function[8]). When the perfect-pairing assumption is not adequate, Eq 8 must be replaced by the more general form

$$\Phi = \sum_j c_j \det\left[\{\text{core}\} \prod_{i=1}^{4} \varphi_{ia}\varphi_{ib}\theta_{Sj}(8)\right] \quad \text{(Eq 9)}$$

where the alternative singlet spin-couplings are taken into account. We refer to the terms of Eq 9 as *valence-bond spin structures*.

In some circumstances more than one set of spatial orbitals must be considered. This means the spatial orbitals must carry an additional index that indicates to which set of spatial orbitals we are referring. Hence, Eq 9 could be rewritten as

$$\Phi_\nu = \sum_j c_j^\nu \det\left[\{\text{core}\} \prod_{i=1}^{4} \varphi_{ia\nu}\varphi_{ib\nu}\theta_{Sj}(8)\right] \quad \text{(Eq 10)}$$

to inform us that only the spatial orbitals labeled by ν are used in the approximate wave function Φ_ν. A wave function more general than Eq 9 is sometimes needed. This may be written as a linear combination of terms such as in Eq 10—i.e.,

$$\Psi = \sum_\nu d_\nu \Phi_\nu \quad \text{(Eq 11)}$$

The terms in Eq 11 are thus *valence-bond orbital structures*, each of which is made up of valence-bond spin structures.

An alternative approach is to describe the wave function as a linear combination of Slater determinants made up of a single set of *orthogonal* spatial orbitals. That is, for N electrons in a singlet state:

$$\Psi = \sum_\nu c_\nu \Phi_\nu^O \quad \text{(Eq 12a)}$$

with

$$\Phi_\nu^O = \sum_j c_j^\nu \det\left[\{\text{core}\} \left(\prod_{i=1}^{N} \phi_{i\nu}\right)\theta_{Sj}\right] \quad \text{(Eq 12b)}$$

and $\langle \Phi_\nu^O | \Phi_\mu^O \rangle = \delta_{\nu\mu}$ and $\langle \phi_{k\nu} | \phi_{\ell\nu} \rangle = \delta_{k\ell}$. The label "O" reminds us that the orbitals are restricted to be orthogonal (i.e., have *zero* overlap). The approach of Eq 11 is vastly more complex mathematically, and for this reason the expansion in Eq 12 is the one typically assumed both in molecular and solid-state work. The ease with which the single-particle basis can be obtained in the latter case is certainly a significant advantage. Furthermore, it might be argued that since either approach is equivalent in the end (i.e., when the expansions are complete), it makes more sense to choose the mathematically more straightforward approach. In fact, however, one always considers only a small fraction of the terms in either expansion, and the more relevant question is: Which series is more rapidly convergent and/or more physically motivated?

If the series in Eq 11 is more rapidly convergent and the individual terms are more easily motivated physically than in the case of Eq 12, the additional complexity may be worth the effort. This, in fact, is the case illustrated in the following discussions on molecules and metal clusters. More specifically, in the following sections we will attempt to show that approximations to the general wave function in Eq 11 lead to results that are: (1) easy to visualize, (2) essentially transferable from one system to another, and (3) more compact than an expansion of the wave function using Eq 12. The ability to construct a three-dimensional spatial picture of the results is a useful feature illustrated in all the examples that follow.

MOLECULES AND METAL CLUSTERS

In order to introduce the valence-bond description of metals it is important to describe some recent advances in our understanding of bonding in molecules and metal clusters deduced from valence-bond calculations. The concepts derived from detailed, correlated calculations on a few prototypic molecules and metal clusters will be discussed because these concepts are directly carried over to our discussion of the bonding in metals. In particular, the consideration of molecules which follows may appear irrelevant at first glance to someone interested in metals and metallurgy. A key concept, however—that of *hybrid orbitals*—is deduced from the discussion. This is followed by a consideration of metal clusters, again a topic somewhat removed from bulk alloys. Yet, a discussion of metal clusters allows another key concept—that of *interstitial electrons*—to be deduced from first principles. It is the integration of these two concepts that we then use to discuss bonding in metals and alloys.

Molecules and Hybrid Orbitals

In some molecules, such as the methane molecule already mentioned, one component of Eq 10 with one spin coupling (resulting in a wave function in the form of Eq 8) suffices to give a reasonably accurate description of the ground state. When each of the spatial orbitals of methane is determined self-consistently from a variational calculation,[9] the results of Fig. 3 are obtained. In this figure, orbital contour plots in the plane of a H-C-H group are shown. The orbitals centered on carbon are essentially perturbed sp^3 hybrid orbitals (*tetrahedral hybrids*), and the orbital centered on the hydrogen atom is a perturbed atomic *1s* orbital. In Fig. 3, the spin-coupling is suggested in the orbital plots by the large overlap of the carbon and hydrogen orbitals forming a particular bond. Thus, the relationship of the schematic in Fig. 2(a) to the detailed numerical results is clear.

As an example of the transferability of results from molecule to solid, a comparison of the carbon-carbon bond in ethane and diamond is instructive. It is

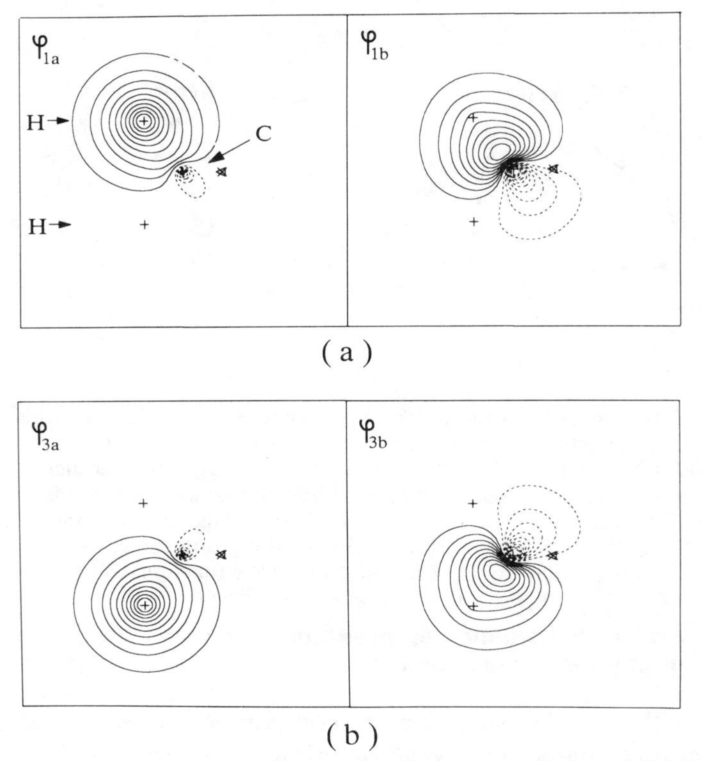

(a) The two orbitals of a singlet-coupled carbon-hydrogen bond in methane. The contours are shown in a plane containing the carbon and two hydrogen atom centers. These centers are indicated by the symbol "+." The remaining two hydrogen centers are not in this plane: one is above the plane ("×") and the other is below the plane ("Δ") shown. The contour interval is constant and the outer contour has the smallest value. Dotted lines indicate negative contours and are artifacts of the restrictions imposed on the wave function (orthogonality of orbitals). Orbital φ_{1a} is essentially a hydrogen *1s* orbital that has polarized toward the carbon, and orbital φ_{1b} is very similar to an sp^3 hybrid orbital, hence the schematic in Fig. 2(a). (b) Contours of another set of carbon-hydrogen bond orbitals of methane. These orbitals have the same shape as the orbitals in (a) except for a rotation, hence the name "equivalent" orbitals.

Fig. 3. Contour plots of variationally obtained orbitals of the methane wave function shown schematically in Fig. 2(a)

known from experiment that the bond length, vibrational force constant, and bond energy of ethane are transferable (to a reasonable degree of accuracy) to the diamond lattice, where they are expressed as the lattice constant, bulk modulus, and cohesive energy, respectively.[10] This can be understood in a simple intuitive fashion if we assemble a collection of carbon atoms with tetrahedral hybrid orbitals and join each carbon to four other carbon atoms. The resulting electronic structure consists of electron pairs, shared between a carbon atom and its nearest-neighbor atoms, as in Fig. 4(a). Thus the wave function of diamond can be

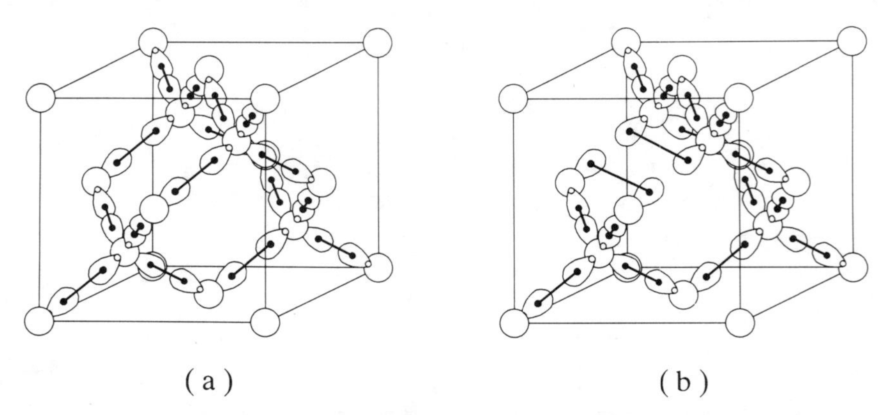

Each carbon atom (positions indicated by circles) has four sp^3 hybrids that are directed toward the carbon atom's four nearest neighbors. (a) Each pair of hybrid orbitals along nearest-neighbor internuclear axes are singlet coupled. This valence-bond structure corresponds to the classical ground-state valence-bond wave function of diamond (*cf.* Eq 8). (b) Alternate spin pairing of the ground-state orbitals. This configuration would be included in a wave function of the form of Eq 9, but would have a very small coefficient.

Fig. 4. Schematic representation of valence-bond structures for diamond

approximated with a single many-electron component (*valence-bond structure*)—i.e., Eq 11 is approximated by a wave function of the form of Eq 8. This is just the classical valence-bond description in which each of a collection of tetrahedral carbon atoms is joined by covalent bonds to its neighbors.

For a more complete wave function, additional valence-bond structures for diamond could be considered. An auxiliary valence-bond structure might differ from the primary structure (primary wave-function component) by having the electrons spin-coupled as in Fig. 4(b). Even though this valence-bond structure has a higher energy than the primary component (Fig. 4a), it is a physical possibility and can be included (together with many other components) if a more accurate description of the wave function is desired. Many aspects of diamond, however, are correctly predicted with a single valence-bond structure.

As examples of multiple bonds in molecules in which only one valence-bond orbital structure is necessary, we consider C_2H_4 (ethylene), C_2H_2 (acetylene), and C_2F_2 (difluoroacetylene). A quick conjecture of the geometric and electronic structure of the first two molecules can be obtained by taking two copies of the methane molecule described above. If one removes two hydrogen atoms from each of the copies and joins the singly occupied orbitals of one carbon atom with the singly occupied orbitals of the other carbon atom, the situation shown in Fig. 5(a) results. Similarly, the ground-state electronic structure of acetylene can be understood by taking two copies of the wave function of methane, removing three hydrogens, and forming three carbon-carbon bonds, as shown in Fig. 5(b).

For C_2F_2, again a wave function of the form of Eq 8 is found [11] to offer a

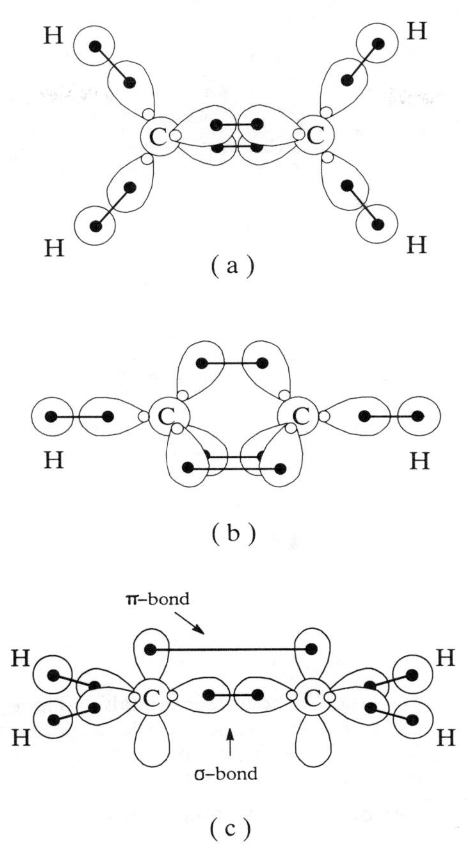

Note the tetrahedral nature of the orbitals associated with carbon. In (a), two sets of carbon hybrid orbitals overlap, forming a double bond, and in (b) three sets of hybrids overlap to form a triple bond. (c) Conventional description of ethylene double bond in terms of hybrid orbitals. Note that in this description there are two types of "hybrid" orbitals: sp^2 orbitals and p orbitals. (See text.)

Fig. 5. Schematic representations of perfect-pairing valence-bond wave functions for ethylene (a) and acetylene (b)

good approximation. A schematic representation of the wave function is shown in Fig. 6(a). The detailed computed contour plots of the orbitals from this form of the valence-bond wave function are shown in Fig. 6(b). Note that all of the orbitals are highly localized and are approximately tetrahedral hybrids. In the schematic of Fig. 6(a), the dots denote that the orbitals are occupied by electrons and the lines denote which orbitals are coupled to form chemical bonds—the convention used in the discussion of other cases above. For the lone pairs of electrons on the F atoms, the orbitals are shown as containing a pair of electrons. In fact, there are two orbitals for each pair—one for each electron—but the angular dependence of the two orbitals in a pair is practically identical, so the simplified representation of Fig. 6(a) is used. In order to obtain the identical description using the assump-

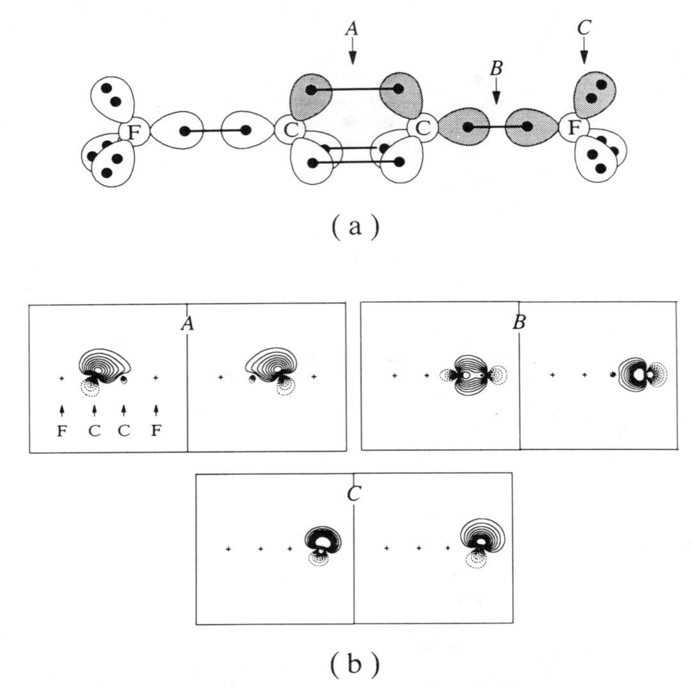

(a) Schematic representation of bond-pair orbitals and *lone-pair* orbitals. (b) Contour plots of some of the variationally obtained orbitals for C_2F_2. *A*, *B*, and *C* correspond to the shaded orbitals of the wave function shown schematically in (a). (See text.)

Fig. 6. The perfect-pairing valence-bond wave function for C_2F_2

tions of Eq 12a, an expansion of 2048 orthogonal determinants would be required. Thus the valence-bond description provides a very compact way to represent the wave function.

In each of the examples above, only a single spatial valence-bond structure (or wave-function component) was used to describe the wave function. In general, several spatial components may be required in order to provide a reasonable approximation. Benzene, for example, requires at least two components (corresponding to the two Kekulé structures[12]) in order to provide a proper wave function (cf. Fig. 7a). A basic feature of quantum mechanics is that if physically indistinguishable alternatives exist, the wave function must be made up of a superposition of those alternatives. Hence, the wave function takes the form of Eq 11, where two separate sets of spatial orbitals are used in the valence-bond structures. In Fig. 7(b) a schematic representation of the carbon-carbon bonding in the benzene ring for one of the two valence-bond structures as obtained from recent calculations[13] is shown. Although benzene has been considered to be the prototypic molecule for π-bonding, the variational principle for the valence-bond approach shows that approximately tetrahedral hybrids leading to equivalent bent-bonds are favored. A cross section of the calculated orbitals forming a double bond in one of the Kekulé structures is shown in Fig. 7(c).

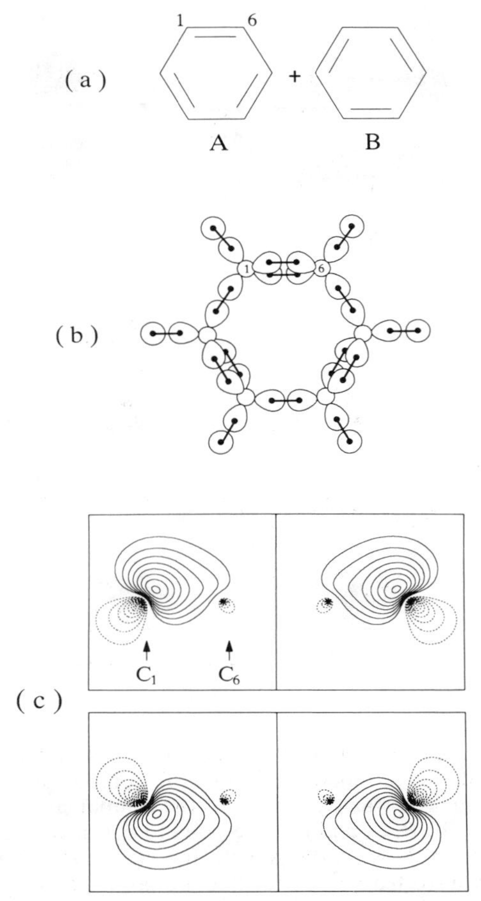

(a) Abstract schematic representation of the two valence-bond structures required to represent the wave function of benzene with tetravalent carbon atoms. (b) Schematic representation of component A of the form of the benzene wave function shown in (a). (c) Contour plots of one of the variationally obtained "double-bond" orbitals (between atoms 1 and 6) of the component of the wave function in (a). (See text.) Note that the orbitals are essentially distorted sp^3 atomic-like orbitals.

Fig. 7. Various representations of the valence-bond wave function of benzene

Another example which requires more than one spatial valence-bond structure to obtain a reasonable approximation is the CO_2 molecule.[14] In this molecule, as in benzene, two terms in Eq 11 are required, each with one spin coupling. A schematic representation of some of the detailed orbitals (Fig. 8a) of the first component is shown in Fig. 8(b). Again, the C and O atoms exhibit localized orbitals which are essentially tetrahedral in nature. The second component of the wave function can be obtained from the first (see Fig. 8a) by a rotation of 90° about the internuclear axis, and the approximate wave function is a coherent superposition of these two alternative structures.

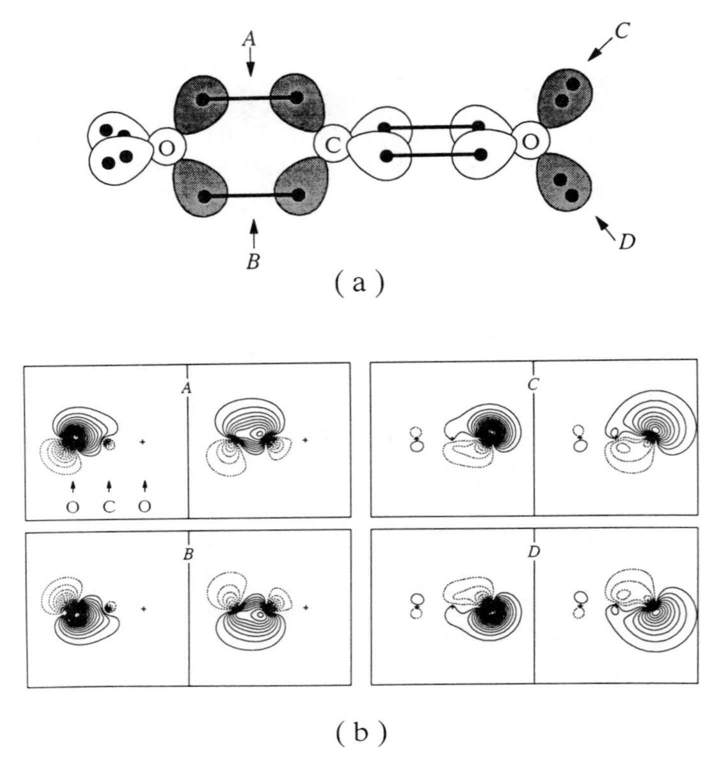

(a) Schematic representation of one of the two equivalent valence-bond structures of the wave function of CO_2. (b) Contour plots of some of the variationally obtained orbitals of the component of the wave function in (a). *A*, *B*, *C*, and *D* correspond to the shaded orbitals of the wave function shown schematically in (a) (see text). The orbitals are essentially distorted sp^3 atomic-like orbitals on both the carbon and the oxygen. Note the similarity between the orbitals associated with carbon here and the orbitals associated with carbon shown in Fig. 1A, 3, 6(b), and 7.

Fig. 8. Valence-bond wave function of CO_2

It is interesting to observe that the usual description for multiple bonds obtained by molecular orbital theory consists of orbitals of σ and π symmetries, which are distinctly different in their spatial descriptions. For example, in the conventional description of ethylene, the double bond is composed of a σ bond and a π bond. The σ bond consists of orbitals that are directed along the internuclear axis, and the π bond consists of orbitals that have zero amplitude along the internuclear axis. σ and π bonds can also be described in terms of hybrids, as shown in Fig. 5(c). Here, the σ bond is made up of sp^2 atomic-like hybrids and the π bond is made up of overlapping atomic-like *p* orbitals. The results presented here from recent studies[11,13-15] which take account of electron correlation do not support such a description. Two equivalent, off-axis or *bent* bonds (Ω-bonds) with the *same symmetry* are energetically preferred for the carbon-carbon double bond of ethylene rather than a σ and a π bond (compare Fig. 5a and c).[15] A similar situation exists for the triple-bond case of C_2F_2, where three equivalent bonds are formed instead of a σ and two π bonds. Even in the multicomponent form of the wave functions required for benzene and CO_2, the individual valence-bond structures have an interpretation in terms of Ω-bonds with tetrahedral hybrid orbitals on the atoms. In any of these cases, forcing the orbitals to become σ- and π-like (by imposing variational constraints) increases the calculated energy of the ground state, demonstrating that it is an inferior solution.[11,13-15]

From the results discussed above and from other calculations, it can be concluded that the carbon atom and other atoms in this row of the Periodic Table (which have only *s* and *p* valence orbitals) can be thought of as forming essentially sp^3 hybrids regardless of their bonding environments. Furthermore, one may use this insight to formulate a simple method for predicting the molecular structure of *sp*-bonded molecules.[16] Namely, to a good approximation, the structure of *sp*-bonded molecules can be thought of as arising from two simple principles: (1) each atom shares electron pairs so as to achieve a closed-shell configuration (the Lewis-Langmuir octet rule), and (2) the atoms arrange themselves so that the electron pairs about each atom are approximately tetrahedrally distributed (to minimize the potential energy). These principles are illustrated in Fig. 9 and 10. In Fig. 9(a), two schematic representations of a C-H bond are shown. In the upper diagram, the correlated nature of the two-electron bond is represented in a manner previously used in describing methane in Fig. 2(a); a shorthand notation is given in the lower diagram. In Fig. 9(b), using the shorthand notation, the many-electron wave function of CH_4 is depicted. Note the tetrahedral distribution of electron pairs about the central carbon atom. In Fig. 9(c), the bonding in C_2H_4 is illustrated. Observe that two electron pairs are shared by the carbon atoms. Figure 9(d) shows the bonding in B_2H_6. The similarity to C_2H_4 is very clear in this representation. The B_2H_6 molecule can be considered to arise from the removal of a proton from each carbon nucleus of C_2H_4, with the protons attaching themselves to the electron pairs shared by the resulting boron atoms. It must be kept in mind what the shorthand notation actually stands for. For example, the electron pair at a bridging H in Fig. 9(d) is shorthand for a situation which is more accurately represented in Fig. 9(e). There are three orbitals, one from each atom, which have two electrons distributed among them. The two electrons are singlet-coupled, and the actual wave function will be a superposition of the various alternatives of distributing two electrons among the three orbitals. Thus, those two electrons will provide the bond between both borons and the hydrogen.

As another example of bonding in boron hydrides, consider the electron distribution and bonding in B_4H_{10} as shown schematically in Fig. 10. Here a further simplification in notation from that used in Fig. 9 is employed. The B-H two-center bonds (involving the four terminal hydrogen atoms) are now represented as dark connecting lines between the atoms. The electron pairs involved in three-center bonds with hydrogen (bridging hydrogen atoms) are not explicitly shown, since the hydrogen atom marks the approximate position of the pair. The hybrids from the boron atoms which share electron pairs are shown as light connecting lines. There is one pair of electrons which is not shared by a proton and forms an ordinary covalent B-B bond. It is shown as a shaded sphere. As described elsewhere,[17] this

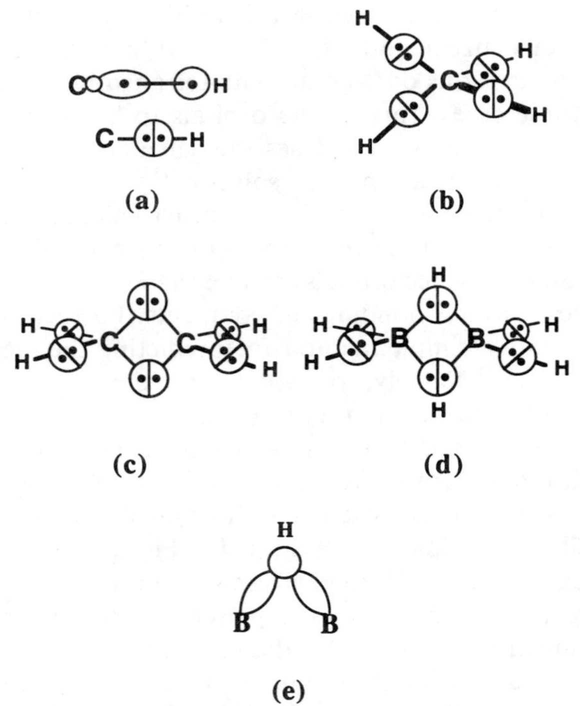

(a) The orbitals of a C-H bond showing "left-right" correlation of the bonding electron pair, together with a shorthand notation employed in the other diagrams. (b) Diagram of the bonding in CH_4 (compare Fig. 2a). (c) Bonding in the C_2H_4 molecule showing how electron pairs are shared between the two C atoms (compare Fig. 5a). (d) Bonding in B_2H_6 showing the close correspondence to the C_2H_4 molecule. (e) The three orbitals among which two electrons are alternatively placed in a "three-center, two-electron bond" (as in B_2H_6).

Fig. 9. Schematic representations of bonding determined from the orbitals of correlated many-electron wave function

approach (i.e., the description of bonding in terms of hybrid orbitals) provides a simple way of understanding the structures of a wide variety of boron hydrides.

The concept of approximate tetrahedral hybrids for describing the bonding of carbon and boron compounds is thus seen to be very useful and general. The fact that boron hydrides are electron-deficient molecules raises the possibility that the concept may be applicable to other electron-deficient materials—for example, metal clusters and bulk metals. This is investigated next.

Metal Clusters and Interstitial Electrons

An important contribution to the understanding of electronic structures in metals has been made by McAdon and Goddard.[18] They studied the electronic structures of many lithium clusters using explicitly correlated many-electron wave func-

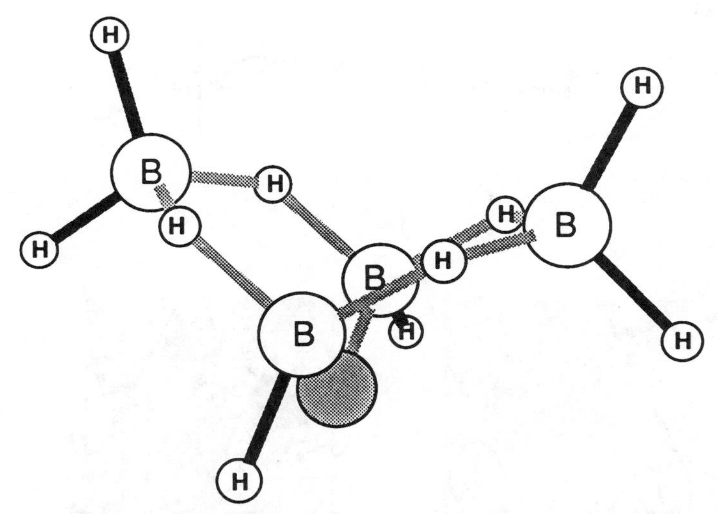

The lighter connecting lines represent tetrahedral hybrid orbitals originating from the boron atoms. The hybrids overlap to share an electron pair (denoted by shaded sphere near bottom of figure); four such electron pairs are also shared with protons (bridging hydrogens) to create three-center, two-electron bonds.

Fig. 10. Schematic representation of B_4H_{10} wave function

tions and demonstrated that electrons in such clusters occupy localized orbitals in interstitial regions between the atoms. It is convenient to think of these occupied interstitial orbitals as *interstitial electrons*.

The concept of an interstitial electron, *per se*, is not new, and in fact predates quantum mechanics. Lorentz used the concept,[19] and many other workers have discussed variations of this idea, including J. J. Thomson,[20] Pauling,[21] Fajans,[22] Bent,[23] Nowotny,[24] Schubert,[25] Johnson,[26] and others. Much of the work resulted in promising, simple ideas which attempted to bridge the gap between the behavior of electrons in metallic conductors and in materials with strong covalent bonds. However, these ideas have not found much application because of the lack of a rigorous foundation and the difficulty in making contact with traditional ideas of metals (e.g., delocalized electrons and Bloch states).

An essential aspect of the McAdon and Goddard work is the way in which the derivation of the concept of the interstitial electron differs from previous studies. For McAdon and Goddard (MG), the concept is derived from the calculated orbitals of a many-electron wave function. Since their results are important to the development presented here, we will discuss briefly some of their principal findings.

MG found that the relative positions of the interstitial electrons depended on the dimensionality of the cluster. Roughly speaking, for 1D chains of atoms, the interstitial electrons were found on sites between two atoms; in 2D hexagonal clusters, the interstitial electrons were localized in triangular faces; and in 3D clusters, the interstitial electrons were found in tetrahedral regions. More precisely, for one-dimensional chains of atoms (Fig. 11a) an orbital localizes between each pair of

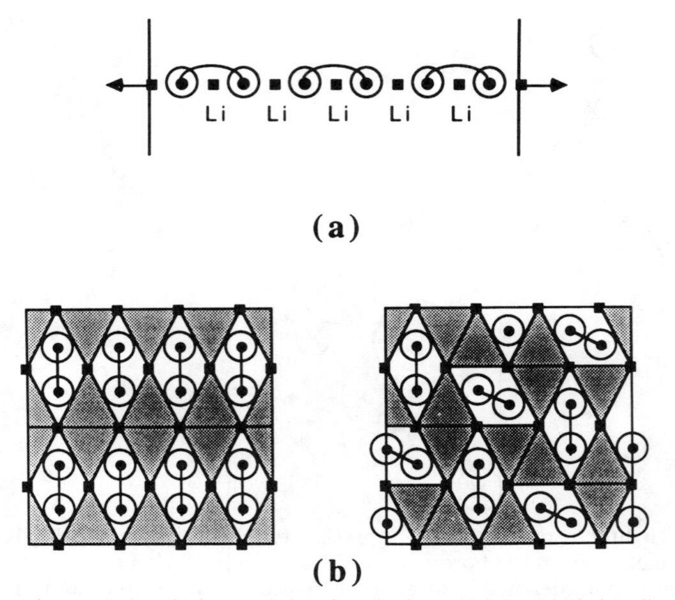

(a) Linear chain of Li atoms showing the interstitial orbitals localized between neighboring atoms with the orbitals occupied by a single electron and singlet coupling between adjacent orbitals. (b) Two different spatial distributions of interstitial orbitals for an ordered, planar array of Li atoms. The different distributions have nearly the same energy. (See text.)

Fig. 11. Schematic representations of one- and two-dimensional lithium clusters

atoms and is occupied by one electron. The electrons in adjacent interstitial orbitals spin-couple to form singlet pairs, but the dominant bonding between atoms is via the one-electron bonds formed by the interstitial electrons. In fact, all the electrons can be high spin-coupled and the chain is still stable with respect to dissociation into lithium atoms! This is in contrast to the situation for normal covalent two-electron bonds—for example, the H_2 molecule discussed above, where the high spin (triplet) state was unbound. The origin of this behavior is the small overlap between the interstitial orbitals of the metal cluster, which results in a small energy difference between the various spin states.

For the one-dimensional chains, all *interstices* are occupied, and thus there are no low-lying excitations (within kT) which could lead to electrical conductivity. However, for two-dimensional sheets there are a variety of valence-bond structures which result in low-lying excited states suitable for conductivity. This is shown schematically in Fig. 11(b). For these planar arrays of atoms, the orbitals localize into triangular interstices among three atoms with one electron in each orbital. Orbitals in adjoining interstices have a small overlap and are singlet-coupled in the ground state as in the 1D chains.

In three-dimensional clusters, it was found that the orbitals localize into tetrahedral interstices with either one or two orbitals in the interstices, depending on the cluster geometry. We consider three of the 3D, 13-atom clusters here. The geometries of these clusters are shown in Fig. 12(a), 13(a), and 14(a). It is convenient to describe the cluster geometries in terms of packing of tetrahedral and octahedral regions. The connecting lines between the atoms make these geometric features more apparent. As in the methane and C_2F_2 examples, we consider the wave function to have only one set of spatial orbitals and one spin pairing.[(27)] Each of the clusters has only six pairs of electrons and thus has a net charge of +1.

First consider the 13-atom fcc cluster (cf. Fig. 12a). This cluster can be thought of as a fragment of an fcc crystal. The central atom is surrounded by eight tetrahedral regions. The six concave regions that expose the central atom from a perspective outside the cluster can be described as halves of octahedral regions. The wave function computed by MG consists of twelve interstitial orbitals which are shown schematically in Fig. 12(b) (only six are shown explicitly, the others are hid-

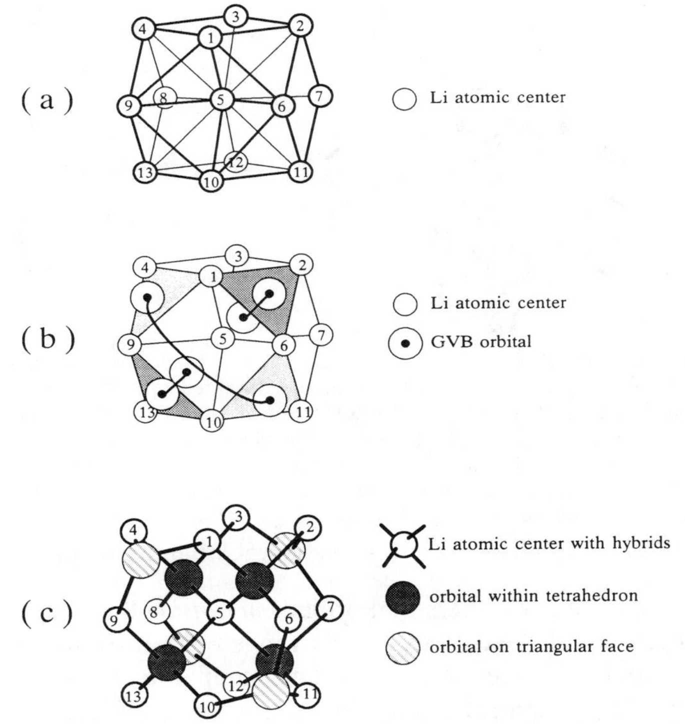

(a) Geometry of the cluster is shown with a perspective view. (b) Schematic representation of variationally obtained orbitals from a wave function of the form of Eq 8 (from Ref 18c). The pair of orbitals associated with the tetrahedra T(1,2,5,6) and T(5,8,10,13) are *inside* the tetrahedron but are shown as separated spheres to indicate that there are two electrons in this region. The centers of the orbitals near the triangular faces t(1,2,6) and t(9,10,13) are *outside* of the cluster. The other two pairs of orbitals are associated with tetrahedra on the other side of the cluster and are not shown for clarity. Note that none of the orbitals is directed along an internuclear axis, hence the name interstitial orbitals. (c) Locations of interstitial orbitals obtained by overlapping tetrahedral hybrids of nearest-neighbor atoms. (See text.)

Fig. 12. Thirteen-atom fcc lithium cluster

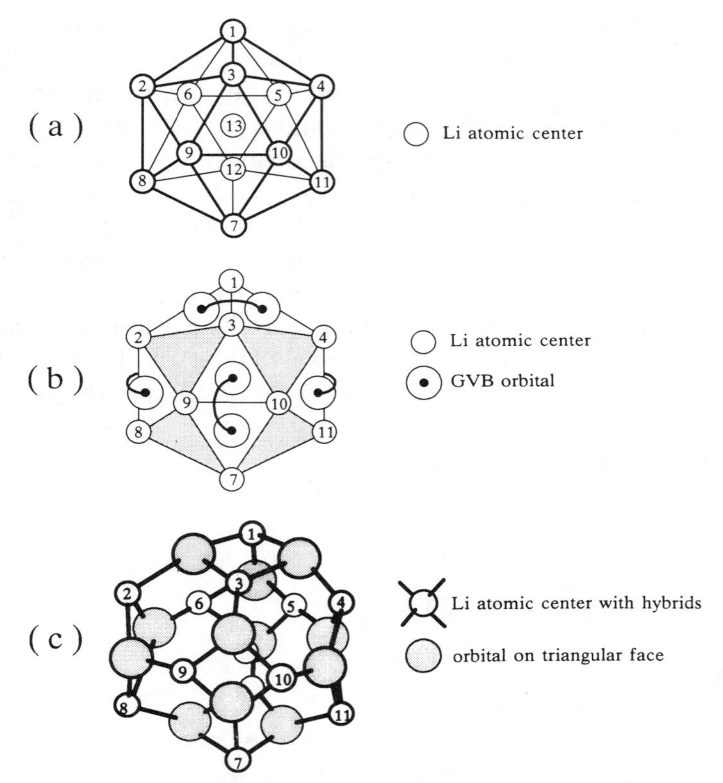

(a) Geometry of the cluster is shown with a perspective view. (b) Schematic representation of variationally obtained orbitals from a wave function of the form of Eq 8 (from Ref 18c). Orbitals associated with tetrahedra on the far side of the cluster are not shown for clarity. As in Fig. 12, none of the orbitals is directed along an internuclear axis. (c) Locations of interstitial orbitals obtained by overlapping tetrahedral hybrids of nearest-neighbor atoms. (See text.)

Fig. 13. Thirteen-atom icosahedral lithium cluster

den from view). Four pairs of interstitial orbitals are *inside* the cluster. These allow four pairs of singlet-coupled electrons to be arranged tetrahedrally around the central atom. These four pairs can be described as occupying four of the eight tetrahedra that surround the central atom. The atoms are labeled in Fig. 12. In the following, the symbol T(i,j,k,ℓ) denotes the tetrahedron defined by the atoms i, j, k, and ℓ and $\frac{1}{2}$O(i,j,k,ℓ,m) denotes the half octahedron defined by the atoms i, j, k, ℓ, and m. The tetrahedra containing correlated pairs of electrons are T(1,2,5,6), T(3,4,5,8), T(5,9,10,13), and T(5,7,11,12). The other four orbitals are singly occupied, and are distributed on the external faces of the four unoccupied tetrahedra—T(1,4,5,9), T(2,3,5,7), T(5,6,10,11), and T(5,8,12,13). The pairing of the electrons in orbitals on the faces is indicated by solid lines connecting the dots. Examples of half octahedra are $\frac{1}{2}$O(1,2,3,4,5) and $\frac{1}{2}$O(1,5,6,9,10).

Note that the valence-bond structure shown in Fig. 12(b) cannot represent a complete wave function for the cluster. First, for the orbitals on the external faces, there are two independent ways to pair the electrons without changing the energy.

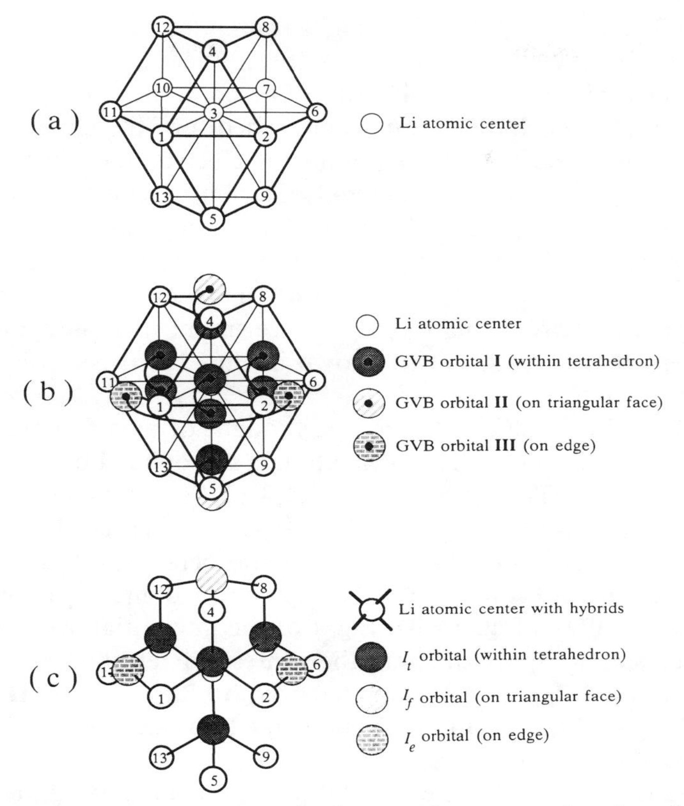

(a) Geometry of the cluster is shown with a perspective view. (b) Schematic representation of variationally obtained orbitals from a wave function of the form of Eq 8 (from Ref 18c). As in Fig. 12 and 13, none of the orbitals is directed along an internuclear axis. (c) Locations of interstitial orbitals obtained by overlapping one set of tetrahedral hybrids of nearest-neighbor atoms. (See text.)

Fig. 14. Thirteen-atom hcp lithium cluster

This results in two valence-bond spin structures for the particular distribution of interstitial orbitals. In addition, by inverting the orbital coordinates and keeping the lithium positions fixed, the interstitial orbitals are redistributed such that the previously unoccupied tetrahedra become occupied and *vice versa*. The combination of two energetically equivalent electron pairings with two energetically equivalent ways of distributing the interstitial orbitals means that a minimum of four valence-bond structures are required to represent an approximate many-body wave function of the cluster.

Next, consider the 13-lithium-atom *icosahedral* cluster (Fig. 13a). This cluster consists of 12 atoms arranged at the corners of an icosahedron and a single atom at the geometric center. This can be thought of as a packing of 20 slightly distorted tetrahedra around the central atom. The results of MG are indicated schematically in Fig. 13(b). The 12 singly occupied interstitial orbitals are arranged in six pairs on the faces of the cluster separated from each other as far as possible. Thus there are 12 triangular faces, denoted by t(i,j,k), such as t(1,3,4), which contain an interstitial electron. These are separated by eight triangular faces, such as t(3,4,10), which are empty. Together they account for the 20 faces of the icosahedron. The singlet-coupled electron pairs are denoted by the connecting lines.

Note that none of the orbitals are *inside* tetrahedra of the cluster. This can be understood in terms of electron correlation—the electron pairs repel each other. If four of the electron pairs were tetrahedrally distributed around the central atom, two of the electron pairs would be forced to occupy regions exterior to the cluster, as in the fcc cluster. These latter pairs of electrons have higher energy than the interior pairs of electrons. Since there is no way to arrange four interior pairs and two exterior pairs of electrons in a way that is consistent with the symmetry of the cluster, the *compromise* arrangement shown in Fig. 13(b) is obtained. We will return to this point later.

Finally, the hcp cluster (Fig. 14a) can be thought of as a fragment of an hcp crystal with an ideal c/a ratio. The central atom is surrounded by eight tetrahedral regions [e.g., T(1,2,3,4), T(1,2,3,5), and T(3,4,8,12)] and six halves of octahedral regions [e.g., $\frac{1}{2}$O(2,3,4,6,8) and $\frac{1}{2}$O(2,3,5,6,9))] as for the fcc cluster, but with a different relative orientation of the tetrahedral and octahedral regions. The wave function computed by MG consists of 12 singly occupied interstitial orbitals which are shown schematically in Fig. 14(b). Eight of the interstitial orbitals (labeled **I**) are *inside* the cluster in approximately the centers of tetrahedral regions [e.g., T(3,6,7,8) and T(3,6,7,9)], two of the interstitial orbitals (labeled **II**) are outside the cluster in three-fold faces [t(4,8,12) and t(5,9,13)], and two of the interstitial orbitals (labeled **III**) are along edges.

In considering the results of these calculations, some interesting questions arise: Can we predict where the interstitial electrons should be without detailed calculations? Can the concept of interstitial electrons be used to think about bulk metals? If so, is it possible to predict where the interstitial electrons should be in the metal? What will be the consequences? For the moment, only the first question is considered.

McAdon and Goddard(18) have presented some rules which are deduced from their results. The rules are:

(α) Orbitals (each with one electron) are localized in different tetrahedral hollows where possible.

(β) If necessary, two electrons may be placed in one tetrahedron, but they must be spin-paired.

(γ) No more than two electrons may be distributed between a pair of face-shared tetrahedra, and these must be spin-paired.

(δ) No more than three electrons may be distributed between a pair of edge-shared tetrahedra.

(ε) No more than four bond pairs of electrons may share one central bulk atom, and no more than three bond pairs of electrons may share one central surface atom.

(η) Additional electrons must be in surface orbitals at edge or face sites that do not share edges with occupied tetrahedra.

These observations suggested that a structure with a total energy even lower than that of the icosahedra might be obtained with some rearrangement of the 13 atoms. In particular, MG were led to consider arrangements of the 13 atoms that maximize the number of tetrahedra under the restriction that the number of tetrahedra shared by any given atom should be relatively small. This restriction was suggested by "the observation that 12 valence electrons cannot all simultaneously bond in an effective manner to one central atom."[18c] This is essentially a consequence of the Lewis-Langmuir octet rule. Based on the rules above, a set of alternative geometries (OPTET clusters) that were anticipated to have lower energies were constructed. As expected, the wave functions of the lower-symmetry OPTET structures had lower energies than the wave functions of the high-symmetry clusters.

While these rules are, perhaps, useful, they are strictly empirical and lack a more fundamental basis. On the other hand, it has been found from our studies that: (1) the tetrahedral hybrids offer a natural way to understand the electron distribution and predict the results of their cluster calculations and (2) tetrahedral hybrids provide new insight about bulk metal alloys. While tetrahedral hybrids can also be used to understand the positions of the interstitial electrons in the OPTET clusters, we will not make direct use of these results later and so our analysis of these clusters is not presented here. We present only the results of our analysis of the high-symmetry clusters. It is important to note, however, that the application of tetrahedral hybrids is *not* restricted to high-symmetry geometries.

Connection Between Interstitial Electrons and Tetrahedral Hybrids

First, consider the Li_{13^+}, fcc cluster. In Fig. 12(c), it is shown how each of the tetrahedral (sp^3) hybrid orbitals (H-orbitals) on the central atom overlap with hybrids from three other atoms so as to share an electron pair. The orientation of the hybrid orbitals is shown by the heavy lines, and the vertex of the orbital overlaps are shown as spheres. By comparing Fig. 12(b) and (c), it can be seen that the interstitial orbital (I-orbital) positions calculated by MG correspond to the vertices of the overlapping H-orbitals. The H-orbitals localized to a particular interstitial region can be thought of as a basis for describing the I-orbitals. Thus, the tetrahedral hybrids provide a simple, direct way of understanding the calculated results (see Fig. 12b) which show that four of the eight tetrahedra about the central atom contain electron pairs that are tetrahedrally oriented about the central atom. The other surface interstitial electrons are each shared by three surface atoms, with each atom contributing one tetrahedral hybrid orbital.

Next we consider the *icosahedral* cluster. The interpretation of a single valence-bond structure in terms of tetrahedral hybrids on the 12 outer lithium atoms is shown in Fig. 13(c). The hybrids on the central atom are not used directly. Each interstitial orbital in this cluster is formed from the overlapping of three H-orbitals, one from each of three adjoining lithium atoms, as for example in t(1,3,4). That is, each I-orbital can be constructed as a linear combination of three H-orbitals.

We can now return to a discussion of why none of the electron pairs are *inside* the cluster. In our previous discussion, we described the distribution as arising from unfavorable contributions to the energy (higher total energy)—namely, electron-pair repulsions. A favorable contribution to the energy (lower total energy) arises from overlapping of the tetrahedral hybrids. The tradeoff between electron-pair repulsion and hybrid overlapping results in the distribution shown in Fig. 13. Some insight into this tradeoff can be obtained by counting the number of hybrids that are actually used in the icosahedral cluster. For the arrangement shown in Fig. 13(c), three of the hybrids (*H*-orbitals) on each of the outer atoms are used, for a total of 36. In order for the *I*-orbitals to exist *inside* the cluster, fewer *H*-orbitals on the 12 exterior atoms would be used. This can be seen by noting the arrangement in the fcc cluster, where only 28 *H*-orbitals are used. This delicate balance between the maximal use of *H*-orbitals and the repulsion between electron pairs will play a key role in our later discussion of the Laves phases.

In the hcp cluster a new and important feature arises. In the two examples above, the GVB wave functions correspond to one of several equivalent valence-bond structures that are required to represent a more accurate wave function. Occasionally, however, the orbitals from a GVB calculation *differ* from the more highly localized orbitals derived from tetrahedral hybrids that we have been using as a basis. This occurs when several valence-bond structures, related to each other by translations in spatial position, are necessary to describe the total wave function. In the case of two valence-bond structures, the GVB orbitals are *smeared* along the direction of the translation that relates the two structures. Such is the case for the hcp cluster, where the orbital arrangement can be understood in terms of tetrahedral hybrids if two arrangements of the hybrids are used. These two structures are shown in Fig. 15(a). In Fig. 15(b), the two valence-bond structures are superimposed, allowing a clearer comparison with Fig. 14(c). In this case, all of the orbitals *inside* the cluster contain one electron. The GVB wave function of MG can be viewed as a *mean field* average of the two structures, which results in the orbital positions shown.

Returning to the linear chain and two-dimensional clusters, one might ask how tetrahedral hybrids can explain the structures. Recall that the tetrahedral hybridization is a consequence of two effects: (1) the stability inherent in forming a closed shell of electrons about each atom (the octet rule) and (2) the minimization of potential energy. If one chooses to arrange atoms in a way such that tetrahedral hybrids cannot be effectively used (as in linear chains or two-dimensional structures), a rehybridization will be required in order to optimize the energy of the structure chosen. It should be clear, however, that for larger clusters the energy of the optimal 1-D and 2-D structures will be higher than for the optimal 3-D structures. This is, in fact, found.[(18)] For the 1-D case, it is easily seen that the best overlapping of orbitals is achieved for *sp*-hybridization, which leaves two orbitals unoccupied on each atom. For the 2-D case, it is the sp^2-hybridization which is most favorable, leaving one orbital/atom unoccupied. The relative energy of the 1-D cluster is less favorable than that of the 2-D cluster. In general, for atoms with *s* and *p* orbitals, the use of approximate hybridizations other than tetrahedral will raise the energy of the molecule or cluster.

To summarize the results for 3-D clusters: we recognize several types of interstitial orbitals in the fcc, icosahedral, and hcp lithium clusters. These arise because

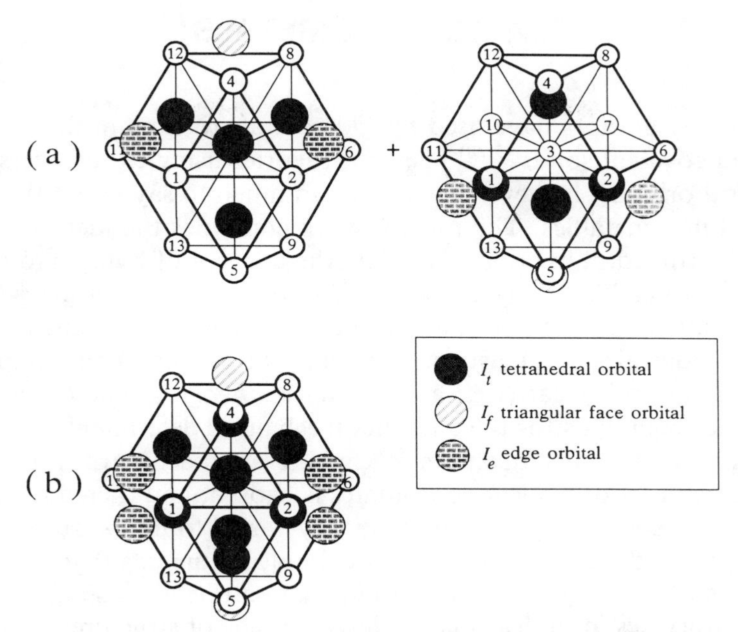

(a) The interstitial orbitals in the structure on the left correspond to the hybrids of Fig. 14(c). The interstitial orbitals on the right are obtained with an alternative orientation of the hybrids. (b) Combined set of interstitial orbitals from (a). Compare with McAdon and Goddard's results (Ref. 18c) shown in Fig. 14(b).

Fig. 15. Location of interstitial orbitals formed from two sets of overlapping tetrahedral hybrids

of the various ways in which tetrahedral hybrid orbitals (H-orbitals) can combine to yield interstitial orbitals (I-orbitals). When two H-orbitals overlap they can form an *edge* interstitial orbital (I_e-orbital); when three H-orbitals overlap they form a triangular face interstitial orbital (I_f-orbital); and when four H-orbitals overlap they form a tetrahedral interstitial orbital (I_t-orbital). These I-orbitals can be used to describe the bonding in the lithium clusters (Fig. 12 to 14) as:

1. The fcc cluster (Fig. 12) is made up of four I_t-orbitals and four I_f-orbitals.
2. The icosahedral cluster (Fig. 13) is made up of 12 I_f-orbitals.
3. The hcp cluster (Fig. 14) is made up of four I_t-orbitals, two I_f-orbitals, and two I_e-orbitals.

These I-orbitals are constructed to contain one electron, but in certain cases, such as the I_t-orbitals of the fcc and hcp clusters, two electrons occupy the tetrahedral interstitial region. A simple approximate way to think of this situation is that such an I_t-orbital contains two electrons. In reality, the situation is somewhat more complicated than this and the H-orbitals, which serve as a basis set to describe the I-orbitals, have to be combined differently (i.e., new linear combinations of the basis functions are needed) to produce two new orbitals to allow for electron correlation effects. For the discussion that follows, we ignore these complications as corrections to the general qualitative picture we present.

SIMPLE BULK METALS

In the past, many authors have used hybrid orbitals to explain the structures of molecules and covalent solids.[5,28] Previous theories using either interstitial electrons or hybrid orbitals, however, have not had much to say about the structures of metals and intermetallics. The present work combines the ideas of interstitial electrons and tetrahedral hybrids, extending the concept of hybrids to bonding in metallic and intermetallic structures. We have seen that the use of hybrids provides a relatively simple means of determining the geometrical arrangements of atoms (*structure*) in molecules and clusters. The utility of an analogous simple scheme for metals and alloys is clear (i.e., a *constructive theory* of alloy structure).

The computational basis for the quantum mechanical determination of structure is energy minimization. Most quantum mechanical approaches use a trial-and-error method—that is, a set of nuclear coordinates is chosen, and then a total energy is computed.[29] The coordinates are systematically varied to minimize the energy. In this situation, there do not appear to be any simple arguments that can be applied to decide the structure *a priori*. Thus any progress toward a simple scheme for rationalizing structures, or at least narrowing the range of structures to be checked, should be welcomed.[30] While it is our general aim to provide a quantitative theory to investigate structural questions of alloys, at the present stage of development the best we can offer is an analysis of the electronic structure and a rationalization of the structure if the atomic arrangement is already known. There are many cases, however, where plausible arguments can be made to predict structure. We believe that our analysis, illustrated below, does provide some insight into the properties of the materials as well as a basis for comparing the electronic structure and properties of different materials.

One of the advantages of a constructive procedure is the insight it provides by employing *building blocks* to construct complicated structures from simpler ones. Just as large molecules can be built from smaller ones and small molecules can be built from atoms, a framework based on localized orbitals allows the transfer of a portion of a wave function—i.e., local orbitals—from one system to another (from molecule to molecule and from molecule to solid). Thus, the tetrahedral nature of the bonding in diamond, cubic-BN, and BeO is not surprising. In fact, there are a large number of III-V and II-VI semiconductors involving only *s* and *p* orbitals in their bonding which have either zinc blende or wurtzite structures (cf. Fig. 16a and b). In these cases, tetrahedral hybrids provide a natural way of describing the bonding. In a simple valence-bond picture, the bonds are formed by overlapping pairs of tetrahedral hybrids from each of the atoms. Discussion of these semiconducting solids in terms of sp^3-hybrids is quite common.[28,31]

In many cases, and particularly for most of the II-VI compounds such as BeO, the bonds are highly polarized, with the electron pairs more tightly drawn to the region of space surrounding the Group VI element. Starting with BeO, which has the wurtzite structure (two interpenetrating hcp lattices), and the valence-bond structure shown in Fig. 17(a), imagine that neutral oxygen atoms are removed from

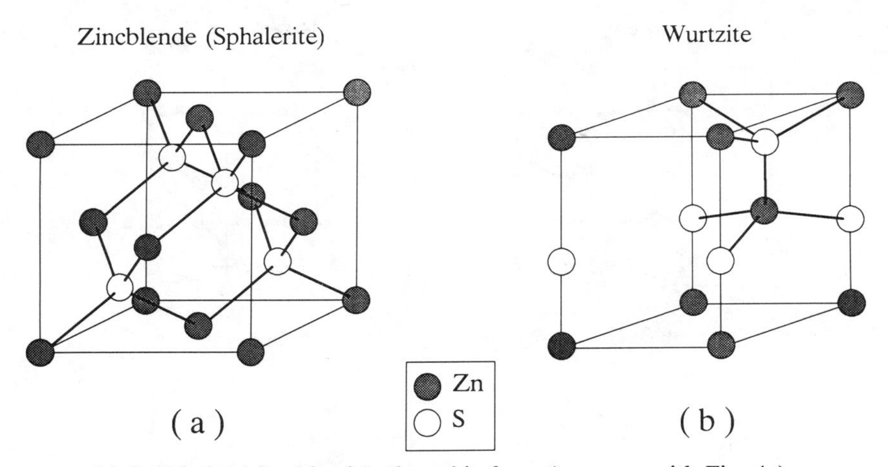

(a) Sphalerite (zinc blende), the cubic form (compare with Fig. 4a).
(b) Wurtzite, the hexagonal form. Note that each atom is tetrahedrally surrounded by four pairs of electrons.

Fig. 16. Simplified schematic representation of valence-bond structure of zinc sulfide

the lattice, leaving two electrons localized in this vicinity. By uniformly compressing the lattice such that the beryllium hybrids overlap each other, an interstitial orbital is formed from the hybrids that can be occupied by the two electrons (Fig. 17b). This results in an hcp metal lattice with electron pairs in tetrahedral interstices of the lattice such that each metal atom has four electron pairs distributed tetrahedrally about it.

There are two tetrahedra per electron pair into which electrons can be placed. Thus, an alternative way of arranging the orbitals that is energetically equivalent to Fig. 17(b) is shown schematically in Fig. 17(c). Furthermore, there are many other valence-bond structures of nearly equal energy that differ from those in Fig. 17(b) and (c) in that the hybrids at only a single atom are modified. Since this represents a relatively small difference in orbital structure, which can be thought of as an *excitation* at a single site, these excited single-site valence-bond structures and multiple-site excitations must be included in a complete description of the wave function. Thus, the two ordered arrays of pairs (Fig. 17b and c) are only a shorthand way of describing a large number of valence-bond structures.

Charge Density in Beryllium Metal

An array of electron pairs as described above can be thought of as a useful starting point for considering the ground-state properties of the bulk metal.[32] Before proceeding down this path, however, it is reasonable to ask if this description is consistent with known facts—for example, the charge distribution of the metal. Fortunately, recent experimental work has produced a high-quality charge density for the case of beryllium metal.[33]

The orbital arrangements shown in Fig. 17(b) and (c) suggest an accumulation of charge in the tetrahedral regions of the crystal at the expense of charge in the

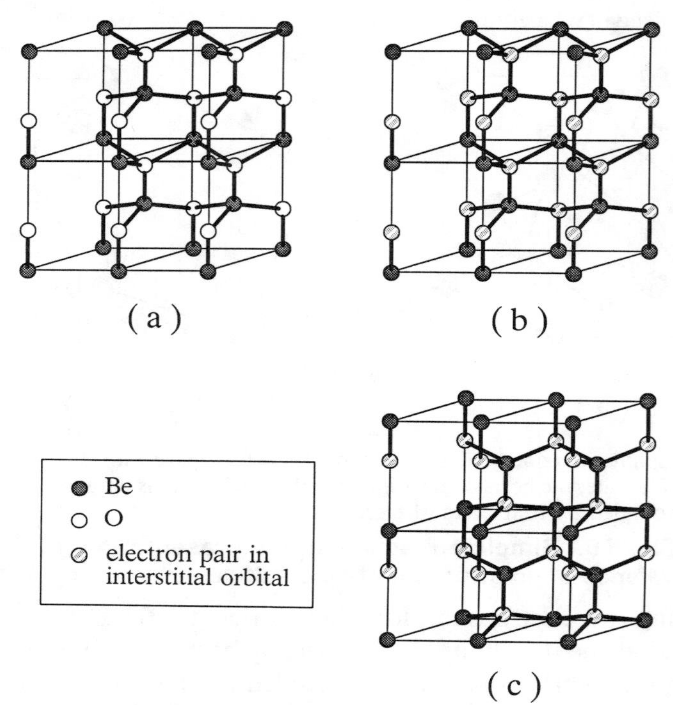

(a) Simplified schematic representation of the valence-bond structure of BeO (as in wurtzite, Fig. 16b). (b) One valence-bond structure of Be with interstitial orbitals, each containing an electron pair. (c) Another valence-bond structure of Be that uses a different orientation of the hybrids (*H*-orbitals) than (b). Here, the orbitals occupy a different set of tetrahedra than in (b). Note that each Be and O in (a) and each Be in (b) and (c) are tetrahedrally surrounded by electron pairs.

Fig. 17. Valence-bond structures of BeO (a) and Be (b and c)

octahedral regions. In the hexagonal plane, this should lead to an accumulation of charge in the triangular faces shared by two tetrahedral regions and a depletion of charge in the triangular faces shared by two octahedral regions. This is, in fact, observed in the experimental charge density shown in Fig. 18(a).

In order to compare the results of localized electrons with this experimental data, it is necessary to describe some recent cluster calculations for beryllium which were used to generate a charge density for the bulk metal.[(34)] The hcp structure can be thought of as being constructed from face-sharing tetrahedra of beryllium atoms separated by face-sharing octahedral voids as shown in Fig. 19(a). Considering just the atoms labeled 1 to 5, a cluster calculation was set up with boundary conditions appropriate to the solid.

The orbitals of the wave function with the pair of electrons in one of the tetrahedral interstices—e.g., that defined by atoms 1, 2, 3, and 4—was determined by a variational calculation. The calculation was repeated for the pair in the tetrahedron defined by atoms 1, 2, 3, and 5. To construct a proper wave function for the

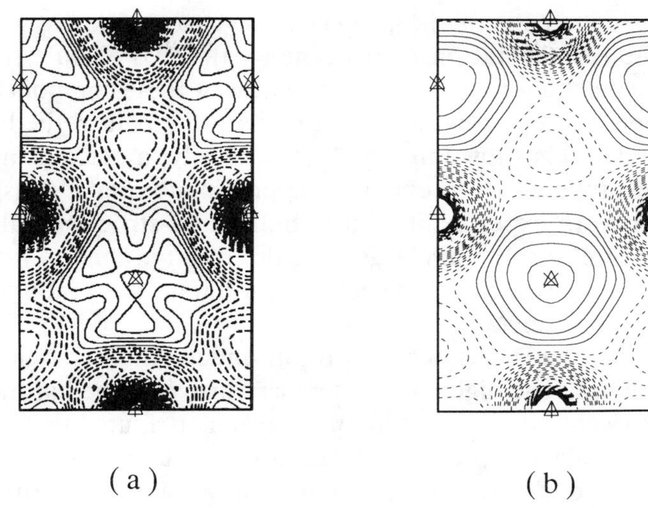

In (a), the contours have positive (solid lines) or negative (dotted lines) values relative to the average density which has been subtracted to show the regions of charge accumulation and depletion. The region of positive contours in the lower half of the figure is bounded by a triangle of atoms that form the shared face of two tetrahedra (e.g., atoms 1, 2, and 3 in Fig. 14a; see also Fig. 19). The region with negative contours in the upper half of the figure is bounded by a triangle of atoms that form the shared face of two octahedra (e.g., atoms 1, 3, and 11 in Fig. 14a; see also Fig. 19).

Fig. 18. (a) Contour plots of (a) experimentally determined Be charge density in the hexagonal plane (from Ref 33) and (b) approximate charge density of bulk Be obtained from the superposition of a periodic array of interstitial electron orbitals (see text)

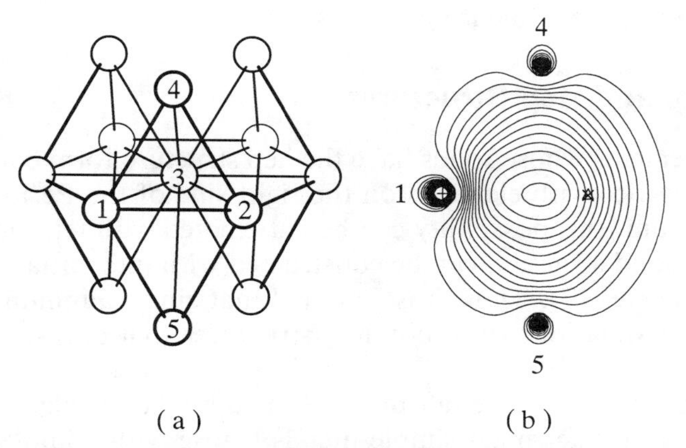

Charge density in (b) was obtained by coherent superposition of orbitals from T(1,2,3,4) and T(1,2,3,5) of (a) (see text).

Fig. 19. Portion of hcp lattice showing cluster used in model calculation (a) and contour plot of charge density of 5-atom cluster in (a) in plane of atoms 1, 3, and 5 (b)

pair of electrons, a coherent superposition of these two alternative structures must be made (as in the representation of benzene by the two Kekulé structures[13]). The charge density resulting from the superposition of these components (one with the pair in the upper tetrahedron and the other with the pair in the lower tetrahedron) was obtained and is shown in Fig. 19(b). Finally, a periodic array of these charge densities was superposed to obtain an approximate charge density (Fig. 18b) in good agreement with experiment[32] and bulk band-structure calculations.[35] Thus, the concept of electron pairs largely localized to tetrahedral interstitial sites is consistent with the known charge density of beryllium, even though it is not the usual way one thinks of metals.

This picture of the bonding is also helpful in understanding some of the properties of bulk beryllium. The high melting temperature, extreme hardness, strength, and diamagnetic susceptibility of beryllium are often attributed to a certain degree of covalency. The singlet-coupled interstitial orbitals suggest an extension to the concept of a covalent bond. An ordinary covalent bond involves the sharing of a pair of electrons between two neighboring atoms. A three-center covalent bond, as in the B_2H_6 examples discussed previously, involves the sharing of two electrons among three neighbors, and a four-center covalent bond involves the sharing of an electron pair among four neighbors, as in the tetrahedrally localized orbitals above. Note the systematic progression from two- to three- to four-center bonds as we move from carbon to boron to beryllium in the Periodic Table. This picture can also explain the metallic conductivity of beryllium. The fact that there are more hybrid orbitals than valence electrons provides the electrons with many spatially different arrangements (valence-bond structures) that have nearly the same energy. Thus, in the presence of an applied electric field, the electrons can easily move from one interstitial region to another without moving into an orbital that is already occupied by another electron. In fact, all structures with metallic conductivity have more hybrid orbitals than valence electrons.

Bonding in Fcc and Hcp Structures

Many metallic elements whose solids have the hcp structure form compounds having the wurtzite lattice with elements on the right side of the Periodic Table. By analogy to beryllium, this possibility can be anticipated with tetrahedral hybrids, where a valence-bond structure can be constructed with interstitial orbitals at the anionic positions (Fig. 17). Thus, it is expected that zinc, cadmium, magnesium, etc., to first approximation, form bonding structures that are similar to that of beryllium.

The energetic preference for tetrahedra, indicated by the calculations on the lithium clusters, suggests that most simple metals (those with completely occupied outermost *d* or *f* shells) should have close-packed structures. McAdon and Goddard[18] proposed this as the reason for the frequent occurrence of the close-packed structures among the elements. Thus the bonding in metals with a single valence electron, such as lithium and copper, is expected to have a close-packed structure with an interstitial electron in alternate tetrahedral sites. In fact, whereas at room temperature the alkali metals are bcc, at low temperatures, bulk lithium *is* in the form of a close-packed structure, a so-called 9R structure.[36] Furthermore, sodium undergoes a transition to a close-packed structure,[37] and there is evidence of precursors to similar transitions in potassium and rubidium.[38] A reason for the bcc form of the alkalis at room temperature will be suggested shortly. In the case of aluminum, the structure is also fcc, but the valence-bond structure is somewhat different. Instead of a single electron in alternate tetrahedra, we must imagine that there are three electrons attempting to fit into alternate tetrahedra. This is accomplished by not combining all four of the *H*-orbitals (which point into a given tetrahedral region) into a single *I*-orbital. Thus, the three electrons are distributed in a more complicated fashion among the four hybrid orbitals. For the purpose of understanding the structure, however, it is convenient to use the shorthand picture of three electrons in a single interstitial region.

Note that as in the case of the hcp structure, the fcc structure also provides two alternative ways of arranging interstitial electron orbitals and, again, the two valence-bond structures should be viewed as a shorthand notation for the multiple alternate ways of arranging the electrons.

It is also possible to provide a simple explanation of the tendency for the fcc structure to exist with metals having one and three valence electrons, such as copper and aluminum, respectively, and the tendency of the hcp structure to exist with metals having two valence electrons. This is accomplished by considering the consequences of electron pairing and the Coulomb repulsion between the metal nuclei. We propose that the driving force for determining the hcp structure is the minimization of electron pair–pair repulsion energy, whereas for the fcc structure the primary force is the minimization of the energy of Coulomb repulsions between the nuclei.

For this discussion we make use of the analogy between fcc and zinc blende and between hcp and wurtzite, with electrons in fcc and hcp corresponding to anions in zinc blende and wurtzite, respectively.* To estimate the electrostatic contributions, we consider ideal lattices of *point charges*. In a wurtzite lattice, which has two hcp sublattices made up of oppositely charged species, the electrostatic attraction between the two sublattices is greater than if the two oppositely charged sublattices were arranged in the cubic zinc blende structure. This is indicated by the relative magnitudes of the Madelung constants, which are proportional to the total electrostatic energy of such arrays of point charges. The wurtzite Madelung constant is 1.641, while the zinc blende Madelung constant is 1.638. Thus the wurtzite arrangement is favored by electrostatic interactions when the oppositely charged species are well approximated by pointlike charges.

In a metal with two valence electrons, the electrons can pair and occupy alternate tetrahedra. The Pauli repulsion between pairs tends to favor maximum distances between pairs. Recall that in small molecules, this is accomplished by

---

*For this discussion, it will be helpful to refer to Fig. 16 and mentally replace S with an *I*-orbital that contains either one, two, or three electrons depending on whether a metal with one, two, or three *sp* valence electrons (such as Cu, Zn or Al) occupies the Zn positions.

orienting the electron pairs in roughly tetrahedral positions. For an fcc or hcp array of metal ions this is accomplished by placing the pairs in an ordered array of alternate tetrahedra. If the electron pairs are relatively stable in the tetrahedra, the hcp array of metal ions is favored according to the argument in the preceding paragraph, because the electrostatic energy is minimized.

In the case of metals with one valence electron per atom, the *ideal* electron arrangement, from a classical electrostatic point of view, is again the placement of electrons in alternate tetrahedra. While the spin pairing in metals with two valence electrons per atom helps to stabilize the positions of the interstitial electrons, in the case of metals with one valence electron per atom, the spin pairing tends to delocalize the electrons. An unpaired electron in one tetrahedron will attempt to singlet-couple with an electron in another tetrahedron. If the tetrahedra are too distant, the pairing energy is less favorable. Thus, there is increased probability that the electrons are paired in neighboring tetrahedra and the screening of the nuclear charge that occurs between second nearest neighbors across the close-packed planes for the hcp structure is not as complete. Therefore the cations assume the fcc structure in order to minimize electrostatic repulsions between the nuclei. A similar argument can be constructed for metals with three valence electrons and, thus, can also be applied to aluminum.

Bcc Metals

From the results described above, close-packed structures (lots of tetrahedra) are, in general, favored by *enthalpy*. However, structure is determined by the *free energy* and the enthalpy is only one part. A study of the thermodynamics of various structures indicated that a high *entropy* is associated with the bcc structure.[39] The authors of that study arrived at this conclusion by estimating the contributions of the low-order terms in a Fourier representation of an expansion of the free energy. They showed that the bcc structure *should* be favored at high temperatures. In fact, the high-temperature structure of most elements up to the melting temperature *is* the bcc structure.

In special circumstances, the bcc structure may be viewed as the result of a process favored by enthalpy. In these cases, the bcc structure results from a resonance between two fcc-like bonding structures. (A detailed discussion of this process is presented in the appendix.) This idea may be useful for understanding the structural phase transitions in some of the elements such as iron.* At low temperatures and high temperatures iron is bcc, but it has a close-packed structure at intermediate temperatures. The arguments presented in the appendix suggest that the bcc structure may be stabilized at low temperatures by resonance energy.

Thus, a simple scenario for the structural phase transitions in iron might be as follows. At low temperatures, the resonance energy stabilizes the bcc structure. This

*Other transition metals, such as Ta and Cr, also exhibit phase transitions in thin films, although this may be related to compound formation with hydrogen, oxygen, or nitrogen. See references in Donohue's book.[10]

occurs because the individual structures are combined coherently (add in phase). As the temperature increases, the lattice expands, weakening the resonating bcc bonds. The entropic contribution to the free energy is still relatively small. At a critical temperature, the resonance energy is not sufficient to offset the loss of energy in the weakened bcc bonds and the structure rearranges such that individual bonds are stronger, to the close-packed structure. Finally, at high temperatures, the entropic contribution dominates, forcing a transformation back to bcc. While this scenario is admittedly speculative and neglects consideration of the *d* orbitals, it may provide a basis for more detailed numerical calculations.

SIMPLE INTERMETALLICS

The notions of semiconductors with two-center covalent bonds and metals with interstitial electron bonds provide a plausible physical picture of the extreme differences in the bonding. Intermetallic compounds, on the other hand, frequently have properties somewhat intermediate between those of a pure metal and those of a nonmetal. Thus, a model that merges these two extremes is needed to understand intermetallic bonding.

The simplest models of the bonding in intermetallic compounds are found in their molecular analogs, the organolithium compounds. The advantage of studying these molecules is that they are small enough to enable detailed calculations to check various bonding ideas. The smallest molecule in this class is represented by CLi_4. This will serve as a prototype for some of the more complex models of intermetallic bonding that we will discuss later.

Naively it might be thought that CLi_4 would have the bonding structure of methane, since lithium, like hydrogen, has only one valence electron. On the other hand, the presence of low-energy *s* to *p* excitations in the lithium atom and the metallic nature of bulk lithium might lead one to expect something different. From the results discussed above on the lithium clusters and our previous discussion of carbon, we might anticipate the following picture of bonding: the lithium nuclei will assume a tetrahedral arrangement, and four lithium valence electrons will be interstitially localized in the four triangular faces of the tetrahedron, forming a metallic cage. The carbon atom will be located in the center of the molecule, and its four valence electrons in sp^3 hybrids will each pair with one of the electrons in the faces of the tetrahedron. A model of this structure is shown in Fig. 20. Recent calculations confirm this picture.[40] Contour plots of one of the four equivalent carbon-lithium bonds are shown in Fig. 21, where it can be seen that the metallic orbitals are polarized toward the carbon atom.

Although the species CLi_4 is known to exist,[41] the structure has not been determined experimentally. On the other hand, the structure of the species $(CH_3Li)_4$ has been determined (cf. Fig. 22a).[42] From calculations,[41] it is found that the bonding in this structure is similar in many respects to the bonding in CLi_4. The results of these calculations are shown schematically in Fig. 22(b). Note again metallic orbitals in each of the four triangular faces of the tetrahedron and that

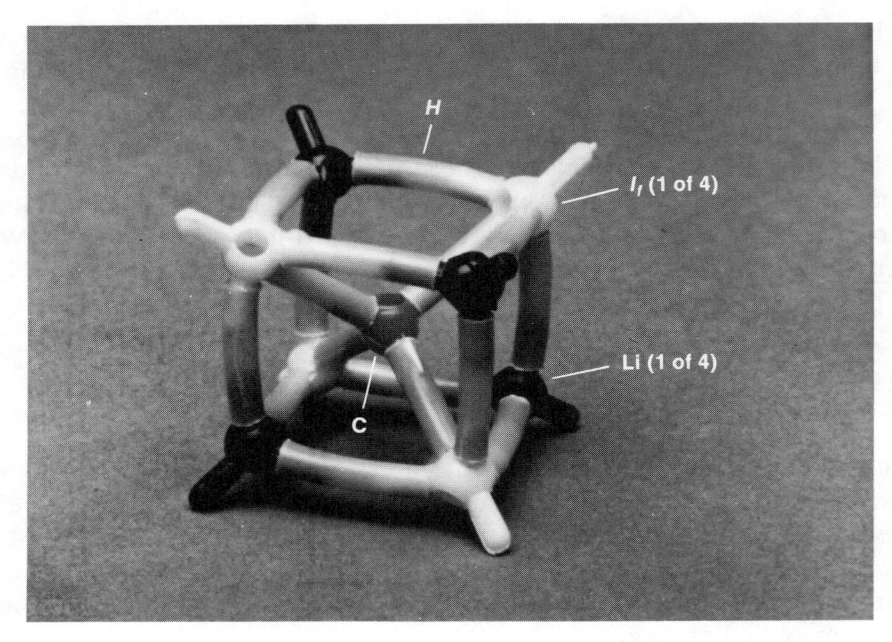

The Li atom centers are represented by the outer dark vertices, the C atom is represented by the center vertex and the centers of interstitial electron orbitals (I_f) are represented by light-colored vertices. The sp^3 hybrid orbitals (H) are represented by connecting lines. The distortion of the tetrahedral angles on the hybrids is partially an artifact of the model components. (See text.)

Fig. 20. Model of CLi_4

each of these orbitals is spin-paired with a tetrahedral hybrid orbital on a carbon atom and polarized toward the carbon atom. The similarity between the two types of hybrid carbon orbitals that bond to hydrogen and the metal cage can be seen by comparing Fig 22(c) and (d).

This picture of the bonding provides a simple way of thinking about the bonding of a wide variety of other organolithium molecules without introducing special forms of carbon, such as hexavalent carbon.[(43)] In fact, there has been a great deal of argument in the chemical literature about the nature of the bonding of these materials and the degree of ionicity or covalency. The present picture offers a simple alternative that provides somewhat greater qualitative flexibility in a situation where the classical concepts of ionic bonding and covalent bonding do not seem to apply. For example, with a slight modification, this bonding scheme provides a simple way of thinking about the bonding of interstitial atoms in transition-metal alloys. First, however, note that the most common interstitial atoms—i.e., oxygen, nitrogen, carbon, and silicon—typically occupy octahedral sites[(44)] rather than the tetrahedral sites suggested by the preceding paragraphs.* That this is not incompatible with the organolithium examples can be understood by considering the geo-

*Note that hydrogen frequently occupies tetrahedral sites. The study of hydrogen trapping and hydrogen embrittlement in metals along the present lines is virtually unexplored. See also Ref 30.

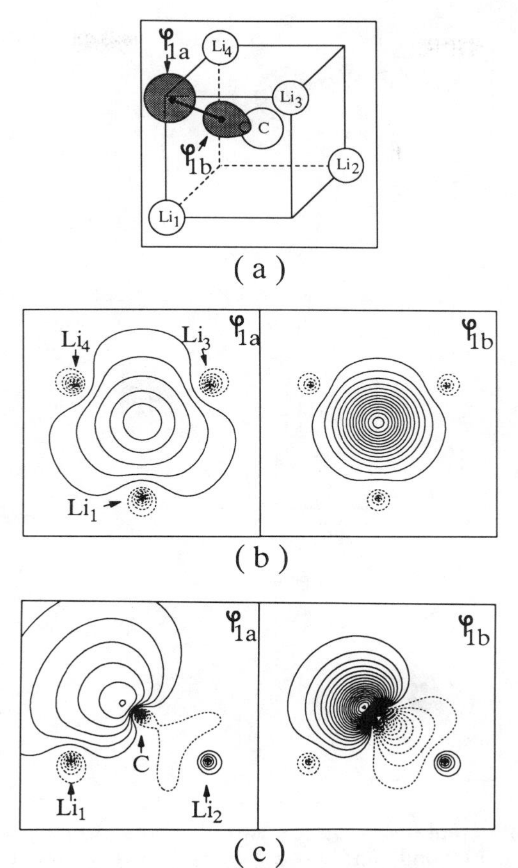

(a) Schematic representation of one of the C-Li bond-pair orbitals consisting of a Li interstitial electron orbital and a tetrahedral hybrid orbital on C. The Li interstitial orbital is composed of sp^3 hybrids on neighboring Li (cf. Fig. 20). (b) Contour plots of variationally obtained orbitals using a wave function of the form of Eq 8 (i.e., one component). The plane of the plot passes through the centers of Li atoms 1, 3, and 4 in (a). Cross sections of the interstitial orbital (φ_{1a}) and the carbon orbital (φ_{1b}) are shown. (c) Another cross section of the interstitial orbital and the carbon orbital showing the C-Li bond. Here, the plane passes through Li atoms 1 and 2 as well as the carbon atom in (a).

Fig. 21. Orbitals of the valence-bond wave function of CLi_4

metric structure. The atomic radii of the transition metals are smaller than lithium and, in the cubic close-packed structures, an octahedra is surrounded by eight tetrahedra. Thus, the more electronegative interstitials can orient their hybrid orbitals through four of the triangular faces of an octahedron into the tetrahedral bonding regions. This is shown schematically in Fig. 23. Since there are two ways of orienting the hybrids through the octahedral faces, the valence-bond wave function contains a superposition of both alternatives. The resonance energy of such a coherent superposition of two alternatives provides additional stability to the

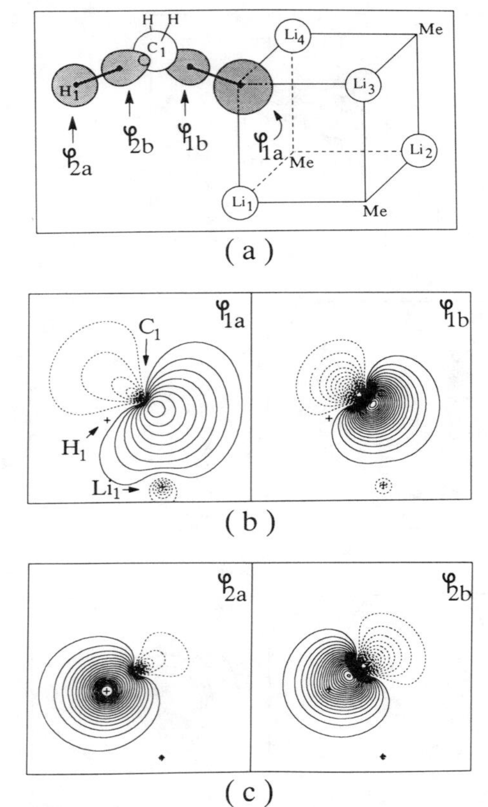

(a) Schematic representation of two pairs of orbitals: φ_{1a}-φ_{1b} describe a C-Li bond, and φ_{2a}-φ_{2b} describe a C-H bond of the methyl group (CH_3). (b) Contour plots showing cross sections of the C-Li bond orbitals highlighted in (a). The contour plane passes through Li atoms 1 and 2 as well as C_1 and H_1 in (a). (c) Contour plots showing cross sections of the C-H bond orbitals highlighted in (a). The contour plane is the same as in (b). Note the similarity between the carbon orbitals in the C-Li bond (φ_{1b}) and in the C-H bond (φ_{2b}).

Fig. 22. Orbitals of $(CH_3Li)_4$ (the structure is described in the text)

bonds. A model of a valence-bond structure showing how the *H*-orbitals on the interstitial atom can form bonds to the surrounding metal atoms is shown in Fig. 24. This modification of the carbon-lithium bonding scheme has many other applications as well. In the next section we discuss the applications to some of the cubic structures listed in Table 1.

Binary and Ternary Cubic Structures

In this section we will describe the valence-bond structures of LiAl, which has the NaTl structure, and several compounds with the Cl_3 structure. These examples illustrate that simply counting valence electrons is not sufficient to determine the

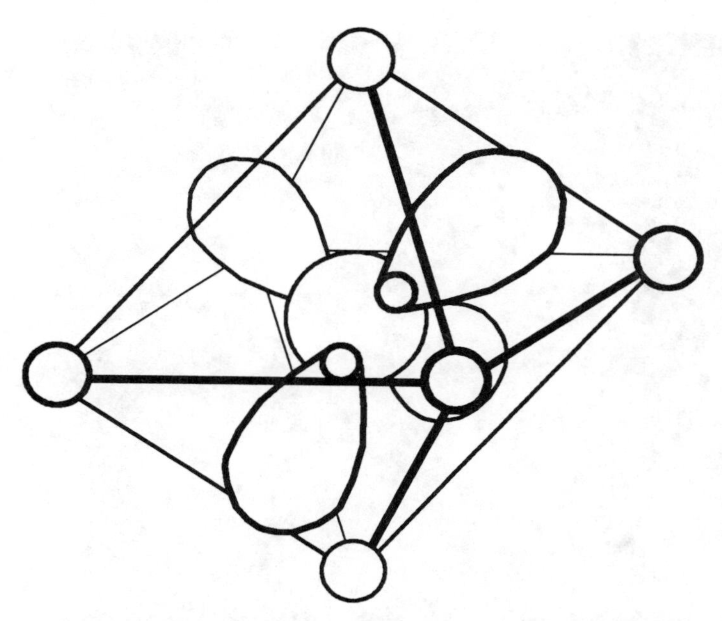

The hybrid orbitals on the interstitial atom are tetrahedrally directed through four of the triangular faces of the surrounding octahedron.

Fig. 23. Schematic representation of one of the valence-bond structures of an interstitial atom at an octahedral site in a transition metal

Table 1.[45] Relationships among various cubic structures indicated by occupation of the *A, B, C* or *D* sites in Fig. 25

E in the table means that the site is vacant (*empty*). The compounds in the last three rows of the table are all examples of the $C1_3$ structure.

Structure	Name	Example	Position *A*	Position *B*	Position *C*	Position *D*
*A*2	bcc	Li	Li	Li	Li	Li
*B*2	CsCl	CsCl	Cs	Cl	Cl	Cs
*B*3	Diamond	C	C	*E*	C	*E*
	Zinc blende	ZnS	Zn	*E*	S	*E*
*B*32	NaTl	LiAl	Al	Li	Al	Li
$L2_1$	Heusler alloys	Cu_2MnAl	Cu	Mn	Al	Cu
DO_3	BiF_3	Li_3Bi	Bi	Li	Li	Li
*B*1	NaCl	NaCl	Na	*E*	*E*	Cl
*C*1	CaF_2	CaF_2	Ca	F	F	*E*
$C1_3$	...	LiMgSb	Mg	*E*	Sb	Li
	...	LiAlSi	Al	*E*	Si	Li
	AgMgAs	CuMgSb	Cu	Mg	Sb	*E*

electronic structure. Before describing these more complicated structures, however, we note their relation to the simple cubic structures that are normally observed for metallic elements, namely fcc and bcc. Recall that fcc can be thought of as a structure with two tetrahedra and one octahedron per atom. These features of the fcc

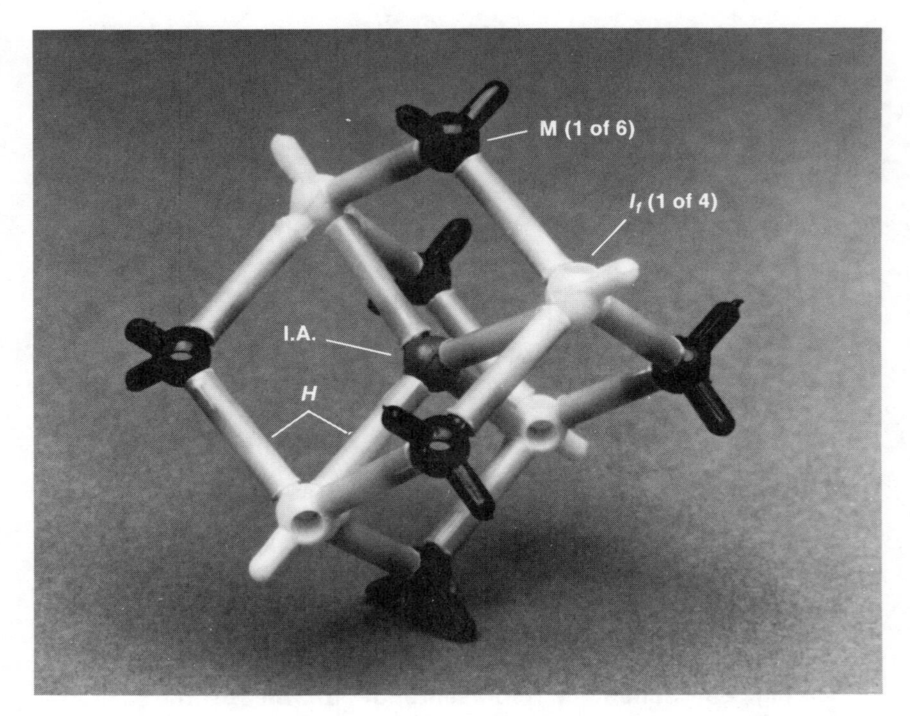

As in Fig. 20, the metal atom centers (M) are represented by the six dark outer vertices, the interstitial atom (I.A.) is represented by the central gray vertex, interstitial electron orbitals are represented by four light-colored, outer vertices (I_f) and sp^3 hybrids are represented by the connecting lines (H). The interstitial orbitals are composed of tetrahedral hybrids on the three surrounding metal atoms.

Fig. 24. Model of the valence-bond structure of Fig. 23

structure can be seen by referring to Table 1 and Fig. 25, where the A positions form an fcc lattice, the B and C positions are located at the centers of tetrahedra (tetrahedral sites) of the A lattice, and the D positions are located at octahedral sites. Because there is a significant degree of ionic bonding in the materials with more complicated structures, it will be necessary to elaborate further on the relationship between ionic bonding and valence-bond structures.

Many intermetallics have a degree of ionic bonding. This qualitative concept is difficult to quantify, but is clear in the extreme case of large charge transfer from one element to another in a compound. The picture of the bonding for CLi_4 and interstitial atoms (Fig. 23 and 24) described above offers a complementary view to the traditional model of ionic bonding in terms of point charges. To further illustrate this complementary nature, consider NaCl—the prototypical system for ionic bonding. The structure of NaCl can be viewed as an fcc sodium lattice with chlorine in octahedral sites (see Table 1 and Fig. 25). If it is imagined that the valence electron of sodium is *transferred* to the chlorine, then the chlorine has four pairs of electrons and the sodium has four empty orbitals. The four chlorine lone pairs are directed tetrahedrally about the chlorine center. It is not unreasonable to

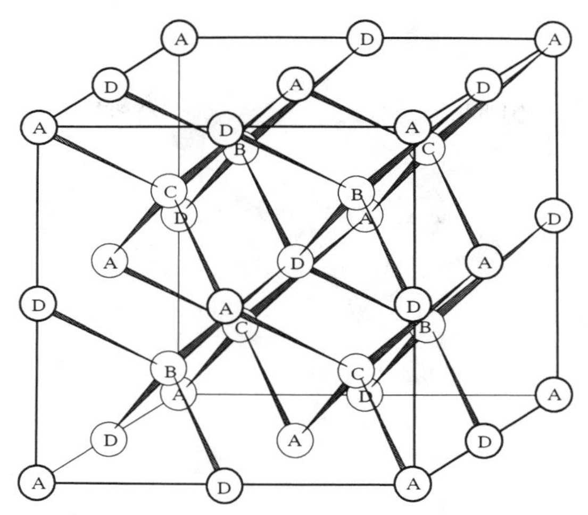

Note that the usual bcc lattice has only one basis atom. This figure is used with Table 1 to illustrate the relationships between many classes of cubic structures.

Fig. 25. Bcc lattice with four basis atoms

imagine that the chlorine lone pairs are directed toward four of the triangular faces of the surrounding octahedral sodium cage, as in Fig. 23 and 24, providing additional stability to the ionic component. (Note that a purely ionic collection of point charges is more stable in the CsCl structure.) As in the example of the atomic interstitials, there are two possible orientations for the lone pairs. Thus, NaCl can be viewed as an fcc sodium lattice with chlorine atomic interstitials.

The NaTl structure is related to the NaCl structure (see Table 1) and is a structure formed by a number of I-III compounds. Here we consider one of them: LiAl. Zintl proposed that each of the lithium atoms transfers its valence electron to an aluminum atom so that each aluminum atom can then form covalent bonds with its four aluminum nearest neighbors.[46] Thus aluminum satisfies the Lewis-Langmuir octet rule. This idea is consistent with the fact that LiAl is a semiconductor and is a useful first approximation to the electronic structure. While the role of tetrahedral hybrid orbitals on aluminum is immediately apparent, the orientation and use of the tetrahedral hybrids on lithium are not so clear. However, by orienting the lithium hybrids toward the nearest-neighbor aluminum, some of the charge can delocalize into the lithium orbitals, allowing lithium also to satisfy the octet rule (cf. Fig. 26). This is the simplest valence-bond structure for this system that is consistent with both physical properties and the charge density obtained by density functional band-structure calculations.[47] Three other compounds that form this structure are LiGa, NaTl, and NaIn. The electronic structure in these compounds is expected to be similar to the electronic structure of LiAl.

A variation of the LiAl structure is found in LiAlSi. In this compound the aluminum and silicon atoms occupy the positions of a zinc blende lattice, and lithium is tetrahedrally arranged around the silicon (see Table 1.) Again, the simplest interpretation of the bonding in this structure is that proposed by Zintl and Brauer.[46] As in LiAl, the lithium in LiAlSi donates its valence electron to the alu-

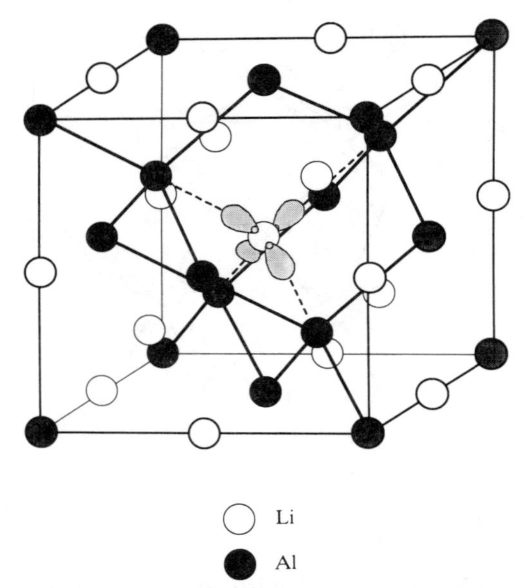

The Al atom hybrids overlap, forming a diamondlike bonding structure. The essentially unoccupied Li hybrids are oriented toward the corners formed by the surrounding tetrahedra of Al atoms. The hybrids on only one Li atom are shown for clarity. (See text.)

Fig. 26. Geometry and valence-bond structure of LiAl

minum, which allows the aluminum to form a covalent bond with each of its four silicon nearest neighbors. The electron affinity of silicon is slightly greater than that of aluminum, however, so the Si-Al bonds will be polarized with greater electron density (negative charge) on the silicon. By using the same argument about the orientation of the lithium hybrids as for LiAl, or an electrostatic argument, it is possible to understand why the lithium prefers the sites around the silicon rather than the sites around the aluminum. The valence-bond structure for this system is shown in Fig. 27.

The charge density that might be expected from such a simple picture of the bonding in these two systems is consistent with recent density functional band-structure calculations.[48] In discussing the results of the calculation, the authors expressed surprise at finding fractional occupancy of lithium *p* states (about 0.5 electron per atom). In our view, because of the hybrids, *p* occupancy is *required* for the bonding.

LiMgSb forms a structure that is similar to LiAlSi, except that magnesium occupies the aluminum positions and antimony occupies the silicon positions. The total number of valence electrons per formula unit is again eight, so it might be anticipated that a similar bonding picture applies. The electron affinity of antimony is much greater than that of either lithium or magnesium, and the ionization potential of lithium is lower than that of either antimony or magnesium. Thus, it is logical to assume that the valence electron is transferred from lithium to antimony, and that the magnesium and antimony species form a network of covalent bonds. The

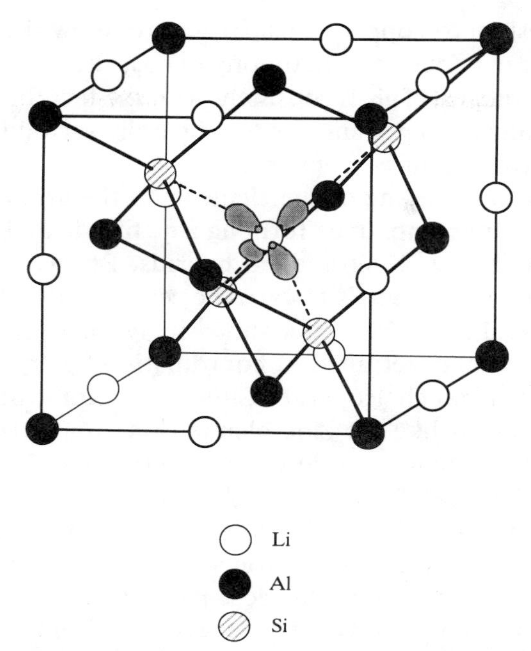

The Al and Si atom hybrids overlap, forming a diamondlike bonding strucutre. As in LiAl, the Li hybrids are essentially unoccupied, but are oriented toward the corners of the more electropositive Si atom. (See text.)

Fig. 27. Geometry and valence-bond structure of LiAlSi

donor atom, lithium, is again surrounded by the element with the greater electron affinity and, as before, all elements attempt to satisfy the octet rule. The valence-bond structure again describes the material as a semiconductor, which is consistent with experimental knowledge.

Assuming that the *d* shell does not participate in the bonding, the compound LiZnAs is isoelectronic with LiMgSb and should form a similar structure with zinc replacing magnesium and arsenic replacing antimony. The electronic structure predicted using the above construction is fully consistent with a charge-density analysis from a detailed density functional band-structure calculation.[(49)] The authors describe the material as a "half ionic and half covalent tetrahedral semiconductor."

Again neglecting the *d* shell, CuMgSb is isoelectronic with LiMgSb and might be expected to form an identical structure, with copper in the positions of the lithium. In fact, the structure is different. In this case, copper and antimony form a cubic zinc blende lattice and magnesium appears to occupy the position of a donor element, surrounded by a tetrahedral arrangement of copper atoms. By considering the relative electron affinities of these elements, it is expected that magnesium will donate charge to copper. Thus, a simple valence-bond structure consists of overlapping tetrahedral hybrid orbitals between copper and antimony. A charge transfer of two electrons from magnesium to the copper and antimony system is not likely because of the high second ionization potential of magnesium. The partial charge transfer from magnesium to copper is insufficient to allow the formation of four covalent bonds between copper and antimony. Thus, there should be many holes in the valence-bond structure. This is consistent with the fact that CuMgSb is metallic, and the donor atoms, magnesium, are tetrahedrally surrounded by the element with the greater electron affinity, copper.

We have seen how the bonding of the elements in the lithium row of the Periodic Table and the binary compounds forming zinc blende and wurtzite structures can be plausibly described with tetrahedral hybrids. But what about the technologically important transition metals toward the center of the Periodic Table? For now, we point to several facts. First, not surprisingly, there exist structures where transition metals do display tetrahedral coordination in crystal structures. For example, iron in the mineral chalcopyrite, $CuFeS_2$, is tetrahedrally coordinated by sulfur.[50] This structure is like the zinc blende (Fig. 16a) structure, except that copper and iron replace zinc in an ordered pattern (see Fig. 28). Secondly, a wide variety of metals and intermetallics with transition metals form cubic structures (see Table 1). Some of these structures will be described below. Finally, the ubiquity of tetrahedral arrangements of metal atoms in alloys, as first pointed out by Kasper,[51] especially in the topologically close-packed structures such as the Laves phases, and in amorphous metals, are very suggestive.[52] In the following sections, we will show how tetrahedral hybrids provide a natural description of these structures, as well as hexagonal and rhombohedral structures such as NiAs.

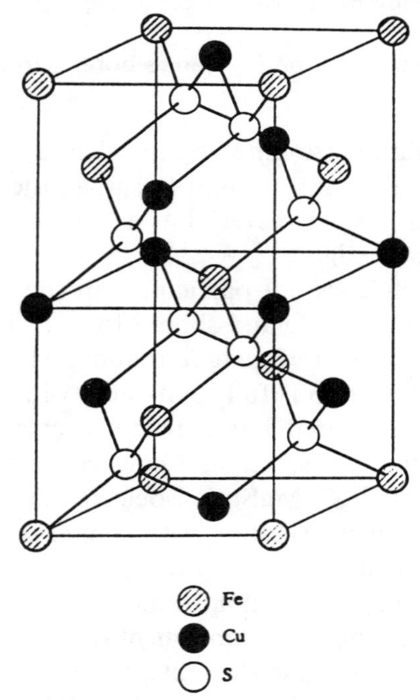

The atoms collectively occupy the positions of a cubic zinc blende lattice, with the ordered sublattices of Fe and Cu replacing Zn.

Fig. 28. Geometry and valence-bond structure of FeCuS

Laves Phases

The three classical Laves phases, represented by $MgCu_2$, $MgNi_2$, and $MgZn_2$ (cf. Fig. 29 and 30) are considered in this section. Collectively, phases of over 200 compounds with the general formula AB_2 have one or more of these structures.[53] It is generally believed that a combination of geometric and electronic factors is associated with the existence of these phases. We will comment on the relationship between these more traditional concepts and those of the present work in the last section of this paper. In this section, we present valence-bond structures for these compounds and use them to explain the observed correlation between the number of valence electrons and the type of structure found in a series of magnesium-based Laves phases with the above structures. Note the distinction among *structure, phase*, and *compound*. Thus, the $MgCu_2$ *structure* refers to the geometry indicated in Fig. 29(a) and 30(a), whereas the $MgCu_2$ *phase* refers to any material with the geometry indicated in Fig. 29(a) and 30(a). The $MgCu_2$ *compound* refers to the particular material composed of magnesium and copper that has the geometry of the $MgCu_2$ *structure*.

The close relationship among the three *structures* can be seen by comparing them along an axis of hexagonal symmetry. For the cubic structure ($MgCu_2$), this is along the (111) direction. Referring to Fig. 29 and 30, where the structures have been divided into *A* and *B* sublattices, respectively, we note several features. First, it can be seen that the *A* atoms (Mg) form tetrahedrally connected networks. In the cubic structure, the magnesium positions can be described as a diamond lattice (Fig. 29a). In the $MgZn_2$ structure, the magnesium positions are best de-

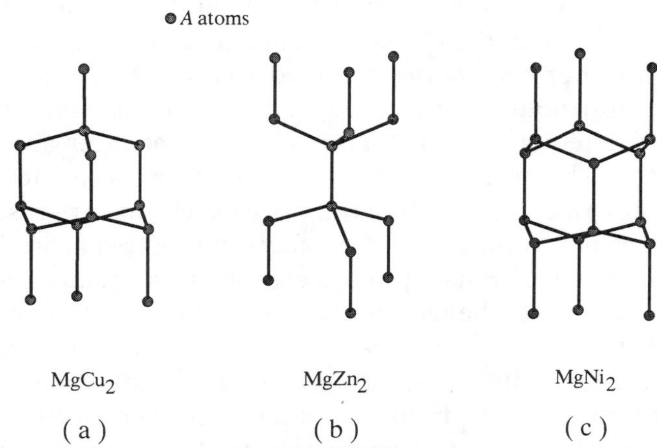

The vertical axis is parallel to the (111) axis in (a) and the hexagonal axes in (b) and (c). These nonprimitive unit cells are chosen to highlight the similarities and differences among the three structures. (a) Diamond-like lattice of the *A* atoms of $MgCu_2$. (b) Wurtzite-like lattice of the *A* atoms of $MgZn_2$. (c) The mixed (diamond and wurtzite) stacking of the *A* atoms in the $MgNi_2$ structure.

Fig. 29. Relative positions of the *A* atoms in the (a) $MgCu_2$, (b) $MgZn_2$, and (c) $MgNi_2$ Laves phase structures in terms of tetrahedral networks

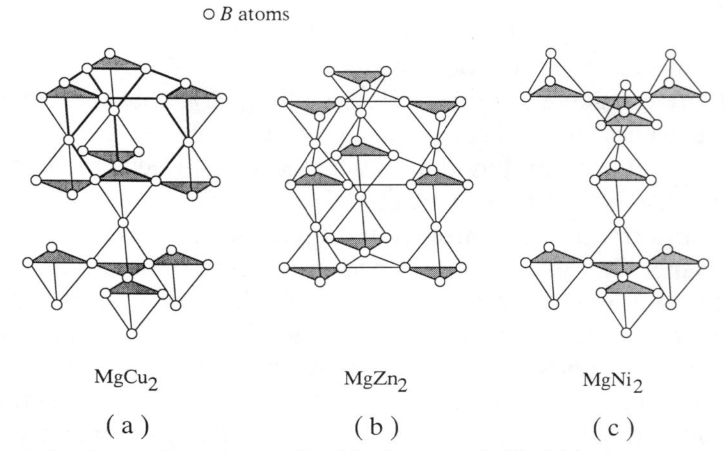

Referring to the region outlined in the upper half of (a), the *A* atoms are approximately centered in a region surrounded by four large hexagonal faces and four small triangular faces. This shape also contains the *A* atoms in (b) and (c).

Fig. 30. Relative positions of the *B* atoms in the (a) $MgCu_2$, (b) $MgZn_2$, and (c) $MgNi_2$ Laves phase structures in terms of stacked tetrahedra using the same orientations and unit cells as in Fig. 29

scribed as a wurtzite lattice where both atom positions are occupied by magnesium (Fig. 29b). The arrangement of the atoms in the $MgNi_2$ structure can be described as a combination of the diamond and wurtzite patterns (Fig. 29c). Secondly, the positions of the *B* atoms are most easily described in terms of a stacking of tetrahedra. In the cubic structure, the *B*-atom tetrahedra stack so that each tetrahedron shares a corner with four neighboring tetrahedra (Fig. 30a). In the $MgZn_2$ structure, the *B*-atom substructure can be thought of as parallel chains of tetrahedra. The tetrahedra are alternately joined along the chain at faces and at corners as shown in Fig. 30(b). The relative orientation of the face-sharing tetrahedra along the chain can be described as an alternating *twist* such that cross sections of the space between the chains in the plane of the face-sharing tetrahedra form an alternating sequence of large hexagonal openings and small triangular openings (cf. Fig. 30). In the $MgNi_2$ structure, the tetrahedra can be divided into two classes in the proportion 1:2 (cf. Fig. 30c). Tetrahedra in the first class, which make up one-third of the total, share a face with one other tetrahedron in the same class and share corners with three other tetrahedra in the other class. The tetrahedra in the second class are connected to other tetrahedra only by sharing corners such that three of the corners are shared with tetrahedra in the same class and one corner is shared with a tetrahedron in the first class. Finally, the *A* and *B* substructures are aligned so that the *A* atoms reside at the centers of the large spaces in the stacks of *B* tetrahedra. The polyhedral shape of the space that encloses each *A* atom and is formed by the *B*-atom tetrahedra is the same in all three cases (cf. Fig. 30). Furthermore, *A* atoms reside on both sides of the large hexagonal cross sections formed by the stacked tetrahedra of *B* atoms.

In order to describe the bonding in these phases, it will be convenient to describe valence-bond structures with tetrahedral hybrids (H-orbitals) first and to discuss the distribution of the valence electrons among the orbitals afterward. This sequence will lead to a simple rationalization of the correlation of the structure with the electron/atom ratio in the Mg-based Laves phases.

Our discussion of the bonding in these phases begins with the cubic *compound* $MgCu_2$, since this will make it easier to build upon the results of the previous section. The copper sublattice of $MgCu_2$ is considered first. In the discussion of bonding in the lithium clusters and in the carbon-lithium clusters above, it was observed that there are two qualitatively different ways to orient the H-orbitals of four tetrahedrally arranged atoms. First, the hybrids can be oriented such that four H-orbitals (one from each of the four atoms) point to the center of the tetrahedron and an I_t-orbital is formed at the interior of the tetrahedral cluster. Second, the H-orbitals can be oriented so that they are all inverted (through their atomic origins) with respect to the first orientation, resulting in four I_f-orbitals being formed—one at each of the tetrahedral faces. An example of the first arrangement is provided by the fcc cluster of Fig. 12(c) with atoms 1, 2, 5, and 6 forming the tetrahedron. An example of the second arrangement is provided by the Li_4 cluster in the CLi_4 molecule of Fig. 20. If it is assumed that four H-orbitals on each copper atom are involved in the bonding, the stacking of the tetrahedra in the copper sublattice shows that both types of orbitals (I_t and I_f) can be formed in alternate tetrahedra. A model of such a valence bond structure is shown in Fig. 31. In this model all of the *upward-pointing* tetrahedra have I_f-orbitals and all of the *downward-pointing* tetrahedra have I_t-orbitals. Note that an equivalent valence-bond structure can be formed with I_f-orbitals in the downward-pointing tetrahedra and I_t-orbitals in the upward-pointing tetrahedra. The larger H-orbitals on the magnesium atoms overlap with each other forming a set of I_e-orbitals* between each pair of nearest neighbor magnesium atoms.

We next consider the distribution of the electrons. This can be obtained by counting interstitial electron orbitals and valence electrons. Each magnesium atom contributes *four* halves of an Mg-Mg I_e-orbital, and two electrons. Each copper atom contributes *one* quarter of a tetrahedron-centered I_t-orbital and *three* thirds to an I_f-orbital. Thus, considering whole numbers of orbitals and electrons from two $MgCu_2$ units, we have: *one* I_t-orbital, *four* I_f-orbitals, and *four* Mg-Mg I_e-orbitals as well as *eight* valence electrons. The greater electronegativity of the copper relative to magnesium leads to the expectation of electron transfer from magnesium to copper. Thus, the I_t-orbitals in the copper tetrahedra are expected to contain two electrons. If *two* valence electrons are placed in the copper tetrahedra, *six* remain to be distributed. These can be easily accommodated by placing one in each of any six of the eight remaining I-orbitals.

The valence-bond structure for the $MgZn_2$ *compound* is different. From our previous discussion of two-electron metals, it is expected that zinc will attempt to form only I_t-orbitals in tetrahedra. A valence-bond structure built from tetra-

*Recall that I_e-orbitals are composed of only two hybrid orbitals.

The Cu atom centers are represented by black vertices, the Mg atom centers are represented by gray vertices, interstitial electron orbitals (I_f, I_t) are represented by light-colored vertices, and atomic hybrids are represented by connecting lines (H). To compare this photograph with Fig. 30(a), notice that the three upward-pointing tetrahedra formed by the Cu atoms (black) in the bottom half of the model correspond to the three upward-pointing tetrahedra in the upper half of Fig. 30(a). In the model shown here the upward-pointing tetrahedra (see Fig. 30a) have face-centered orbitals (I_f-orbitals) formed from three hybrids (H-orbitals) on the surrounding Cu atoms. The downward-pointing tetrahedra have tetrahedron-centered orbitals (I_t-orbitals) formed from four tetrahedral hybrids (H-orbitals). The hybrid orbitals (H-orbitals) of Mg overlap, forming interstitial orbitals that are directed along the Mg-Mg internuclear axis. (See text.)

Fig. 31. Model of a valence-bond structure of $MgCu_2$ in terms of tetrahedral hybrids, showing the alternation in the bonding of the tetrahedra

hedral hybrids is shown in Fig. 32. As before, there are I_e-orbitals between magnesium pairs, formed by magnesium hybrids. It can also be seen that alternate zinc tetrahedra contain I_t-orbitals. Notice that there are also I_e-orbitals *between* the chains formed by overlapping zinc H-orbitals of neighboring chains. In this valence-bond structure, the I_e-orbitals are formed instead of the I_f-orbitals of the $MgCu_2$ structure. In addition, alternate magnesium positions are octahedrally surrounded by these zinc I_e-orbitals. Furthermore, the Mg-Mg I_e-orbitals pass through four of the triangular faces of the octahedral cage of Zn-derived I_e-orbitals.

The valence electrons can be distributed as follows. Each magnesium atom contributes *four halves* of an Mg-Mg I_e-orbital. Each zinc atom contributes *one-quarter* of an I_t-orbital and *three halves* of an I_e-orbital. For convenience, we again consider the whole numbers of orbitals and electrons from two $MgZn_2$ units. Two $MgZn_2$ units contribute *one* I_t-orbital, *six* Zn-derived I_e-orbitals, and *four* Mg-Mg I_e-orbitals as well as *twelve* valence electrons. The greater electronegativity of the zinc relative to magnesium leads to the expectation of electron transfer from magnesium to zinc. Thus, the I_t-orbitals in the zinc tetrahedra are expected to contain two electrons as in bulk zinc metal. If *two* valence electrons are placed in the zinc tetrahedra, *ten* remain to be distributed. These can be precisely accommodated by placing *one* electron in each of the *four* Mg-Mg I_e-orbitals and each of the *six* Zn-Zn I_e-orbitals.

Just as the *crystal structure* of $MgNi_2$ is related to the crystal structures of $MgCu_2$ and $MgZn_2$, the *valence-bond structure* of $MgNi_2$ is related to the two valence-bond structures just discussed. A model of the valence-bond structure of $MgNi_2$ built from tetrahedral hybrids is shown in Fig. 33. Consistent with the previous structures, the face-sharing B-atom tetrahedra form either I_t- or I_e-orbitals, while the corner-sharing tetrahedra form either I_t- or I_f-orbitals. As in the previous structures, exactly *half* of the B-atom tetrahedra have I_t-orbitals (*downward-pointing* tetrahedra in the model of the valence-bond structure shown). One-third of the remaining tetrahedra have I_e-orbitals and the other two-thirds have I_f-orbitals (cf. Fig. 33).

To describe the electron distribution among the orbitals in $MgNi_2$ it is convenient to consider six AB_2 units. As in the previous structures, each magnesium atom contributes *two* Mg-Mg I_e-orbitals, for a total of *twelve* orbitals for *six* $MgNi_2$ units. The *twelve* nickel atoms divide into two classes: four nickel atoms each contribute one-quarter of an I_t-orbital and three halves of an I_e-orbital for a total of *one* I_t-orbital and *six* I_e-orbitals; the remaining eight nickel atoms each contribute *one-quarter* of an I_t-orbital and *three-thirds* of an I_f-orbital for a total of two I_t-orbitals and eight I_f-orbitals. Thus, for six $MgNi_2$ units, there are *twelve* Mg-Mg I_e-orbitals, *three* Ni I_t-orbitals, *six* Ni I_e-orbitals, and *eight* Ni I_f-orbitals. Assuming that two electrons can be placed in the I_t-orbitals and one in each of the remaining types of I-orbitals, this structure can comfortably contain 32 electrons for every six $MgNi_2$ units.

We are now in a position to discuss the observed correlation between valence electron concentration and structure for the Mg-based Laves phases. It has been

Similar to Fig. 31, the Zn atom centers are represented by black vertices, the Mg atom centers are represented by gray vertices, interstitial electron orbitals (I_e, I_t) are represented by light-colored vertices, and atomic hybrids are represented by connecting lines (H). Here the upward-pointing tetrahedra (see Fig. 30b) have tetrahedron-centered orbitals (I_t-orbitals) formed from four hybrids (H-orbitals). The downward-pointing tetrahedra have edge-centered orbitals (I_e-orbitals) formed from two hybrids (H-orbitals) on the surrounding Zn atoms. The hybrid orbitals (H-orbitals) of Mg overlap, forming interstitial orbitals that are directed along the Mg-Mg internuclear axis. (See text.)

Fig. 32. Model of a valence-bond structure of $MgZn_2$ in terms of tetrahedral hybrids, showing the alternation in bonding of the tetrahedra

Similar to Fig. 31 and 32, the Ni atom centers are represented by black vertices, the Mg atom centers are represented by gray vertices, interstitial electron orbitals (I_e, I_t, I_f) are represented by light-colored vertices, and atomic hybrids are represented by connecting lines (H). The three upward-pointing tetrahedra formed from the Ni (black) atoms in the bottom half of the model correspond to the upward-pointing tetrahedra in the lower half of Fig. 30(c). Only one upward-pointing tetrahedron is shown in this region in Fig. 30(c) for clarity. The upward-pointing tetrahedra in the model (see Fig. 30c) have either face-centered orbitals (I_f-orbitals) formed from three hybrids (H-orbitals), as in $MgCu_2$, or edge-centered orbitals (I_e-orbitals) formed from two hybrids (H-orbitals), as in $MgZn_2$. The downward-pointing tetrahedra have tetrahedron-centered orbitals (I_t-orbitals) formed from four hybrids (H-orbitals) on the surrounding Ni atoms. As in Fig. 31 and 32, the hybrid orbitals (H-orbitals) of Mg overlap, forming interstitial orbitals that are directed along the Mg-Mg internuclear axis. (See text.)

Fig. 33. Model of a valence-bond structure of $MgNi_2$ in terms of tetrahedral hybrids, showing the alternation in bonding of the tetrahedra

found that the existence of the three structures is approximately correlated with the ratio of the number of valence electrons to the number of atoms. With ordinary valences assumed for copper (one), magnesium (two), zinc (two), silicon (four), and aluminum (three), it has been observed that the $MgCu_2$ phase is found in alloys with electron/atom ratios between 1.33 and 1.75, the $MgNi_2$ phase is found for electron/atom ratios between 1.78 and 1.90, and the $MgZn_2$ phase is found for electron/atom ratios above 1.95.[(54)] Before continuing, we must comment on $MgNi_2$. The above observations suggest that each nickel atom contributes between one and two electrons. This nonintegral valency of nickel occurs frequently in compounds and seems inconsistent with the sequence across the Periodic Table—gallium (three), zinc (two), copper (one)—which implies that nickel should have a valence of *zero*. Nickel differs from copper and zinc, however, in that in nickel there is a very small difference in energy between the state with one *sp* valence electron (d^9s^1) and the state with two *sp* valence electrons (d^8s^2). In fact, the ground state of atomic nickel is d^8s^2, while in bulk nickel metal, the ground state is essentially d^9s^1. For the isolated atom, the difference in energy between the d^8s^2 and d^9s^1 states is very small (0.025 eV).[(55)] Thus, in particular, a nickel atom might contribute one or two valence electrons to the interstitial bonding electrons of an alloy. This is strongly influenced by interactions with the other elements in its environment.

We now continue our discussion of the correlation of structure with valence electron concentration. The observations are much less mysterious, if, instead of the electron/atom ratio, the electron/I-orbital ratio is used. This leads to the fractional numbers of electrons per atom because there are several types of orbitals. Since the valence-bond structures provide a way to count the number of I-orbitals for a given number of atoms, we will be able to relate the results back to the electron/atom ratio. In order to more easily compare the results for the three structures, we consider the number of orbitals per six AB_2 units (18 atoms).

The numbers of different I-type orbitals per six AB_2 units for the three structures are tabulated in Table 2. We have uniformly assumed that each I_t-orbital can accommodate two electrons and each of the remaining types of orbitals can contain a single electron. It is seen that the ideal electron/atom ratio (see Table 2) for the $MgCu_2$ *structure* is 1.66. This is intermediate in the range 1.33 to 1.75. The $MgCu_2$ *compound* has the electron/atom ratio 1.33. Similar counts of the ideal e/atom ratio lead to 1.78 electrons per atom for the $MgNi_2$ *structure*, and 2.0 electrons per atom for the $MgZn_2$ *structure*. These numbers also hold for the $MgNi_2$ and $MgZn_2$ *compounds*, respectively. Note that the $MgNi_2$ results suggest that, on average, each nickel atom contributes between one and two *sp* valence electrons to the interstitial electron bonding. More precisely, approximately one-third of the nickel atoms contribute one *sp* valence electron and the other two-thirds contribute two valence electrons. This is consistent with our comments above about the number of *sp* valence electrons in nickel. The results in Table 2 also suggest the following. The number of electrons in the $MgCu_2$ *compound* is far below the ideal value, so it is expected that atoms with *sp* valence greater than that of copper, such as aluminum or silicon, can substitute for copper over a wide range and

Table 2. Number of interstitial electron orbitals of different types for six AB_2 units in the Mg-based Laves phase structures (see text)

Orbital type (atom type)	AB_2 structure type		
	$MgCu_2$	$MgNi_2$	$MgZn_2$
I_e(A)	12	12	12
I_t(B)	3	3	3
I_e(B)	0	6	18
I_f(B)	12	8	0
Total	27	29	33
Ideal e/a ratio(a)	30/18 =1.66	32/18 =1.78	36/18 =2.00

(a) Ideal number of electrons is computed from the number of orbitals with the following expression: $2I_t(B) + I_e(B) + I_f(B) + I_e(A)$. (See text.)

that the ternary compound will retain the $MgCu_2$ structure. As the number of electrons increases, so that the electron/atom ratio approaches the ideal value for the $MgNi_2$ *structure*, the *B*-atom-derived *I*-orbitals exterior to the tetrahedra change from I_f-orbitals to I_e-orbitals and the number of hybrids changes from three per *I*-orbital to two per *I*-orbital. This accomplishes at least three things. First, the reorientation of the hybrids allows better overlap between them, forming a better bond; secondly, the number of *I*-orbitals is increased so that more electrons can be accommodated; and thirdly, for a given number of electrons, there is a greater separation between them. Thus, it appears that as the number of valence electrons is increased, the structure adjusts to correlate the electrons, keeping them as far apart as possible. As the electron/atom ratio increases further, this *conversion* process continues until the $MgZn_2$ *structure*, which can accommodate the greatest number of *sp* valence electrons of all of the three structures, is obtained.

There are, of course, Laves phases other than the Mg-based ones described above. For example, KNa_2, $CaLi_2$, $CaMg_2$, and $CaAl_2$ all have the hexagonal, $MgZn_2$ structure. It is tempting to suppose that the charge densities in these materials can be understood from the orbitals of the $MgZn_2$ structure described above, except, perhaps, with a different occupation of the orbitals. The difference in orbital occupation arises from differences in the relative electron affinities and ionization potentials of the atoms. Thus, it is expected that the charge densities can be understood from the orbitals of the valence-bond structure of $MgZn_2$ if the direction of charge transfer is carefully considered. For example, while in $MgZn_2$, it is reasonable to assume that charge is transferred from the *A* atoms to the *B* atoms, in $CaLi_2$ it can be expected that the direction of charge transfer is reversed—i.e., from lithium to calcium.

Such a description of the bonding in these structures would remain little but an idle exercise if it were not for the availability of detailed charge densities. Band structure/local density functional calculations have been performed for several

Laves phase compounds. While the energies and fine details of the charge density provided by these calculations are not usually reliable, gross features of the charge density are usually correct. In addition, for many systems, these calculations are the only practical means of obtaining a detailed charge density. Calculations have been published[56] for KNa_2, $MgZn_2$, $CaLi_2$, $CaMg_2$, and $CaAl_2$. It is possible, in fact, to give a reasonable explanation for the gross features of the charge density in terms of the orbital structure proposed above for $MgZn_2$ with careful consideration given to charge transfer. This will be discussed elsewhere.[57]

NiAs Structure

Many technologically important materials have the NiAs structure or a close derivative of it, such as an NiAs lattice with vacancies. In this structure, the nickel forms a simple hexagonal lattice and the arsenic forms an hcp lattice that is commensurate with the nickel lattice[58] (cf. Fig. 34). The structures are aligned so that each arsenic atom is surrounded by nickel atoms at the corners of a triangular prism. It is noticed that while this structure is frequently found in metal alloys, only transition-metal atoms are observed on the simple hexagonal lattice. This suggests that *d* electrons may be important for bonding in the structure and that the tetrahedral hybrids alone may not be sufficient to describe the electronic structure; we will return to this point in the next paragraph. Nevertheless, a tentative model for the NiAs bonding structure consists of tetrahedral hybrids where the Ni *H*-orbitals are oriented so that one is directed along the *c*-axis toward a nickel atom in another hexagonal layer. Each of the other three hybrids creates interstitial electron orbit-

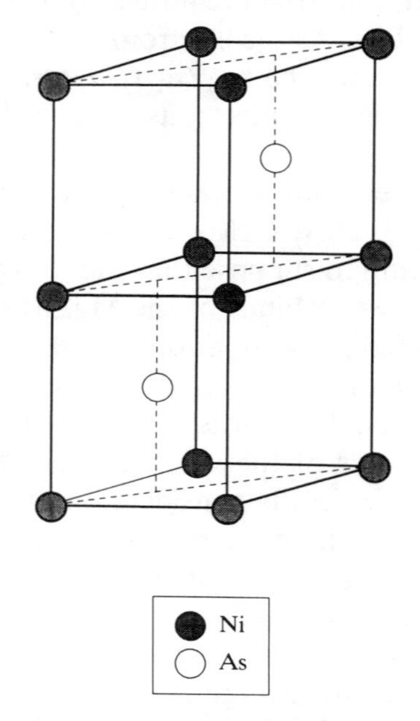

Fig. 34. Unit cell of NiAs structure

als by overlapping with the hybrids from nearest-neighbor nickel atoms in the hexagonal layers. The arsenic hybrids are oriented such that one is directed perpendicular to the hexagonal plane and the other three hybrids are directed toward the nickel interstitial orbitals. This valence-bond structure is shown in Fig. 35. A complementary valence-bond structure can be formed from the one shown by a coordinate transformation that involves a reflection through the hexagonal planes.

This model introduces several new features. First, there are three electrons in the interstitial regions in six orbitals. This greater-than-unity electron/atom ratio is consistent with the high conductivity of NiAs and the high conductivities of many other materials with this structure. Secondly, the nickel hybrids directed along the c-axis do not bond to sp^3 hybrids on the nickel neighbors in this direction. Hence the bonding is possibly ionic; or, more likely, overlap with a d_{z^2} orbital occurs, forming one-electron bonds. The interaction of a bonding orbital with d electrons would explain the absence of a ferromagnetic moment in FeS which also has this structure, and would also explain the appearance of ferromagnetism when there are sulfur vacancies.

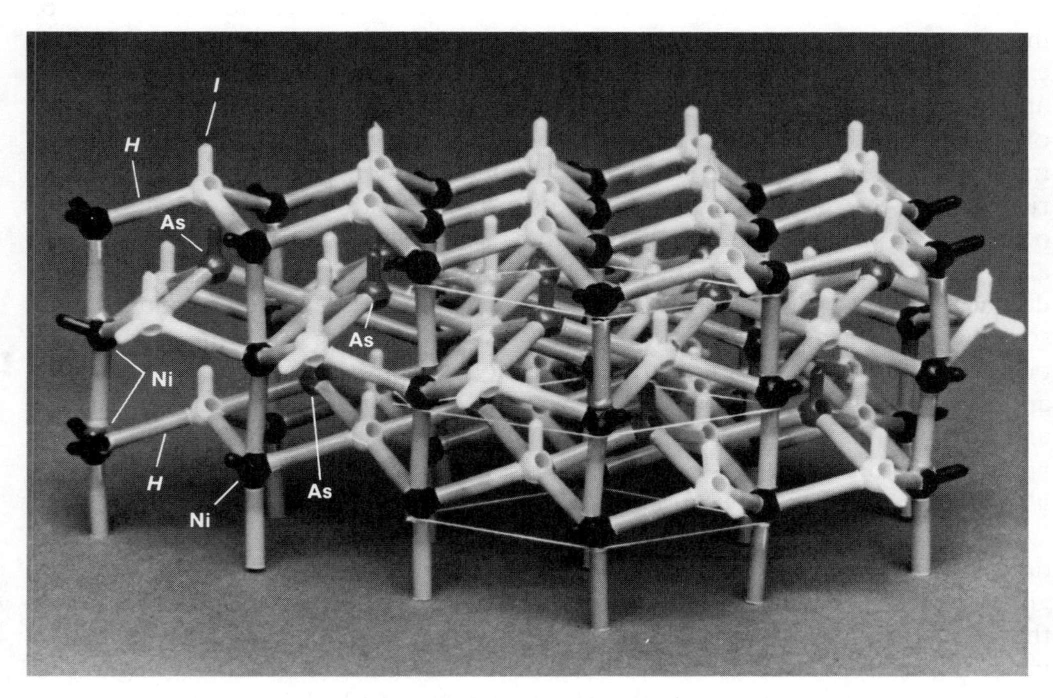

The Ni atom centers are represented by black vertices, the S atom centers are represented by gray vertices, the interstitial electron orbitals are represented by light-colored vertices, and atomic hybrids are represented by connecting lines. The interstitial orbitals are formed from the overlap of three Ni atom sp^3 hybrids (H-orbitals). The remaining Ni atom hybrid is directed toward an Ni atom in the layer below it. (See text.)

Fig. 35. Model of a valence-bond structure of NiAs in terms of tetrahedral hybrids

SUMMARY AND CONCLUSIONS

We have attempted to show that the fundamental property of a metal is reflected in the existence of interstitial electron orbitals. While this result derives from the attempt to understand the results of highly sophisticated quantum mechanical calculations, the concept is quite simple. The simplicity of the concept, however, belies its apparent utility, for when combined with other physical principles, it can be used to rationalize the structures of many of the common simple metals and their alloys. Furthermore, while structure is perhaps the most significant aspect of the description of a material, qualitative features of various electronic properties are readily visualized from a description of the electronic structure in terms of interstitial orbitals. For example, it is often possible to determine whether a magnetic moment is compatible with a given structure and a given number of valence electrons, or whether the electrical conductivity is metallic or semiconducting.

While the conceptual relationship between localized bonding in real space and electronic properties has been emphasized in the past, the rigorous description of interstitial electron orbitals was not feasible. Not only have recent calculations explicitly "crystallized" such an ephemeral concept, but for *sp*-metals nature has apparently been so kind as to permit a description of interstitial orbitals in terms of tetrahedral hybrids! Thus in one step many of the guiding principles and much intuition that have been developed for covalent bonding in molecules can be carried over to the metallic world. The best example of this is the Lewis-Langmuir octet rule. At the same time, use of tetrahedral hybrids in both metals and nonmetals allows the development of a picture of the interactions between metals and nonmetals that supplements the simple idea of an ionic bond and the popular, but somewhat obscure, notion of a bond represented in terms of molecular orbital occupation numbers. We have discussed models of metal/nonmetal bonding in applications to several intermetallic compounds and interstitial atoms in metals, and briefly commented on the application to bonding in ionic materials. Though applications to bonding of small molecules on surfaces were not discussed here[59] it is clear that such an application could extend to the structure at grain boundaries.

The broad applicability of modern valence-bond theory as indicated by the topics developed in the present work may leave the impression that perhaps all metallurgical problems will eventually be understood by the valence-bond approach. While the present work clearly invalidates the opposite claim, several factors make the initial impression unlikely. First, historical precedent suggests that no single view will be the most useful in all cases. Secondly, the level of detail required for a valence-bond description, though considerably simplified here for the case of the interstitial electrons, is still too great for large-scale numerical simulations. Finally, the developments of the line of thought discussed in the present work are too new to benefit from the enlightenment that comes with age and are likely to undergo some changes. Thus, the development of other theoretical approaches will be important. In particular, those that can provide reasonably accurate numerical results will be helpful in refining the qualitative ideas presented here. As for the experimental side, careful charge densities and magnetic characterizations would be particularly useful.

Finally, we remark on some of the more traditional ideas from the metallurgical literature used to understand the structures of metals, intermetallics, and alloys. Laves has identified several *main principles*[60] that influence crystal structure. These are:

1. The space (filling) principle
2. The (high) symmetry principle
3. The (high dimension) connection principle.

He also describes several *factors* which counteract the structure principles and these are briefly listed here:

1. Filling of open shells (stability of closed shells)
2. Temperature (entropy)
3. Competition between long and short forces (electrostatic vs. exchange)
4. Size factor (stronger bonds form when orbitals are of similar size)
5. Electrochemical factor (ionization potential and electron affinity).

While the counteracting factors are based directly on chemical or physical considerations, which are indicated in parentheses, the structure principles remain to be explained in detail. Once the physical basis for the structure principles is established, the balance of the counteracting effects of the various factors listed above will become clearer. We believe that an approach such as that described here, based on valence-bond orbitals, is the first step toward understanding the connection between electronic physics and the structure principles. In rough terms according to this view, the space-filling principle is related to electron sharing and the filling of open shells, the high-symmetry principle is related to either resonance or entropy, and the connection principle is related to the size factor, which is related to the fact that exchange terms are most favorable when orbitals are the same size. A more complete discussion will have to await further experience and development of the concepts presented here.

APPENDIX: BONDING IN BCC METALS

In this appendix, we describe a way of understanding the bonding in bcc metals in terms of resonance. Before describing the bonding in the bcc structure, however, we first discuss how a high-symmetry structure is stabilized by the resonance of two low-symmetry valence-bond structures using the molecular example of benzene.

The wave function of benzene is well represented as a superposition of two Kekulé structures.[13] If it were possible to isolate one of the Kekulé structures of the wave function, the coordinates of the carbon atoms would have to adjust to minimize the energy. A typical C-C double-bond distance (e.g., in ethylene) is shorter than the C-C distance in benzene and a typical C-C single-bond distance (e.g., in diamond) is longer than the C-C distance in benzene, so the system would lose its hexagonal symmetry. Thus, it is the resonance which *stabilizes* the intermediate structure with the six equal C-C distances that is observed.

In a cubic crystal, a real analog to such a fictitious molecular transformation is the Bain transformation.[61] This is a crystallographic distortion that relates the fcc and bcc structures. This can be described as a shearing distortion along the (110) direction.[62] There are several ways of applying this distortion to bring about the same result, but our point can be made by considering just two ways. Starting with an fcc lattice, we apply a shearing distortion along the (110) direction until we arrive at a bcc structure. If this distortion is continued further, a new fcc lattice is formed. This is shown schematically in Fig. 36(a). An individual atom has different sets of neighbors in the two fcc lattices, and the bonds an atom can form in one of the fcc lattices are different from the bonds it can form in the other. If we imagine the two bonding structures to contribute equally to the bcc wave function, the gain in energy due to the resonance of the two valence-bond structures may prevent the structure from actually performing a microscopic shear. Thus, from the point of view of interstitial orbitals, the bcc lattice offers a greater number of options, or local resonance structures, than the close-packed lattice (cf. Fig. 36b). Though an individual bond in the bcc structure is not as strong as an individual bond in the close-packed structures (because of reduced overlap) it is pos-

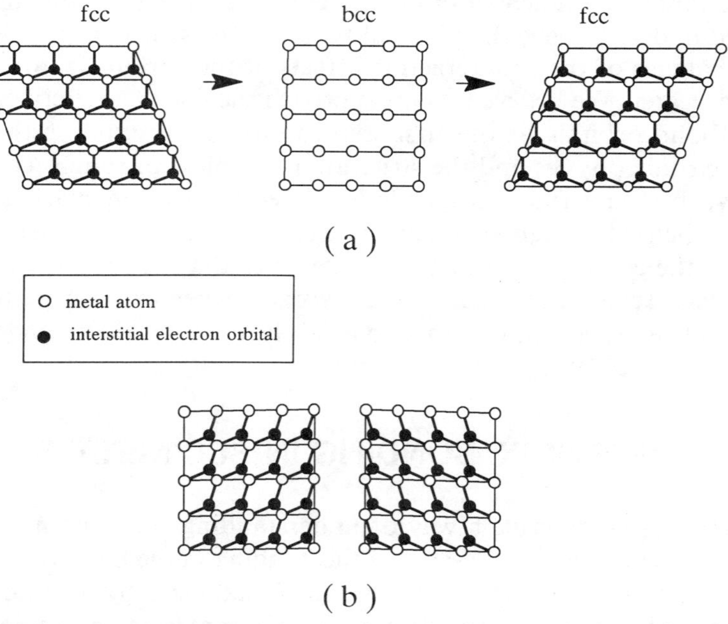

(a) Side view along (110) axis, showing a continuous transformation from the fcc to bcc to fcc structure (Bain transformation; see text). Interstitial electron orbitals formed from overlapping tetrahedral hybrids indicated by dark circles in fcc structures. The bonding in the bcc stage is not shown. (b) Bonding in bcc structure derived from resonance of fcc-like bonding structures shown in (a). (See text.)

Fig. 36. Schematic representation of relationship between bcc and fcc structures

sible for the resonance energy to be large enough to more than compensate for the energy loss in the individual bonds (just as in benzene). If this occurs, the bcc structure will be energetically favored over the fcc structure.

There are other ways of forming sets of alternative valence-bond structures for the bcc structure in terms of tetrahedral hybrids. A second set of alternative valence-bond structures is apparent from Fig. 25 and Table 1. In Table 1, it is observed that the bcc structure results if the four distinct sites in Fig. 25 contain the same atomic species. Figure 25 suggests that this structure can be thought of as two interpenetrating diamond lattices, namely a diamond lattice formed from sites *A* and *C* and a diamond lattice formed from sites *B* and *D*. It is also possible to form interpenetrating diamond lattices by associating sites *A* and *B* and associating sites *C* and *D*. The two ways of forming interpenetrating sublattices form a second set of alternative valence-bond structures. The importance of this alternative set of valence-bond structures for a particular material, relative to the set of valence-bond structures described in the preceding paragraph, depends on the number of valence electrons. It is expected that this latter set of valence-bond structures is more important (contributes a larger percentage to the total wave function) in metallic alloys with an average of 3 to 4 *sp* valence electrons per atom. Conversely, the valence-bond structures described in the preceding paragraph are expected to be more important in materials with fewer *sp* valence electrons.

ACKNOWLEDGMENTS

The authors would like to thank Peter Schultz for a critical reading of the manuscript and helpful discussions.

This work was supported in part by the National Science Foundation MRL program under grant No. DMR-8519095 at the Laboratory for Research on the Structure of Matter, University of Pennsylvania and in part by the Office of Naval Research.

REFERENCES

1. W. Hume-Rothery and G.V. Raynor, *The Structure of Metals and Alloys*, 3rd Ed., The Institute of Metals, London, 1954
2. A.H. Cottrell, *Theoretical Structural Metallurgy*, 2nd Ed., Camelot Press, Ltd., London, 1955
3. J.C. Slater, *Theory of Alloy Phases*, ASM, Cleveland, 1956, p. 1; R.E. Rundle, *Intermetallic Compounds*, edited by J.H. Westbrook, John Wiley & Sons, Inc., New York, 1965; see also C. Kittel, *Introduction to Solid State Physics*, 3rd Ed., John Wiley & Sons, Inc., New York, 1968, and N.W. Ashcroft and N.D. Mermin, *Solid State Physics*, Holt, Rinehart, and Winston, New York, 1976
4. W. Heitler and F. London, *Z. Phys.*, Vol 44, 1927, p. 455
5. L. Pauling, *The Nature of the Chemical Bond*, 3rd Ed., Cornell Univ. Press, Ithaca, NY, 1960

6. L. Pauling, *J. Am. Chem. Soc.*, Vol 53, 1931, p. 1367
7. J.C. Slater, *Phys. Rev.* Vol 37, 1931, p. 481
8. W.J. Hunt, P.H. Hay, and W.A. Goddard, III, *J. Chem. Phys.*, Vol 57, 1972, p. 738; R.C. Ladner and W.A. Goddard, III, *J. Chem. Phys.*, Vol 51, 1969, p. 1073; F.W. Bobrowicz and W.A. Goddard, III, in *Methods of Electronic Structure Theory, Modern Theoretical Chemistry Vol. 3*, edited by H.F. Schaefer, III, Plenum, New York, 1977, Chapter 4
9. The results were obtained using the GVB2P5 program: R.A. Bair, W.A. Goddard, III, A.F. Voter, A.K. Rappé, L.G. Yaffe, F.W. Bobrowicz, W.R. Wadt, P.J. Hay, and W.J. Hunt (unpublished); see also: R.A. Bair, Ph. D. Thesis, California Institute of Technology, 1980 (unpublished)
10. The bond length, force constant, and bond energy of the C-C bond in ethane are: 1.535 Å, 27.3 eV/Å^2, and 3.64 eV, respectively. The same parameters for the C-C bond in diamond when converted from the lattice-constant, bulk-modulus, and cohesive-energy parameters of bulk diamond are: 1.545 Å, 36.4 eV/Å^2, and 3.67 eV. Note that the bond-energy and bond-length parameters for diamond and ethane differ by less than 1%. Similar results hold for germanium and silicon and their ethane analogs. Ethane data from: G. Herzberg, *Electronic Spectra of Polyatomic Molecules*, D. Van Nostrand Co., Princeton, 1966—bond length and bond energy; and T.L. Cottrell, *The Strengths of Chemical Bonds*, 2nd Ed., Butterworths Publications, Ltd., London, 1958. Diamond data from: J. Donohue, *The Structures of Elements*, Wiley, New York, 1974—lattice constant; M. Kittel, *Introduction to Solid State Physics*, 3rd Ed., John Wiley & Sons, Inc., New York, 1968, Chapter 4—bulk modulus; *The NBS tables of chemical thermodynamic properties*, D.D. Wagman, W.H. Evans, V.B. Parker, R.H. Schumm, I. Halow, S.M. Bailey, K.L. Churney, and R.L. Nuttall, American Chemical Society, Washington, 1982—cohesive energy.
11. R.P. Messmer and P.A. Schultz, *Phys. Rev. Lett.*, Vol 57, 1986, p. 2653
12. A. Kekulé, *Justus Liebigs. Ann. Chem.*, Vol 162, 1872, p. 77
13. P.A. Schultz and R.P. Messmer, *Phys. Rev. Lett.*, Vol 58, 1987, p. 2416
14. R.P. Messmer, H.-J. Freund, R.C. Tatar, and P.A. Schultz, *Chem. Phys. Lett.*, Vol 126, 1986, p. 176
15. W.E. Palke, *J. Am. Chem. Soc.*, Vol 108, 1986, p. 6543
16. R.P. Messmer (to be published). See also: N. Sidgwick and H. Powell, *Proc. R. Soc. London, Ser. A*, Vol 176, 1940, p. 153; R. Gillespie and R. Nyholm, *Quart. Rev. Chem. Soc.*, Vol 11, 1957, p. 339; R. Gillespie, *J. Chem. Educ.*, Vol 43, 1970, p. 18, and Vol 51, 1974, p. 367
17. R.C. Tatar and R.P. Messmer (to be published)
18. (a) M.H. McAdon and W.A. Goddard, III, *Phys. Rev. Lett.*, Vol 55, 1985, p. 2563; (b) M.H. McAdon and W.A. Goddard, III, *J. Non-Cryst. Sol.*, Vol 75, 1985, p. 149; (c) M.H. McAdon and W.A. Goddard, III, *J. Phys. Chem.*, Vol 91, 1987, p. 2607
19. H.A. Lorentz, *Theory of Electrons*, Teubner, Leipzig, 1906
20. J.J. Thomson, *The Electron in Chemistry*, The Franklin Institute, Philadelphia, 1923
21. L. Pauling and B. Kamb, *Proc. Natl. Acad. Sci. USA*, Vol 83, 1986, p. 3569; L. Pauling and B. Kamb, *Proc. Natl. Acad. Sci. USA*, Vol 82, 1985, p. 8286; B. Kamb and L. Pauling, *Proc. Natl. Acad. Sci. USA*, Vol 82, 1985, p. 8284; L. Pauling, *J. Solid. State. Chem.*, Vol 54, 1984, p. 297; L. Pauling, *Proc. Roy. Soc. A*, Vol 196, 1949, p. 343; L. Pauling, *Phys. Rev.* Vol 54, 1949, p. 899
22. K. Fajans, *Ceramic Age*, Vol 54, 1949, p. 288
23. H. Bent, *J. Chem. Educ.*, Vol 42, 1965, p. 348
24. H. Nowotny, F. Holub, and A. Wittmann, *The Physical Chemistry of Metallic Solutions and Intermetallic Compounds*, N.P.L. Symposium No. 9, Chem. Pub. Co., New York, 1960, p. 366

25. K. Schubert, *Intermetallic Compounds*, edited by J.H. Westbrook, John Wiley & Sons, Inc., New York, 1965, Chapter 6. See also references therein.
26. O. Johnson, *J. Chem. Soc. Jpn.*, Vol 45, 1972, pp. 1599 and 1607; Vol 46, 1973, pp. 1919, 1923, and 1929
27. See Ref 18 for a more accurate description of the wave function.
28. W.A. Harrison, *Electronic Structure and the Properties of Solids*, W.A. Freeman and Co., San Francisco, 1980
29. See, for example, the references in M.H. Kang, R.C. Tatar, E.J. Mele, and P. Soven, *Phys. Rev. B*, Vol 35, 1987, p. 5457
30. M.I. Baskes, C.F. Melius, and W.D. Wilson, *Hydrogen Effects in Metals*, edited by I.M. Bernstein and A.W. Thompson, AIME, New York, 1981
31. See, for example, Parthé, *Intermetallic Compounds*, edited by J.H. Westbrook, John Wiley & Sons, Inc., New York, 1967, Chapter 11
32. R.P. Messmer, *Solid State Commun.*, Vol 63, 1987, p. 405; R.P. Messmer, *Phys. Scripta* (in press)
33. L. Massa, M. Goldberg, C. Frishberg, R.F. Boehme, and S.J. La Placa, *Phys. Rev. Lett.*, Vol 55, 1985, p. 622
34. R.C. Tatar and R.P. Messmer (to be published)
35. M.Y. Chou, P.K. Lam, and M.L. Cohen, *Phys. Rev. B*, Vol 28, 1983, p. 4179
36. H.G. Smith, *Phys. Rev. Lett.*, Vol 58, 1987, p. 1228. See also R. Berliner and S.A. Werner, *Phys. Rev. B*, Vol 34, 1986, p. 3586; G. Ernst, C. Artner, O. Blaschko, and G. Krexner, *Phys. Rev. B*, Vol 33, 1986, p. 6465
37. O. Blaschko and G. Krexner, *Phys. Rev. B*, Vol 30, 1984, p. 1667
38. J.A. Wilson and M. de Podesta, *J. Phys. F: Met. Phys.*, Vol 16, 1986, p. L121
39. S. Alexander and J. McTague, *Phys. Rev. Lett.*, Vol 41, 1978, p. 702
40. R.P. Messmer and R.C. Tatar (submitted for publication.)
41. F.J. Landro, J.A. Gurak, J.W. Chinn, Jr., and R.J. Lagow, *J. Organometal. Chem.*, Vol 249, 1983, p. 1; C.H. Wu and H.R. Ihle, *Chem. Phys. Lett.*, Vol 61, 1979, p. 54
42. E. Weiss and G. Henken, *J. Organometal. Chem.*, Vol 23, 1970, p. 265
43. See, for example, G.A. Olah, G.K. Prakash, R.E. Williams, L.D. Field, and K. Wade, *Hypercarbon Chemistry*, John Wiley & Sons, Inc., New York, 1987
44. C.S. Barret and T.B Massalski, *Structure of Metals*, McGraw-Hill, New York, 1966, p. 235
45. See also Table 8 in A.E. Dwight, *Intermetallic Compounds*, edited by J.H. Westbrook, John Wiley & Sons, Inc., New York, 1967; and Table 1 of Ref 48
46. E. Zintl and G. Brauer, *Z. Phys. Chem. Abt. B*, Vol 20, 1933, p. 245
47. N.E. Christensen, *Phys. Rev. B*, Vol 32, 1985, p. 207
48. N.E. Christensen, *Phys. Rev. B*, Vol 32, 1985, p. 6490
49. S.-H. Wei and A. Zunger, *Phys. Rev. Lett.*, Vol 56, 1986, p. 528
50. L. Pauling and L.O. Brockway, *Z. Krist.*, Vol 82, 1932, p. 188
51. J.S. Kasper, *Theory of Alloy Phases*, ASM, Cleveland, 1956, p. 264
52. Much progress has been made in describing very complicated alloy structures (e.g., β-brass, σ-phases, χ-phases, etc.) in terms of regular polyhedra, especially in terms of tetrahedra and octahedra. See S. Andersson, *Structure and Bonding in Crystals II*, edited by M. O'Keeffe and A. Navrotsky, Academic Press, New York, 1987, p. 233, as well as the references therein. Work is in progress that utilizes Andersson's description of complex crystal structures to obtain valence bond structures.
53. See the references in *Structure of Metals*, C.S. Barret and T.B. Massalski, McGraw-Hill, New York, 1966, Chapter 10, p. 256
54. See Table 4 in J.H. Wernick, *Intermetallic Compounds*, edited by J.H. Westbrook, John Wiley & Sons, Inc., New York, 1967, p. 202; also, F. Laves and H. Witte, *Metallwirtschaft*, Vol 14, 1935, p. 645

55. C. Moore, *Atomic Energy Levels*, Nat. Stand. Ref. Data Ser., Nat. Bur. Stand., 35/V.II, 1971
56. J. Hafner, *J. Phys. F.: Met. Phys.*, Vol 15, 1985, p. 1879
57. R.C. Tatar and R.P. Messmer (to be published)
58. R.W.G. Wycoff, *Crystal Structures*, Vol 1, John Wiley & Sons, Inc., New York, 1963
59. R.P. Messmer and P.A. Schultz (to be published in *J. Phys. Chem.*)
60. F. Laves, *Intermetallic Compounds*, edited by J.H. Westbrook, John Wiley & Sons, Inc., New York, 1967, Chapter 8
61. E.C. Bain, *Trans. AIME*, Vol 70, 1924, p. 25
62. C. Zener, *Phys. Rev.*, Vol 71, 1947, p. 846; see also discussion in C. Kittel, *Introduction to Solid State Physics*, 2nd Ed., John Wiley & Sons, New York, 1956, end of Chapter 4

3

The Theory of Metallic Alloys

J.S. FAULKNER
Florida Atlantic University

The purpose of the work described here is to develop a theory for the electronic states in substitutional solid-solution alloys and to use this theory to explain properties of alloys that are of interest to materials scientists. As will be seen, from the references cited, these developments have resulted from the efforts of many theorists over a period of almost two decades.

Let us define precisely the kind of solid that we will consider. A solid-solution alloy has a well-defined crystal structure, just like a metal. However, atoms of two or more species are distributed randomly on the lattice sites. In a binary alloy, for example, we will assume that the probability of finding an A atom on a given site is C_A, while the probability of finding a B atom on that site is C_B. In this example, C_A is the concentration of A atoms in the alloy and C_B is the concentration of B atoms. Such an alloy has no short-range order in the sense that the Warren-Cowley short-range-order (SRO) parameters are zero.

There are almost no alloys in nature for which the measured SRO parameters will all be zero. Our model is a useful abstraction, however, because it serves as the reference state in the thermodynamics of alloys. We will demonstrate that calculations on such models can be used to predict the SRO parameters of real alloys.

THEORETICAL CONSIDERATIONS

The theory for calculating the electronic states in pure metals and ordered intermetallic compounds was understood in principle in the 1930's and has been applied to many systems since the 1950's. These calculations make use of the density functional theory (DFT) of Hohenberg, Kohn, and Sham to express the many-electron wave function in terms of one-electron wave functions $\Psi_{\vec{k}}^{\alpha}(\vec{r})$. These functions are solutions of the one-electron Schroedinger equation

$$\left[-\frac{h^2}{2m}\nabla^2 + V(\vec{r})\right]\Psi_{\vec{k}}^{\alpha}(\vec{r}) = E_{\alpha}(\vec{k})\Psi_{\vec{k}}^{\alpha}(\vec{r}) \quad \text{(Eq 1)}$$

where $V(\vec{r})$ is the one-electron potential that is defined in the DFT, and $E(\vec{k})$ is the one-electron eigenvalue. Because of the long-range order in metals and intermetallic compounds, the potential is invariant under translation through a lattice vector $\vec{R}_n$

$$V(\vec{r} + \vec{R}_n) = V(\vec{r}) \quad \text{(Eq 2)}$$

Whenever the function $V(\vec{r})$ in a differential equation such as Eq 1 has the property shown in Eq 2, it is possible to prove Bloch's theorem, which states that the solutions $\Psi_{\vec{k}}^{\alpha}(\vec{r})$ have the property

$$\Psi_{\vec{k}}^{\alpha}(\vec{r} + \vec{R}_n) = e^{i\vec{k}\cdot\vec{R}_n}\Psi_{\vec{k}}^{\alpha}(\vec{r}) \quad \text{(Eq 3)}$$

where $\vec{k}$ is a real vector. For a fixed value of $\vec{k}$, there is an infinity of $E_{\alpha}(\vec{k})$. The different eigenvalues are distinguished by the band index α. The $E_{\alpha}(\vec{k})$ in a given band are continuous functions of $\vec{k}$. The results of a band-theory calculation of $E_{\alpha}(\vec{k})$ and $\Psi_{\vec{k}}^{\alpha}(\vec{r})$ are frequently summarized by plotting the $E_{\alpha}(\vec{k})$ for values of $\vec{k}$ along certain symmetry axes.

The density functional theory can be used in calculating the electronic states of alloys, so there are one-electron wave functions and eigenvalues for that case. Solid-solution alloys have no long-range order, however, so the one-electron potential, $V(\vec{r})$, does not satisfy Eq 2. From this it follows that there is no Bloch's theorem for an alloy. The one-electron eigenvalues are not related to a $\vec{k}$ vector.

Since the band theory used to calculate the electronic states of metals and intermetallic compounds is not applicable to alloys, it was necessary to develop an entirely new set of techniques to solve the one-electron Schroedinger equation for these systems. The search for such techniques started in the 1930's, but they only began to show signs of success when it was realized in the 1950's that attention should be focused on the one-electron Green's function $G(E,\vec{r},\vec{r}')$ rather than on the Schroedinger equation. The reason is that there is a statistical aspect to the alloy problem in addition to the quantum mechanical aspect. It is impossible to know which of the 10^{23} lattice sites in a binary alloy are occupied by A atoms and which ones are occupied by B atoms. Only the concentrations C_A and $C_B = 1 - C_A$ are known. It is therefore necessary to consider averages over an ensemble made up of all alloys that have the given concentration. The average of a Schroedinger equation leads to no meaningful quantities, but the average of a Green's function $G(E,\vec{r},\vec{r}')$ is meaningful.

The only piece of mathematics that must be understood is the way that observable quantities are calculated from the Green's function. One example is the density of states

$$\rho(E) = -\frac{1}{\pi}\int \mathrm{Im}G(E,\vec{r},\vec{r})\,d\vec{r} \qquad \text{(Eq 4)}$$

which is the number of one-electron eigenvalues of the alloy that fall between E and E + dE divided by the interval dE. Another example is the density of the electronic charge in the alloy

$$Q(\vec{r}) = \frac{e}{4\pi}\int_{-\infty}^{E_f} \mathrm{Im}G(E,\vec{r},\vec{r})\,dE \qquad \text{(Eq 5)}$$

The point is that the observable is linearly related to the Green's function. It follows that inserting the ensemble-average Green's function $\langle G(E,\vec{r},\vec{r}')\rangle$ on the right side of Eq 4 and 5 will yield the ensemble average of the density of states $\langle \rho(E)\rangle$ and charge density $\langle Q(\vec{r})\rangle$. Similarly, ensemble averages of all relevant observables can be calculated from $\langle G(E,\vec{r},\vec{r}')\rangle$.

Of course, it would be impossible to calculate the ensemble-averaged Green's function exactly, so we must look for approximations to $\langle G(E,\vec{r},\vec{r}')\rangle$. Experience has shown that the best tractable approximation to $\langle G(E,\vec{r},\vec{r}')\rangle$ is the function $G_C(E,\vec{r},\vec{r}')$ that is given by the coherent-potential approximations (CPA). There were several precursors to the CPA, but the approximation was stated most clearly by Soven[1] in 1967. We will not try to explain the mathematics of the CPA in detail, but merely give the flavor of the approximation. Much more about this and other questions can be found in a review article I published a few years back.[2]

The function $G_C(E,\vec{r},\vec{r}')$ is calculated from the standard formula for the Green's function for a period solid. The mathematical model has the same set of lattice sites as the physical alloy, but the atomic potentials on both the A and B sites are replaced by an effective potential $V_C(E,\vec{r})$ that is complex and energy-dependent. No true atomic potential can be either complex or energy-dependent.

The equation for the CPA effective potential $V_C(E,\vec{r})$ is obtained by writing the formula for the CPA Green's function $G_C(E,\vec{r},\vec{r}')$ which contains the, as yet unknown, $V_C(E,\vec{r})$. We then define two scattering functions $\hat{t}_A(E,\vec{r},\vec{r}')$ and $\hat{t}_B(E,\vec{r},\vec{r}')$ by

$$\hat{t}_A = (V_A - V_C)(1 + G_C\hat{t}_A) \quad \hat{t}_B = (V_B - V_C)(1 + G_C\hat{t}_B) \qquad \text{(Eq 6)}$$

where we are using abbreviations like

$$(V_A - V_C)G_C\hat{t}_A = \int [V_A(\vec{r}) - V_C(E,\vec{r})]G_c(E,\vec{r},\vec{r}_1)\hat{t}_A(E,\vec{r}_1\vec{r}')\,d\vec{r}_1 \qquad \text{(Eq 7)}$$

The Green's function for an electron traveling in a lattice that has a $V_C(E,\vec{r})$ on every site except for the central site, and has a potential $V_A(\vec{r})$ on that site, is

$$G_A = G_C + G_C\hat{t}_A G_C \quad \text{(Eq 7a)}$$

while

$$G_B = G_C + G_C\hat{t}_B G_C \quad \text{(Eq 8)}$$

is the Green's function when the lattice has $V_B(\vec{r})$ on the central site. The theory requires that $V_A(\vec{r})$ and $V_B(\vec{r})$ should be the potential functions that describe the interaction of the electron with a real A or B atom.

In terms of these quantities, the CPA condition that defines $V_C(E,\vec{r})$ may be stated in two ways. The first way of stating it is

$$C_A G_A + C_B G_B = G_C \quad \text{(Eq 9)}$$

while the second is

$$C_A\hat{t}_A + C_B\hat{t}_B = 0 \quad \text{(Eq 10)}$$

When these equations are written out in all their gory detail as in Eq 7, they lead to an integral equation for a scattering function $t_C(E,\vec{r},\vec{r}')$ that is defined by

$$t_C = V_C(1 + G_C t_C) \quad \text{(Eq 11)}$$

The CPA equation for $t_C(E,\vec{r},\vec{r}')$ [or, equivalently, $V_C(E,\vec{r})$] must be solved iteratively. That means we first guess a form for t_C that can be used to calculate a new t_C. The new t_C is substituted back into the defining equation, and the process is repeated until there is no further change upon iteration. In addition to E, $\vec{r}$, and $\vec{r}'$, the solution t_C depends on $C_A, C_B, V_A(\vec{r})$ and $V_B(\vec{r})$.

An additional complication of the CPA equation arises from the fact that atomic potentials $V_A(\vec{r})$ and $V_B(\vec{r})$ are supposed to be calculated from the exact electronic wave functions in the alloy, which means the alloy problem must be solved before they can be obtained. It follows that these functions must also be found by a self-consistent procedure. This is sometimes called a self-consistent KKR-CPA calculation because it borrows much from the Korringa-Kohn Rostocker (KKR) band-theory technique. The first such calculations were done in 1983[(3)], and the technical details that must be understood in order to carry them out are explained in references 4 and 5.

The important insights into properties that come from this alloy theory can be appreciated without understanding the details of the CPA calculation, but there are some points that should be understood. First, the only parameters that must be given as input to the calculation are the concentrations, C_A and C_B, and the atomic numbers of the A and B atoms, Z_A and Z_B. The crystal structure and lattice constant are normally chosen by the calculator, although the equilibrium values for these quantities can be obtained from total-energy calculations based on the CPA. The next point is that these calculations are the equivalents of many band-theory calculations. In this day of ever-improving computers, the complexity of the calculation does not limit its usefulness. The time is not far off when a KKR-CPA calculation can be carried out on a micro-computer.

At the present time, most of the KKR-CPA calculations have been carried out for binary metallic alloys. There is no difficulty in principle in treating ternary, quaternary, or more complex substitutional alloys, but the calculations become more time-consuming. The KKR-CPA has been used to treat substoichiometric compounds[6,7,8] such as palladium hydride (PdH_x) where x is less than one. Such systems may be treated as alloys for which the A atom is an occupied site and the B atom is an empty site. These techniques have been applied less extensively to semiconductor alloys.[9] Other systems without long-range order are the amorphous metals and alloys. The KKR-CPA cannot be used for such systems because they do not have well-defined crystal structures. Efforts have been made to adapt the ideas of the CPA to the treatment of such systems,[10] but they have not been completely successful.

THE ELECTRONIC STRUCTURE OF ALLOYS

The results of the first CPA calculation on a real alloy system that led to a definitive comparison of theory with experiment[11] are shown in Fig. 1. The copper-nickel alloy system on which these calculations were done is a particularly convenient test case for the CPA. The alloys have the face-centered-cubic structure over the entire concentration range, and samples can be prepared with little short-range order by quenching from high temperatures. The curves in the left panel show density-of-states functions $\rho(E)$ that were done using the virtual crystal approximation (VCA). In these calculations, the Green's function that is used in Eq 4 is calculated for an ordered crystal with the potential

$$\bar{V}(\vec{r}) = C_A V_A(\vec{r}) + C_B V_B(\vec{r}) \quad \text{(Eq 12)}$$

on every lattice site. The curves in the right panel show a set of CPA calculations in which a somewhat simplified version of $G_C(E,\vec{r},\vec{r}')$ was used in Eq 4. The curves in the center panel are measurements of the energy distribution of electrons emitted in an ultraviolet photoemission experiment.

The density-of-states functions calculated with the VCA are only very slight improvements over the ones that can be obtained from the well-known rigid-band approximation (RBA). In the RBA it is assumed that the distributions of energies in the conduction bands of all transition metals and alloys are the same, so that there is a universal density-of-states function $\rho_0(E)$ for all alloys in this class. If

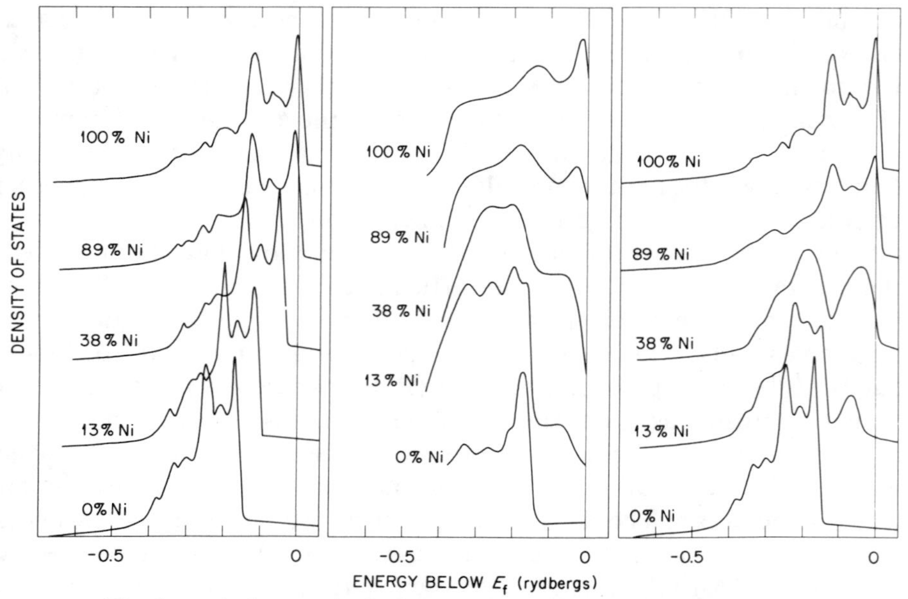

The theoretical curves in the left panel are calculated with the virtual-crystal approximation and are similar to the results of the rigid-band model. Those in the right panel are calculated with the coherent-potential approximation. The experimental curves in the center panel are the ultraviolet photoemission spectra reported in Ref 12.

Fig. 1. Density of states versus energy for a series of copper-nickel alloys

N_A is the number of electrons in the conduction band of metal A and N_B is the corresponding number for metal B, the number of electrons in the alloy is

$$\bar{n} = C_A N_A + C_B N_B \qquad \text{(Eq 13)}$$

The only difference in the distribution of electrons for alloys with different concentrations is in the position of the Fermi energy, which is given by

$$\int_0^{E_f} \rho_0(E)\, dE = \bar{n} \qquad \text{(Eq 14)}$$

The $\rho(E)$ curves in the left panel of Fig. 1 show this effect. There is one more electron in the conduction band of copper than in the conduction band of nickel. The Fermi energy is well above the peaky structure caused by the d-bands in copper, but it lies in the d-bands for nickel. According to the VCA or RBA, the Fermi energy will move continuously toward the d-bands as nickel replaces copper in the d-bands. Since the Fermi energy has a fixed position in the figure, the d-bands are seen to move continuously up.

As can be seen from Fig. 1, the CPA gives a very different picture of the density-of-states functions for copper-nickel alloys. When a little bit of nickel is alloyed with copper, a bump appears slightly above the copper d-bands. The positions of the leading edges of the copper d-bands remain fixed relative to the Fermi energy. As more nickel is added to the alloy, the bump grows and begins to take on a peaky structure, rather as a tree grows from a seed. The copper d-bands soften and become less distinguishable.

The energy-distribution curves measured in the UV photoemission experiment[12] are primarily a reflection of $\rho(E)$, although they are modified by secondary reflections and matrix-element effects. It can be seen by comparing the center panel in Fig. 1 with the panels on either side that the CPA agrees much better with experiment than does the VCA or RBA. In particular, the leading edges of the copper d-bands show no tendency to move toward the Fermi energy with increasing nickel content, which is in violent disagreement with the rigid-band approximation. The peak above the d-bands predicted by the CPA can also be seen.

The results in Fig. 1 demonstrate two points. First, the CPA gives very different predictions for the electronic states of an alloy than does the rigid-band model that has been used so often in textbooks on materials science. Second, the CPA picture is the one that agrees with experiment. The copper-nickel system has been studied experimentally many times since these calculations were published in 1971.[13] As the experimental results are improved, they look even more like the CPA curves in the left panel of Fig. 1.

The curves in Fig. 2 show the kind of agreement between theory and the results of photoemission experiments that is expected today.[14] The experiment is called ARUPS, angular-resolved ultraviolet photoemission spectroscopy. The theoretical and experimental curves for three different copper-palladium alloys are seen

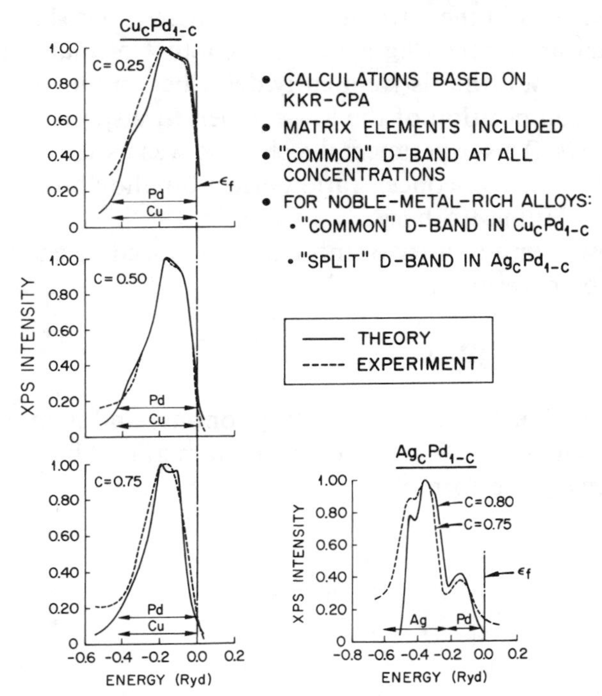

Fig. 2. Calculated and measured XPS-spectra

to agree almost exactly. The theoretical curves are not just density-of-states functions. The CPA scattering functions t_C are inserted into a model that tracks the motion of the photoemitted electron from the instant it is excited in the solid until it reaches the counter.

The experimental and theoretical results for a silver-palladium alloy are also shown in Fig. 2. This curve looks a lot like the density-of-states function of the copper-nickel system. It is surprising that the copper-palladium results are so different, since the elements involved are from the same two columns of the Periodic Table. The first CPA calculation on the 25% copper-palladium alloy showed a second peak above the d-band as in the silver-palladium case. As the calculation was iterated to self-consistency, the extra peak moved under the d-band. This is a dramatic illustration of the importance of iterating the atomic potentials to self-consistency.

THE SOLUTION OF HISTORIC PROBLEMS

Since the early 1970's the predictions of the CPA have been compared many times with the results of such physics experiments as photoemission, positron-annihilation, and soft x-ray emission or absorption. This body of work has demonstrated the ability of the CPA to predict the electronic states of solid-solution alloys in great detail with no adjustable parameters. However, such experiments do not appear to be relevant to the kinds of problems dealt with by most materials scientists. As I stated in the introduction, our interest has shifted toward dealing with problems that are interesting within the context of materials science.

An experimental result that is frequently described in materials science texts is the variation with composition of the low-temperature specific heat coefficient in Hume-Rothery alloys. This experimental technique was used in the earliest attempts to shed light on the electron-concentration effect on alloy phase stability that was first enunciated by Hume-Rothery.

It has been known for a century or more that the specific heat of a metal or alloy is related to the temperature T by

$$C = \gamma T + bT^3 \qquad \text{(Eq 15)}$$

at low temperature. The T^3 term is due to phonons, and is present in insulators as well as metals. The term γT is present only in metals. The coefficient γ is now known to be given by the formula

$$\gamma = \hat{\gamma}(1 + \lambda) \qquad \text{(Eq 16)}$$

where $\hat{\gamma}$ is

$$\hat{\gamma} = \frac{1}{3}\pi^2 k_B^2 \rho(E_F) \qquad \text{(Eq 17)}$$

in which k_B is the Boltzmann constant and $\rho(E_F)$ is the density of states at the Fermi energy. The parameter λ in Eq 16 is an enhancement factor that may arise from several sources, the most important one being the electron-phonon interaction.

The experimental measurements[15] of γ as a function of concentration are shown for the typical Hume-Rothery alloy system copper-zinc by the empty circles and triangles in the top panel in Fig. 3. The abscissa in this figure is the electron concentration or number of electrons per atom, conventionally written e/a. If it is assumed that copper has one electron per atom in the conduction band and zinc has two, e/a is one plus the concentration of zinc in the alloy.

It was thought originally that the upward slope of the experimental curve in Fig. 3 is physically reasonable because, according to the free-electron theory of metals, the Fermi surface of copper should be spherical. The rigid-band model would predict that the addition of electrons to the system by alloying with zinc will cause the spherical Fermi surface to expand until it touches the hexagonal faces of the Brillouin zone. This picture, which would lead to an increase in γ with e/a because $\rho(E_F)$ is proportional to the area of the Fermi surface, was used by Jones[16] to calculate the values of e/a at which the phase boundaries should appear in Hume-Rothery alloys.

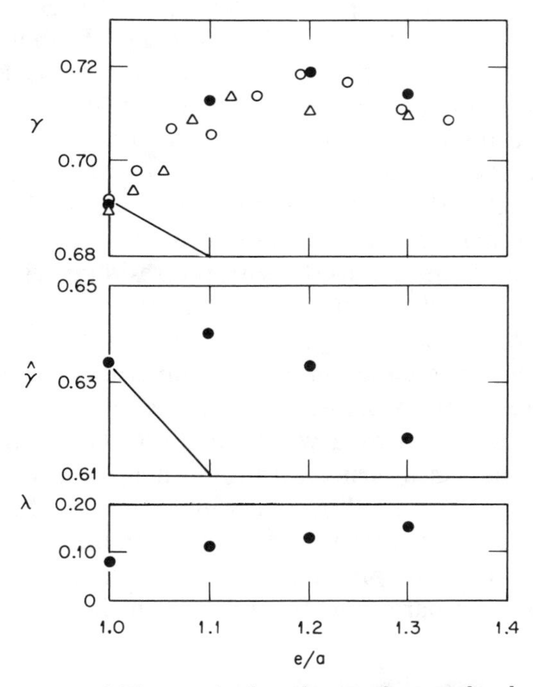

Fig. 3. (Top) Theoretical and experimental values of the specific-heat coefficient γ for alloys of zinc in copper as a function of electron concentration. (Middle) Theoretical calculations of γ ignoring electron-phonon enhancement. (Bottom) The electron-phonon enhancement factor.

It is a pity that this simple and internally consistent picture of the copper-zinc alloy system is completely incorrect. It was discovered in the early 1960's that the Fermi surface of pure copper is far from spherical, and that it indeed makes contact with the hexagonal faces of the Brillouin zone. The reason for this deviation from the expected free-electron behavior is that the d-electrons lie in the conduction band. The electronic states of pure copper are known quite well from band-theory calculations. The density-of-states function $\rho(E)$ has a negative slope in the neighborhood of $E = E_F$. The rigid-band model predicts that $\hat{\gamma}$ should decrease rapidly when zinc is alloyed with copper, as shown by the line in the middle panel in Fig. 3.

The change in the electron-phonon parameter λ with alloying can be estimated from the deviations from Matthiesen's rule using the resistivity data on alloys. This parameter increases rapidly with e/a as shown in the bottom panel of Fig. 3. In spite of this, the specific heat constant γ calculated from the rigid-band $\hat{\gamma}$ using Eq 16 is a linearly decreasing function of e/a. This function is shown by the straight line in the top panel in Fig. 3, and it disagrees completely with the experimental data.

The obvious thing to do at this stage is to use the coherent-potential approximation to calculate $\rho(E_F)$ for the copper-zinc alloy. The values of $\hat{\gamma}$ shown by the large dots in the center panel in Fig. 3 are the result of such a calculation[17] for alloys containing 10, 20, and 30% zinc. The coefficients γ that are obtained from these $\hat{\gamma}$ values using Eq 16 and the λ values described previously are shown in the top panel in Fig. 3. In looking at these results it should be borne in mind that there are absolutely no adjustable parameters in the theory.

The excellent agreement of the CPA values for γ with experiment is simply an illustration of the way that the modern theory of alloys can be used to sort out a classic problem that materials scientists had been wrestling with for many years. Another such classic problem has to do with the residual resistivity of alloys.

It has been known since the 19th century that the resistivity of a pure metal approaches zero as the temperature approaches zero, but that the resistivity of an alloy approaches some finite value ρ_R as the temperature approaches zero. The disappearance of the resistivity of a metal is explained by band theory since an electron in a Bloch state described by a wave function like the one in Eq 3 travels through the lattice without scattering with a momentum proportional to $\vec{k}$. The finite residual resistance ρ_R of an alloy is a consequence of the fact that the wave functions are not related to a $\vec{k}$-vector. A simple argument based on first-order perturbation theory leads to the prediction that the residual resistivity for a given alloy system should have a parabolic dependence on the concentration C of one of the constituents:

$$\rho_R = AC(1 - C) \qquad \text{(Eq 18)}$$

Experiments by such pioneers as Matthiesen and Linde confirmed that Eq 18 is roughly correct for most alloy systems, but there are some exceptions.

One of the best-known exceptions to Eq 18 is the silver-palladium alloy system, for which ρ_R is not symmetrical about C = 0.5. The peak in the residual-resistivity function occurs at a silver concentration of about 0.4%, and the resistivity is lower in the silver-rich range than would be expected from Eq 18. In an effort to explain these data, Mott[16] used his rigid-band model and introduced another theoretical postulate for transition metals that has been used in many other applications, the S-D model. Butler and Stocks[18] used a simplified theory of resistivity and a CPA calculation of the electronic states to calculate ρ_R for the silver-palladium system. They were able to explain the asymmetry in the function, and to obtain reasonable quantitative agreement with experiments. Later, Butler[19] derived a theory for the residual resistivity that makes essentially the same assumption as the CPA. This theory has been used[20] to explain the residual resistivity of silver-palladium and many other alloy systems in great detail and without the use of any adjustable parameters.

A detailed analysis of the modern theory of residual resistivity demonstrates again that the rigid-band model is not reliable for most alloy systems. It also shows that the conclusions drawn from the S-D model are not correct. Many materials scientists may not feel this chipping away at some of the simple and intuitive models that they have relied on is a positive contribution to their field. On the other hand, it is necessary to clear away misconceptions before progress can be made. It is to be hoped that, as time goes on, simple and intuitive pictures will be distilled from the CPA machinery that will be useful in analyzing the properties of alloys without large computations.

PREDICTIONS ABOUT STRUCTURE

Although an understanding of the specific heats and resistivities of alloys is useful to materials scientists, there is more interest in mechanical properties such as strength and ductility. For hundreds of years metalworkers have achieved the mechanical properties that they desired by controlling the microstructures of their alloys with the help of such time-honored techniques as heating, beating, and quenching. It is only in recent years that this activity has been put on a more scientific basis. It is now known that the information that is needed in the design of a treatment for achieving a given microstructure is data on the equilibrium and nonequilibrium structures that can occur in the alloy system. The equilibrium data are normally depicted in a phase diagram. A major goal of modern alloy theory is to calculate parameters that can be used in thermodynamic models of the phase stability of alloys.

As an example of this kind of study, let us consider the electron-diffraction pattern on a copper-palladium alloy[21] shown in Fig. 4. The sets of sharp spots in the diffuse scattering region between the large Bragg peaks are indicative of concentration waves with wave vectors $\vec{K}_i$ that all have the same magnitude, $|\vec{K}_i| = K$. They look a lot like the superlattice peaks in the diffraction pattern of a system that has a long-period superlattice structure such as, for example, the copper-gold system. However, the spacing between the peaks in the copper-palladium system

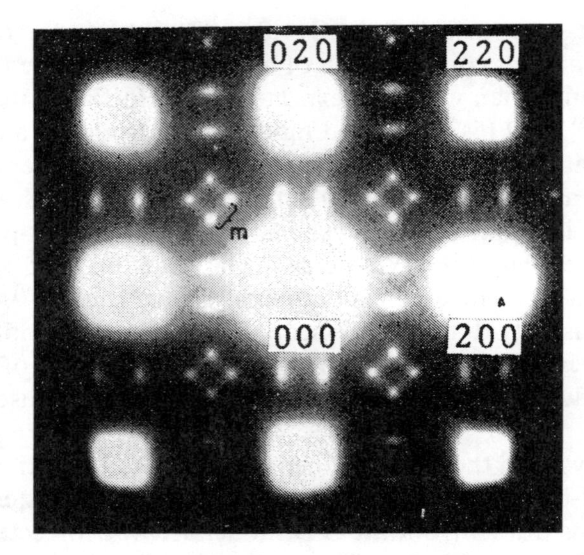

Fig. 4. Electron-diffraction pattern in a copper-palladium alloy (from Ref 21)

varies smoothly and continuously with concentration. This means that the wavelength of the concentration wave is not commensurate with the crystal structure because if it were commensurate for the concentration C, it would not be for $C + \delta C$. An incommensurate concentration wave cannot be interpreted as a superlattice unless one includes the trivial case of a superlattice with an infinitely large unit cell.

The length of the side of the square defined by the spots in the four-fold clusters that was measured experimentally for alloys of many different concentrations is plotted in Fig. 5. An obvious question is, what determines the wavelengths of these concentration waves? It is known that short-range forces that could arise from size effects or interatomic potentials can lead to complicated superlattices, but only a long-range effect can explain an incommensurate concentration wave.

Physicists have observed incommensurate waves in solids before. For example, there are incommensurate magnetic waves in some magnets, and incommensurate charge-density waves in low-dimensional solids. In all of these cases, the wavelength is related to a dimension of the Fermi surface. More specifically, there must be rather large areas on the Fermi surface that are flat and parallel to each other. The vectors that are perpendicular to these planes and connect them are the wavevectors $\vec{K}_i$ of the incommensurate waves.

The calculated values for the separation of the concentration-wave spots for three copper-palladium alloys are plotted with the experimental data in Fig. 5. These points were obtained by Gyorffy and Stocks,[22] who used the KKR-CPA to calculate the Fermi surfaces of the alloys. The flat parts of the calculated Fermi surfaces are such that the connecting vectors are in the (110) directions. The excellent agreement between the theoretical and experimental separations proves that the wavelengths of the concentration waves do arise from a Fermi-surface effect.

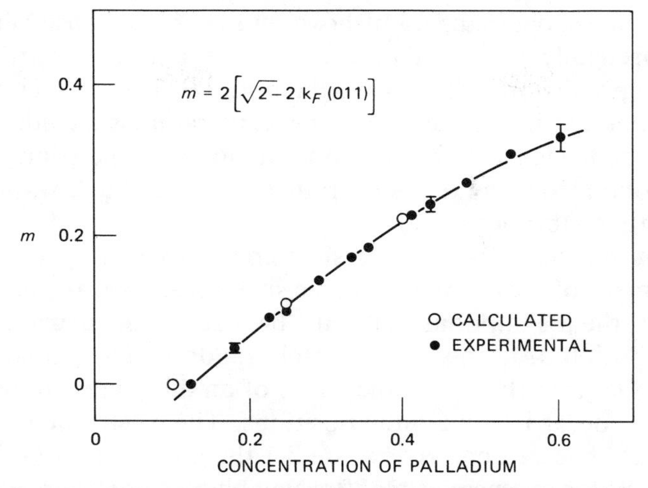

Fig. 5. Calculated and experimental values of the separation of the peaks in the diffuse scattering from copper-palladium alloys

Actually, Moss[23] had predicted that the wavelengths of concentration waves should be related to Fermi-surface dimensions, but he could not demonstrate this effect because the only theory of electronic states available to him was the rigid-band model. The accuracy of the predictions in Fig. 5 is even more impressive when it is recalled that there are no adjustable parameters in the calculations.

The theory of Gyorffy and Stocks[22] goes far beyond the prediction of the wavelengths of the concentration waves. Using a classical density-functional theory for a lattice gas model of the alloy, they developed a technique for calculating a correlation function $S^{(2)}(\vec{k})$ directly from the KKR-CPA. The correlation function or its Fourier transform $S^{(2)}(\vec{R}_n - \vec{R}_m)$ describes the correlations in the distribution of the A and B atoms on the lattice sites. Indeed, the short-range-order parameter that is proportional to the diffuse scattering intensity is given by

$$\alpha(\vec{k}) = [1 - C(1 - C)S^{(2)}(\vec{k})]^{-1} \quad \text{(Eq 19)}$$

The function $S^{(2)}(\vec{k})$ has been calculated for the copper-palladium system. Using this function, it has been possible to predict the diffuse scattering pattern in its entirety, not just the positions of the peaks. These predictions agree with the experimental measurements.

The KKR-CPA has been used[24] to calculate $S^{(2)}(\vec{k})$ for the palladium-rhodium alloy system. The short-range-order parameter $\alpha(\vec{k})$ is very different for this case, having peaks at the same points in $\vec{k}$-space as the Bragg peaks. The physical significance of this is that the alloy is unstable against phase separation. That is, the equilibrium distribution of the atoms on the lattice sites is such that there are regions in which the percentage of palladium atoms is much greater than random and regions in which the percentage of rhodium atoms is greater. It follows that there is a miscibility gap in the solid solution. This can be seen from the palladium-rhodium phase diagram shown in Fig. 6. The miscibility gap can be traced out theoretically by calculating $S^{(2)}(\vec{k})$ as a function of temperature. These theoretical predictions are also shown in Fig. 6. The open circles are on the theoretical spinodal that is obtained when the temperature is included only through the temperature dependence of the entropy of mixing. The triangles are on the spinodal that is obtained when the temperature is included in a way that simulates the effects of lattice vibrations.

It is surprising that the CPA has had such great success in predicting the equilibrium arrangement of atoms on the lattice sites since, as was pointed out in the beginning, one of the assumptions that must be made in the deviation of the CPA is that the distribution of atoms is completely random. The explanation for this comes from the fact that the electronic states of an alloy are almost independent of the short-range order in the atomic positions. The electronic states calculated with the CPA can thus be analyzed to predict the atomic configuration that the alloy will have in its lowest energy state. Presumably this procedure would not work so well if the equilibrium atomic configuration were too different from the random arrangement for which the CPA calculations are carried out.

The successes that have been obtained so far in the use of the CPA to predict the equilibrium phases of an alloy system should encourage theorists to continue their efforts in this area. Time constraints have made it impossible to mention another active area of research in which the CPA has proved its usefulness—the study of magnetism at finite temperatures.[25] The proper inclusion of the magnetic contribution to phase stability will, for the first time, provide a fundamental explanation for the behavior of steels, the most important class of metallic alloys.

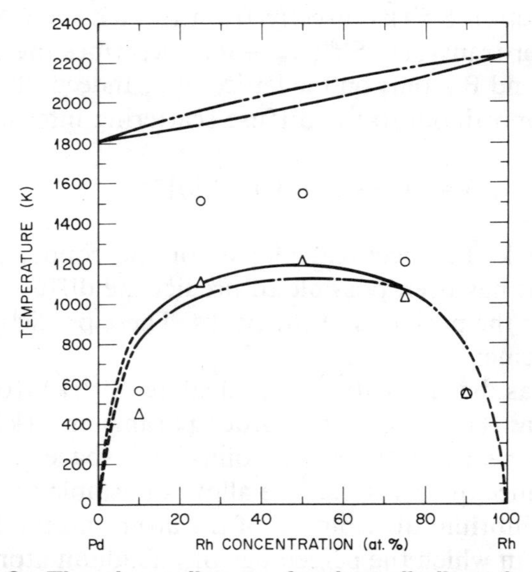

Fig. 6. The phase diagram for the palladium-rhodium alloy system. The triangles and circles define two calculated spinodals.

REFERENCES

1. P. Soven, *Phys. Rev.*, Vol 156, 1967, p. 809
2. J.S. Faulkner, *Prog. Mater. Sci.*, Vol 27, 1982, p. 1
3. H. Winter and G.M. Stocks, *Phys. Rev.*, Vol B27, 1983, p. 882
4. B.L. Gyorffy and G.M. Stocks, "Electrons in Disordered Metals and at Metallic Surfaces," in NATO-ASI Series B: *Physics*, Vol 42, edited by P. Phariseau, B.L. Gyorffy, and L. Scheire, Plenum Press, New York, 1979, p. 89
5. G.M. Stocks and H. Winter, "The Electronic Structure of Complex Systems," in NATO-ASI Series B: *Physics*, Vol 113, edited by P. Phariseau and W.M. Temmerman, Plenum Press, New York, 1984, p. 463
6. J.S. Faulkner, *Phys. Rev.*, Vol B13, 1976, p. 2391
7. D.A. Papaconstanlopoulos and A.C. Switendick, *Phys. Rev.*, Vol B32, 1985, p. 1289
8. G. Schadler, P. Weinberger, A. Gonis, and J. Klima, *J. Phys.*, Vol F15, 1985, p. 1675
9. A. Chen, G. Weisz, and A. Scher, *Phys. Rev.*, Vol B5, 1972, p. 2897; L. Kleinman, *Phys. Rev.*, Vol B35, 1987, p. 3854
10. L.M. Roth, *Phys. Rev.*, Vol B11, 1975, p. 3769
11. G.M. Stocks, R.W. Williams, and J.S. Faulkner, *Phys. Rev.*, Vol B4, 1971, p. 4390
12. D.H. Seib and W.E. Spicer, *Phys. Rev.*, Vol B2, 1970, p. 1676
13. N.J. Schevchik and C.M. Penchina, *Phys. Stat. Sol.*, Vol 70, No. 6, 1975, p. 619
14. H. Winter, P.J. Durham, W.M. Temmermann, and G.M. Stocks, *Phys. Rev.*, Vol B33, 1986, p. 2370
15. B.W. Veal and J.A. Rayne, *Phys. Rev.*, Vol 130, 1963, p. 2156; U. Mizutani, S. Noguchi, and T.B. Massalski, *Phys. Rev.*, Vol B5, 1972, p. 2057
16. N.F. Mott and H. Jones, *The Theory of Metals and Alloys*, Clarendon, Oxford, 1936
17. J.S. Faulkner and G.M. Stocks, *Phys. Rev.*, Vol B23, 1981, p. 5628
18. G.M. Stocks and W.H. Butler, *Phys. Rev. Lett.*, Vol 48, 1982, p. 55
19. W.H. Butler, *Phys. Rev.*, Vol B31, 1985, p. 3260
20. J.C. Swihart, W.H. Butler, G.M. Stocks, D.M. Nicholson, and R.C. Ward, *Phys. Rev. Lett.*, Vol 57, 1986, p. 1181
21. K. Olishima and D. Watanabe, *Acta. Crystallogr.*, Section A29, 1973, p. 520
22. B.L. Gyorffy and G.M. Stocks, *Phys. Rev. Lett.*, Vol 50, 1983, p. 374
23. S.C. Moss, *Phys. Rev. Lett.*, Vol 22, 1969, p. 1108
24. D.M. Nicholson, G.M. Stocks, F.J. Pinski, and D.D. Johnson (to be published)
25. J. Staunton, B.L. Gyorffy, G.M. Stocks, and J. Wadsworth, *J. Phys.*, Vol F16, 1986, p. 1761

4 Dimensional Analysis of the Crystal Structures of Intermetallic Phases

W.B. PEARSON
Department of Physics
University of Waterloo

I shall begin by putting the cart before the horse, since I shall first give an example of an explanation of other physical properties that can arise from the so-called dimensional analysis of a series of intermetallic phases with a given crystal structure. Then I shall go into the reasons for, and previous hindrances to, such analyses, and how the difficulties can be avoided. Finally, I shall discuss the findings from three such analyses to indicate some of the types of new information to which they may give rise.

There is some surprising information in a recent paper by Pinto *et al.*[1] on the magnetic structures of a few ternary RM_2X_2 phases with the tetragonal $BaAl_4$ or $ThCr_2Si_2$ type structure. In the phases they discuss, R is a heavy lanthanide from gadolinium to thulium; M is a 3*d* element, iron, cobalt, nickel, or copper; and X is silicon or germanium. The phases $TbCo_2Si_2$ and $DyCo_2Ge_2$ have antiferromagnetic Néel temperatures of 46 K and 16 K, respectively, but their lattice parameters give axial ratios, *c/a*, that are severally identical to five figures at measurement temperatures of 300 K and 4.2 K, even though the magnetic-structure parameters are $\vec{k}$: (0,0,1) and $\vec{\mu}//\vec{c}$ in each case.

In all previously known examples of such antiferromagnetic ordering of uniaxial crystals, or crystals that become uniaxial as a result of antiferromagnetic ordering, there is a considerable change of axial ratio on proceeding to lower temperatures below T_N. The so-called dimensional analysis gives the reason for the unique behavior of these two phases of cobalt.[2]

We have room-temperature unit-cell parameters (and a means of getting the *z* parameters of the X atoms) of ten nearly complete series of RM_2X_2 phases with this structure, where R is a rare earth, M is manganese, iron, cobalt, nickel, or copper, and X is silicon or germanium. Analysis of these ten series of cell dimensions (by means that we shall discuss in due course) leads to the data shown in Fig. 1, where the axes are respectively the slopes of the observed cell dimensions *a* and *c* versus D_R, the CN12 diameter of the rare-earth atom, for the ten series of alloys as severally M and X remain fixed and R is varied. On this diagram are plotted lines representing the conditions $f_{R-X} = 0$ and $f_{R-M} = 0$, where f_{i-j} is the slope of the straight line of $(R_i + R_j) - d_{i-j}$ with D_R. R_i, R_j are the atomic radii for CN12 of the *i*, *j* components and d_{i-j} are the observed distances between the *i* and *j* components in the series of phases as the rare earth is changed and M and X remain fixed.

Since the R–M and R–X contacts point in different directions in the structure, it is seen that the condition $f_{R-M} = f_{R-X} = 0$ given by the intersection of the two lines defines a rigid framework for the structure, so that a line $f_{c/a} = 0$ representing no change of axial ratio with change of D_R must also pass through the point of intersection of these two lines (Fig. 1). On Fig. 1 we have also plotted points representing data for nine of the ten series of alloys. Note that points for the two cobalt series lie just about at the intersection of these three lines. Thus for the two cobalt series a rigid framework is maintained in the structure as D_R is changed. Furthermore, as $TbCo_2Si_2$ and $DyCo_2Ge_2$ are each part of the whole, their alloys cannot undergo change of c/a as the temperature changes or after antiferromagnetic ordering, thus accounting for the results of Pinto *et al.* There must therefore be large magnetostrictive forces in these alloys below T_N.

We shall return to the results summarized in Fig. 1 later after explaining why we have analyzed the data in this way. Here we just observe that the conditions $f_{R-X} = 0$ and $f_{R-M} = 0$ indicate that R–X and R–M contacts, respectively, control the unit-cell dimensions. However, while we have the figure here we shall note

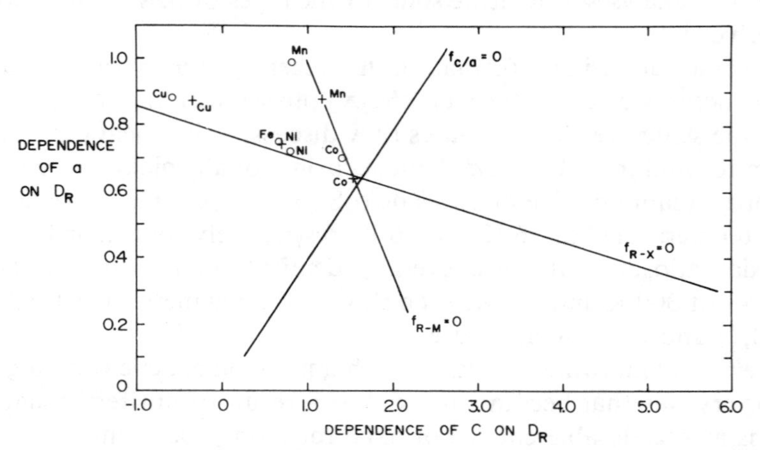

The symbols represent the observed behavior of series of alloys RM_2Ge_2 (o) and RM_2Si_2 (+), the M components being indicated.

Fig. 1. Lines representing $f_{R-X} = 0$, $f_{R-M} = 0$, and $f_{c/a} = 0$ (see text) plotted on axes representing slopes of cell dimensions *a* and *c* with D_R, for series of alloys as M and X components are kept fixed and R, the rare-earth atom, is changed

that the iron, nickel, and copper alloys all lie close to the line $f_{R-M} = 0$, whereas the manganese alloys lie on the line $f_{R-N} = 0$. The cobalt alloys, which lie close to the intersection of the two lines, have antiferromagnetic ordering of the type we have discussed. Copper alloys (well away from the $f_{R-M} = 0$ line) have a different but distinct type of antiferromagnetic ordering. Iron and nickel alloys on the line for $f_{R-X} = 0$, but not so far from that for $f_{R-M} = 0$, were found also to undergo a transition to antiferromagnetism but with incommensurate propagation vectors, so here we see an example of one of the interesting intermediate states in physics. The magnetic properties of any manganese alloy have yet to be reported, but since the alloys lie firmly about the line for $f_{R-M} = 0$ we expect them to be distinct and different from those of iron, cobalt, nickel, and copper alloys. Thus we may expect a correlation between the observed magnetic properties and the interatomic contacts that control the unit-cell dimensions as the diameter of the rare-earth atom is changed.

THE METHOD OF DIMENSIONAL ANALYSIS

Now the thing that we are all interested in is understanding the chemistry and physics of metals and alloys, which generally is a much more difficult subject than understanding inorganic and organic materials—even the crystallography of alloys with relatively complex structures is generally more difficult. I shall discuss one means of entry to the gaining of such understandings. Its great attraction is that it uses the one physical property that we are certain to have for all intermetallic phases once their crystal structures are known: the unit-cell dimensions. However, in the past there have been hindrances to pursuing this line of investigation that have prevented any real progress. Let us consider these hindrances.

Most prominent is the coordination problem since the size of a metallic atom, insofar as it can be characterized, depends on its coordination number. Thus problems arise when an atom in a structure has several different coordination numbers, or when an atom such as magnesium in the Laves phase $MgCu_2$ has 16 neighbors, four of which may be much closer relative to the atomic diameters than the other 12. Attempts to handle such problems have been made by deriving different sets of empirical radii for nonspherical surroundings as in phases with tetrahedrally close-packed structures (Shoemaker and Shoemaker[3]), or by deriving effective "fractional" coordination numbers to represent the atomic surroundings (e.g., Carter[4]).

However, practical sets of radii such as those of the Shoemakers (or of Geller[5] or Pauling[6] for phases with the β–W or Cr_3Si type structure) do not permit further *a priori* interpretations based on the physical properties (sizes) of the elements themselves. Secondly, as far as I am aware, nobody devising schemes for calculating "fractional" coordination numbers has ever related the results effectively to *a priori* elemental diameters.

Another hindrance to obtaining useful analytical results must be ascribed to the great influence that the late Professor Laves had. Even though in 1956[7] he had more understanding of the geometrical properties of intermetallic phases than anybody else, his use of the radius ratio as an analytical tool was sterile, because of the intractable nature of ratios—any relationships found in terms of radius ratios cannot be further developed in terms of the sizes of the individual atoms.

There is also the question of the absolute size of atoms, which strictly is undefinable: What *a priori* radius sum $(R_i + R_j)$ should be related to an observed interatomic distance d_{i-j}? As we shall show, this problem can be avoided by always considering *slopes*, rather than absolute values. The radius-ratio problem is easily avoided by seeking relationships in terms of individual radii R_i, R_j, or diameters, rather than the ratio R_i/R_j. Finally, since coordination number is a *situation* rather than a fundamental physical property of the elements composing a phase with a given structure, and because a situation that is inconvenient is best avoided, this is exactly how we handle the coordination problem, by avoiding it.

Consider Pauling's[8] equation:

$$R_{(1)} - R_{(n)} = 0.3 \log n$$

where n equals valency divided by C.N. Let us assume for the moment that this equation holds for intermetallic phases. Now for a series of intermetallic phases, MN_x, with a given cubic structure, if there are sufficient phases formed by the same element M, we can plot the cell edge a against the diameter of the N atom that is varied (and vice versa) and obtain a simple relationship between a and D_M and D_N since experience shows that linear plots are obtained. For the diameters of the M and N atoms, any self-consistent set of radii can be used, but for intermetallic phases, the CN12 radii of Teatum *et al.*[9] are most appropriate since they are derived directly from the elemental structures, and thus represent a physical property quite independent of the structure considered. If the CN's in the structure considered are not 12, or if the neighbors of the atoms are at various distances, correcting the diameters from CN12 to whatever CN may be appropriate (it matters not) only involves adding or subtracting a constant term to all D_M and D_N values, provided that Pauling's equation holds. This would only move the linear plots up or down on the diagram. Therefore, provided that subsequent analysis deals with *slopes*, we can use elemental diameters for CN12, regardless of the actual CN's in the structure, and so the coordination problem is avoided. Thus we consider the dependence of the dimensional parameter, a, on D_M or D_N, rather than comparing it directly with values calculated from absolute values of D_M or D_N.

But, you will say, Pauling's equation is known not to hold, for example, in relating CN8 of the bcc or CsCl structure to CN12 of the fcc structure where Teatum *et al.*[9] have shown that a percentage law holds. Furthermore, it is also known that it does not hold in changing diameters of rare-earth and actinide metals with valency. Nevertheless, practical experience shows that linear plots of a are obtained against diameters for CN12, at least within the accuracy of available experimental data that are drawn from numerous sources and exhibit scatter due to differing systematic errors, purities, and uncertainties of alloy composition.

So far we have only considered phases with noncomplex cubic crystal structures.

In structures of greater complexity and/or lower symmetry which may have variable atomic parameters (x, y, z) that are not known, the unit-cell volume or atomic volume can be plotted against D_M, D_N, etc. However, such plots are generally analytically uninformative—at best they give information on overall energy considerations. More information can be gained from plots of the unit-cell parameters a, c, etc. against D_M, D_N, etc.

For example, if for binary phases MN_x with a uniaxial crystal structure we plot a (also c) against D_M for a series of alloys with fixed N and changing M components, we obtain a series of parallel lines separated according to the D_N values of the N components to which they apply. If then we plot the a values obtained from these lines at some arbitrary D_M value (say 3.6 Å) against the D_N values of the N components involved, all data lie on a single straight line (Fig. 2). From these plots we obtain an equation representing the dependence of a (also c) on D_M and D_N that well reproduces the observed a (also c) values of all MN_x phases with the structure. The dependences of a (also c) on D_M and D_N in these equations permit the identification of the arrays of atoms in the structure that control the unit-cell dimensions.[10] However, even in structures of very moderate complexity this may not be sufficiently diagnostic. Provided that any variable atomic parameters are known, the most diagnostic plots are those involving the various interatomic distances that are less than or not much greater than the appropriate atomic radius sums. Generally we plot the CN12 radius sums minus the observed interatomic distances, $(R_i + R_j) - d_{i-j}$ versus $D_i, D_j, \ldots$, the diameter of the atom that is changed in the series of phases considered; such plots are also linear with

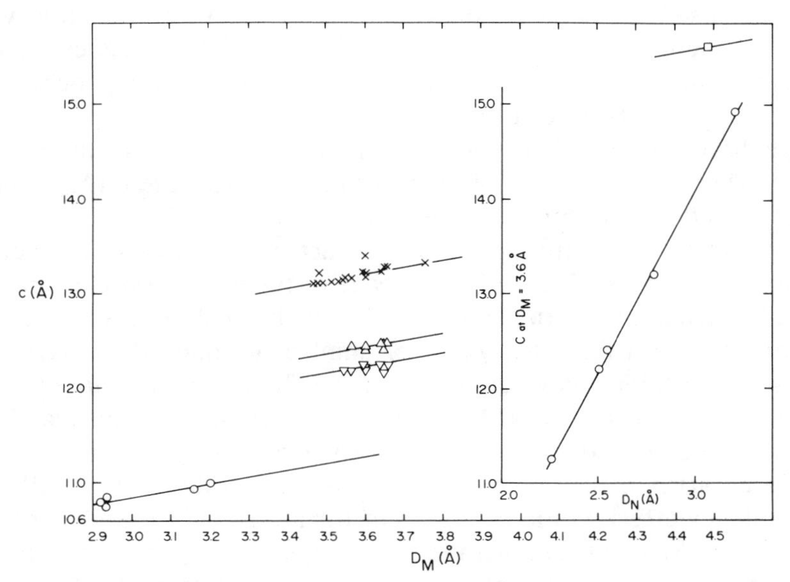

N = Be (o), Co (▽), Fe (△), Zn (×), and Mg (□). Inset shows c at a D_M value of 3.6 Å vs. D_N for the five N components.

Fig. 2. Plot showing unit-cell edge c versus D_M for M_2N_{17} phases with the rhombohedral Th_2Zn_{17} structure (hexagonal cell)

$D_i, D_j, \ldots$ Now, as we shall explain, we have a means of finding out which interatomic contacts are responsible for changes in unit-cell dimensions as one of the component atoms is changed. We consider only slopes; the absolute values are unimportant at this stage of the analysis.

Consider the ten series of phases with the tetragonal $BaAl_4$ or $ThCr_2Si_2$ (RM_2X_2) structure referred to earlier. Plots of $\Delta_{i-j} = (R_i + R_j) - d_{i-j}$ for the various interatomic contacts of interest in each of these ten series of phases are linear with D_R as the rare-earth atom is changed, as shown in Fig. 3 for the five series of germanium phases. Furthermore, it is seen in Fig. 3 that the slope, f_{i-j}, of one (or more) of these lines is approximately zero.

A slope f_{i-j} of zero indicates that there is no change of $(R_i + R_j) - d_{i-j}$ as D_R changes; therefore it is presumed that that particular interatomic contact, i–j, controls the unit-cell dimensions as D_R changes.

Note that $f_{R-X} = 0$ for the iron, nickel, and copper phases, whereas for the manganese phases $f_{R-M} = 0$ (for both silicon and germanium series), so the R–X contacts control the unit-cell dimensions in the former case and the R–M contacts in the latter, and the usual differences in behavior of manganese alloys compared with those of iron, cobalt, and nickel are to be noted. The interesting case of the cobalt phases has already been discussed, and the information on the ten diagrams of the type shown in Fig. 3 has been summarized in the plot of Fig. 1. The a cell parameter increases with D_R for all ten series of phases, as does the c parameter, except for the two series of copper alloys, where it decreases notably as D_R increases! Nevertheless, the R–X contacts control the cell dimensions as with the iron and nickel series of phases for which c increases with D_R. However, Fig. 1 indicates that there is nothing strange in such behavior; if R–X contacts control the cell dimensions, a negative dependence of c on D_R must occur when the dependence of a on D_R exceeds a certain value.

We have discussed these alloys as an example of how such analyses can be carried out.[11] There is much more of interest in phases with the $BaAl_4$ structure that we cannot consider here, but which is discussed in Ref 12.

We have now to draw attention to an artifact, known as a valency effect, that may enter these analyses. The CN12 diameters of the elements are determined from the elemental structures on the assumption that they exhibit a specific valency in the structure. However the valency and/or number of bond orbitals used by an element in an intermetallic phase considered may differ from that/those in its elemental structure from which its CN12 diameter was determined. This may lead to the recognition of a so-called valency effect, provided that the element concerned is involved in an array of contacts that controls the cell dimension(s).[13]

Such a valency effect would be apparent for atoms N in the plots of a (also c) against D_N for MN_x phases with a given uniaxial crystal structure, already described above, if the a (also c) values read at an arbitrary D_M value and plotted against D_N are not collinear, but separate into a series of parallel lines according to the Group number of the N atoms. In some cases, for example, it may be observed that points for phases formed by N atoms of Groups II to IV are collinear, but those formed by N atoms of Groups V and VI lie severally on separate

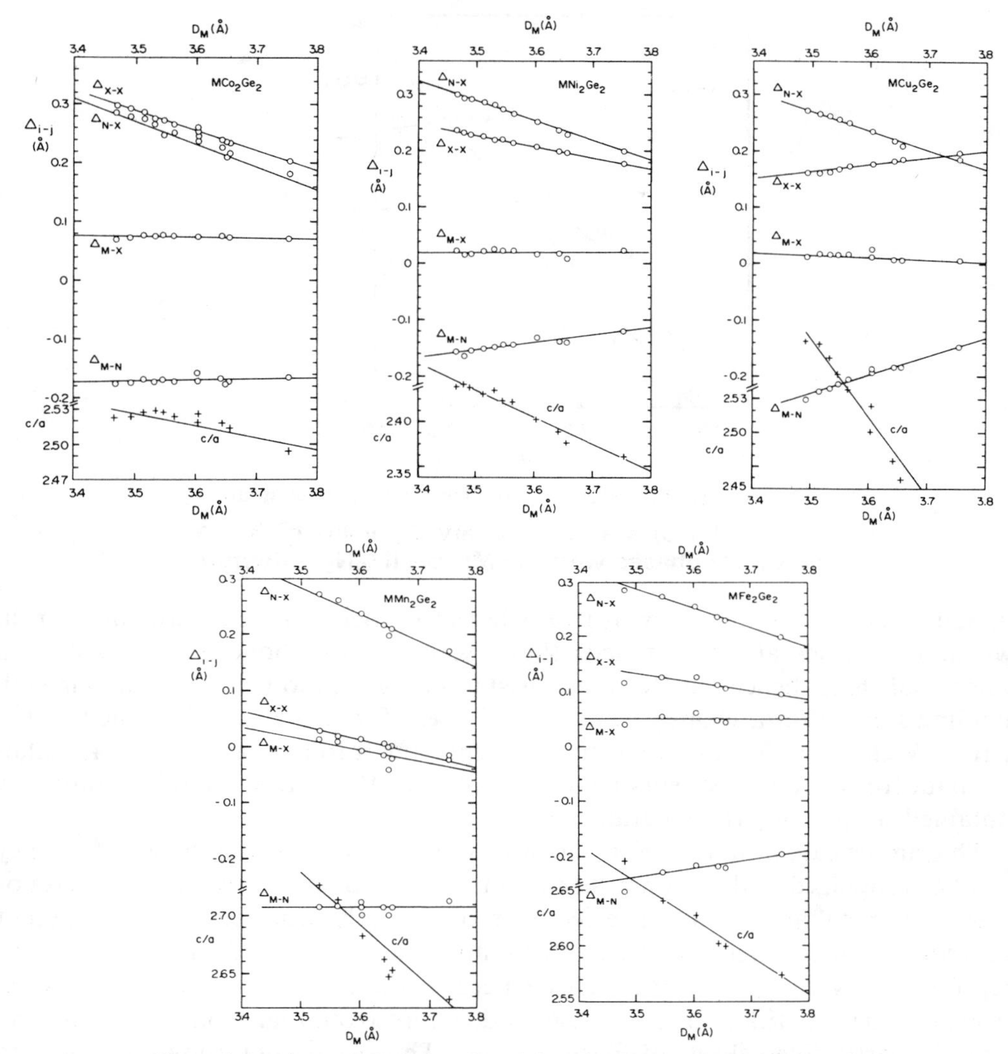

The axial ratios, c/a, of the phases are also shown as functions of D_M. *Note particularly that on this figure the rare earth is accorded the letter M whereas in the text and in Fig. 1 it has the letter R; also the transition metal is accorded the letter N whereas in the text and in Fig. 1 it has the letter M.*

Fig. 3. Plots of Δ_{i-j} vs. D_M for MN_2Ge_2 phases of Mn, Fe, Co, Ni, and Cu with the $BaAl_4$-type structure

parallel lines. Such is the behavior found for phases with the hexagonal Mn_5Si_3 (M_5N_3) structure shown in Fig. 4, where, in plots of a (at an arbitrary D_M value of 3.2 Å) against D_N, phases formed by N atoms of Groups II to IV (lead excepted) are collinear, but phases formed by N atoms of Group V lie on a lower, but parallel line. The reason for this is that the N components of Groups II to IV exhibit the same valencies in phases with the Mn_5Si_3 structure as in their elemental structures, as indeed they must. However the Group V elements use five or six

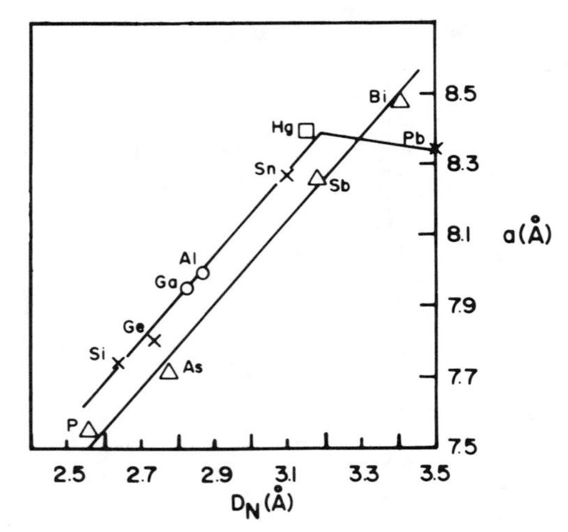

The various N components are identified on the diagram.

Fig. 4. Plot of *a* at an arbitrary D_M value of 3.2 Å vs. D_N for phases with the Mn_5Si_3 (M_5N_3) structure

bonding orbitals for a valency of five, whereas in their elemental structures (from which D_N values are determined) they use only three bonding orbitals for a valency of three; hence the D_N values used in Fig. 4 are too large. Calculating with Pauling's equation a change in D_N of −0.13 Å for a change of valency of the Group V elements from three to five and applying it to the data of Fig. 4, makes the data for all N components from Groups II to V collinear. Similar results are obtained on plotting the *c* parameters.

Thus in the case of normal elements with specific valencies, the change of valency can be recognized and its consequences interpreted quantitatively. In the case of transition-metal alloys, valency effects can be similarly identified, but they cannot generally be interpreted quantitatively because transition-metal valency cannot be represented by a single number. This is because the cohesion resulting from electrons in *d* bands (filled, or otherwise) is different from that from the outer electrons in partly filled sheets of Fermi surface. Thus we do not know the extent to which the valency effects result from changes in the number of cohesive electrons in *d* orbitals or from the number of outer electrons in partly filled sheets of Fermi surface, or both. We can only observe evidence that the CN12 diameters of the transition metals of different Groups bear a different relationship to each other in phases with the structure considered, compared with that found in their elemental structures. In derived equations that express the dependence of *a*, *c*, etc. on D_M and D_N we can allow for the valency effects by adding an empirical term, *S*, where *S* generally has integral values for transition metals of Groups IV, V, VI . . . .

Here we have only been able to sketch a brief outline of the basic effects that can be observed if the valency and/or number and distribution of bonding orbitals of atoms differs in a structure examined, compared with that in the elemental structure from which its CN12 diameter was derived. Further details are given in Ref 10 and 13.

EXAMPLES OF THE USE OF DIMENSIONAL ANALYSIS

Bcc or CsCl Structure Phases of Cu, Ag and Au

Now we shall discuss dimensional analyses of phases with two other structure types to indicate the types of information that may arise from such analyses.

You will be aware of such expressions for the total energy of a phase as:

$$U = U_0 + U_E + U_{bs}$$

(compare, for example, Heine and Weaire[14]) where U_0 is a structure-independent energy; U_E is a structure-dependent energy and can be regarded as a chemical energy arising particularly from the nearest neighbors of the atoms; and U_{bs} is the band-structure energy, also structure-dependent. Normally in intermetallic phases U_{bs} is such a small fraction of the structure-dependent energy (say 5%) that structural changes as a function of electron concentration (the diagnostic parameter for band-structure-energy effects) are not observed. Only in special cases, such as the polytypic structures studied by Sato and coworkers,[15] are structural changes at specific electron concentrations observed. The reason for this is that the nearest-neighbor surroundings of the atoms in the polytypic structural changes remain the same. Therefore there is little change in the structure-dependent chemical energy term, U_E, and changes in U_{bs}, that depend here on changes of second- or third-nearest neighbors, become observable as a result of structural changes at specific electron concentrations. The one exception to this discussion is the series of Hume-Rothery α, β, γ, etc. phases of copper, silver, and gold alloys with the succeeding B Group metals in which, despite changing composition and hence near-neighbor surroundings, structural changes between successive structures occur at specific electron concentrations. The reason for this is that with the filled *d* bands of copper, silver, and gold, all of the outer electrons are in partly filled sheets of Fermi surface, so that U_{bs} and its change with structure is a much larger fraction of the total energy than it is in intermetallic phases generally.

The bcc or ordered CsCl structure β phases of copper, silver, and gold with the following B Group metals are such electron phases. However, there are also CsCl-structure phases of copper, silver, and gold, with lithium, magnesium, scandium, yttrium, and the rare earths that are not electron phases since they (except magnesium) occur at electron concentrations other than 1.5 e/a, which is the limiting electron concentration of the β electron phases. Indeed, they are normal "chemical" intermetallic phases in which the chemical energy terms are dominant.

Calling all of these phases AB with copper, silver, or gold as A, we plot *a* versus the average D_B in Fig. 5. Note that the electron phases of silver and gold are collinear with no distinction of cell dimensions for silver and gold, as expected, since their CN12 diameters are, respectively, 2.890 and 2.884 Å. However, there is a notable difference (that averages 0.051 Å) in the cell dimensions of the "nonelectron" phases of silver and gold with lithium, magnesium, scandium, yttrium, and

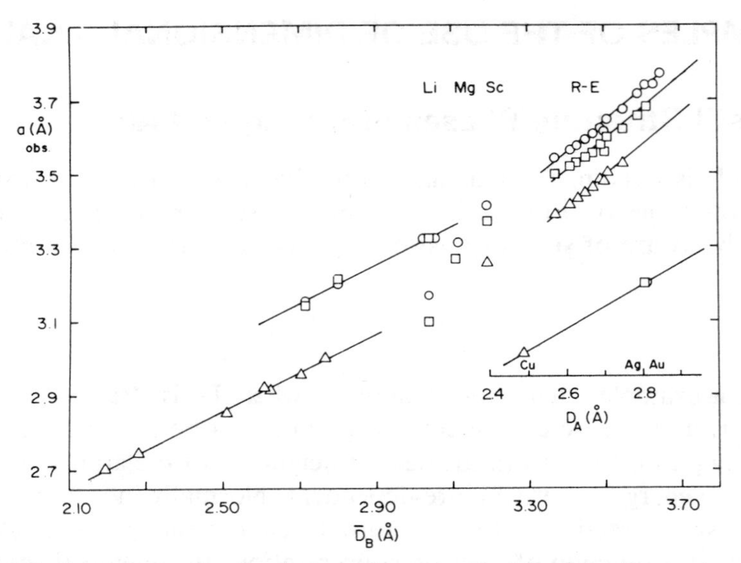

The inset shows, for the electron phases (on the left of the diagram), the variation of a with D_A at an arbitrary D_B value of 2.8 Å. The plots are constructed using CN8 diameters obtained from CN12 diameters with the relationship of Ref 9. Plots using CN12 radii are very similar in appearance.

Fig. 5. Plot of *a* vs. average diameter D_B of B component for AB phases with bcc or CsCl structures formed by A components Cu (△), Ag (○), and Au (□).

the rare earths, the silver phases having the larger unit cells. This results from the much larger electronegativity value of 2.3 for gold compared with 1.8 for silver,[16] which is a factor of importance in these "chemical energy" phases, but not in the electron phases where the band-structure energies are relatively much more important.

Indeed, if one applies the old Schomaker-Stevenson equation (17)

$$R_{(A-B)} = R_{(A)} + R_{(B)} - 0.09\ |x_A - x_B|$$

and $R_{(A-B)} = (a\sqrt{3})/2$ for the CsCl structure to calculate the difference in unit-cell size of the silver and gold phases with the CsCl structure that is expected to result from their difference in electronegativity, x, one obtains a predicted shortening of 0.052 Å for the cell edges of the gold phases compared with the silver phases formed with the same B components. Such close agreement with the average observed shortening of 0.051 Å is of course fortuitous, but it does serve to confirm our arguments concerning the relative importance of the band-structure energy and the structure-dependent chemical energy in this group of phases with the same crystal structure. Indeed, here is a further example of the possible usefulness of dimensional analysis.

Phases With the H-Phase, $AlCr_2C$ Structure

We shall now discuss some results of dimensional analysis of ternary phases with the hexagonal H-phase $AlCr_2C$ (NM_2C) structure. These are generally regarded as interstitial phases with carbon or nitrogen sitting at the center of an octahedron of transition-metal atoms (Fig. 6). The structure has two degrees of freedom: c/a and the z parameter of the transition-metal atoms that allows them to move in the [0001] direction. Our interest originally was to discover the mode of possible distortion of the transition-metal octahedron under operation of these two variables—a feature that does not occur in the cubic interstitial structures since everything scales with a.

In this analysis we have plotted both a and c as functions of D_M as M is changed and N and C are fixed. Figure 7 is an example of the former, where for simplicity we show data for phases with N = Al, Tl, and S only. Note that the data lie on straight parallel lines, positioned in some way related to the diameter of the N atom. Note also an apparent valency effect involving the transition metal M component in the data for N = Al where the comparison can be made. Similar results, but with different slopes, are obtained on plotting c versus D_M. When,

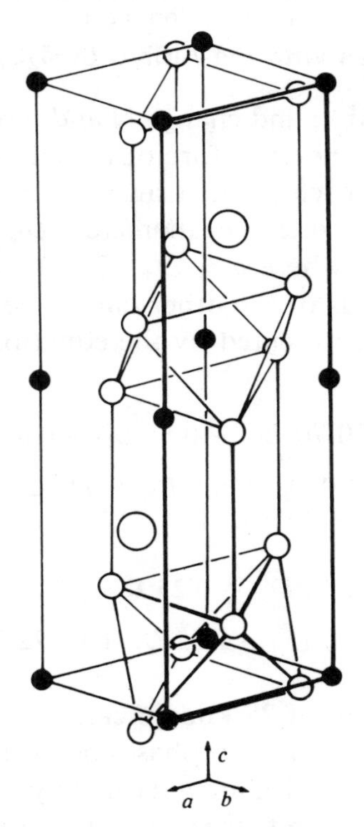

The large circles are the Al atoms.

Fig. 6. Diagram of the $AlCr_2C$ structure, showing the octahedra of transition-metal atoms (○), surrounding the carbon atoms (●).

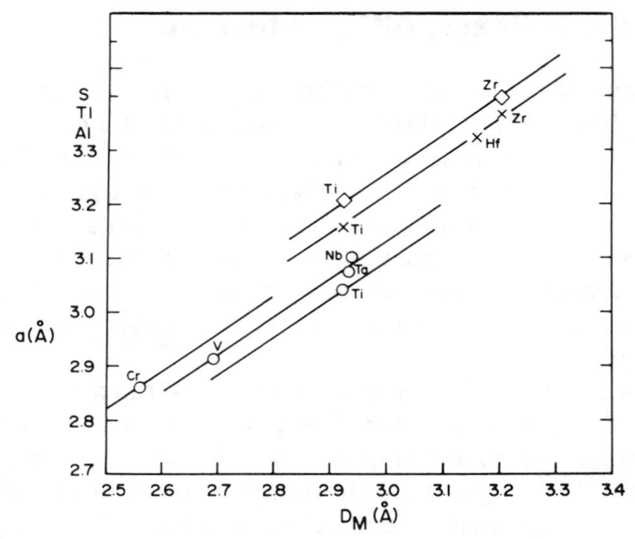

Plot of a vs. D_M for carbide phases of N = Al (○), Tl (×), and S (◇) (only) as M is changed. A valency effect is apparent for the transition metals, since their phases with Al lie on separate lines according to the group number of the transition metals.

Fig. 7. Phases with the $AlCr_2C$ (NM_2C) structure

however, we keep M and C fixed and change N and plot both a and c against D_N, remarkable results are found, never before observed in any dimensional analysis (Fig. 8). Again, the data for each of the transition metals lie on different parallel lines placed relative to their different diameters D_M, and the dramatic change of slope occurs in the region where $a \approx D_N$. The N = Sn alloys that lie in this region do not follow the trends of the other data. The data for the a and c plots against D_M and D_N can be represented by the equations:

$$\text{When } a > D_N \begin{cases} a = 0.70\, D_M - 0.65\, D_N + 0.019S + 2.833 & \text{(Eq 1)} \\ c = 3.0\, D_M + 6.5\, D_N + 0.072S - 13.798 & \text{(Eq 2)} \end{cases}$$

$$\text{When } a < D_N \begin{cases} a = 0.70D_M + 0.25\, D_N + 0.019S + 0.273 & \text{(Eq 3)} \\ c = 3.0\, D_M - 1.65\, D_N + 0.072S + 10.812 & \text{(Eq 4)} \end{cases}$$

that reproduce the a dimension of 28 known carbide phases to within 0.5% and the c dimension to within 0.25%, the tin phases being excepted. Also excepted are data for the three N = S phases which are indicated by further analysis as exhibiting a valency of six in these phases. In these equations S represents the valency effect in respect to the transition metals. It has values, respectively, of 0, 3, and 5 for the Group IV, V, and VI M components. These values are selected so as to make the valency effect zero for the Group IV metals, as shown in further analysis that cannot be discussed here.

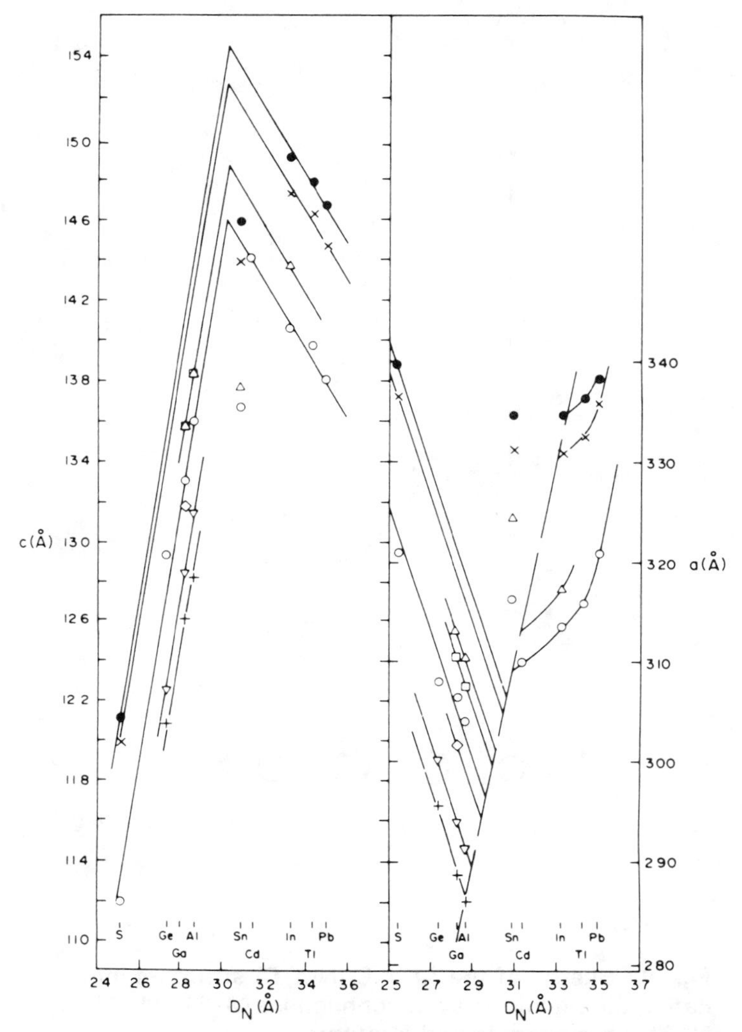

Plots of a and of c vs. D_N for the carbide phases as M is kept constant and N is changed. The broken line in the right-hand plot represents $a = D_N$. M components are Cr (+), V (▿), Mo (◇), Ti (○), Ta (□), Nb (△), Hf (×), and Zr (●).

Fig. 8. Phases with the $AlCr_2C$ (NM_2C) structure

Since the M—M separation in (0001) planes is always greater than D_M, the a dependence on 0.7 D_M and the c dependence on 3.0 D_M are accounted for by the resolved components of the zig-zag array of atoms —M—N—M—M—N—M— (possibly with M—C—M replacing M—M) that runs throughout the structure (Fig. 9) and thus controls the unit-cell dimensions. However, the c dependences on −1.65 D_N and most particularly on 6.5 D_N must be accounted for especially; there is no array of N atoms in the structure that could possibly lead to a c dependence on 6.5 D_N. Such a dependence must therefore be the result of a dimensional adjustment in the structure that does not arise directly from the N atoms themselves.

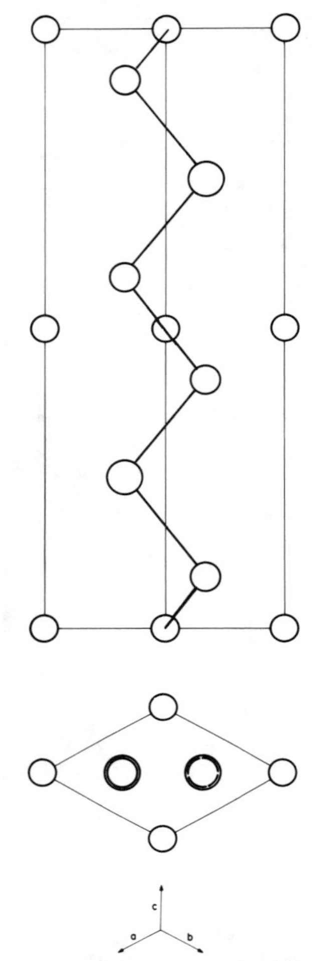

Fig. 9. Sketch of the $AlCr_2C$ (NM_2C) structure indicating the zig-zag array of connections M–N–M–M–N–M– between M and N atoms

In order to see if and how the transition-metal octahedra distort, we examined the NTi_2C series of phases, since this was the largest series with the most information on the atomic parameter z. We found (Fig. 10) that the M–M edges ($\neq a$) that did not lie in (0001) planes remained constant (range $\approx$ 2.89 to 2.97 Å) and essentially equal to the interatomic distances in elemental hexagonal titanium (2.89 and 2.95 Å). That this should prove to be the invariant feature of the structure as the diameter of the N atom increases from about 2.5 to 3.5 Å is interesting since it is the one condition that would maintain three-dimensional d bands of titanium similar to those of its elemental structure. This condition is essential to true Hägg interstitial phases, which the titanium phases are with $R_C/R_{Ti} < 0.59$.

Now we can show how this invariant feature of the structure is responsible for the dependences of c on D_N mentioned above. The M–M ($\neq a$) distance, $d =$

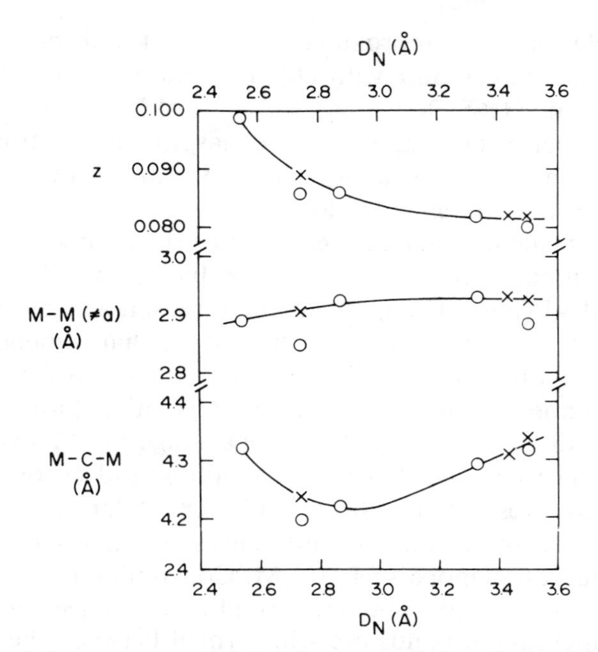

Plots of the z parameter of the Ti atoms and the calculated M–M ($\neq a$) and M–C–M distances versus D_N. Points (o) represent data with experimentally determined z values. Points (×) represent data with smoothed z values.

Fig. 10. NTi_2C phases with the $AlCr_2C$ structure.

$(a^3/3 + 4z^2c^2)^{1/2}$, so that, for example, a change in c must be compensated by changes in a and/or z to keep d constant. Note, however, that we have already shown that the M–N distances are important in controlling the cell dimensions. These distances also involve a, c, and z, the latter in a $(1/4 - z)$ relationship, so a change in z has the opposite effect on the M–N distances as on the M–M ($\neq a$) distances. Although the M–N distance changes appreciably with z, it varies little as a and c (which are coupled) change widely at a constant z and M–M ($\neq a$) distance.

Thus, maintaining M–M ($\neq a$) distance constant and an acceptable M–N distance prescribes the value of z closely, but the coupled values of a and c can still vary widely, while satisfying these conditions. This possible range of a and c values (e.g., large a and small c, or vice versa) causes a wide range in the unit-cell volume which is greatest for large a and small c. Since the unit-cell volume has to have an equitable value to give a low energy for the unit cell, it is this that ultimately determines the relative values of a and c.

Thus the conditions of invariant M–M ($\neq a$) distance and equitable M–N distances and unit-cell volumes provide the driving force to control the cell dimensions as the diameter of N changes. This leads to the dependence of c on 6.5 D_N when $a > D_N$. In the region where $a < D_N$, z is approximately constant (Fig. 10). The triangular, 3^6, networks of N atoms in (0001) planes now in part control a

since $d_{N-N} = a$. Because z is approximately constant and the M–M ($\neq a$) distance is invariant, an increase in a with D_N must cause a decrease in c, and hence a dependence of c on $-1.65\ D_N$ is observed.

I have given a rather brief summary of the reasons for the strange dimensional behavior of these phases, but the analysis leads to further results of interest (see Ref 18) that we can only indicate in closing.

The final result is calculation and prediction of the number of electrons transferred to carbon 2p bands or in nonbonding d bands for alloys of M atoms of Groups IV, V, and VI (Fig. 11) and an apparent reason why no phases have been found with M atoms of Group III. I am still hoping that somebody will confirm or refute these predictions with band-structure calculations for the phases.

Finally, I should note that for the true Hägg interstitial phases of titanium, hafnium, zirconium, niobium, and tantalum, the average M–M ($\neq a$) distances are approximately equal to D_M, but for the vanadium and chromium phases, the average distances are, respectively, 0.111 and 0.187 Å larger, and it is clear that the diameter of the carbon atom, C, influences the invariant M–M distances through the octahedron diagonals M–C–M. Concomitantly, the vanadium and chromium phases are not true Hägg interstitial phases since for them $R_C/R_M >$ 0.59; whether elementlike d bands are still formed between the transition-metal atoms under these conditions is uncertain.

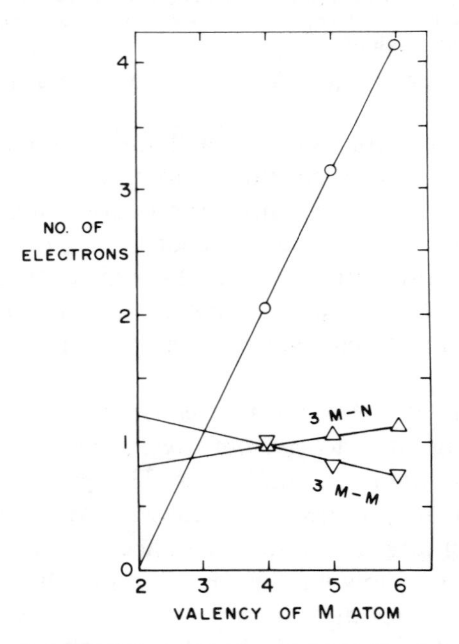

Fig. 11. Calculated average number of valency electrons involved in three M–M bonds (▽), three M–N bonds (△), and transferred to carbon 2p bands or in nonbonding d bands (○) vs. M atom valency for NTi_2C, NV_2C, and NCr_2C phases in the region where $a > D_N$

We have sought to show how dimensional analysis can give information and understanding of other physical properties such as magnetic and energy-band effects, besides information on the more obvious features such as the interatomic contacts and/or interactions that control the unit-cell dimensions, or account for phases in which the axial ratio has a constant value, such as those with the Th_2Zn_{17} structure[19] that we did not mention.

REFERENCES

1. H. Pinto, M. Melamud, M. Kuznietz, and H. Shaked, *Phys. Rev.*, Vol B31, 1985, p. 508
2. W.B. Pearson, *Phys. Rev.*, Vol B33, 1986, p. 3545
3. C.B. Shoemaker and D.P. Shoemaker, in *Developments in the Structural Chemistry of Alloy Phases*, edited by B.C. Giessen, Plenum, New York, 1969, pp. 107-139
4. F.L. Carter, *Acta Crystallogr.*, Vol B34, 1978, p. 2962
5. S. Geller, *Acta Crystallogr.*, Vol 9, 1956, p. 885
6. L. Pauling, *Acta Crystallogr.*, Vol 10, 1957, p. 374
7. F. Laves, in *Theory of Alloy Phases*, ASM, Cleveland, 1956, p. 124
8. L. Pauling, *J. Amer. Chem. Soc.*, Vol 69, 1947, p. 542
9. E. Teatum, K. Gschneidner, and J. Waber, Los Alamos Laboratory Research Report LA 2345, U.S. Dept. of Commerce, Washington, 1960
10. W.B. Pearson, *Philos. Trans. Roy. Soc., London*, Vol 298, 1980, p. 415
11. W.B. Pearson and P. Villars, *J. Less-Common Met.*, Vol. 97, 1984, p. 119
12. W.B. Pearson, *J. Solid State Chem.*, Vol 56, 1985, p. 278
13. W.B. Pearson, *J. Less-Common Met.*, Vol 72, 1980, p. 107
14. V. Heine and D. Weaire, in *Solid State Physics*, Vol 24, Academic Press, New York, 1970, p. 268
15. H. Sato and R.S. Toth, *Bull. Soc. Fr. Miner. Crist.*, Vol 91, 1965, p. 557; H. Sato, R.S. Toth, G. Shirane, and D.E. Cox, *J. Phys. Chem. Solids*, Vol 27, 1966, p. 413; H. Sato, R.S. Toth, and G. Honjo, *J. Phys. Chem. Solids*, Vol 28, 1967, p. 137; H. Sato and R.S. Toth, *J. Phys. Chem. Solids*, Vol 29, 1968, p. 2015
16. L. Pauling, *The Nature of the Chemical Bond*, 3rd Ed., Cornell University Press, Ithaca, N.Y., 1960; M. Haissinskij, *J. Phys. Radium*, Vol 7, 1946, p. 7; W. Gordy and W.J.O. Thomas, *J. Chem. Phys.*, Vol 24, 1956, p. 439
17. V. Schomaker and D.P. Stevenson, *J. Amer. Chem. Soc.*, Vol 63, 1941, p. 37
18. W.B. Pearson, *Acta Crystallogr.*, Vol A36, 1980, p. 724
19. W.B. Pearson, *Acta Crystallogr.*, Vol B35, 1979, p. 1329

5

Microscopic Theories of Alloy Phase Equilibrium

C. SIGLI
Cegedur Pechiney
Centre de Recherche et Developement de Voreppe
J.M. SANCHEZ
Henry Krumb School of Mines
Columbia University

The computation of temperature-composition phase diagrams, besides being potentially valuable in the design and development of new materials, provides an exacting test for *ab initio* and phenomenological theories of alloy phase stability. Phase diagrams often exhibit a great variety of equilibrium and metastable phases competing closely for stability within given temperature and composition ranges. Although knowledge of the free energy of the system is clearly sufficient to address the question of stability, free-energy calculations from first principles represent a formidable task. Nevertheless, during the last decade we have witnessed a great deal of activity by theorists and experimentalists—using a variety of empirical, phenomenological, and *ab initio* methods—which have set the principles governing alloy phase stability on solid ground.

The empirical approach proposed by Kaufman and Nesor,[1] presently practiced by a large number of investigators throughout the world, has resulted in an extensive thermochemical data base for binary and ternary alloys. The philosophy of the approach consists of fitting simple functions to available thermochemical data and to known phase-diagram information, a procedure that generally produces reasonably consistent values for the thermodynamic potentials. In some instances these thermodynamic functions may be extrapolated to unknown temperature and composition regions in order to make specific predictions on phase stability. Despite its practical usefulness, this strictly empirical approach to phase-diagram characterization does not provide any insight into the underlying physical reasons for alloy phase stability which, invariably, must be sought in the electronic structure as well as in the correct description of the statistical mechanics problem at finite temperatures.

One of the most significant recent developments in alloy theory has been the implementation by several investigators of *ab initio* quantum mechanical methods for studying the electronic structure and, in particular, for calculating the total energy at zero temperature of pure solids,[2-4] relatively simple compounds,[5-7] and disordered alloys.[8-10] By reproducing a wide range of physical properties within a few per cent of the experimental values, these quantum mechanical total-energy calculations have provided almost conclusive evidence in favor of the theoretical approximations—most notably the local-density approximation—which of necessity are involved in the treatment of the many-body electron problem in metals, semiconductors, and alloys. At the same time, the observed structural and alloying trends have been extensively investigated and satisfactorily explained by means of simple microscopic models, such as tight-binding Hamiltonians for transition-metal alloys.[11-20] Although *ab initio* and model Hamiltonian calculations have clearly different goals, both approaches complement each other very nicely: the former provides a solid quantitative basis for specific alloy systems whereas the latter has been most valuable in establishing the general trends and in identifying the key factors that control alloy formation. It is generally the case that, for most alloy systems, fully *ab initio* calculations of equilibrium phase diagrams are beyond present theoretical and computational capabilities whereas model Hamiltonian calculations lack quantitative reliability. However, the result of *ab initio* and model Hamiltonian calculations, when combined with accurate statistical mechanics methods, provide important guidelines for a sound phenomenological model of alloy phase stability. We will explore such a phenomenological model in the sections headed "Enthalpy of Formation of Transition-Metal Alloys" and "Phenomenological Computation of Phase Diagrams," with the emphasis placed on how electronic alloy theory can guide us in the global description of stable and metastable equilibrium.

As already mentioned, the statistical mechanics treatment of alloys at finite temperatures plays an important role in the theoretical description of phase equilibrium. Owing to the simplicity of the subregular solution or the Bragg-Williams models, these approximations are widely used for the description of the configurational free energy. However, numerous recent studies using Monte Carlo simulations[21-23] as well as the Cluster Variation Method (CVM)[24-28] have conclusively shown the inadequacy of the single-site mean-field approximation (subregular solution and Bragg-Williams). The shortcomings of the strict mean-field approach stems primarily from the fact that short-range order (SRO), which is neglected in the single-site approximation, has profound qualitative and quantitative effects on phase stability.

In the next section we present a cluster theory of the configurational thermodynamics of alloys. The description provides the formal framework for the treatment of SRO effects, in both the configurational energy and entropy, in terms of pair and multisite correlation functions. In the subsequent section we make contact with microscopic electronic theories via a simple tight-binding description of the d-electrons in transition-metal alloys. In particular, we present the argument, first advanced by Gautier, Ducastelle, and coworkers,[14-16] that the energy of alloy formation may be accurately written as the sum of the energy of a disordered reference state plus an ordering energy which incorporates short-ranged pair and, possibly, multisite interactions. Finally, in the last section, we incorporate the main features of the electronic energy of alloy formation into a phenomenological CVM model of phase equilibrium. As an example, we present the calculated equilibrium and metastable phase diagrams for the Al-Li system and discuss the results in the light of available experimental data.

CONFIGURATIONAL THERMODYNAMICS

In this section we discuss briefly a general formalism for the description of the configurational thermodynamics of alloys. For the sake of simplicity we consider only binary systems, although the theory can be easily extended to multicomponent alloys.[29] The configuration of a crystalline binary alloy is described in terms of spin or occupation numbers σ_p at each lattice site p and taking, respectively, values +1 and −1 for components A and B. Any configuration of the system is then fully specified by the N-dimensional vector $\boldsymbol{\sigma} = \{\sigma_1, \sigma_2, \ldots \sigma_N\}$, where N is the number of lattice points. In general, one is confronted with the problem of describing functions that depend explicitly on the occupation variables σ_p, such as the energy of formation and entropy. In order to provide an unambiguous description of such functions, it is convenient to introduce an orthogonal functional basis in configurational space. Although in the thermodynamic limit the dimension of the complete orthogonal basis is infinite, judicious choice of the basis functions allows us to obtain accurate approximations to the thermodynamic potentials in terms of a subset of finite dimension.

We begin by defining the scalar product in multidimensional configurational space between two arbitrary functions, $f(\boldsymbol{\sigma})$ and $g(\boldsymbol{\sigma})$, by

$$\langle f, g \rangle = (1/2^N)\mathrm{Tr}^{(N)} f(\boldsymbol{\sigma}) \cdot g(\boldsymbol{\sigma}) \qquad \text{(Eq 1)}$$

where $\mathrm{Tr}^{(N)}$ designates the trace over the 2^N configurations.

A complete orthogonal functional basis is given by the identity plus the set of cluster functions Φ_α formed by the products of the occupation numbers σ_p on all possible clusters of points in the lattice[29]:

$$\Phi_\alpha = \sigma_p \sigma_{p'} \ldots \sigma_{p''} \qquad \text{(Eq 2)}$$

where α refers to the cluster defined by lattice points $\{p, p', \ldots, p''\}$. Due to the orthogonal character of the basis set, any function of configuration $f(\boldsymbol{\sigma})$ may be written [29]:

$$f(\boldsymbol{\sigma}) = f_0 + \sum f_\alpha \Phi_\alpha \qquad \text{(Eq 3)}$$

where

$$f_0 = \langle f, 1 \rangle = (1/2^N)\mathrm{Tr}^{(N)} f(\boldsymbol{\sigma}) \qquad \text{(Eq 4)}$$

and

$$f_\alpha = \langle f, \Phi_\alpha \rangle = (1/2^N)\mathrm{Tr}^{(N)} f(\boldsymbol{\sigma}) \cdot \Phi_\alpha(\boldsymbol{\sigma}) \qquad \text{(Eq 5)}$$

In particular, it follows from Eq 3 that the configurational energy of the alloy, $H(\boldsymbol{\sigma})$, can always be written in terms of effective pair and multisite interactions:

$$H(\boldsymbol{\sigma}) = h_0 + \sum_\alpha h_\alpha \Phi_\alpha(\boldsymbol{\sigma}) \qquad \text{(Eq 6)}$$

where the effective interactions h_α are calculated using Eq 5.

Likewise, the reduced-density matrix or probability distribution for a given cluster α can be written as:

$$\rho_\alpha(\boldsymbol{\sigma}_\alpha) = (1/2^{m_\alpha}) \left[1 + \sum_\beta \langle \Phi_\beta \rangle \Phi_\beta(\boldsymbol{\sigma}_\beta) \right] \qquad \text{(Eq 7)}$$

where m_α is the number of points in α, $\boldsymbol{\sigma}_\alpha$ is a m_α-dimensional vector describing the configuration of cluster α, and $\langle \Phi_\alpha \rangle$ are multisite correlation functions defined by:

$$\langle \Phi_\alpha \rangle = \mathrm{Tr}^{(\alpha)} \rho_\alpha(\boldsymbol{\sigma}_\alpha) \Phi_\alpha(\boldsymbol{\sigma}_\alpha) \qquad \text{(Eq 8)}$$

The reduced-density matrices ρ_α can also be described in terms of renormalized cluster Hamiltonians H_α defined by:

$$\rho_\alpha(\boldsymbol{\sigma}_\alpha) = (1/Z_\alpha)\exp[-H_\alpha(\boldsymbol{\sigma}_\alpha)/kT] \qquad \text{(Eq 9)}$$

where Z_α is a normalization constant and where the H_α are given in terms of renormalized multisite interactions $h_\beta^{(\alpha)}$ by:

$$H_\alpha(\boldsymbol{\sigma}_\alpha) = \sum h_\beta^{(\alpha)} \Phi_\beta(\boldsymbol{\sigma}_\beta) \qquad \text{(Eq 10)}$$

A key element of the CVM is the fact that we can define a new set of irreducible cluster Hamiltonians $\hat{H}_\alpha$ such that the renormalized Hamiltonians H_α are given by the sum of the $\hat{H}_\beta$ over all subclusters β of α:

$$H_\alpha(\boldsymbol{\sigma}_\alpha) = \sum \hat{H}_\beta(\boldsymbol{\sigma}_\beta) \qquad \text{(Eq 11)}$$

In particular, this transformation allows us to write the configurational energy of the system, which corresponds to the renormalized Hamiltonian of the infinite system, as:

$$H(\boldsymbol{\sigma}) = \sum_{\beta} \hat{H}_{\beta}(\boldsymbol{\sigma}_{\beta}) \qquad \text{(Eq 12)}$$

Although Eq 12 is exact, it has no practical use since the summation involves an infinite number of clusters β. However, a useful approximation is obtained if we neglect the irreducible cluster Hamiltonians $\hat{H}_{\beta}$ for clusters larger than a maximum cluster γ. This approximation, which leads to results identical to the CVM introduced by Kikuchi in 1951,[30] allows us to write the configurational Hamiltonian in the following manner:

$$H(\boldsymbol{\sigma}) = \sum_{\beta}{}' \hat{H}_{\beta}(\boldsymbol{\sigma}_{\beta}) = \sum_{\beta}{}' a_{\beta} H_{\beta}(\boldsymbol{\sigma}_{\beta}) \qquad \text{(Eq 13)}$$

where the sum is now restricted to clusters smaller than or equal to the maximum cluster γ, and where the a_{β} are geometrical coefficients given by:

$$\sum{}' a_{\beta} = 1 \qquad \text{(Eq 14)}$$

Equation 14, one for each cluster α contained in the maximum cluster γ, follows directly from Eq 11 and 13.

Equations 13 represent a set of self-consistency relations for the renormalized interaction $h_{\beta}^{(\alpha)}$ (see Eq 10) that must be solved as a function of temperature and composition. With the help of Eq 6, 9, and 10, the self-consistency relations (Eq 13) may be written as:

$$h_{\alpha} = -kT \sum_{\beta} a_{\beta} \langle \ell n\, \rho_{\beta}, \Phi_{\alpha} \rangle = -kT \sum a_{\beta} h_{\beta}^{(\alpha)} \qquad \text{(Eq 15)}$$

Once the renormalized interactions $h_{\beta}^{(\alpha)}$ are obtained from the solution of Eq 15, the free energy of the system is given by:

$$F = -kT \sum_{\alpha}{}' a_{\alpha}\, \ell n\, Z_{\alpha} \qquad \text{(Eq 16)}$$

where the cluster-partition functions Z_{α} follow from Eq 9.[29]

ENTHALPY OF FORMATION OF TRANSITION-METAL ALLOYS

Several of the first-principles electronic theories of binary alloys include SRO (short-range order), either explicitly or implicitly, in the calculation of the enthalpy of formation. For example, Connolly and Williams,[31] following the description of the configurational energy given by Eq 6 and assuming the existence of a finite set of multisite interactions h_α, used the results of *ab initio* total-energy calculations of ordered compounds in order to estimate the pair and many-body interactions h_α. These interactions could then be used to describe the enthalpy of formation ΔH of disordered alloys. This approach relies strictly on short-range interactions for the description of ΔH at a fixed alloy volume. However, a configuration-independent effective-medium contribution to ΔH follows from the concentration dependence of the equilibrium molar volume of the alloy.[32]

An alternative approach, proposed by Gautier and coworkers,[14-16] consists of calculating ΔH for the alloy with SRO by extrapolation from the random system (rather than extrapolating from the ordered compounds as done by Connolly and Williams). The method of Gautier and coworkers is based on the calculation of ΔH for a random alloy using the coherent-potential approximation (CPA) followed by a perturbation expansion of ΔH in terms of short-range-order parameters. This technique, known as the generalized perturbation method (GPM), is used to calculate short-range interactions (pair and many-body), which contribute to ΔH in addition to the mean-field CPA energy of the random alloy.

The GPM has been studied extensively in the tight-binding (TB) approximation and has been very valuable in elucidating general trends in the alloying behavior of the transition metals. However, some of the additional approximations usually made in actual implementations of the GPM, such as neglecting the effects of charge transfer and the off-diagonal disorder in the TB Hamiltonian, generally preclude a quantitative comparison between the calculated and the experimentally determined ΔH.

The cluster Bethe lattice method (CBLM) has also been used recently together with the CVM in order to calculate the enthalpy of formation and the associated phase diagrams for Cr-Mo, Cr-W, and Mo-W binary alloys.[18] These calculations were carried out using a self-consistent Hartree-Fock Hamiltonian in the TB approximation with a six orbital basis per atom (one s- and five d-orbitals). Despite the fact that the CBLM relies on a rather severe topological approximation that replaces the real crystal lattice with a Cayley tree, the results of the calculations were in general agreement with experiment. The CBLM results also showed that, in order to achieve the accuracy in ΔH needed for phase-diagram calculations, one must include off-diagonal disorder in the TB Hamiltonian and, in addition, carry out a self-consistent treatment of charge transfer.

Recently, we have used a microscopic theory based on the TB approximation for the calculation of the enthalpy of formation of transition-metal alloys.[19,20] In our analysis, the topological approximation of the CBLM is lifted using the CPA to calculate the enthalpy of formation of the random alloy, and SRO is treated using the GPM. Furthermore, we include off-diagonal disorder in the TB Hamiltonian, and the effects of charge transfer in the random alloy are treated self-consistently within the Hartree-Fock approximation.

Within the CPA-GPM approach, the enthalpy of formation of an alloy is written as the sum of the random alloy enthalpy of formation (ΔE_{rand}) plus the ordering energy (ΔE_{ord}) which includes both short- and long-range order:

$$\Delta H = \Delta E_{rand} + \Delta E_{ord} \quad \text{(Eq 17)}$$

The random alloy enthalpy is calculated within the coherent potential approximation, whereas the ordering energy is obtained by perturbing the energy of the alloy about the random state. The ordering or configurational energy takes the form of a cluster expansion involving concentration-dependent effective interactions for pairs, triplets, etc.[14-16] The GPM results indicate that these interactions are short-ranged: for the case of the nonmagnetic transition metals, the leading contributions are given by pair interactions which extend to first nearest neighbors in the fcc lattice and to first and second nearest neighbors in the bcc lattice.[14-16] In what follows, only pair interactions will be included in the expression of the ordering energy. Thus, the expectation value of the ordering energy then takes the form:

$$\Delta E_{ord} = (1/2) \sum_k \omega_k V_k (\xi_2^{(k)} - \xi_1^2) \quad \text{(Eq 18)}$$

where ω_k and V_k are, respectively, the coordination number and the effective interaction for the kth pair, and where the pair correlation function $\xi_2^{(k)}$ and the point correlation function ξ_1 are defined by:

$$\xi_2^{(k)} = \langle \sigma_p \sigma_{p+k} \rangle \quad \text{(Eq 19)}$$

and

$$\xi_1 = \langle \sigma_p \rangle \quad \text{(Eq 20)}$$

where the brackets represent configurational averages.

The enthalpy of formation of the random alloy, ΔE_{rand}, depends only on concentration and follows directly from the self-consistent CPA treatment of the model TB Hamiltonian. In general, ΔE_{rand} can be accurately described by means of a polynomial expansion in the point correlation ξ_1:

$$\Delta E_{rand} = (1 - \xi_1^2) \left[\sum_{m=0}^{m} h_n \xi_1^n \right] \quad \text{(Eq 21)}$$

The expansion of Eq 21 is such that ΔE_{rand} vanishes when ξ_1 equals 1 (pure component A) or -1 (pure component B).

In order to calculate ΔE_{rand}, one needs to compute the density of states of the pure elements. For that purpose, a tight-binding Hamiltonian is used with a basis of five d-orbitals per atom. Accordingly, the contributions of s- and p-electrons to the enthalpy of formation are neglected. This approximation is expected to work relatively well for the nearly half-filled d-band alloys for which the d-electron contribution to ΔH dominates. The hopping integrals for the pure elements are determined using canonical Slater-Koster parameters[33] according to the prescription of Harrison.[34] The Slater-Koster parameters between unlike atoms in the alloy are estimated, following Shiba,[35] as the geometric mean of those for the pure metals. The on-site energy ϵ_i^0 for pure metal i is taken to be equal to the d-energy level in the excited free-atom configuration $s^1d^{(z_i-1)}$, where z_i is the total number of valence electrons. This configuration is close to the $s^{1.3}d^{(z_i-1.3)}$ configuration predicted by band-structure calculations for the pure metals.[36] In all our calculations, we have taken the number of d-electrons per i-atom to be equal to z_i − 1.3. The on-site energy ϵ_i, at a site occupied by element i in the alloy, is determined self-consistently using Eq 14:

$$\epsilon_i = \epsilon_i^0 + U(n_i - n_i^0) \qquad \text{(Eq 22)}$$

where U is an effective direct-exchange Coulomb integral, $n_i^0 = z_i - 1.3$ is the number of d-electrons in the pure metal, and n_i is the actual d-electron count for species i in the alloy. A value of U = 3 eV has been used for all elements. In Eq 22 we consider only the intra-atomic effects of charge transfer on the on-site energies. This is consistent with the fact that, in our calculations, electronic self-consistency is implemented only for the random alloy for which interatomic corrections to the on-site energy vanish.[17,18] The input parameters of the microscopic theory—i.e., the d-band width W_d, the on-site energy ϵ^0, and the number of d-electrons—are given in Table 1.

The local densities of states for the pure metals are obtained using five levels of the recursion method, which guarantee that the first 11 moments of the density of states are reproduced exactly.[37,38] For a random alloy, the local densities of states are obtained using the CPA. In general, the total energy of the alloy is given by three terms: (1) the one-electron energy, obtained from the calculated density of states; (2) the intra- and interatomic electron-electron energies, which are subtracted from the energy of the alloy since they are counted twice in the one-electron term; and (3) the ion-ion electrostatic energy. For a random alloy, however, the ion-ion energy and the interatomic electron-electron correction cancel each other, and, thus, these terms do not appear explicitly in ΔH. The enthalpy of formation of the random alloy ΔE_{rand} is obtained, as usual, by subtracting from the total energy of the alloy the energies of the pure elements weighted by their concentrations.

The effect of SRO on the enthalpy of alloy formation is described in terms of effective pair interactions (EPI) calculated using the GPM.[14-16] In the alloy with SRO, we also include a small contribution to the EPI arising from the net charge transfer between the different atomic species. This screened electrostatic interaction is characterized by an interatomic potential V, which is taken to be equal to 0.4 eV.[19,20] At a given temperature and concentration, the amount of SRO is obtained self-consistently using the CVM described in the previous section.

In the present tight-binding approximation, the enthalpy of formation of an alloy

Table 1. Input parameters of the microscopic theory(a)

Parameter	Sc (hcp)	Ti (hcp)	V (bcc)	Cr (bcc)	Fe (bcc)	Co (hcp)	Ni (fcc)	Cu (fcc)	
n^0	1.7	2.7	3.7	4.7	6.7	7.7	8.7	9.7	
ϵ^0	−1.51	−2.10	−2.63	−3.14	−4.07	−4.51	−4.93	−5.34	
W_d	5.13	6.08	6.77	6.56	4.82	4.35	3.78	2.80	
	Y (hcp)	Zr (hcp)	Nb (bcc)	Mo (bcc)	Tc (hcp)	Ru (hcp)	Rh (hcp)	Pd (fcc)	Ag (fcc)
n^0	1.7	2.7	3.7	4.7	5.7	6.7	7.7	8.7	9.7
ϵ^0	−1.77	−2.50	−3.22	−3.95	−4.68	−5.42	−6.17	−6.92	−7.68
W_d	6.59	8.37	9.72	9.98	9.42	8.44	6.89	5.40	3.63
	Hf (hcp)	Ta (bcc)	W (bcc)	Re (hcp)	Os (hcp)	Ir (fcc)	Pt (fcc)	Au (fcc)	
n^0	2.7	3.7	4.7	5.7	6.7	7.7	8.7	9.7	
ϵ^0	−2.18	−2.86	−3.56	−4.26	−4.98	−5.70	−6.44	−7.18	
W_d	9.56	11.12	11.44	11.02	10.31	8.71	7.00	5.28	

(a) The parameters given here are the d-band width, W_d (in eV), the d-energy level of the free atom in the configuration ($s^1d^{z_i-1}$), ϵ^0 (in eV), and the number of d-electrons, n^0. For all elements, we have used U = 3 eV and V = 0.4 eV.

(AB) depends on three parameters characterizing the pure elements: the diagonal disorder ($\Delta\epsilon = \epsilon_A^0 - \epsilon_B^0$), the off-diagonal disorder ($\Delta W = W_A/W_B$), and the difference in the number of d-electrons between A and B ($\Delta n = n_A^0 - n_B^0$). Note that the other input parameters, the effective intra-atomic Coulomb integral (U) and the interatomic potential (V), are fixed, respectively, to 3 eV and 0.4 eV. The diagonal disorder and the difference in the number of d-electrons represent negative contributions to the enthalpy of formation, whereas the off-diagonal disorder represents a positive contribution to the enthalpy of formation.

Off-diagonal disorder and the self-consistent effect of charge transfer have often been assumed to contribute negligibly to the enthalpy of formation. Our results indicate, however, that both these effects are as important as that of diagonal disorder. To illustrate this point, the enthalpy of formation of the random Cr-Mo alloy has been calculated at the equiatomic concentration neglecting electronic self-consistency and off-diagonal disorder effects; a value of −0.137 eV is then obtained for ΔE_{rand}. The negative sign of ΔE_{rand} indicates an ordering tendency in the Cr-Mo system which is contrary to experiment.[39] By including electronic self-consistency in the calculation but still neglecting off-diagonal disorder, we obtain a value of ΔE_{rand} equal to 0.016 eV. In this case, the sign is consistent with experiment but the calculation is in poor quantitative agreement with the experimental measurement of 0.075 ± 0.010 eV reported in Ref 39. Finally if both effects are taken into account, a value for ΔE_{rand} of 0.062 eV is obtained in fair agreement with experiment.

In what follows, we present the enthalpy of formation and the EPI's calculated for the bcc alloys obtained by mixing elements belonging to the VB and VIB columns in the Periodic Table (V, Nb, Ta, Cr, Mo, and W). At the equiatomic concentration, these alloys have an average number of d-electrons equal to 3.7, 4.2, or 4.7. We will also present the enthalpies of formation of alloys obtained by mixing the hexagonal metals, titanium, zirconium, and hafnium, with the fcc metals, nickel, palladium, and platinum. At the equiatomic concentration, the number of d-electrons of these alloys is equal to 5.7. For these alloys, the density of states of the hcp structure will be approximated by that of the fcc structure with the same nearest-neighbor distance.

The calculated enthalpies of formation for the random bcc alloys (ΔE_{rand}) and the corresponding first- and second-nearest-neighbor pair interactions (V_1 and V_2) are shown in Fig. 1 to 4. Available experimental measurements are also shown in

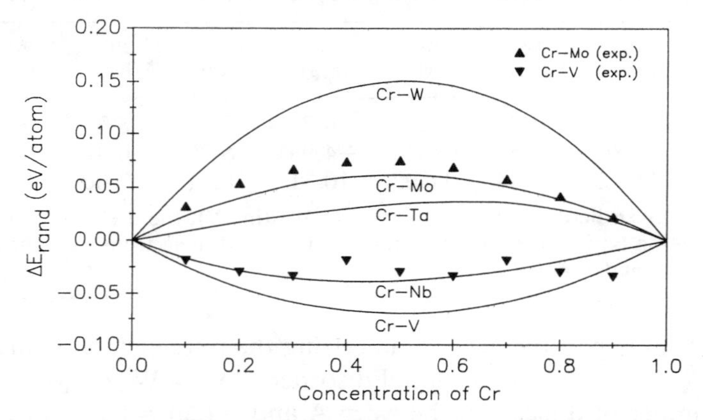

Fig. 1A. Calculated enthalpies of formation for random bcc binary alloys of Cr with Mo, W, V, Nb, and Ta (full lines). Available experimental data for Cr-Mo[39] and Cr-V[69] are indicated by closed symbols.

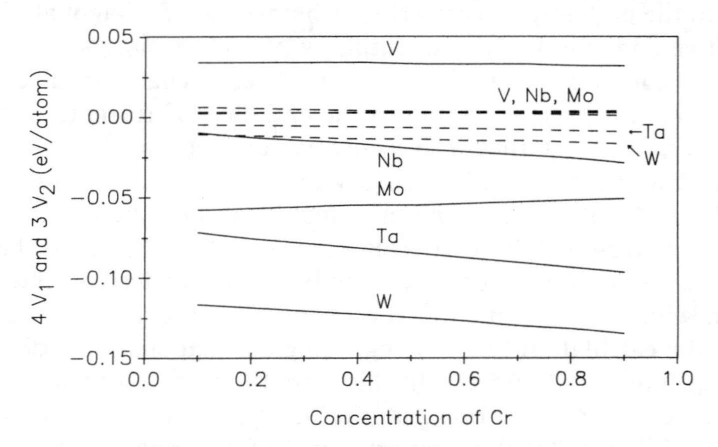

Fig. 1B. Calculated effective pair interactions for first (full lines) and second (dashed lines) nearest neighbors for bcc binary alloys of Cr with Mo, W, V, Nb, and Ta

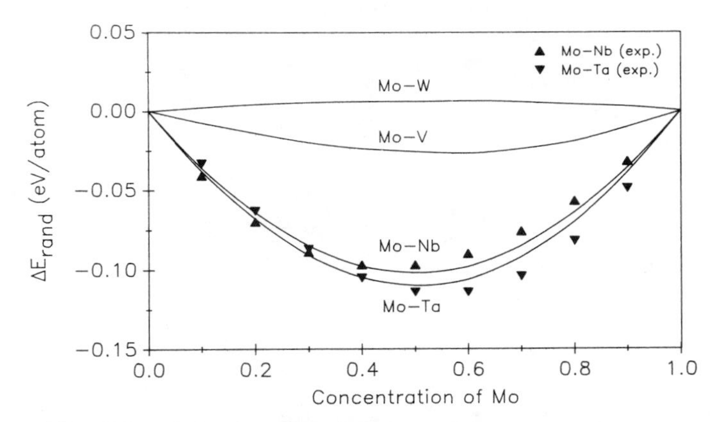

Fig. 2A. Calculated enthalpies of formation for random bcc binary alloys of Mo with W, V, Nb, and Ta (full lines). Available experimental data for Mo-Nb[70] and Mo-Ta[71] are indicated by closed symbols.

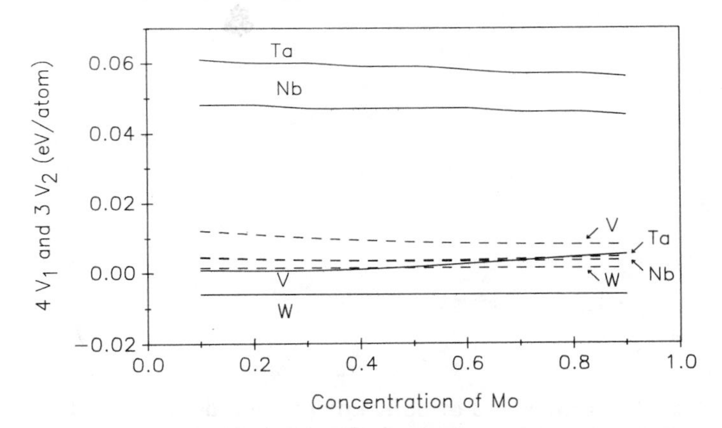

Fig. 2B. Calculated effective pair interactions for first (full lines) and second (dashed lines) nearest neighbors for bcc binary alloys of Mo with W, V, Nb, and Ta

these figures. We find good overall agreement between the calculations and the experimental data.

The effective pair interactions (EPI) shown in Fig. 1 to 4 are almost independent of concentration due, primarily, to the fact that the difference in the number of d-electrons Δn is small for this set of alloys. The enthalpy of formation of the random alloys and the EPI's can be used, together with the CVM, in order to calculate the configurational free energy and degree of SRO as a function of temperature and concentration. Furthermore, if we neglect the vibrational entropy of formation, the calculated configurational free energy may be used to construct temperature-composition phase diagrams. We will discuss here those systems for which experimental enthalpies of formation are available (i.e., Cr-Mo, Cr-V, Mo-Nb, Mo-Ta, and Ta-W). The experimental data were obtained at high temper-

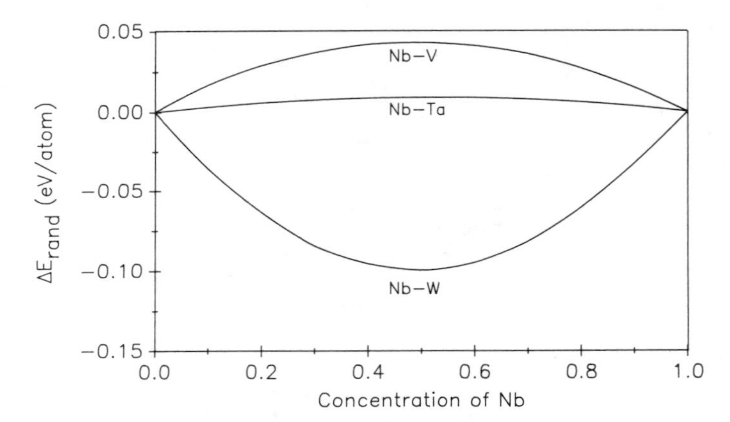

Fig. 3A. Calculated enthalpies of formation for random bcc binary alloys of Nb with W, V, and Ta

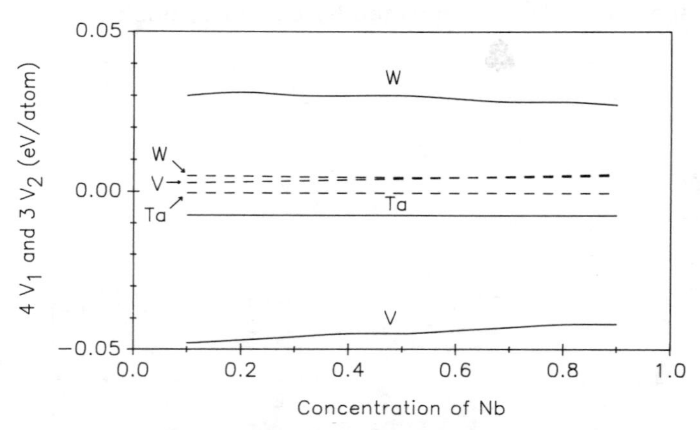

Fig. 3B. Calculated effective pair interactions for first (full lines) and second (dashed lines) nearest neighbors for bcc binary alloys of Nb with W, V, and Ta

atures (above 1200 K) in the bcc solid solution. At these temperatures, our calculations indicate that the SRO contribution (i.e., the ordering energy) to the enthalpy of formation is approximately one-tenth of ΔE_{rand}. As explained below, the ordering energy becomes more significant at lower temperatures.

The first-nearest-neighbor pair interaction (V_1) is negative for the Cr-Mo system, indicating a clustering tendency, whereas the sign of V_1 for Mo-Nb, Mo-Ta, Cr-V, and Ta-W is positive, indicating an ordering tendency. The model predicts the presence of a miscibility gap for Cr-Mo below 1310 K, which is in fair agreement with the experimental results of Kubaschewski and Chart,[39] indicating the presence of a miscibility gap below 1200 K. For the W-Ta, Ta-Mo, Mo-Nb, and Cr-V systems, we predict the existence of a CsCl (B_2) ordered phase below 920, 1040, 800, and 590 K, respectively. These results are consistent with the complete solubility found in these alloys above 1200 K. To the authors' best knowledge, how-

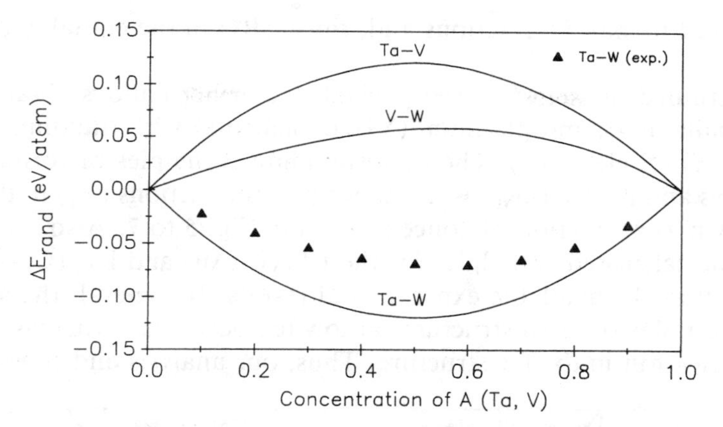

Fig. 4A. Calculated enthalpies of formation for the random bcc V-W system and for alloys of Ta with W and V (full lines). Available experimental data for Ta-W[72] are indicated by closed symbols.

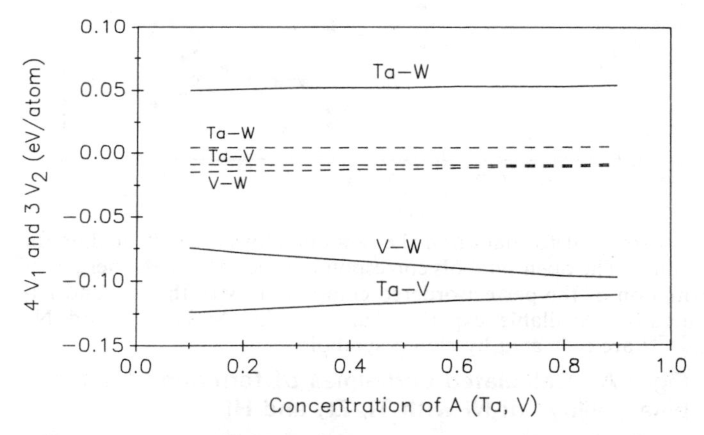

Fig. 4B. Calculated effective pair interactions for first (full lines) and second (dashed lines) nearest neighbors for the V-W system and for bcc binary alloys of Ta with W and V

ever, no experimental evidence is currently available concerning the existence of a low-temperature B_2 phase.

Williams, Gelatt, and Moruzzi[5] have used the local-density approximation together with a muffin-tin potential in order to predict the enthalpy of formation of the MoNb compound in the CsCl structure. Our result (−146 meV), obtained by adding ΔE_{rand} (102 meV) and ΔE_{ord} (−44 meV), is in good agreement with their calculation (−152 meV). Our analysis, however, enables us to describe the disordered phase and to calculate a disordering temperature of 800 K for the compound MoNb in the CsCl structure. Note that the ordering energy of MoNb (CsCl) is about one-third of the total enthalpy of formation, which underlines the important role played by pair interactions and, thus, SRO in the enthalpy of alloy formation.

We have studied the set of closed-packed (fcc or hcp) alloys obtained by mixing nickel, palladium, and platinum (VIIIB elements) with titanium, zirconium, and hafnium (IVB elements). The corresponding enthalpies of formation of the random alloys and the first-nearest-neighbor pair interactions (V_1) in the fcc structure are shown as a function of concentration in Fig. 5 to 7. Also shown in these figures are the calculated enthalpies for the $L1_0$ (CuAu) and $L1_2$ (Cu_3Au) ordered phases as well as the available experimental results. In general, the alloys under study have complex ordered structures at low temperatures, whereas our calculations are carried out in the fcc structure. Thus, our analysis and comparison with

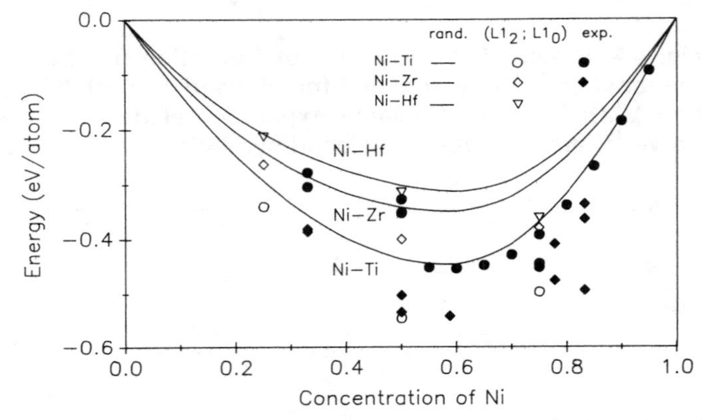

The energies of formation of the random alloys are indicated by the full lines. The open symbols correspond to the calculated energies of formation of the perfect ordered compounds with the $L1_0$ and $L1_2$ structures. Available experimental data for Ni-Ti[73-76] and Ni-Zr[76-78] are indicated by closed symbols.

Fig. 5A. Calculated enthalpies of formation for fcc binary alloys of Ni with Ti, Zr, and Hf

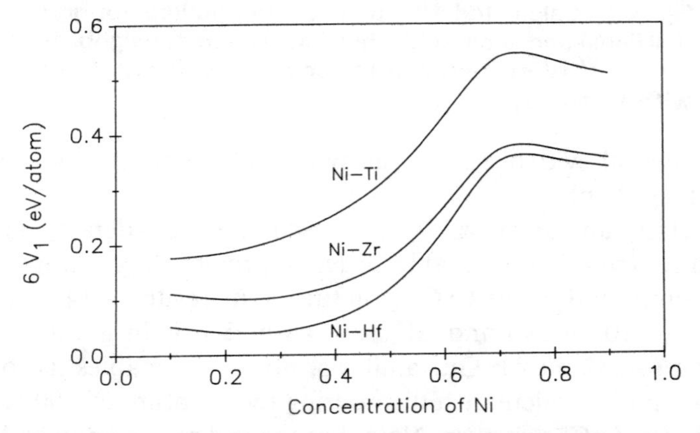

Fig. 5B. Calculated first-nearest-neighbor effective pair interactions for fcc binary alloys of Ni with Ti, Zr, and Hf

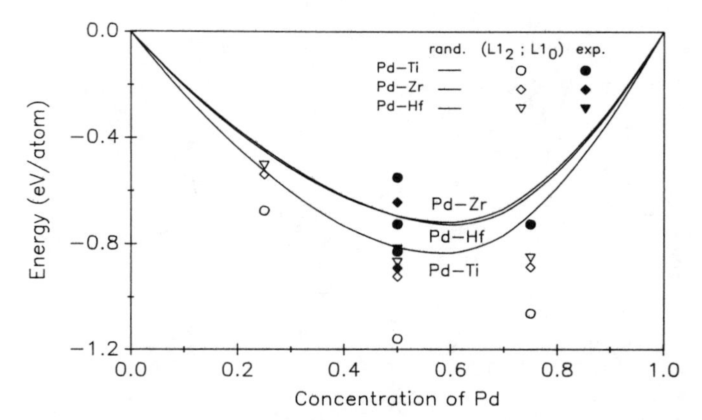

The energies of formation of the random alloys are indicated by the full lines. The open symbols correspond to the calculated energies of formation of the perfect ordered compounds with the $L1_0$ and $L1_2$ structures. Available experimental data for Pd-Ti,[79-81] Pd-Zr[79,80] and Pd-Hf[79] are indicated by closed symbols.

Fig. 6A. Calculated enthalpies of formation for fcc binary alloys of Pd with Ti, Zr, and Hf

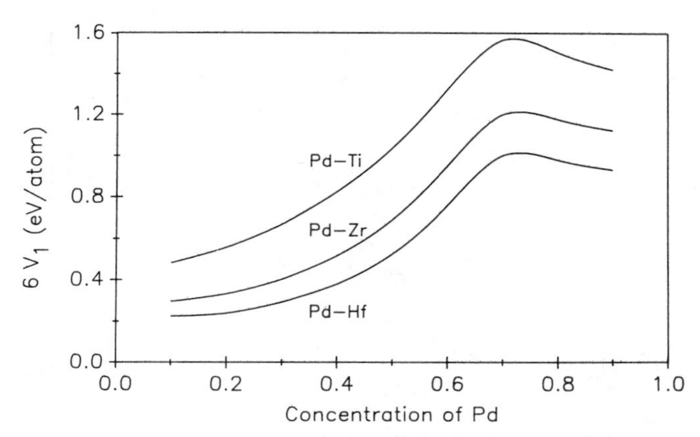

Fig. 6B. Calculated first-nearest-neighbor effective pair interactions for fcc binary alloys of Pd with Ti, Zr, and Hf

the experimental data neglects the structural energy involved in going from the fcc to the actual complex ordered structures. One can notice that, for a given alloy, the scattering of the experimental data is generally larger than the ordering energy. Within this uncertainty, our results agree well with experiment. Note that the enthalpies of formation of these alloys are much more negative than those obtained for the bcc alloys shown in Fig. 1 to 5. This is explained well by the fact that the $\Delta\epsilon$ and Δn values for the closed-packed alloys are larger than those for the bcc alloys studied here.

The Pt-Ti system provides an interesting test case for the present tight-binding model since the ordered Pt_3Ti phase having the Cu_3Au structure ($L1_2$) exists

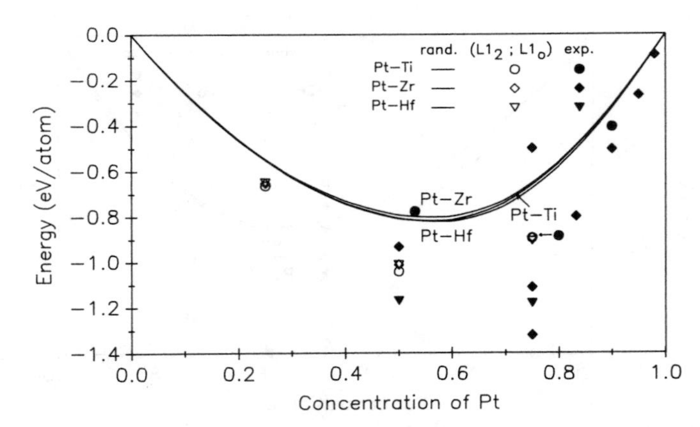

The energies of formation of the random alloys are indicated by the full lines. The open symbols correspond to the calculated energies of formation of the perfect ordered compounds with the $L1_0$ and $L1_2$ structures. Available experimental data for Pt-Ti,[41,79] Pt-Zr,[79,82-84] and Pt-Hf[79,82,83] are indicated by closed symbols.

Fig. 7A. Calculated enthalpies of formation for fcc binary alloys of Pt with Ti, Zr, and Hf

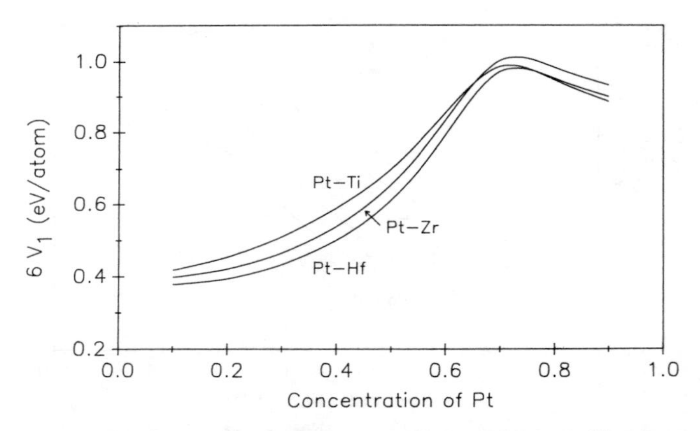

Fig. 7B. Calculated first-nearest-neighbor effective pair interactions for fcc binary alloys of Pt with Ti, Zr, and Hf

experimentally[40] and its enthalpy of formation has been measured[41] (see Fig. 7A). Since no structural energy is involved in the formation of Pt_3Ti, we expect the model to describe accurately the energy of formation for this phase. Furthermore, Ref 41 provides a measured enthalpy of formation for the disordered fcc solid solution (A1). Accordingly, the accuracy with which the model describes SRO effects can be tested in the Pt-Ti alloys by comparing the calculated and experimental enthalpies of formation of the A1 and $L1_2$ phases. We have used the CVM to obtain equilibrium thermodynamic potentials for the A1 and $L1_2$ phases at the experimental concentration and temperature. The results of the calculations are compared with the experimental data in Table 2. The good agreement between cal-

Table 2. Comparison of the calculated enthalpies of formation for the Pt_3Ti phase and the fcc solid solution (A1) with the available experimental data[41]
Energies are in eV.

Phase	Calculated random alloy enthalpy, ΔE_{rand}	Calculated ordering energy, ΔE_{ord}, at T = 1300 K	Calculated enthalpy of formation at T = 1300 K	Measured enthalpy of formation at T = 1300 K
Pt_3Ti ($L1_2$)	−0.653	−0.245	−0.898	−0.886 ± 0.034
A1 (c_{Pt} = 0.9)	−0.318	−0.034	−0.352	−0.407 ± 0.034

culated and experimental results indicates that ordering effects are well described by the theory, at least for the case of simple ordered phases such as the $L1_2$ structure. For the other close-packed alloys investigated here, the detailed analysis of the results is complicated by the scattering of the experimental data and by the structural energy associated with the complex ordered structures actually observed. Nevertheless, within the scattering of the measurements, our results are in good agreement with the experimental data.

The variation of the first-nearest-neighbor EPI (V_1) with concentration shown in Fig. 5B, 6B, and 7B is qualitatively the same for all the IVB-VIIIB alloys: the magnitude of V_1 increases with the concentration of the VIIIB element up to about 0.75, followed by a slight decrease for higher concentrations. Experimentally, the dependence of V_1 on concentration is not known. However, the phase diagrams of these alloys[42] show that the congruent point temperatures between ordered and liquid phases increase as the content of the VIIIB element increases. A plausible reason for this behavior is the increase of V_1 (or equivalently of the ordering energy) with the content of the VIIIB element. As a result, the ordered phases are more stable relative to the liquid phases for high contents of the VIIIB element.

We have shown in this section that it is possible to understand the ordering tendency of transition-metal alloys on the basis of a model tight-binding Hamiltonian. Furthermore, accurate enthalpies of formation can be predicted for alloys having a half-filled d-band at the equiatomic concentration. However, this microscopic theory cannot be easily extended to a general binary alloy since we have neglected magnetic interactions as well as the effects of s- and p-electrons on the enthalpy of formation. Although in principle these effects can be included, the number of parameters entering the description of the model Hamiltonian increases substantially. In particular, there are numerous and essentially unknown intra- and interatomic potentials which greatly limit the quantitative reliability of the theory. An alternative approach of considerable practical value consists of using the general form for the enthalpy of alloy formation suggested by the CPA-GPM results, followed by an optimization of the energy parameters (i.e., h_n and V_k in Eq 18 and 21) using known experimental thermodynamic data as well as a few experimental points in the phase diagram. This approach is described in the next section. As an example, we present the calculated stable and metastable temperature-composition phase diagram for the Al-Li system.

PHENOMENOLOGICAL COMPUTATION OF PHASE DIAGRAMS

Most investigations of phase equilibria are based on phenomenological models which rely heavily on existing phase diagrams and thermochemical data. A semiempirical approach along such lines has been successfully implemented by Kaufman and Nesor,[1] who developed an extensive free-energy database for transition-metal alloys. The general procedure consists of using a subregular-solution model to fit experimental thermodynamic data and available phase-diagram information. In addition to equilibrium free energies, the database provides lattice-stability energies for the pure elements in their metastable phases. Although some of Kaufman's lattice-stability parameters are in sharp disagreement with first-principle calculations,[43-45] recent reassessments by Miodownik[46] have produced a set of values in closer agreement with theory. Despite such discrepancies, the overall agreement between experimental phase diagrams and those obtained from the semiempirical free energies is generally good, with the notable exception of ordered phases which are treated by Kaufman and coworkers as stoichiometric compounds. The subregular-solution model and its generalization to ordered compounds, the Bragg-Williams approximation, has also been used extensively to characterize alloy free energies and to compute equilibrium phase diagrams.[47-51] This semiempirical approach tends to produce a more realistic description of nonstoichiometric ordered compounds due to the improved treatment of the configurational entropy of ordered phases by means of sublattices.[47,49]

A characteristic of all free-energy functions based on the subregular-solution model and/or the Bragg-Williams approximation is that SRO is not explicitly included in the configurational entropy. It must be emphasized that SRO has a significant effect on phase equilibria although its contribution to the alloy's total free energy of mixing is usually small. Thus, SRO effects are commonly incorporated into the Bragg-Williams models by means of a phenomenological expansion of the excess free energy in powers of temperature and composition. This empirical approach to the description of the configurational free energy has, however, some important limitations. Among the most significant of such limitations is the fact that the configurational entropy cannot be properly approximated by a polynomial expansion over extended temperature and composition ranges. In addition, the Bragg-Williams approximation, when applied to a simple fcc model alloy with nearest-neighbor interactions, fails to describe general features of the equilibrium phase diagram.[24] Consequently, free-energy functions obtained by fitting equilibrium data to binary alloys cannot be extrapolated with confidence to treat, for example, metastable phases or multicomponent systems.

We have recently developed a phenomenological approach based on the CVM for the description of solid-solid and solid-liquid phase equilibria in binary alloys.

This approach—which has been used to investigate the Ni-Al,[52] Ni-Cr,[53] and Al-Li[54] systems—emphasizes the description of SRO in the solid phases and enables a more accurate prediction of metastable phase equilibria. In this section we review the salient features of the phenomenological approach and present the main results obtained for the Al-Li system.

In general, the total free energy per lattice point takes the form:

$$F = 1/2[(F_A + F_B) + (F_B - F_A)\xi_1] + \Delta H - T(\Delta S_{conf} + \Delta S_{vib}) \qquad \text{(Eq 23)}$$

where F_i is the free energy of a pure element (i = A or B), T is the temperature, and ΔS_{conf} and ΔS_{vib} are, respectively, the configurational entropy and the vibrational entropy of formation.

For simplicity we assume that the free energy of a pure element is given by the expression:

$$F_i = H_i - TS_i \qquad \text{(Eq 24)}$$

where H_i and S_i are assumed to be temperature-independent. The form of the alloy enthalpy of formation is suggested by the results obtained by the CPA-GPM calculations and is given by Eq 21. For a solid phase, the effect of SRO in the configurational entropy is expressed using the CVM formalism, whereas for the liquid phase SRO is neglected. The vibrational entropy of formation is assumed to be SRO-independent. A polynomial expansion similar to the one given in Eq 21 for ΔE_{rand} is therefore used to express ΔS_{vib}:

$$\Delta S_{vib} = (1 - \xi_1^2)\left[\sum_{m=0}^{m} s_n \xi_1^n\right] \qquad \text{(Eq 25)}$$

The input parameters for each phase (h_n, s_n, V_k) are obtained by means of a CVM least-square fit to available thermochemical data and to a few points in the phase diagram. In general, the least-square fitting procedure cannot be reliably carried out with respect to the pair interactions V_k, except in those cases where a congruent order-disorder temperature is experimentally known. Thus, the quantum mechanical calculations outlined in the previous section are particularly valuable in order to obtain an initial estimate for the effective pair interactions. Furthermore, since the least-square fit to a given phase diagram is not unique, the calculations also serve to increase the reliability of the final set of energy parameters.

As an example, we present the calculated stable and metastable phase diagram for the binary Al-Li system. A comprehensive review of the thermodynamic data and equilibrium temperature-concentration phase diagram for Al-Li alloys can be found in Ref 55 and 56. The different stable phases appearing in the updated phase diagram in Ref 55 are the liquid phase, the fcc Al-rich solid solution (α), the bcc Li-rich solid solution, the ordered AlLi phase (β), which has a NaTl-type structure, the Al_2Li_3 phase (γ) based on a rhombohedral structure and reported to have a very narrow range of solubility,[57] and the Al_4Li_9 phase (δ) based on a monoclinic structure below 548 K and reported by Myles *et al.* to transform to a different, yet undetermined, structure (δ') above 548 K.[58] This last structure will not be described in our analysis.

The parameters required for describing the Al-Li system are listed in Table 3, and were obtained using 11 points in the phase diagram and three measured thermodynamic potentials (ΔS and ΔH for the liquid phase, and ΔH for the AlLi phase). Using the energy parameters of Table 3, the entire equilibrium phase diagram has been calculated and it is compared in Fig. 8 with the experimental points taken from Ref 55. As expected, a very good overall agreement has been obtained between experimental and calculated equilibrium concentrations. In particular, the calculated α solvus line compares very well with the different experimental concentrations shown in Fig. 8.

The activities of lithium in the α-β two-phase region have been determined by several investigators[58-60] at different temperatures, and have been found to be very consistent.[60] A comparison of the experimental data with values obtained from the present thermodynamic description is given in Fig. 9. As can be seen, the temperature dependence of the lithium activities is reproduced well by the model. This agreement is of particular interest since these experimental lithium activities have not been used to calculate the input parameters.

A metastable Al_3Li (α') phase having an $L1_2$ structure is also reported to exist

Table 3. Parameters characterizing the Li-Al system(a)

	Enthalpy/entropy parameters						
	E_{rand}, kcal/g-at.			ΔS_{vib}, cal/g-at./K		E_{ord}, kcal/g-at.	
Phase	h_0	h_1	h_2	s_0	s_1	V_1	V_2
Fcc (A1)	−1.64	−0.25	−0.27	−1.2	0.0	0.83	...
Al_3Li	−1.64	−0.25	−0.27	−1.2	0.0	0.83	...
AlLi	−3.83	−0.10	0.00	−2.5	0.0	0.76	0.76
Liquid	−2.70	−1.24	0.00	−1.5	−0.5	...	...

Free-energy parameters(b)		
Compound	A_c, kcal/g-at.	B_c, cal/g-at./K
Al_2Li_3	−7.41	5.62
Al_4Li_9	−8.53	8.90

Lattice-stability parameters(b)				
	Aluminum		Lithium	
Structure	H_{Al}, kcal/g-at.	S_{Al}, cal/g-at./K	H_{Li}, kcal/g-at.	S_{Li}, cal/g-at./K
Fcc	−2.560	−2.750	−0.427	−1.710
Bcc	−0.150	−1.600	−0.717	−1.580
Liquid	0.000	0.000	0.000	0.000

(a) The point-correlation function ξ_1 is defined as $c_{Li} - c_{Al}$. (b) Reference structure: liquid.

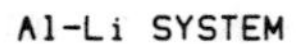

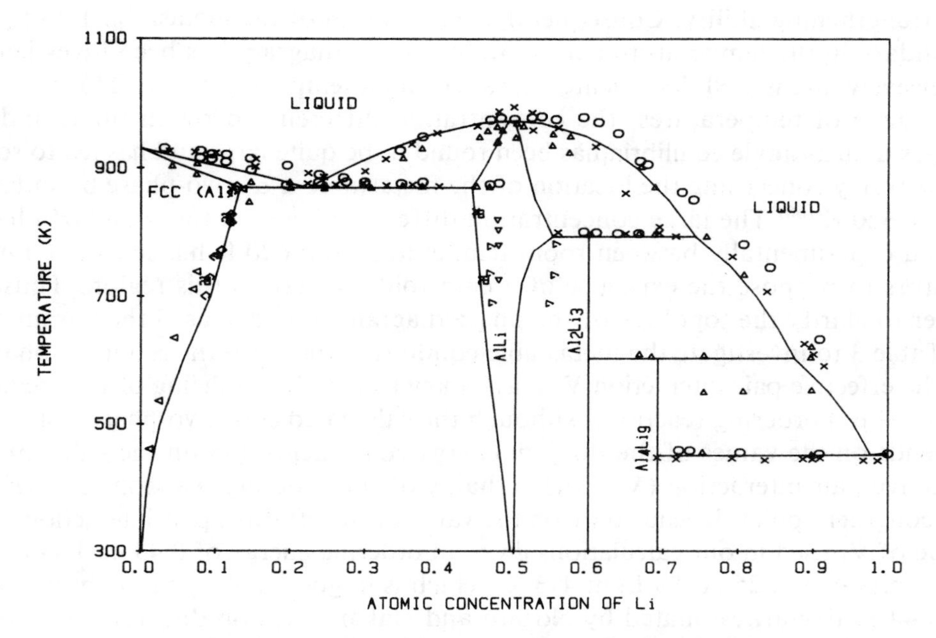

Fig. 8. Comparison between the calculated phase diagram for the Al-Li system (full lines) and equilibrium concentrations compiled in Ref 55

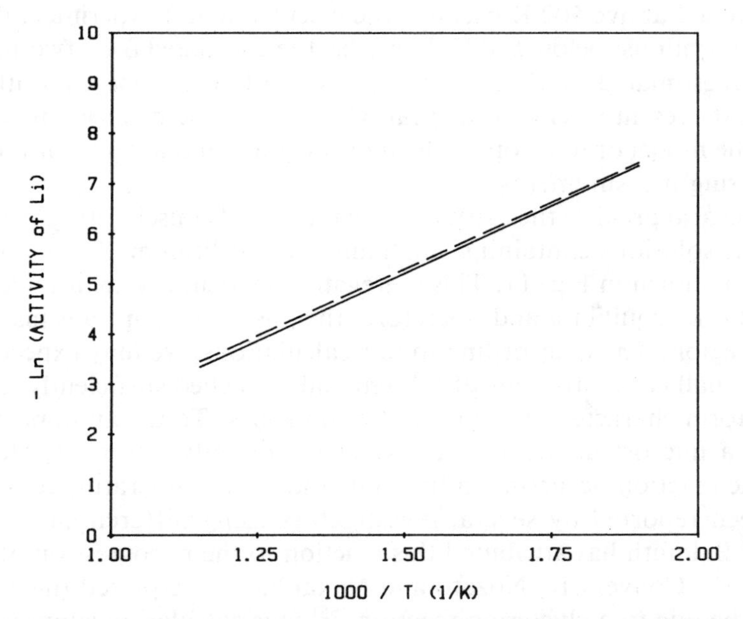

Fig. 9. Calculated (full line) and experimental[60] (dashed line) activities of lithium as a function of temperature at the α-β coexistence line

in the Al-Li system. This phase plays a significant role in Al-Li alloys because of its strengthening ability. Consequently, the location of the metastable two-phase boundary in the temperature-concentration phase diagram has been investigated extensively and is well documented between room temperature and 620 K.[61-65] In this range of temperatures, the concentration difference between the α and α' phases in metastable equilibria has been found to be quite large and has led to some controversy concerning the location of the metastable α-α' two-phase boundaries above 620 K.[66] The large concentration difference between the α and α' phases found experimentally between room temperature and 620 K has led some investigators to propose the existence of a eutectoid reaction in this region. Thus, in order to clarify the topology of the phase diagram, we have used the parameters of Table 3 to investigate the metastable equilibrium between the α and α' phases.

The effective pair interaction V_1 plays a key role in the modeling of the metastable ($\alpha \rightarrow \alpha'$) ordering reaction. Although the calculated α-α' two-phase boundary depends on the values of the random energy coefficients and on the value of the effective pair interaction (V_1), the enthalpy of the ordering reaction ($\alpha \rightarrow \alpha'$) at the congruent point depends only on the value of the effective pair interaction. The value of V_1 used in our calculations gives an ordering energy of 0.43 kcal/g-at. for the α phase with 25 at. % Li at 473 K, which is in good agreement with the value of 0.44 kcal/g-at. estimated by Nozato and Nakai based on differential thermal analysis.[65]

The calculated metastable α-α' two-phase boundaries are shown in Fig. 10. Due to their metastable character, they fall inside the α-β two-phase region. Good agreement is obtained above 500 K between the calculated and experimental[61,62] α-α' solvus concentrations; below 500 K, however, the calculated α-α' two-phase region is wider than estimated by Ceresara *et al.*[64] using low-angle x-ray scattering measurements. At present it remains unclear whether the discrepancies are due to deficiencies in the model or to errors in the interpretation and analysis of the low-angle x-ray scattering measurements.

Our model also predicts the existence of a metastable miscibility gap below 400 K for Al-Li fcc solutions containing small amounts of lithium. The computed miscibility gap is shown in Fig. 11. This segregation tendency is metastable relative to the α-β and α-α' equilibria and, therefore, the miscibility gap falls inside the α-α' two-phase region. Thus, according to our calculations, we may expect that Al-Li alloys with small concentrations of lithium and quenched sufficiently fast will segregate and form characteristic Guinier-Preston zones. To the authors' best knowledge, such a microstructure has not yet been directly observed. However, an endothermic reaction occurring in the same range of temperature and concentration has been reported by several investigators using differential thermal analysis.[65,67,68] Balmuth has attributed this reaction to the retrogression of fine Al_3Li precipitates.[67] Conversely, Nozato and Nakai have interpreted the endothermic reaction to be due to a clustering reaction.[65] Our calculation supports the latter interpretation. Furthermore, recent differential scanning calorimetry experiments by Papazian, Sigli, and Sanchez,[68] illustrated in Fig. 12, indicate that the mea-

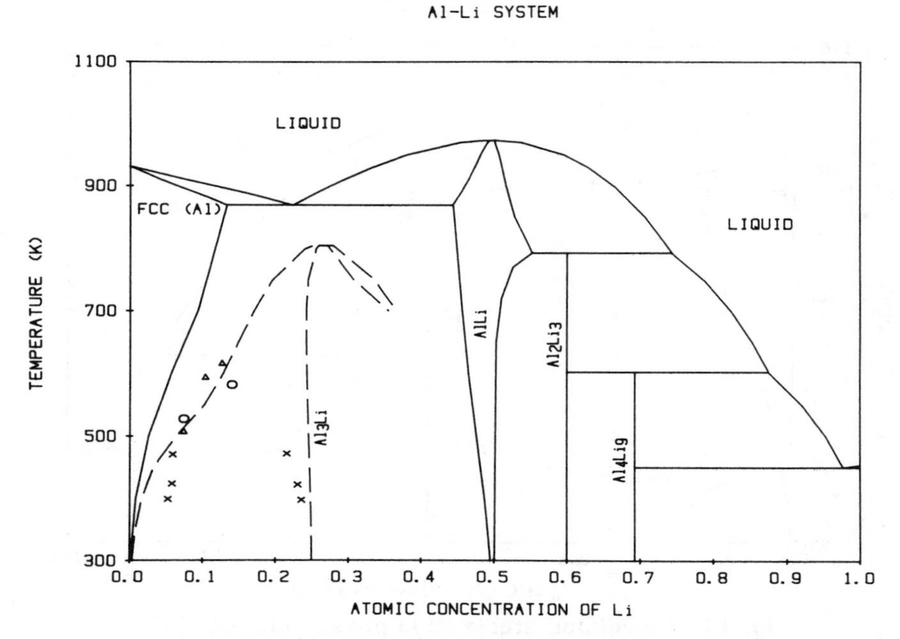

Fig. 10. Calculated equilibrium Al-Li phase diagram (full lines) and calculated metastable α-α' two-phase boundary (dashed lines). Experimental solvus concentrations are shown by triangles,[61] circles,[62] and crosses.[64]

sured dissolution peak temperatures as a function of the lithium content in the alloy are in good agreement with the miscibility gap predicted by the theory.

SUMMARY

Short-range order in alloys plays an important role in the description of phase equilibria and, as such, must be properly accounted for in calculations of both the energy of formation and the configurational entropy. We have shown that accurate values for the enthalpy of formation as a function of SRO may be obtained for transition-metal alloys having approximately a half-filled d-band at the equiatomic concentration. The calculations were made using a model tight-binding Hamiltonian together with the CPA-GPM method. However, the application of the theory to other alloys will require that s- and p-electrons as well as magnetic interactions be taken into account. In particular, there is a need for first-principle calculations of the different intra- and interatomic potentials which control charge transfer in the alloy and can, therefore, greatly affect the values of the heat of formation.

At a given concentration and temperature, the degree of SRO in an alloy is

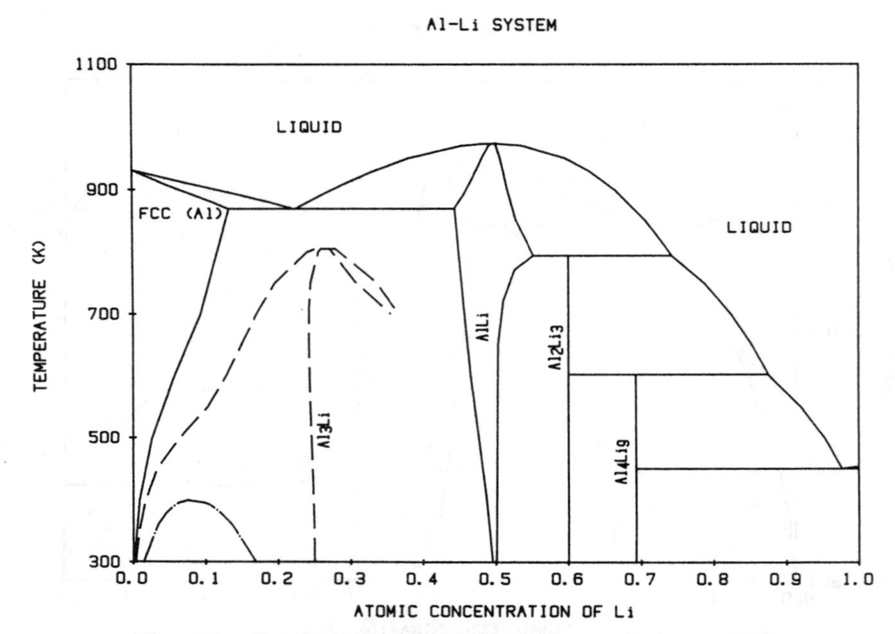

Fig. 11. Calculated stable Al-Li phase diagram (full lines), metastable α-α' two-phase boundary (dashed lines), and metastable miscibility gap for supersaturated fcc solid solutions (dash-dot line)

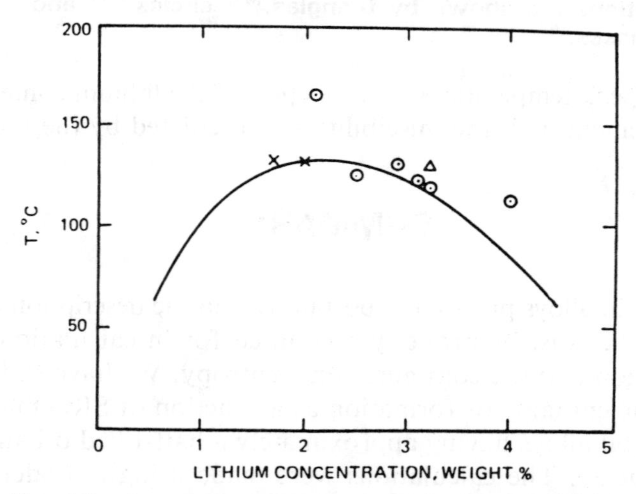

Fig. 12. Comparison between the calculated metastable miscibility gap and the dissolution peak temperatures observed in various Al-Li alloys[65,88,89]

directly related to the magnitude of the effective cluster interactions. A microscopic theory such as the CPA-GPM together with a tight-binding Hamiltonian provides valuable qualitative and, in some instances, quantitative information about these cluster interactions. In particular, the CPA-GPM results indicate that the magnitude of cluster interactions decreases as the size of the cluster increases. In the case of nonmagnetic transition-metal alloys, for example, the leading contributions to the ordering energy are generally given by pair interactions which extend to first nearest neighbors in the fcc lattice, and to first and second nearest neighbors in the bcc lattice.

The problem of phase stability at finite temperatures can be treated accurately and in a relatively straightforward manner by means of the CVM. In particular, the CVM, used in conjunction with a phenomenological model for the alloy internal energy based on the GPM-CPA, gives a reliable description of free energies in binary alloys for both stable and metastable phases. For example, in the Al-Li system we have computed the metastable A1-Al_3Li two-phase boundaries. The model also predicts a second level of metastability within the A1-Al_3Li two-phase region which seems to be confirmed by experiment. Namely, a segregation tendency is predicted to occur within the Al-rich fcc solid solution below 400 K. However, more experimental work is needed in order to deny or confirm this prediction.

ACKNOWLEDGMENTS

This work was funded by the National Science Foundation under grant DMR-85-10594. J.M. Sanchez thanks the ALCOA Foundation for the award of an ALCOA Science Support Grant. The authors express special thanks to Dr. R.J. Rioja for helpful comments on the Al-Li system.

REFERENCES

1. L. Kaufman and H. Nesor, *Met. Trans.*, Vol 5, 1974, p. 1623; *CALPHAD*, Vol 2, 1978, p. 337
2. V.L. Moruzzi, J.F. Janak, and A.R. Williams, in *Calculated Electronic Properties of Metals*, Pergamon Press, 1978
3. O.K. Andersen, O. Jepsen, and D. Glötzel, *Highlights of Condensed-Matter Theory*, Proceedings of the International School of Physics Enrico Fermi, North-Holland, Amsterdam, 1985
4. M.T. Yin and M.L. Cohen, *Phys. Rev. Lett.*, Vol 45, 1980, p. 1004
5. A.R. Williams, C.D. Gelatt, and V.L. Moruzzi, *Phys. Rev. Lett.*, Vol 44, 1980, p. 429
6. C.D. Gelatt, A.R. Williams, and V.L. Moruzzi, *Phys. Rev.*, Vol B27, 1983, p. 2005
7. R.E. Watson, L.H. Bennett, J.W. Davenport, and M. Weinert, *Computer Modeling of Phase Diagrams*, edited by L. Bennett, The Metallurgical Society, Warrendale, PA, 1986
8. J.S. Faulkner, *Prog. Mater. Sci.*, Vol 27, Pergamon Press, London, 1982, pp. 1-187
9. H. Winter and G.M. Stocks, *Phys. Rev.*, Vol B27, 1982, p. 882
10. H. Winter, P.J. Durham, and G.M. Stocks, *J. Phys.*, Vol F14, 1984, p. 1047
11. F. Ducastelle, *J. Physique* (Paris), Vol 31, 1970, p. 1055
12. D.G. Pettifor, *J. Phys.*, Vol F7, 1977, p. 613
13. D.G. Pettifor, *CALPHAD*, Vol 1, 1977, p. 305
14. F. Gautier, J. Van der Rest, and F. Brouers, *J. Phys.*, Vol F5, 1975, p. 1884
15. G. Treglia, F. Ducastelle, and F. Gautier, *J. Phys.*, Vol F8, 1978, p. 1437

16. A. Bieber, F. Gautier, G. Treglia, and F. Ducastelle, *Sol. St. Comm.*, Vol 39, 1981, p. 149
17. M.O. Robbins and L.M. Falicov, *Phys. Rev.*, Vol B29, 1984, p. 1333
18. R.J. Hawkins, M.O. Robbins, and J.M. Sanchez, *Phys. Rev.*, Vol B33, 1986, p. 4782; *Sol. Stat. Comm.*, Vol 55, 1985, p. 253
19. C. Sigli, M. Kosugi, and J.M. Sanchez, *Phys. Rev. Lett.*, Vol 57, 1986, p. 253
20. C. Sigli and J.M. Sanchez, *Acta Metall.*, Vol 36, 1988, p. 367
21. K. Binder, *Phys. Rev. Lett.*, Vol 45, 1980, p. 811
22. K. Binder, J.L. Lebowitz, M.K. Phani, and M.H. Kalos, *Acta Metall.*, Vol 29, 1981, p. 1655
23. H.T. Diep, A. Ghazali, B. Berge, and P. Lallemand, *Europhys. Lett.*, Vol 2, 1986, p. 603
24. D. de Fontaine and R. Kikuchi, Nat. Bur. Stand., Report No. SP-496, 1978, p. 999
25. J.M. Sanchez and D. de Fontaine, *Phys. Rev.*, Vol B21, 1980, p. 216
26. J.M. Sanchez, W. Teitler, and D. de Fontaine, *Phys. Rev.*, Vol B26, 1982, p. 1456
27. T. Mohri, J.M. Sanchez, and D. de Fontaine, *Acta Metall.*, Vol 33, 1985, pp. 1171-1185
28. C.M. van Baal, *Physica* (Utrecht), Vol 64, 1973, p. 571
29. J.M. Sanchez, F. Ducastelle and D. Gratias, *Physica*, Vol 128A, 1984, p. 334
30. R. Kikuchi, *Phys. Rev.*, Vol 81, 1951, p. 988
31. J.W.D. Connolly and A.R. Williams, *Phys. Rev.*, Vol B27, 1983, p. 5169
32. A.E. Carlsson and J.M. Sanchez, *Sol. Stat. Comm.*, Vol 65, 1988, p. 527
33. J.C. Slater and G.F. Koster, *Phys. Rev.*, Vol 94, 1954, p. 1498
34. W.A. Harrison, *Electronic Structure and the Properties of Solids*, W.H. Freeman, San Francisco, 1980
35. H. Shiba, *Progr. Theor. Phys.*, Vol 46, 1971, p. 77
36. R. Nieminen and C. Hodges, *J. Phys.*, Vol F6, 1976, p. 573
37. R. Haydock, V. Heine, and M.J. Kelly, *J. Phys. C.*, Vol 5, 1972, p. 2845
38. R. Haydock, V. Heine, and M.J. Kelly, *J. Phys. C.*, Vol 8, 1975, p. 2591
39. O. Kubaschewski and T.G. Chart, *J. Inst. Met.*, Vol 93, 1964-1965, p. 329
40. P. Villars and L.D. Calvert, in *Pearson's Handbook of Crystallographic Data for Intermetallic Phases*, Vol 3, ASM, Metals Park, OH, 1985, p. 3059
41. P.J. Meschter and W.L. Worrell, *Met. Trans.*, Vol 7A, 1976, p. 299
42. *Metals Handbook*, 8th Ed., Vol 8, ASM, Metals Park, OH, 1973
43. H.L. Skriver, *Phys. Rev.*, Vol B31, 1985, p. 1909
44. D.G. Pettifor, *J. Phys.*, Vol C3, 1970, p. 367
45. D.G. Pettifor, *CALPHAD*, Vol 1, 1977, p. 305
46. A.P. Miodownik, *Computer Modeling of Phase Diagrams*, edited by L. Bennett, The Metallurgical Society, Warrendale, PA, 1986
47. M. Hillert and L.I. Staffanson, *Acta Chem. Scand.*, Vol 24, 1970, p. 3618
48. M. Hillert and M. Jarl, *CALPHAD*, Vol 2, 1978, p. 227
49. B. Sundman and J. Agren, *J. Phys. Chem. Solids*, Vol 42, 1981, p. 297
50. S. Hertzman and B. Sundman, *CALPHAD*, Vol 6, 1982, p. 67
51. A.F. Guillermet, *CALPHAD*, Vol 6, 1982, p. 127
52. C. Sigli and J.M. Sanchez, *Acta Metall.*, Vol 33, 1985, p. 1097
53. C. Sigli, Ph.D. Dissertation, Columbia University in the City of New York, 1986
54. C. Sigli and J.M. Sanchez, *Acta Metall.*, Vol 34, 1986, p. 1021
55. A.J. McAlister, *Bull. Alloy Phase Diagrams*, Vol 3 (No. 2), 1982, p. 177
56. R.P. Elliott and F.A. Shunk, *Bull. Alloy Phase Diagrams*, Vol 2 (No. 3), 1981, p. 353
57. C.J. Wen, B.A. Boukamp, R.A. Huggins, and W. Weppner, *J. Electrochem. Soc.*, Vol 126, 1979, p. 2258
58. K.M. Myles, F.C. Mrazek, J.A. Smaga, and J.L. Settle, Proc. Symp. and Workshop on Adv. Battery Res. and Design: U.S. ERDA Report ANL-76-8, March, 1976

59. N.P. Yao, L.A. Heredy, and R.C. Saunders, *J. Electrochem. Soc.*, Vol 118, 1971, p. 1039
60. E. Veleckis, *J. Less-Common Met.*, Vol 73, 1980, p. 49
61. B. Noble and G.E. Thomson, *Met. Sci. J.*, Vol 5, 1975, p. 144
62. D.B. Williams and J.W. Edington, *Met. Sci. J.*, Vol 9, 1975, p. 529
63. S. Ceresara, G. Cocco, G. Fagherazzi, and L. Schiffini, *Philos. Mag.*, Vol 35, 1977, p. 373
64. S. Ceresara, G. Cocco, G. Fagherazzi, A. Giarda, and L. Schiffini, *La Metall. Italiana*, Vol 1, 1978, p. 20
65. R. Nozato and G. Nakai, *Trans. J.I.M.*, Vol 18, 1977, p. 679
66. F.W. Gayle and J.B. Vander Sande, *Bull. Alloy Phase Diagrams*, Vol 5, 1984, p. 19
67. E.S. Balmuth, *Scripta Met.*, Vol 18, 1984, p. 301
68. J.M. Papazian, C. Sigli, and J.M. Sanchez, *Scripta Met.*, Vol 20, 1986, p. 201
69. A.T. Aldred and K.M. Myles, *Trans. Met. Soc. AIME*, Vol 230, 1964, p. 736
70. S.C. Singhal and W.L. Worrel, *Met. Trans.*, Vol 4, 1973, p. 1125
71. S.C. Singhal and W.L. Worrel, in *Metallurgical Chemistry*, edited by O. Kubaschewski, Her Majesty's Stationary Office, London, 1972, pp. 65-74
72. S.C. Singhal and W.L. Worrell, *Met. Trans.*, Vol 4, 1973, p. 895
73. R. Hultgren, P.D. Desai, D.T. Hawkins, M. Gleiser, and K.K. Kelley, in *Selected Values of Thermodynamic Properties of Binary Alloys*, ASM, Metals Park, OH, 1973, p. 192
74. Yu.O. Esin, M.G. Valishev, A.F. Ermakov, O.V. Gel'd, and M.S. Petrushevskii, *Russ. J. Phys. Chem.*, Vol 55, 1981, p. 421
75. G.A. Levshin and V.I. Alekseev, *Russ. J. Phys. Chem.*, Vol 53, 1979, p. 437
76. J.C. Gachon and J. Hertz, *CALPHAD*, Vol 7, 1983, p. 1
77. M.P. Henaff, C. Colinet, A. Pasturel, and K.H.J. Buschow, *J. Appl. Phys.*, Vol 56, 1984, p. 307
78. G.A. Levshin, V.I. Alekseev, G.B. Petrov, and V.I. Polikarpov, *Dokl. Akad. Nauk. SSSR*, Vol 269, 1983, p. 870
79. J.C. Gachon, J. Charles, and J. Hertz, *CALPHAD*, Vol 9, 1985, p. 29
80. P. Steiner and S. Hufner, *Acta Metall.*, Vol 29, 1981, p. 1885
81. U.V. Choudary, K.A. Gingerich, and L.R. Cornwell, *Met. Trans.*, Vol 8A, 1977, p. 1487
82. P.J. Meschter and W.L. Worrell, *Met. Trans.*, Vol 8A, 1977, p. 503
83. V. Srikrishnan and P.J. Ficalora, *Met. Trans.*, Vol 5, 1974, p. 1471
84. R.S. Carbonara and G.D. Blue, *High Temp. Sci.*, Vol 3, 1971, p. 225

6

Theory of Ordering

PHILIP C. CLAPP
Department of Metallurgy and
Institute of Materials Science
University of Connecticut

A useful alloy ordering theory is one which is able to predict ordering phase transitions (their temperatures, compositions, and structures) from a limited amount of experimental information. In fact, the day is approaching when perhaps only the atomic numbers of the constituents and their relative proportions will be required (referred to hereafter as "first-principles calculations"). In the meantime this review offers a guide to a variety of theoretical approaches which may be used to accomplish many of the same objectives.[1,2]

Most theories of alloy ordering are based on the Ising model,[3] which had originally been created to describe magnetic ordering. The statistical mechanics is virtually identical, and the energy of any imaginable binary alloy configuration is approximated by this model as:

$$E(\sigma) = \frac{1}{2} \sum_{i,j}^{N} V_{ij} \sigma_i \sigma_j \qquad \text{(Eq 1)}$$

where $\sigma_i = (+1, -1)$ if atom (A,B) is on site i of N sites in the lattice. This expression may be trivially extended to cover the case of multicomponent alloys with any number of independent chemical species.[1]

One normally assumes that the V_{ij} values are at least temperature-independent and possibly composition-independent, although this latter assumption rarely stands up to detailed examination. Presuming that the V_{ij} values are known from theory or experiment, the energy of any specific configuration of the alloy is calculable from Eq 1. In principle, the free energy may then be determined by the standard statistical mechanical formula involving the partition function:

$$Z = \sum_{(\sigma)} e^{-\beta E(\sigma)}; \ \beta = 1/kT \qquad \text{(Eq 2)}$$

$$F = -kT \ln Z \qquad \text{(Eq 3)}$$

This would provide the free energy of the alloy as a function of temperature, thereby permitting a prediction of a number of thermodynamic properties and the phase diagram.

In practice, this calculation has rarely been carried out because of its inherent complexity. Some of the more notable successful examples are as follows. One example is the first calculation by Ising(3) for a one-dimensional chain of atoms with nearest-neighbor interactions, which, however, did not show any ordering transition. Onsager(4) exactly calculated the critical properties near the ordering transition of a two-dimensional square lattice with nearest-neighbor interactions. More recently, Wilson(5) has developed a very powerful approach called "Renormalization Group Theory" which can calculate critical properties of three-dimensional lattices with short-range interactions near their ordering transitions.

Even though these calculations are "exact" within the framework of the Ising model, it still may be asked how they can be related to the properties of real alloy systems. Certainly a first concern would be to include longer-range interactions beyond nearest neighbors. A second concern would be to find a way of including the effect of atomic displacements off the regular lattice sites due to atoms of different sizes trying to fit on the same lattice. These "size-effect" displacements (which are measurable by diffraction methods) depend on the local atomic configuration and can lead to quite large elastic energy contributions. A third concern involves ways in which many-body interactions, volume forces, etc., which are predicted to have some importance by first-principles calculations, could be included in the configurational energy. Only when all of these factors have been adequately included can it be expected that highly accurate ordering and phase-diagram predictions will be forthcoming.

This review details the general progress that has been made to date toward this overall goal. To make the picture clearer, it is natural to divide the task into its component parts. One major endeavor has been the determination of the lowest-energy ordered structures, given a basic lattice and a set of V_{ij} values. This is, in effect, the determination of the phase diagram at 0 K. A second major effort is directed at determining the effects of temperature on the relative stability of different orderings. This necessitates an evaluation of the entropy, in addition to the energy, as a function of temperature in some fashion. Last to be discussed here are the current efforts to go beyond the Ising model in the various directions already cited.

LOWEST-ENERGY ORDERED STRUCTURES

An early method for determining the ground state of fcc or bcc lattices at AB or A_3B compositions was developed by Clapp and Moss[6] for V_{ij} values as long ranged as third neighbors. They found it convenient to use the Fourier transform of the V_{ij} values, which can be written as:

$$V(q) = (1/2N) \sum V_{ij} e^{iq \cdot R_{ij}} \quad \text{(Eq 4)}$$

where R_{ij} is the vector separation between sites i and j.

They found that the minima of V(q) predict the locations of the diffraction peaks of the minimum-energy ordered structure (if such a structure can be constructed at that composition). Based on this approach they produced maps predicting the lowest-energy ordered structures for varying ratios of the V_{ij} values. Examples are shown in Fig. 1 and 2.

This approach was generalized and extended by Richards, Allen, and Cahn[7,8] to deal with binary alloys of any composition on fcc or bcc lattices. This led to ground-state diagrams of the sort shown in Fig. 3. Kanamori, Kudo, and Katsura[9,10] carried this work even further to include interactions up to fourth nearest neighbors and also considered hcp lattices up to second neighbors. Very little work of this sort has been done for more complex lattice types, but since a majority of ordering systems fall into the simpler lattice categories, most cases of interest can be dealt with.

Perhaps the most exciting change in predicting the relative energies of different ordered configurations has recently come from the much higher accuracy now pos-

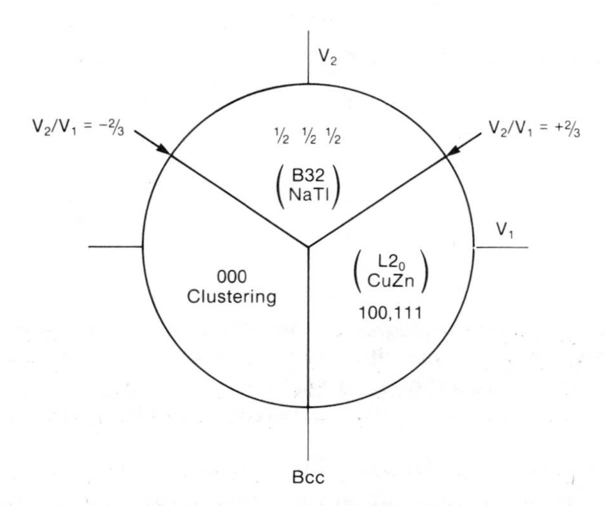

Fig. 1. Ordering diagram (from Ref 16) for a bcc lattice at AB or A_3B composition with only first- and second-neighbor interactions obtained by locating the minima of V(q)

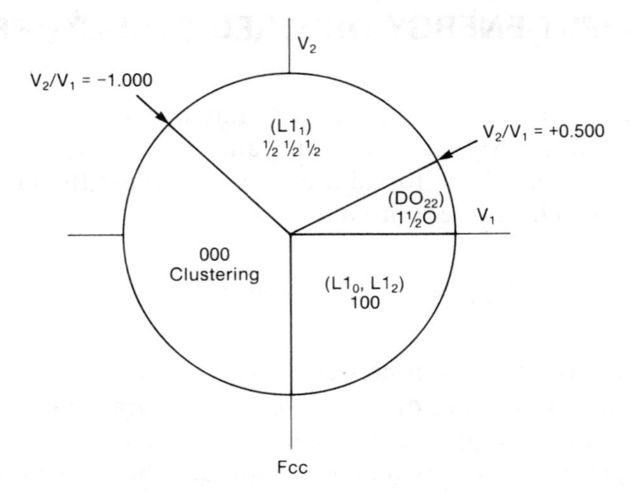

Fig. 2. Ordering diagram (from Ref 16) for a fcc lattice at AB or A_3B composition with only first- and second-neighbor interactions based on locations of the minima of V(q)

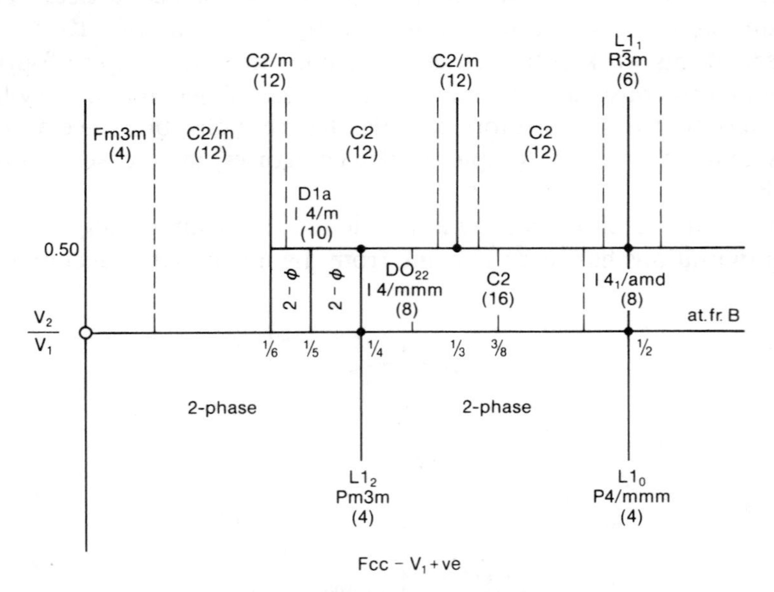

Fig. 3. Ordering diagram (from Ref 7) for a fcc lattice at any binary composition with only first- and second-neighbor interactions. In this particular diagram the first-neighbor interaction is restricted to be positive.

sible with first-principles calculations. These involve calculating the total lattice energy for a given configuration, including the energies of all the electrons. The input is simply the atomic number of the chemical species involved and their relative proportions, as well as the type of lattice the atoms are to be arranged upon. The actual method of performing the calculation may use what has come to be known as the "density functional method," possibly combined with older methods of approximately solving a many-body Schrodinger equation such as the ASW method, the KKR-CPA method, or the LCAO method. The output can be designed to yield the electronic density of states, bulk modulus, phonon-dispersion curves, equilibrium lattice parameters, elastic constants, shape of the Fermi surface, and what is most important here, the total lattice energy of a chosen configuration as a function of lattice parameter, all at 0 K of course.

An impressive example is the calculation by Connolly and Williams[11] of the copper-silver alloy system. These calculations utilized the density functional approximation combined with the ASW method. Some of their results are shown in Fig. 4, which shows the total lattice energy of each of a set of ordered alloys on an fcc lattice of either the $L1_2$ or $L1_0$ structure at different stoichiometric compositions in addition to the limiting pure-element cases. It may be seen from this diagram that a two-phase mixture of the pure elements will always have a lower energy than any of the ordered structures. Thus their calculation predicts that the Cu-Ag system should be a phase-separating one (into the two terminal solid solutions) at low temperatures over the entire composition field, which agrees very well with accepted data.

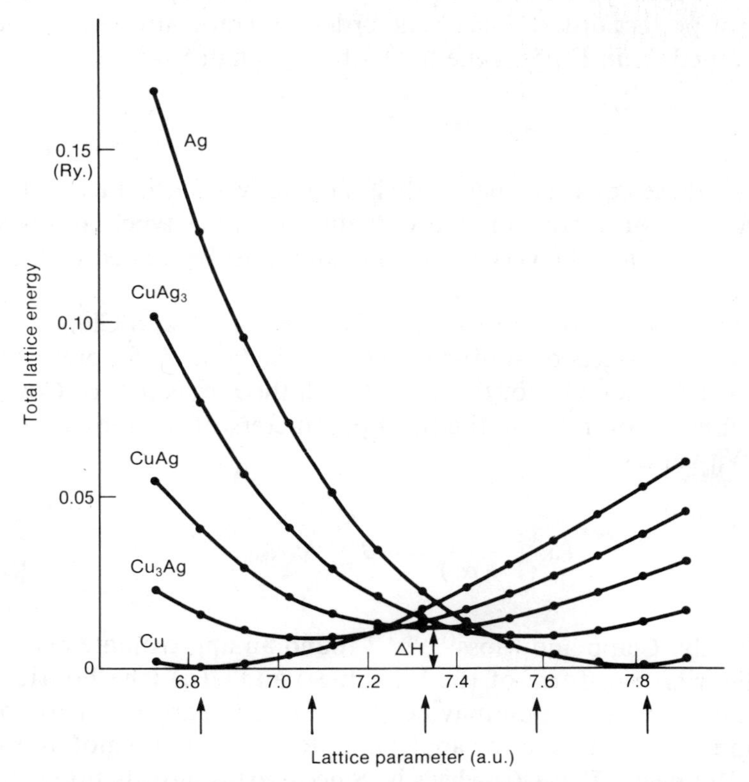

Fig. 4. Total lattice energy versus fcc lattice parameter first principles density functional calculation of the Cu-Ag system at specific stoichiometries based on the work reported in Ref 11[12]

ORDERING AT FINITE TEMPERATURES

The simplest and probably still the most useful ordering theory for the purposes of obtaining a quick, approximate result is known variously as the Weiss molecular field theory, the Bragg-Williams theory, or the mean field approximation (MFA). This is generally considered to be the zeroth-order approximation in that it is the lowest-order calculation which is capable of predicting an ordering transition and structure. Brout[13] has discussed the MFA's accuracy and limitations in some detail. For the purposes of this review it is sufficient to quote the main result of the MFA for "simple" lattices, which is the self-consistent equation for the long-range-order (LRO) parameter S:

$$S = \tanh\{-V(0)S/2kT\} \qquad \text{(Eq 5)}$$

where V(0) is the value of Eq 4 evaluated for $q = 0$, k is the Boltzmann constant, and T is the absolute temperature. S may be defined as the fraction of atoms on their correct sublattice less the fraction of atoms on the wrong sublattice, referred to the ideal perfectly ordered structure. Thus for complete, random disorder, $S = 0$; and for perfect order, $S = 1$. The ordering temperature T_c predicted by the MFA is obtained from Eq 5 in the limit of very small S—i.e.,

$$T_c = -V(0)/2k \qquad \text{(Eq 6)}$$

The beauty of these equations lies in their simplicity and the fact that they can be used for any range of interaction and virtually any lattice type. However, like anything beautiful they may be very misleading and must be treated with considerable caution.

The next-higher order of approximation (see Brout[13] and Clapp[14] for discussions) includes the effects of short-range order (SRO) and can predict the degrees of both SRO and LRO. One of the best of such theories is that of Cowley,[15] who developed an equation relating the SRO parameters of an alloy at any temperature to the V_{ij} values—i.e.,

$$\ell n \frac{(1 + \alpha_{ij})}{(1 - \alpha_{ij})} = -\beta \sum_{m}^{N} V_{im}\alpha_{mj} \qquad \text{(Eq 7)}$$

Subsequently, Clapp and Moss[14,16,17] found an approximate relation between V(q) (the Fourier transform of the V_{ij} values) and $\alpha(q)$ (the Fourier transform of the α_{ij} values). This relation may be classed as a linearized version of Cowley's equation, and is easier to use for analyzing SRO data, but is not appropriate for describing LRO below T_c, as Cowley's is. Since $\alpha(q)$ is directly proportional to the measured short-range-order diffuse intensity obtained by x-ray or neutron scattering, the Clapp-Moss approximation made it possible to relate the data directly to the interaction energies (the V_{ij}'s) in any alloy system in a relatively simple and straightforward manner. This has been done for a substantial number of alloy systems to date. The equation is given as follows:

$$\alpha(q) = \frac{C}{1 + 2C_A C_B \beta V(q)} \quad \text{(Eq 8)}$$

Krivoglaz[18] independently derived an equation similar to Eq 8 with the difference that C was taken to be unity for all cases, whereas Clapp and Moss used it as a self-consistent normalization term that always ensures that the integral of the SRO diffuse intensity is unity over a complete Brillouin zone, as it must be. Thus Krivoglaz's equation generally violates that normalization condition except at very high temperatures.

A match between Eq 8 using only first- and second-neighbor V_{ij} values and the measured SRO intensity for Cu_3Au at 405 °C is shown in Fig. 5. It can be observed that the fit is quite reasonable. Wilkins[19] showed that the approximation errors in Eq 8 could be reduced by fitting to the V_{ij} ratios. Such a fit for the copper-gold system is illustrated in Fig. 6. Equation 8 may also be used to predict the configuration of the impending ordered structure and the temperature at which it will appear once the V_{ij} values have been determined by fitting Eq 8 at some higher temperature. This is accomplished by determining the temperature for which the denominator of Eq 8 vanishes and the places in q-space that show pronounced maxima just above that ordering temperature, again according to Eq 8. These positions are Eq 8's predictions of where the Bragg peaks of the new ordered structure will appear and will obviously coincide with the points in q-space where V(q) has its minima.

The prediction of the ordering temperature given by Eq 8 turns out to be no better than that of the MFA, but does have the advantage of predicting the ordered structure, as well as predicting the degree of SRO at any temperature above the ordering temperature. In some cases the minima of V(q) may occur only at the origin of q-space, which means that the system is tending toward segregation or phase separation, rather than ordering. A case study in point was the example of copper-nickel alloys,[17] which was later verified experimentally.

Recent work of considerable interest by Wadsworth, Gyorffy, and Stocks,[20] using a first-principles KKR-CPA calculation to determine the diffuse SRO scattering that should be produced by the Fermi surface in copper-palladium alloys, obtained a remarkably good fit to experimental data over a range of compositions. Furthermore, they were able to analyze their results in the form of an equation relating $\alpha(q)$ to a two-body response function which can be regarded as a generalization of the two-body interaction V(q) in Eq 8. Thus it seems that "new wine can be put into old skins."

The cluster variation method (CVM) has been used increasingly in the last decade or so to calculate entropies and free energies of systems with relatively short-range interactions. Kikuchi[21] pioneered this method, with subsequent improvements provided by Sanchez and de Fontaine[22] (among others). The input in this method is a specification of the lattice type, the lattice composition, and the val-

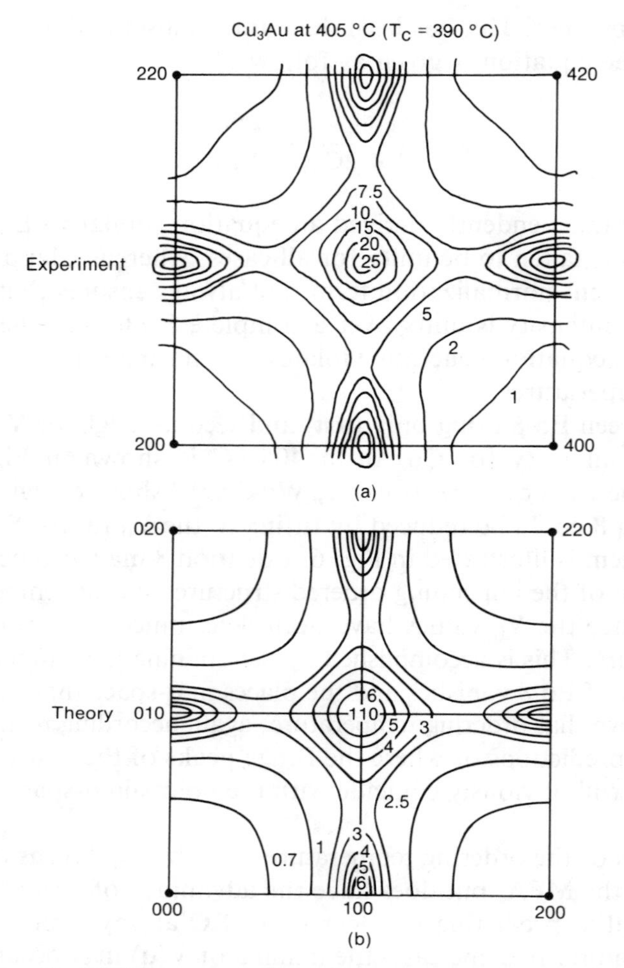

Fig. 5. Diffuse scattering measurements of Cu_3Au at 405 °C of J.M. Cowley[15] compared to Clapp-Moss theory (from Ref 17)

ues of V_{ij} as a function of distance in the lattice. The output, after substantial calculation on a high-speed computer, is the free energy as a function of temperature of different ordered structures as well as the disordered phase. Detailed predictions of the phase diagram are then possible. The reverse process of proceeding from the experimental data to the V_{ij} values using a fluctuation formulation of the CVM developed by Sanchez[23] has recently been applied to several specific cases and compared with Eq 8.[24,25]

However, probably the most spectacular breakthrough in this problem area has very recently been accomplished by Gerold, Kern, and Schweika[26,27] with what they have called the "Inverse Monte-Carlo" (IMC) method. They have shown that they can extract V_{ij} values with an accuracy of better than 10% from artificially generated test data for a system having known values of first- and second-neighbor interactions. They have also demonstrated an impressive ability to extract V_{ij} val-

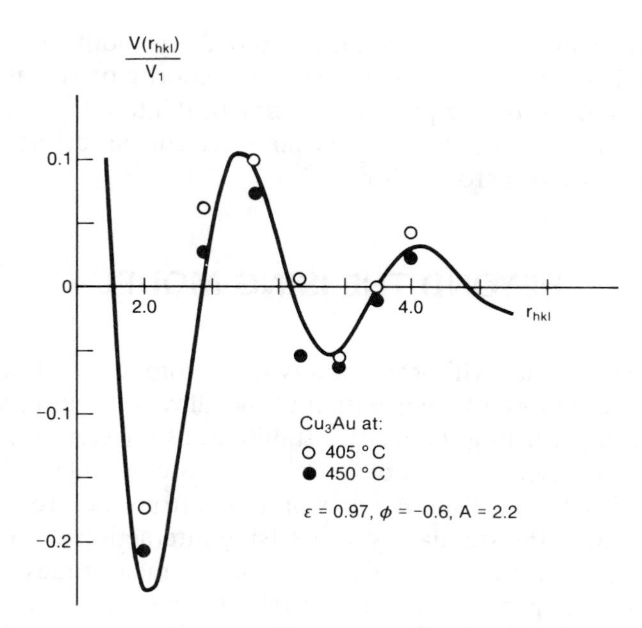

Fig. 6. Plot of ratio of higher-neighbor interactions to first-neighbor interaction in Cu_3Au derived from experiments at two temperatures demonstrating only weak temperature dependence of pair interactions (from Ref 19). Solid curve is a best fit to a theoretical oscillating potential form.

ues up to 8th neighbors or more from real experimental data for a variety of alloy systems. Once the V_{ij} values have been obtained at some temperature above the ordering transition, the normal Monte-Carlo (M-C) method can be used to shuffle atoms around on a suitably large lattice array at any desired temperature until "equilibrium" is achieved, thus providing a detailed atomic picture of the stable phase (or phases) at each temperature simulated. This gives a very precise image of the SRO and LRO with a high degree of accuracy, and so can map out the phase diagram at the compositions for which experimental input was available. If the V_{ij} values were composition-independent (which one assumes only at great peril), the IMC combined with the MC would be sufficient to map out the entire phase diagram of a binary alloy system from one careful experimental measurement of the SRO diffuse scattering intensity at one composition and temperature! Of course all this requires the services of a high-speed computer, but that is a facility that is becoming almost "standard equipment."

Within the framework of the Ising model, for cases where the ordering transition is higher than first order (i.e., no latent heat), an extremely powerful statistical mechanical analysis known as "renormalization group theory" was developed mainly by Wilson, Kadanoff, and Fisher.[5,28,29] Because of its asymptotically exact accuracy in predicting many thermodynamic characteristics very near the critical transition temperature (T_c), such as critical exponents, it has attracted a great deal of interest, including a Nobel prize. Unfortunately, as yet it cannot tell us very much about phase characteristics well away from T_c or about ordering transitions involving a latent heat, which represents the large majority of actual cases. Accordingly, most of the alloy ordering processes of practical interest fall outside the present scope of this theory, but there are very intensive current efforts to extend it to the class of first-order transformations.

BEYOND THE ISING MODEL

The elastic distortions that will occur in any alloy composed of atoms of different sizes produce an important contribution to the alloy configurational energy that must be included in predicting the relative stabilities of different phases. Cook and de Fontaine[30] addressed this problem in a microscopic elasticity theory that includes terms which couple displacements of atoms from their regular lattice positions and also includes the standard site-site Ising interactions. Within the framework of their theory, and depending on the relative magnitudes of the different energy terms, one may predict an ordering transformation, spinodal decomposition, or "continuous ordering." The latter term describes a process in which ordering and phase separation occur concurrently. Because there are three sets of interaction parameters (instead of one set in the Ising model), the necessary parameters of this theory are much more difficult to extract from experimental data. Nevertheless, one example of this exercise is now extant with the study of Au-40Ni by Wu and Cohen.[31,32)

Machlin[33-37] has taken a different approach to the same distortion problem by developing a set of semiempirical intermetallic potentials that will include both the Ising (chemical) energy and the distortion (elastic) energy in one interaction. Using reasoning based on both theoretical and experimental considerations, he found that the most generally satisfactory "universal" form for this interaction was:

$$V_{ij}^{\alpha\beta} = \frac{-A^{\alpha\beta}}{(R_{ij})^4} + \frac{B^{\alpha\beta}}{(R_{ij})^8} \qquad \text{(Eq 9)}$$

where A and B are constants dependent on the alloy species α and β, and R_{ij} is the interatomic separation.

The adjustable quantities were determined from measured lattice-parameter and cohesive-energy data of the pure elements and a chosen set of intermetallic compounds. Machlin was then able to make extensive predictions of the lattice parameters, the cohesive energies, and the relative stabilities of various other intermetallic cubic phases, such as fcc, bcc, $L1_2$, $L1_0$, A15, C15, etc., and also some noncubic phases. These interactions may then be used in any one of the thermodynamic theories to calculate the phase diagram as a function of temperature and composition. Although this approach is very attractive, it is difficult to assess the accuracy of the result in any particular case or to relate it to the more fundamental first-principles calculations that are now becoming increasingly available. One may therefore regard it as a useful semiquantitative device providing guidelines for the alloy designer, but substantial allowance must be made for error.

Many-body interactions (involving three or more atoms simultaneously) have been shown to exist in alloys either on the basis of fundamental calculations (such as pseudopotential methods) or from fits to experimental phonon-dispersion curves. They are generally weak in comparison with the pairwise interactions but may be important in particular cases. As a result, some effort has been made to include their effects in the calculation of configurational energies. A natural extension of the Ising model to include them would be:

$$E\{\sigma\} = \frac{1}{2}\sum_{i,j}^{N} V_{ij}\sigma_i\sigma_j + (1/3!)\sum_{ijk}^{N} V_{ijk}^{(3)}\sigma_i\sigma_j\sigma_k + (1/4!)\ldots. \qquad \text{(Eq 10)}$$

Using such a configurational energy form, de Fontaine and Kikuchi[38] assumed that the sums would only be taken over nearest neighbors but could include pairwise, three-body, and four-body interactions in their treatment of the copper-gold system. They then employed the CVM to calculate the entire phase diagram with but two adjustable parameters that made the ordering temperatures at the Cu_3Au and CuAu compositions match the correct values. Their CVM calculation of the phase diagram, shown in Fig. 7, compares quite favorably with the presently accepted diagram shown in Fig. 8.

Returning to the first-principles calculation of Connolly and Williams,[11] they have analyzed their results in a form analogous to Eq 10, but have included a one-body and a no-body term as well. The one-body term helps to account for composition-dependent energy terms, and the no-body term reflects volume-dependent energies. Again, they restricted consideration to nearest-neighbor interactions but included up to four-body interactions. The result of their analysis is shown in Fig. 9, and it may be noted that the three-body and four-body interac-

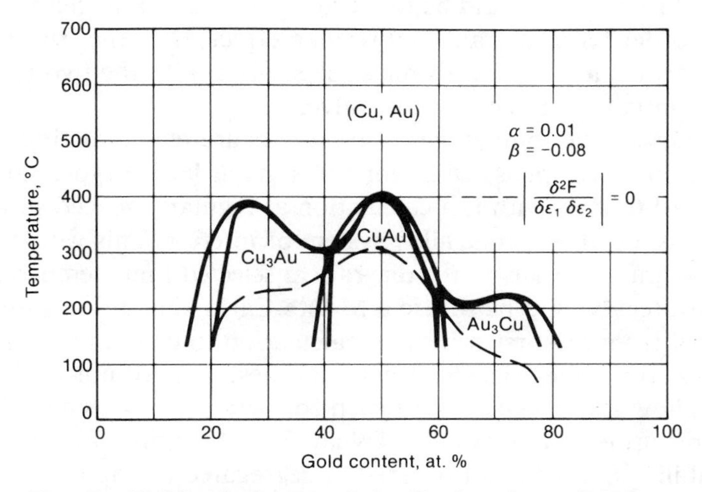

Fig. 7. CVM calculation of the Cu-Au phase diagram (from Ref 38)

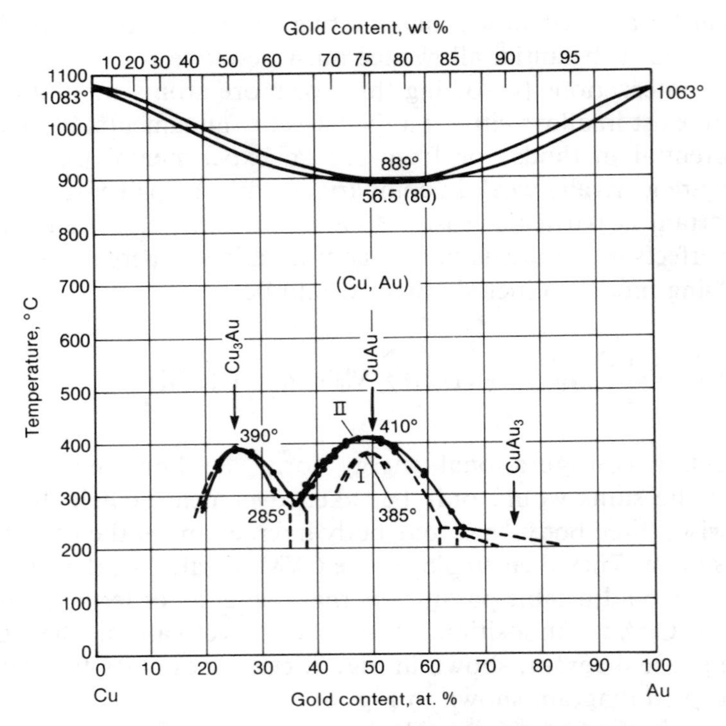

Fig. 8. Cu-Au phase diagram[39]

tions are indeed substantially weaker than the two-body interactions. Since the first-neighbor pair interaction (V_1) is positive, and V_2 is considered negligible in their analysis, based on this interaction alone the ground-state diagram of Fig. 2 would indicate an ordering reaction. If the system were confined to a fixed lattice parameter at the CuAg minimum, this would be a correct conclusion. However, once this constraint is released (as it would be in nature) and the strain energies associated with the lower-order terms are allowed to take effect, then the true energy minimum of the system is achieved with phase separation into the two terminal solid solutions of essentially pure copper and silver.

This result suggests the following scheme for accurately predicting the ordering transitions and complete phase diagram for a given lattice type, say fcc. What would be required to start with is a calculation at regularly spaced stoichiometries similar to that just cited for some alloy system of interest. This would provide the 0 K fcc phase diagram. To obtain the diagram at selected finite temperatures, probably the best strategy would be to use a Monte-Carlo simulation using the interactions extracted from the first-principles calculation to determine configurational energies again at the same selected stoichiometries. The Monte-Carlo procedure would have to allow static relaxation of each configuration by using either an additional M-C subroutine or a Molecular Dynamics subroutine. In this way the relative phase stabilities of different ordered or segregated configurations on the fcc lattice could be established at a grid of temperature-composition points and inter-

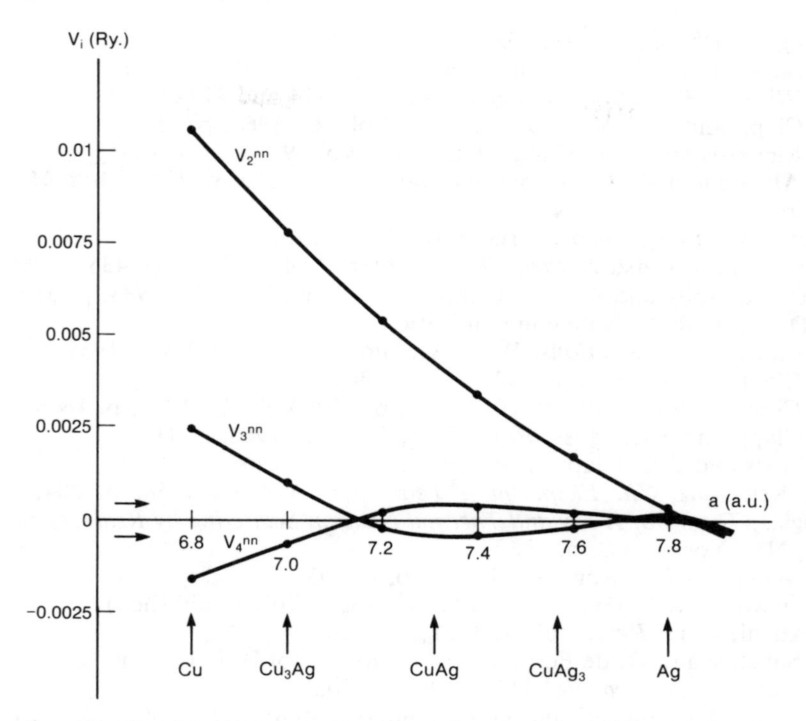

Fig. 9. Nearest-neighbor 2-, 3- and 4-body interactions estimated from a first principles density functional calculation of the Cu-Ag system at specific stoichiometries based on the work reported in Ref 11[40]

polations made to flesh out the entire diagram. Any doubtful interpolations could be checked with additional M-C runs at intermediate points.

One very important advantage of the proposed scheme, as compared with direct experimental investigation, is that it also gives much information about possible metastable phases. Of course, this does not yet provide the complete phase diagram of a given alloy system, because generally other lattice types must also be considered. To include specific alternatives (such as bcc, hcp, etc.), the entire procedure would need to be repeated for each one. Although this may appear to be a very expensive and time-consuming procedure at present, the rapid increase in computing power should make this approach quite feasible in the near future.

REFERENCES

1. P.C. Clapp, *Materials Research Society Symposia Proceedings*, Vol 39, 1985, p. 31 (The present article is an updated and expanded version of this earlier review.)
2. An excellent general review of many of the topics of this paper is given in: D. de Fontaine, *Sol. Stat. Phys.* Vol 34, 1979, p. 73

3. E. Ising, *Z. Physik*, Vol 31, 1925, p. 253
4. L. Onsager, *Phys. Rev.*, Vol 65, 1944, p. 117
5. K.G. Wilson, *Phys. Rev.*, Vol B4, 1971, pp. 3174 and 3184
6. P.C. Clapp and S.C. Moss, *Phys. Rev.*, Vol 171, 1968, p. 754
7. M.J. Richards and J.W. Cahn, *Acta Met.*, Vol 19, 1971, p. 1263
8. S.M. Allen and J.W. Cahn, *Scripta Met.*, Vol 7, 1973, p. 1261; *Acta Met.*, Vol 20, 1972, p. 423
9. J. Kanamori, *Prog. Theor. Phys.*, Vol 35, 1966, p. 66
10. T. Kudo and S. Katsura, *Prog. Theor. Phys.*, Vol 56, 1976, p. 435
11. J.W.D. Connolly and A.R. Williams, *Phys. Rev.*, Vol B27, 1983, p. 5169
12. J.W.D. Connolly (private communication)
13. R. Brout, *Phase Transitions*, W.A. Benjamin, Inc., New York, 1965
14. P.C. Clapp, *Phys. Lett.*, Vol 13, 1964, p. 305
15. J.M. Cowley, *Phys. Rev.*, Vol 77, 1950, p. 669; Vol 120, 1960, p. 1648
16. P.C. Clapp and S.C. Moss, *Phys. Rev.*, Vol 142, 1966, p. 418
17. S.C. Moss and P.C. Clapp, *Phys. Rev.*, Vol 171, 1968, p. 764
18. M.A. Krivoglaz, *Zh. Eksperim. i Teor. Fiz.*, Vol 34, 1958, p. 204; also M.A. Krivoglaz, *Theory of X-ray and Thermal Neutron Scattering by Real Crystals*, Plenum Press, New York, 1969, p. 166
19. S.W. Wilkins, *Phys. Rev.*, Vol B2, 1970, p. 3935
20. J. Wadsworth, B.L. Gyorffy, and G.M. Stocks (to be published)
21. R. Kikuchi, *Phys. Rev.*, Vol 81, 1951, p. 988
22. J.M. Sanchez and D. de Fontaine, *Phys. Rev.*, Vol B17, 1978, p. 2926
23. J.M. Sanchez, *Physica*, Vol 111A, 1982, p. 200
24. T. Mohri, J.M. Sanchez, and D. de Fontaine, *Acta Met.*, Vol 33, 1985, p. 1463
25. D. de Fontaine, A. Finel, and T. Mohri, *Scripta Met.*, Vol 20, 1986, p. 1045
26. V. Gerold and J. Kern, *Acta Met.*, Vol 35, 1987, p. 393
27. W. Schweika, Ph.D. Thesis, Stuttgart, 1987
28. K.G. Wilson and M.E. Fisher, *Phys. Rev. Lett.*, Vol 28, 1972, p. 240
29. L.P. Kadanoff, *Physics*, Vol 2, 1966, p. 263
30. H.E. Cook and D. de Fontaine, *Acta Met.*, Vol 17, 1969, p. 915; Vol 18, 1970, p. 189
31. T.B. Wu, J.B. Cohen, and W. Yelon, *Acta Met.*, Vol 30, 1982, p. 2065
32. T.B. Wu and J.B. Cohen, *Acta Met.*, Vol 31, 1983, p. 1929
33. E.S. Machlin, *Acta Met.*, Vol 22, 1974, p. 95
34. E.S. Machlin, *Acta Met.*, Vol 22, 1974, p. 109
35. E.S. Machlin, *Acta Met.*, Vol 22, 1974, p. 367
36. E.S. Machlin, *Acta Met.*, Vol 22, 1974, p. 1433
37. E.S. Machlin, *Acta Met.*, Vol 24, 1976, p. 543
38. D. de Fontaine and R. Kikuchi, in "Applications of Phase Diagrams in Metallurgy & Ceramics," NBS Special Publication 496, edited by G.C. Carter, 1978, p. 999
39. M. Hansen, *Constitution of Binary Alloys*, McGraw-Hill, New York, 1958, p. 199
40. J.W.D. Connolly (private communication)

7

Alloying of Aluminum: Development of New Aluminum Alloys

EDGAR A. STARKE, JR.
School of Engineering and Applied Science
University of Virginia

Strong, lightweight materials are necessary ingredients for efficient powered flight, and age-hardenable aluminum alloys have been the predominant choice for airframe construction for over 50 years. The selection of materials for today's aircraft depends primarily on strength-to-weight ratio; durability, including corrosion resistance and damage tolerance; and economic considerations which include producibility, material cost, maintainability, and availability. Historically, aluminum alloys have been leaders in satisfying these requirements and currently are the dominant materials used for airframes (see Table 1). However, because of

Table 1. Materials used in aircraft. From Peel.

Material	Amount used, % of total weight
Transports	
Aluminum alloys	75 to 81
Titanium alloys	2 to 6
Steel	10 to 14
Composites/fiberglass	0.5 to 1
Fighters and Trainers(a)	
Aluminum alloys	60 to 70
Titanium alloys	10 to 15
Steel	10

(a) Some aircraft—e.g., the Harrier—contain large amounts of carbon fiber composites approaching 35%.

increased performance requirements, conventional aluminum alloys are being challenged by other materials, particularly titanium and carbon-fiber-reinforced epoxy-matrix composites.

Although about 60% of the structures of aircraft that operate in the Mach 1.5 to 2.0 range consist of aluminum alloys, the skin temperatures produced at higher Mach numbers prohibit the use of conventional age-hardenable aluminum products. Consequently, titanium alloys, which retain their strength to much higher temperatures, are logical alternatives. Since weight is the major driver in performance, lightweight, high-stiffness materials such as epoxy-matrix composites appear attractive when they can satisfy the environmental restrictions. However, since these materials can be expensive and often difficult to fabricate, improving the competitiveness of aluminum alloys has been the topic of recent research and development efforts.

Numerous studies have been conducted during the past several years to identify the property improvements that would have major impacts on improving the structural efficiency of aerospace vehicles. The result of one study is presented graphically in Fig. 1 and shows that for this particular aircraft a reduction in density is three to ten times more effective in reducing structural weight than an increase in strength, damage and durability tolerance assessment (DADTA) (a fatigue and fracture toughness parameter), or modulus of elasticity. The effects of the various engineering parameters will change in relative importance depending on both the mission of the vehicle and the governing failure mode (tension, compression, or another mode). Consequently, in order to improve the competitiveness of aluminum, the focus has been on developing new aluminum alloys with lower density, higher stiffness, improved damage tolerance, and improved high-temperature properties.

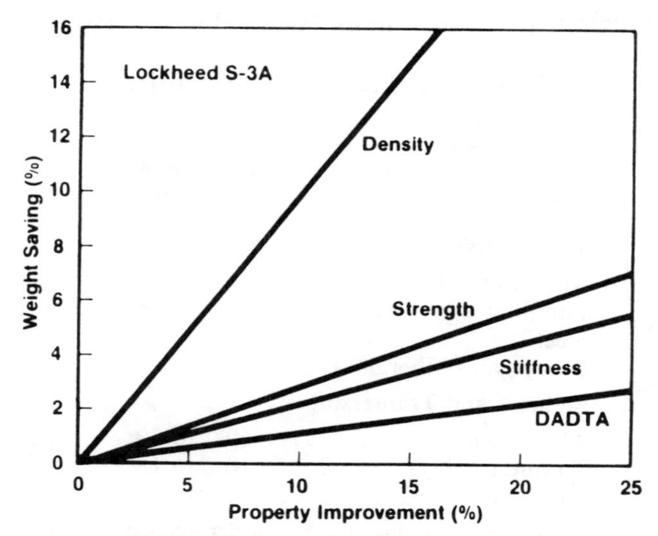

Fig. 1. The influence of various properties on structural weight savings in the S-3A aircraft which is typical of many studies. Courtesy of R.E. Lewis.

BACKGROUND

Aluminum alloys are classified as heat treatable or non-heat-treatable, depending on whether or not they respond to precipitation hardening. The heat treatable alloys contain elements that decrease in solubility with decreasing temperature and in concentrations that exceed their equilibrium solid solubilities at room and moderately higher temperatures. The most important alloying elements in this group include copper, magnesium, zinc, and, more recently, lithium. A normal heat treatment cycle includes a solutionizing soak at a high temperature to maximize solubility, followed by rapid cooling or quenching to a low temperature to obtain a solid solution supersaturated with both solute elements and vacancies. During aging at either room or some intermediate temperature, the solute atoms cluster together, forming preprecipitates which are coherent with the matrix and are often referred to as Guinier-Preston zones after the two researchers who confirmed their existence in 1938. The research by Guinier and Preston followed the discovery of age hardening in 1906 by Alfred Wilm, and the explanation of the phenomenon by Merica, Waltenberg, and Scott in 1919.

Precipitation in Aluminum Alloys

During aging, the type and rate of the decomposition process depend on whether aging is carried out below or above a critical temperature which is defined by a metastable solvus line on the equilibrium diagram (Fig. 2). The solid solution is metastable with respect to Guinier-Preston zones below this temperature, and the zones do not form above the critical temperature. The critical temperature may be

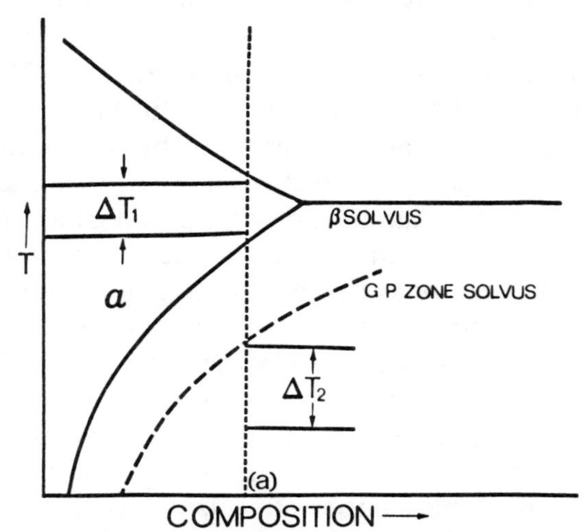

For composition (a), delta T1 temperature for solution heat treating, delta T2 temperature for precipitation heat treatment.

Fig. 2. Phase diagram of a hypothetical system A-B showing the beta solvus and GP-zone solvus

displaced to higher temperatures when the vacancy concentration is increased. The vacancies are believed to accelerate the formation of clusters during quenching and/or to act as aggregates which nucleate the new phase. The decomposition of supersaturated solid solutions below either critical temperature normally occurs by the following sequence: Supersaturated solid solution → clusters → transition structure → final structure.

In precipitation-hardening alloys, three types of interfaces may develop during nucleation and growth of strengthening precipitates: coherent, semicoherent, and incoherent interfaces. No significant difference in atomic positions exists across a coherent interface, although a slight difference in lattice parameter may result in local elastic strains required to maintain coherency. If the coherency strains become excessive, which usually happens as aging proceeds, the coherent interface may be replaced by a periodic array of edge dislocations or structural ledges resulting in a semicoherent interface. The coherent and semicoherent particles may form in a variety of shapes—e.g., spheres, cubes, disks, and needles—depending on the misfit between particles and the matrix and on the interfacial energy. The shape that occurs generally minimizes the total energy associated with the formation of the particle. For example, when the atomic misfit is very small, spherical particles are formed because this shape has the lowest surface-area-to-volume ratio. However, when the atomic misfit is large, the particles generally assume the shapes of disks or needles in order to minimize the strain energy. Cubes are an intermediate state and have not, thus far, been observed in aluminum alloys. The third type of interface is the incoherent interface, which has a degree of disorder comparable to that of a high-angle grain boundary. Dispersions, which usually refer to nonmetallic particles, normally have incoherent interfaces. They are often produced by precipitation from the melt, eutectoid decomposition, internal oxidation, or mechanical mixing of particulate followed by powder metallurgy consolidation.

Often a precipitate-free zone (PFZ) is observed adjacent to a high-angle grain boundary, as shown in the transmission electron micrograph (TEM) of Fig. 3. The PFZ may be caused by two different phenomena. In the first, the aging temperature is above the critical temperature determined by the equilibrium concentration of vacancies but below that determined when excess vacancies are present, and vacancies have been lost to the grain boundary prior to aging. In this case the PFZ is called a vacancy-depleted PFZ. In the second phenomenon, normal nucleation and growth of the equilibrium phase occurs at the grain boundary, depleting the adjacent region of solute, and thus producing a solute-depleted PFZ. This is the type shown in Fig. 3. Both types of PFZ's may be minimized by lowering the aging temperature, which increases the solute supersaturation and thus the driving force for homogeneous decomposition, and decreases diffusion rates which decreases the nucleation and growth of the equilibrium precipitates. Consequently, age-hardening alloys are frequently given double aging treatments: a low-temperature age followed by a higher-temperature age. The low-temperature age increases the number density of preprecipitates and minimizes the formation of PFZ's, and the higher-temperature age accelerates the growth of the precipitates and often their transition to an intermediate precipitate. This allows the desired strength level to be obtained in the minimum amount of time.

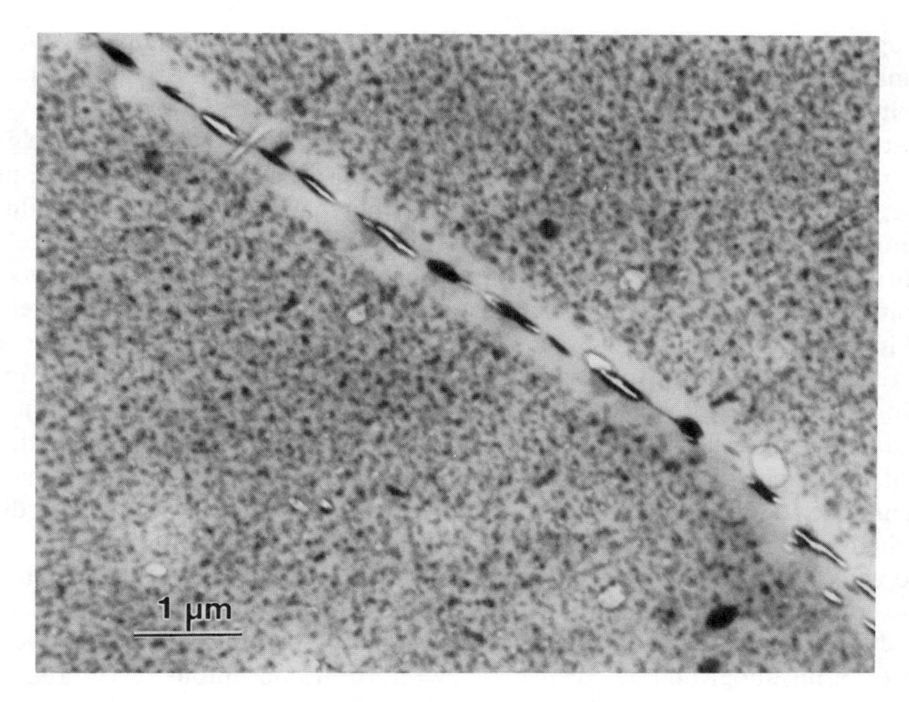

Fig. 3. Transmission electron micrograph showing precipitate-free zone (PFZ) adjacent to a high-angle grain boundary in an aged Al-Zn-Mg alloy

PFZ's almost always have a deleterious effect on the properties of aluminum alloys. Since they are free of precipitates, they are softer than the surrounding matrix and, under an applied stress, can be regions of strain localization. This may lead to premature fracture under either monotonic or cyclic loading. In addition, PFZ's are believed to increase the susceptibility of age-hardenable aluminum alloys to stress-corrosion cracking. The grain-boundary precipitates which produce the PFZ's can also play an important role in the failure process of both mechanically and environmentally induced fracture. Consequently, processing procedures—e.g., low-temperature aging and/or deformation prior to aging—are frequently employed to minimize grain-boundary precipitation and PFZ formation. The dislocations induced during the deformation often serve as nucleating sites for precipitates in competition with the grain boundaries.

Trace elements may often be present in aluminum alloys, either as impurities or by design, and can frequently have an effect on the precipitation process. Although the influence of trace elements depends on the particular element and alloy system, they usually modify the precipitation process by one or more of the following mechanisms:

1. They interact with vacancies and alter the kinetics of the precipitation process.
2. They change the interfacial energy of the clusters and/or precipitates and may raise or lower the critical temperature and alter the nucleation frequency.
3. They may alter the type of precipitate formed.

Although trace elements have been added to some alloy systems to improve machinability, they are normally added to aid in the nucleation of the strengthening precipitates.

Most commercial alloys are designed to have precipitates that control the grain structure. Aluminum combines readily with the transition metals, chromium, manganese, zirconium, etc., to form intermetallic phases with little or no solubility in the aluminum matrix. Because of their slow diffusivity in aluminum, and since they are normally present in low concentrations, these elements usually form very small precipitates, less than 1 μm in size, during either solidification or ingot preheating. This fine distribution of precipitates, called dispersoids, delays or prevents static recrystallization during processing, and, because they are strung out in the working direction (a process called mechanical fibering), they aid in retaining the elongated or "pancake"-shape grains that develop during working. The mean free path of the dispersoids is greater in the working direction than in directions perpendicular to it. Thus, recrystallized grains can grow faster in the working direction. Such elongated grains impart anisotropic behavior in the material.

The effectiveness of a particular dispersoid in controlling the grain structure depends on its size, spacing, and coherency. Small particles (normally less than 0.4 μm in diameter) retard continuous recrystallization by pinning subgrain boundaries and preventing subgrain coalescence. They retard discontinuous recrystallization by pinning migrating high-angle boundaries—a process described by Zener. Zener suggested that a grain would cease to grow when its radius, R, was approximately equal to the particle radius, r, divided by the particle volume fraction, f:

$$R(\text{crit}) = (3\beta/4)(r/f) \qquad \text{(Eq 1)}$$

where β is a constant approximately equal to one. Equation 1 suggests that small, closely spaced particles have the greatest effect in retarding recrystallization. Coherent dispersoids (e.g., Al_3Zr) are more effective than the incoherent dispersoids (e.g., $Al_{12}Mg_2Cr$ and $Al_{20}Mn_3Cu_2$) in inhibiting boundary migration, because the precipitate/matrix interface must either change from coherent to semicoherent or incoherent as the boundary passes the dispersoid or, alternatively, dissolve and reprecipitate after the boundary passes through its local. Both processes require considerable energy.

The type of dispersoid—i.e., coherent or incoherent—can also affect the "quench sensitivity" of an age-hardenable alloy. As mentioned previously, age-hardenable alloys must be quenched from a relatively high temperature to maximize the supersaturation of the alloying elements that will participate in formation of the strengthening precipitates during subsequent aging. The quenching medium may be forced air, cold or boiling water, or some other liquid. The "quench sensitivity" describes the tolerance of a particular alloy to the slower quenching media. During slow cooling, some of the solute may precipitate out as coarse particles, which reduces the degree of supersaturation and adversely affects the alloy's fracture toughness and ductility. Quench sensitivity is higher in the more concentrated alloys and is enhanced in alloys containing incoherent particles whose interfaces serve as nucleation sites for heterogeneous precipitation. Consequently, alloys that contain chromium and manganese as dispersoid-forming elements are more quench sensitive than those that contain zirconium, which forms a coherent dispersoid.

Strengthening in Aluminum Alloys

The increase in strength of age-hardenable aluminum alloys over that of the pure metal or of solid-solution alloys is due to the interaction of dislocations with preprecipitates and transition precipitates which are formed during aging below the critical temperature. The strengthening contribution from shearable particles is related to the particles' intrinsic properties and the volume fraction of particles and can be represented by an equation of the form:

$$\tau = cf^m r^p \qquad \text{(Eq 2)}$$

where τ is the critical resolved shear stress, c is an alloy constant that depends on the particular strengthening mechanism (i.e., coherency, surface, chemical, stacking-fault, and/or modulus hardening), f is the volume fraction, and r is the particle radius. Coherency hardening is associated with the elastic coherency stresses surrounding a particle that does not exactly fit the matrix. Surface hardening is associated with the energy necessary to create additional particle/matrix interfaces when the particle is sheared by the dislocation. Chemical or order hardening is associated with the additional work required to create an internal interface or antiphase boundary in the case of internally ordered particles. Stacking-fault hardening is associated with the difference in stacking-fault energy between the matrix and the particle. Modulus hardening is associated with the difference in modulus between the matrix and the particle. The exponents m and p in Eq 2 are always positive, and the strength increases with both volume fraction and particle size.

Aluminum alloys also contain particles that are incoherent, either due to the transformation from the transition phase to the equilibrium phase during aging, or due to the presence of dispersoid-forming elements (e.g., chromium and manganese). The recent advent of rapid-solidification processing technology has broadened the range of aluminum alloys that can be strengthened by dispersoids. Incoherent particles, or very large coherent particles, are not sheared but are looped or bypassed by dislocations during plastic deformation. This strengthening mechanism is called Orowan hardening, and the critical resolved shear stress can be represented by:

$$\tau = \frac{0.8G \cdot b \cdot f^{1/2}}{2\pi[2(1-\nu)r]^{1/2}} \ell\text{n} \frac{1.6}{r_0} r \qquad \text{(Eq 3)}$$

where G is the shear modulus, b is the Burgers vector, ν is Poisson's ratio, and r_0 is the inner cutoff radius of the dislocation. Here the strength decreases with increasing particle size.

As mentioned above, depending on the size, spacing, and degree of coherency, the precipitates are either sheared or looped and bypassed by the dislocations during plastic deformation. The particular deformation mechanism controls the distribution of strain and can affect the fracture properties of aluminum alloys. When particles are sheared by moving dislocations, the strengthening mechanisms associated with the particles are reduced (the size of the particle on the glide plane is reduced). This results, successively, in a local decrease in resistance to further dislocation motion, concentration of slip, and destruction of the strengthening agents.

Duva and coworkers have recently derived a quantitative indicator for strain localization in age-hardened alloys containing coherent precipitates. They consider both the degradation of strength with slip and the strengthening associated with dislocation pile-ups and take the number of dislocations N that pass on a typical slip plane from the time deformation begins until local slip ends to be an indicator of slip localization. The larger the slip intensity as measured by N, the coarser the deformation expected.

By considering the strengthening effects of both the second-phase particles and the dislocation pile-ups at the grain boundaries, and assuming spherical particles of equal size, two expressions are derived for N, one corresponding to initial softening and one corresponding to initial hardening. When softening occurs initially:

$$N = f^{1/2} r_p^{1/2} r_G \frac{C_p}{C_B b} \quad \text{(Eq 4)}$$

where f is the volume fraction of the particles, r_p is the particle radius, r_G is the grain radius, b is the Burgers vector, C_p is a constant that depends on the intrinsic properties of the particles, and C_B is a constant that depends on the elastic properties of the bulk material. The slip intensity, N, increases with the square roots of both the particle volume fraction and the particle radius and increases linearly with the grain radius. When hardening occurs initially:

$$N = (2/b) r_p + (2/b) f r_G [C_p/4C_B]^2 \quad \text{(Eq 5)}$$

For this case, the slip intensity increases linearly with particle radius and volume fraction and increases as the square of the grain radius. These expressions predict that the propensity for strain localization increases during aging, up to the point when the deformation mode changes from shearing to looping and bypassing of the particles. They also predict that strain localization would increase with increasing grain size. Incoherent strengthening precipitates which are looped and bypassed by dislocations tend to homogenize deformation. Increasing the homogeneity of deformation normally improves ductility and stress-corrosion resistance, but may decrease fracture toughness and fatigue-crack-propagation resistance.

Properties of Conventional Aluminum Alloys

Chemical composition and processing control the microstructure and thus the physical, mechanical, and corrosion properties of heat treatable aluminum alloys. Since these properties can be varied greatly, it is not feasible to cover all possibilities here, and the reader is referred to *Aluminum Standards and Data*, published by the Aluminum Association. Typical mechanical properties of a selected group of heat treatable alloys frequently used in aircraft structures are listed in Table 2. The toughness levels of age-hardenable aluminum alloys generally decrease as the strength level is raised by alloying and heat treatment. This trend is illustrated in Fig. 4(a) for the 2*xxx* and 7*xxx* alloys. Once peak strength is reached for a particular alloy, some improvement in toughness may be obtained by overaging. However, the overaged strength-toughness curves always lie below the underaged strength-toughness curves, as illustrated in Fig. 4(b). This is most likely due to the presence of coarse grain-boundary precipitates and PFZ's in the overaged condition, which lead to an increased incidence of low-energy intergranular fracture.

Aluminum and its alloys may fail by intergranular stress-corrosion cracking while under stress and exposed to aggressive environments, which may include water vapor, aqueous solutions, organic liquids, and liquid metals. Stresses sufficient for the initiation and growth of stress-corrosion cracks may be well below those necessary for gross yielding. In general, aluminum alloys are most susceptible near or slightly below peak strength when planar slip and strain localization are most pronounced. Overaging is often used for improving the stress-corrosion resistance of some 7*xxx* alloys, but is impractical for 2*xxx* alloys; for the latter materials, stress corrosion must be controlled by alloy chemistry or by using a barrier which isolates the environment from the susceptible material.

Table 2. Typical mechanical properties of heat treatable aluminum alloys used in aircraft structures

Alloy and temper	Strength, ksi(a)		Elongation, % in 50 mm or 2 in.		Hardness, HB(b)	Ultimate shear strength, ksi(a)	Endurance limit, ksi(a)	Modulus of elasticity, 10^6 psi(c)
	Ultimate	Yield	1/16-in.-thick specimen	1/2-in.-diam specimen				
2024-O	27	11	20	22	47	18	13	10.6
2024-T3	70	50	18	...	120	41	20	10.6
2024-T4, T351	68	47	20	19	120	41	20	10.6
2036-T4	49	28	24	...	...	...	18	10.3
2117-T4	43	24	...	27	70	28	14	10.3
2219-O	25	11	18	...	...	...	...	10.6
2219-T81, T851	66	51	10	...	...	...	15	10.6
6061-O	18	8	25	30	30	12	9	10.0
6061-T6, T651	45	40	12	17	95	30	14	10.0
7049-T73	75	65	...	12	135	44	...	10.4
7050-T7651	80	71	...	11	...	47	...	10.4
7075-O	33	15	17	16	60	22	...	10.4
7075-T6, T651	83	73	11	11	150	48	23	10.4
7178-O	33	15	15	16	...	...	...	10.4
7178-T6,T651	88	78	10	11	...	...	...	10.4
7178-T76,T7651	83	73	...	11	...	...	...	10.3

(a) To convert ksi to MPa, multiply by 6.8948. (b) 500-kg load, 10-mm ball. (c) To convert 10^6 psi to GPa, multiply by 6.8948.

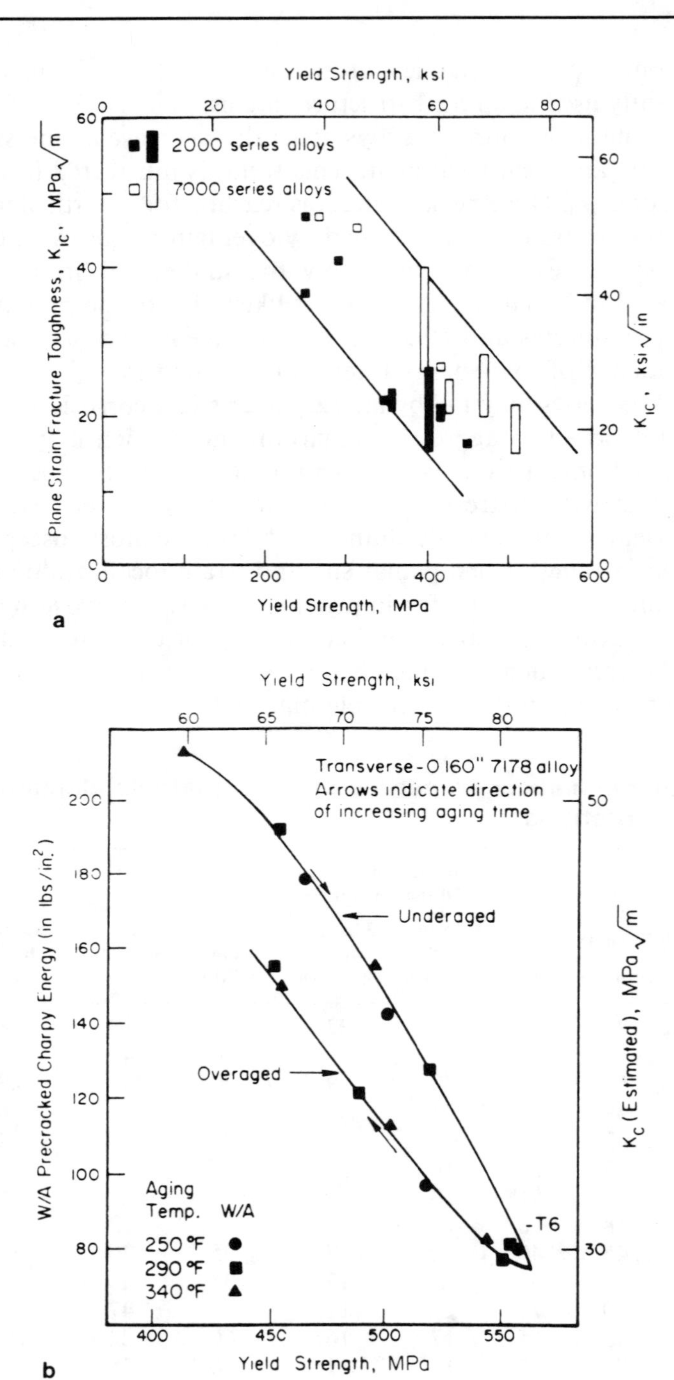

Fig. 4. The relationship between strength and toughness (a) for various 2*xxx* and 7*xxx* alloys and (b) for alloy 7178 as a function of aging. After Hahn and Rosenfield.

ADVANCED ALUMINUM AIRCRAFT ALLOYS

Low-Density Aluminum Alloys

Reducing the density of structural materials has been shown to be the most efficient way to achieve minimum weight in aerospace structures. Consequently, there has been a major effort in developing low-density aluminum alloys. Lithium is the lightest metallic element and has a significant effect on density when alloyed with aluminum, producing a 3% decrease for every 1 wt% added up to the limit of solid solubility. In addition, a concomitant increase in elastic modulus has been reported for both binary and complex alloy systems. The only other alloying element that has this combination of positive effects is beryllium.

The potential advantages of lithium additions to aluminum were first recognized by LeBaron, who obtained a patent on the Al-Cu-Li-X alloys in 1945. Subsequently, the Al-Cu-Li alloy 2020 was commercially produced and in 1957 was used on the United States Navy RA 5C Vigilante; however, concerns about the alloy's brittle behavior, and production problems, thwarted further use, and alloy 2020 was withdrawn from commercial production in the 1960's. More recently, rising fuel costs and the demand to produce more fuel-efficient and higher-performance aircraft have led to renewed interest in Al-Li alloys.

When Al-Li alloys containing more than 1 wt% lithium are quenched from the single-phase field and aged at a temperature considerably below the solvus, homogeneous precipitation of the metastable phase delta prime (Al_3Li) occurs. The metastable precipitates are coherent with the matrix, and, although they can be sheared by moving dislocations, they impede their motion and, consequently, greatly improve the strength over that of the unalloyed aluminum. During aging, heterogeneous precipitation of the equilibrium delta (AlLi) phase also occurs, normally at grain boundaries. The AlLi precipitates consume the lithium from the surrounding region and produce a lithium-depleted PFZ adjacent to the grain boundary. The three microstructural features just described are shown in the bright-field transmission electron micrograph of Fig. 5(a), and sheared Al_3Li is shown in the dark-field TEM of Fig. 5(b).

As described previously, precipitate shearing reduces the strengthening and leads to strain localization, as shown in Fig. 6(a). The PFZ adjacent to the grain boundary is also weaker than the matrix and can likewise be a region of concentrated slip (Fig. 6b). Both types of concentrated slip produce stress concentrations at grain boundaries and at grain-boundary triple junctions and low-energy intergranular or intersubgranular fracture. Grain-boundary precipitates, which are the cause of the PFZ's, also have a major detrimental effect on fracture resistance since they are the sites for microvoid nucleation, which occurs at small macroscopic strains when deformation is localized in the PFZ. The two failure modes are shown schematically in Fig. 7(a) and by the scanning electron micrograph (SEM) of Fig. 7(b). In addition to these types of strain-localization problems, the old 2020 alloy had a large volume fraction of coarse constituents and intermetallic particles. These

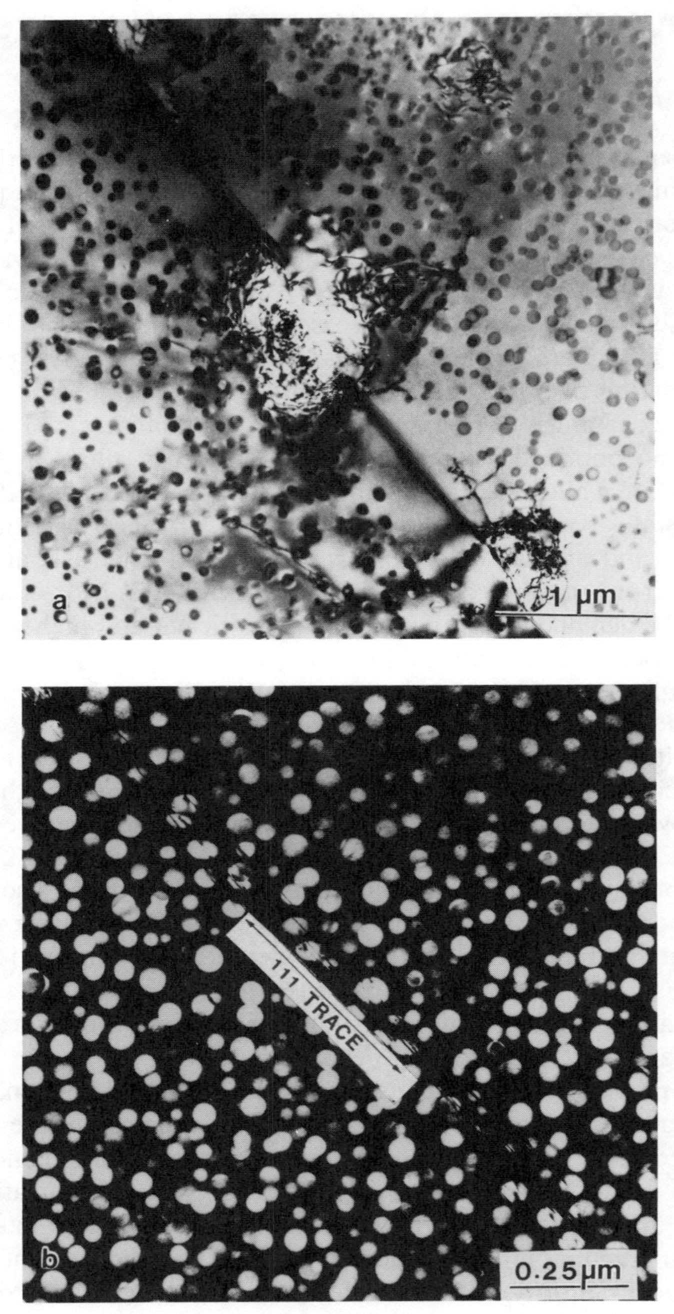

(a) Bright-field image showing Al_3Li in matrix, AlLi at grain boundary and PFZ. (b) Dark field image showing sheared Al_3Li that occurs during deformation.

Fig. 5. Transmission electron micrographs of an Al-Li alloy aged 24 h at 363 K. Courtesy of W.A. Cassada.

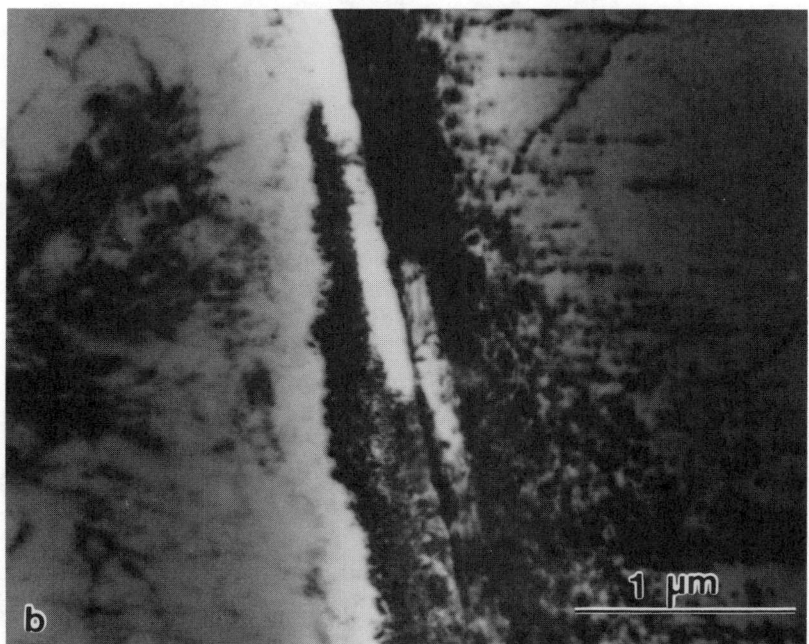

(a) Underaged, showing localized deformation in the matrix. (b) Overaged, showing localized deformation in the PFZ.

Fig. 6. Transmission electron micrographs of deformed Al-Li

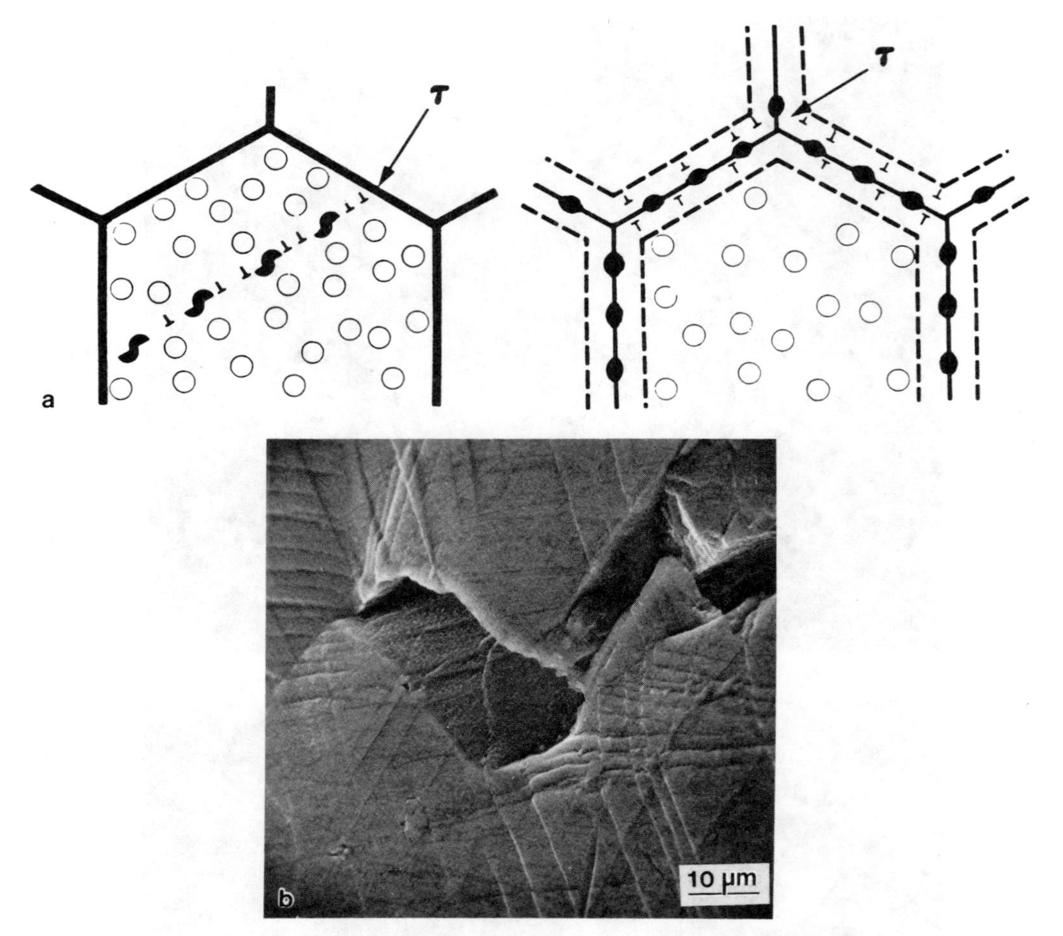

Fig. 7. (a) Schematic representation of deformation leading to fracture in Al-Li alloys and (b) scanning electron micrograph showing intense planar slip and intergranular fracture in an Al-Li binary alloy

coarse particles have been shown to have an adverse effect on fracture toughness, and numerous investigators have demonstrated that decreasing the iron content of high-strength aluminum alloys decreases the volume fraction of large insoluble phases and increase the toughness.

Various modifications in alloy chemistry and fabrication techniques have been used in an attempt to improve the ductility and fracture toughness of Al-Li alloys while maintaining high strength. Essentially, three problems have been addressed: (1) the strain localization in the matrix; (2) coarse precipitates along grain boundaries, their associated PFZ, and strain localization in the PFZ; and (3) the adverse effects of coarse constituents and intermetallic particles. Strain localization due to particle shearing can be minimized by overaging to produce incoherent precipitates, by reducing the grain size, or by adding alloying elements which promote homogeneous deformation. Overaging is impractical in Al-Li alloys because the coherency strains of the Al_3Li strengthening precipitates are small, and coherency loss does not occur during aging. In addition, precipitation of the equilibrium AlLi, or other equilibrium phases, at grain boundaries results in problem 2 above—i.e., the formation of wide PFZ's at grain boundaries.

Since the deformation behavior of age-hardened alloys is determined by the nature of the interaction of dislocations with the strengthening precipitates, it can be modified by changes in the type, size, coherency, and distribution of the precipitates present. Consequently, alloying is a potentially attractive method of eliminating strain-localization effects in Al-Li alloys. Control of alloy chemistry may also be used to eliminate problem 3 above—i.e., the effects of coarse constituents and intermetallic particles. Cassada, Shiflet, and Starke showed that the ductility of Al-Li alloys could be significantly increased by the addition of an element that formed small incoherent precipitates which dispersed slip. They added 0.2 wt% germanium to an Al-2Li alloy, and, during aging, the germanium coprecipitated with the Al_3Li phase, forming very small rod-shape particles (see Fig. 8a). During deformation these small incoherent particles were looped and bypassed by dislocations (Fig. 8b and c), thus minimizing the strain localization associated with the shearable Al_3Li particles. The tensile properties of the alloys studied are given in Fig. 9 as a function of aging time at 473 K. SEM's of the fracture surfaces of the Al-Li binary and the Al-2Li-0.2Ge alloy (Fig. 10) show that homogenizing the deformation decreased the incidence of intergranular fracture and was responsible for the increase in elongation to failure from 3 to 10%.

A reduction in grain size decreases the slip length and reduces the stress concentrations at grain boundaries and grain-boundary triple junctions when deformation is localized within the matrix or PFZ. Small grains also enhance multiple slip at low strains, producing more homogeneous deformation. Both of these effects have been shown to increase the ductility in age-hardened aluminum alloys containing coherent matrix precipitates and PFZ's.

Many programs which focused on the development of Al-Li-X alloys utilized the basic principles mentioned above. A variety of production methods including rapid solidification and powder metallurgy consolidation, mechanical alloying, and ingot casting, have been used. Rapid solidification and mechanical alloying were used as a means of reducing the grain size and extending solid solubility; but simply reducing the grain size did not have a significant effect on ductility, and the advantages of extended solubility were negated by elevated-temperature excursions during processing. Consequently, once problems associated with ingot casting were overcome, ingot casting appeared the most feasible method for the production of large plate and extrusion products and is being used for the alloys that are currently in commercial production. However, both rapid solidification and mechanical alloying may offer advantages for other types of alloys (for example, those that contain elements that are insoluble at high temperatures) or for special product forms (such as large forgings), and this will be discussed later.

The effects of a variety of alloying additions on the microstructure and properties of Al-Li alloys have been studied, as well as the effects of certain types of impurities. Iron and silicon contents have been kept to a minimum in order to

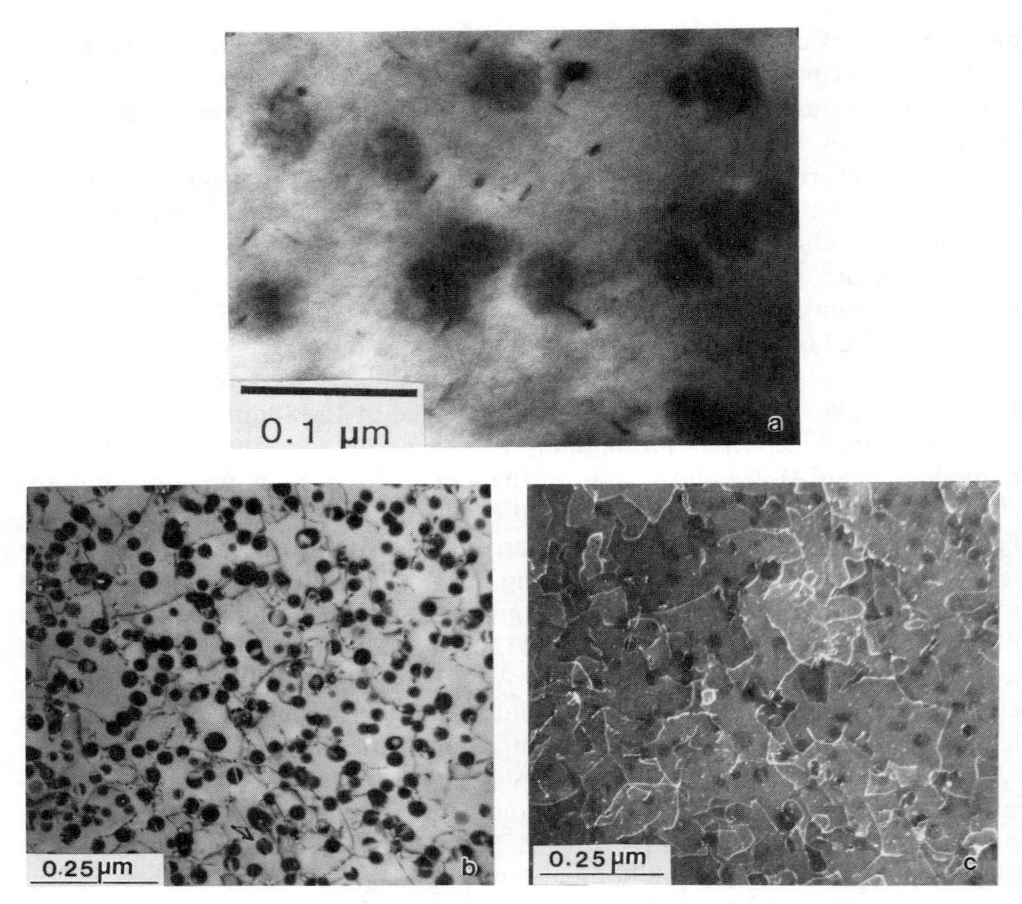

(a) Bright-field image of rod-shape germanium particles along with Al_3Li precipitates in sample aged 9 h at 473 K. (b) Bright-field image of sample aged 48 h at 473 K and stretched 2%. (c) Dark-field image of (b) showing homogeneous distribution of dislocations in matrix along with small germanium particles.

Fig. 8. Transmission electron micrographs of an Al-2Li-0.2Ge alloy

reduce the detrimental effects of constituent particles. The dispersoid-forming element zirconium has been shown to be preferred over both manganese and chromium when fracture toughness, quench sensitivity, and inhibition of recrystallization are considered, and has been selected for use in the new Al-Li-X alloys being considered for commercialization. The alloying elements that form incoherent precipitates, such as the germanium addition mentioned, may homogenize slip but when present in small quantities have no significant effect on strength, and these simple alloys do not meet the strength requirements of structural applications. If dispersoid elements are added in large quantities, they form very coarse particles during ingot casting and have an adverse effect on fracture behavior. Consequently, most programs are focused on alloying additions that coprecipitate with the Al_3Li phase, and, depending on the processing, precipitate up to the grain boundaries,

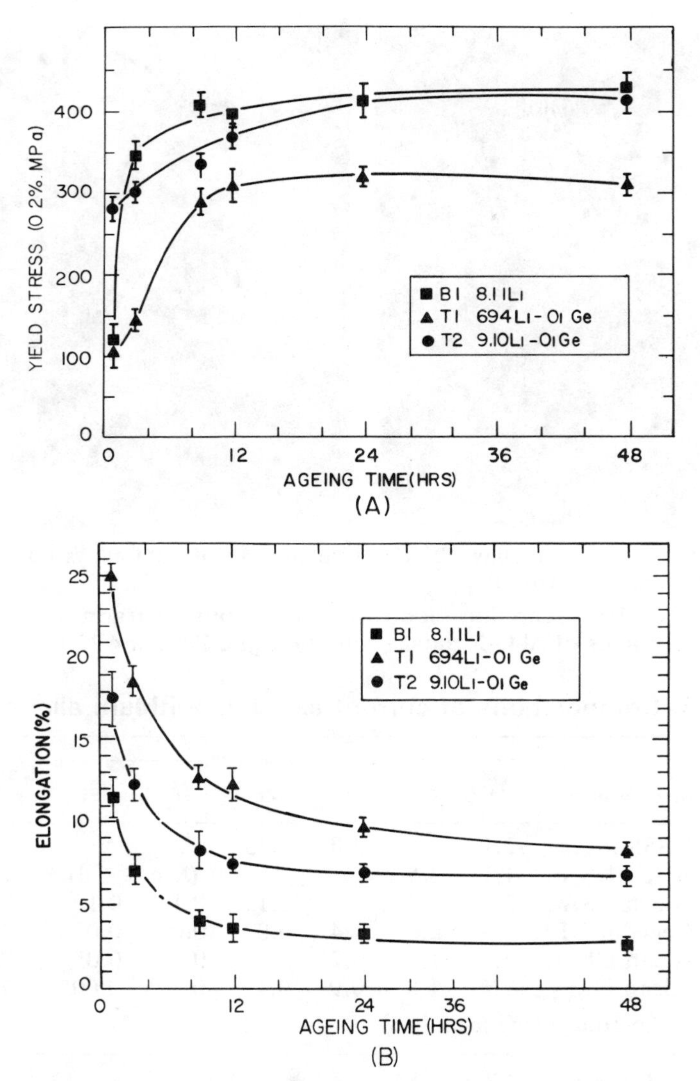

(a) Yield stress plotted as a function of aging time. (b) Elongation corresponding with (a) plotted as a function of aging time.

Fig. 9. Variation in the tensile properties of three Al-Li-X alloys with aging time at 473 K

thus minimizing the PFZ. Copper and magnesium have been shown to be the most attractive additions. The compositions of old and new commercial Al-Li-X alloys are given in Table 3.

Copper has a very positive effect on the strength of Al-Li alloys and, depending on the composition, may coprecipitate in the matrix with the Al_3Li phase as Al_2CuLi (T1) and/or Al_2Cu (theta prime). Often, as is the case for the new alloy 2090, all three phases can occur as shown in the TEM micrograph of Fig. 11. The partially incoherent T1 and theta prime phases may reduce the extent of strain localization; however, their effectiveness in homogenizing strain depends on their

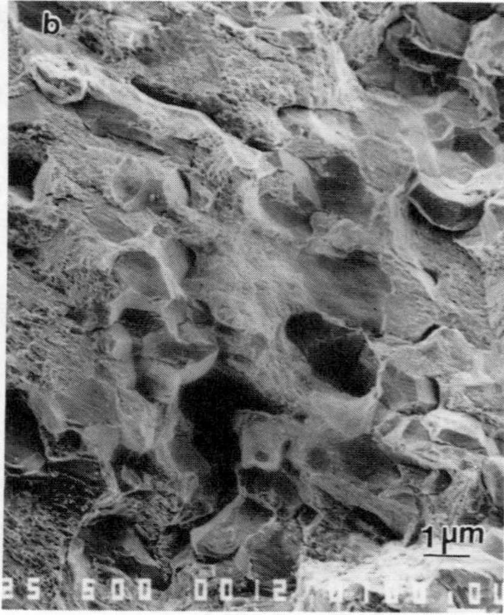

(a) Alloy B1. (b) Alloy T2. The yield strength of both alloys was approximately 410 MPa.

Fig. 10. Scanning electron micrographs of fracture surfaces of Al-Li-X tensile samples aged 24 h at 473 K

Table 3. Compositions of current aluminum-lithium alloys

Alloy	Manufacturer	Composition, wt% Li	Cu	Mg	Zr	Fe	Si	SG(a), $g \cdot cm^{-3}$
1420	USSR	2.1	...	5.0	0.15	...	...	2.50
2020	Alcoa USA	1.1	4.5	...	...	(Mn + Cd)		2.71
2090	Alcoa USA	2.2	2.7	...	0.12	0.12	0.10	2.60
2091	Cegedur RF	2.0	2.0	1.4	0.07	0.04	0.03	2.58
8090	Alcan UK	2.5	1.3	0.7	0.12	0.10	0.05	2.53
8091	Alcan UK	2.6	1.8	0.9	0.12	0.10	0.005	2.54

(a) SG = specific gravity (density).

size and distribution. These parameters are normally controlled by the amount of deformation prior to aging and by the aging temperature and time.

The primary role of deformation is to increase the dislocation density and thereby the number of nucleating sites for heterogeneous precipitation. The effectiveness of the process is related to the interfacial strains of the precipitates, and the process is highly effective for precipitates having large interfacial strains—e.g., T1 and theta prime. Figure 12(a) shows the effects of different amounts of deformation prior to aging on the number density of the T1 precipitates formed during aging for various times at 190 °C. Since strength is related to the precipitate structure, there is a corresponding effect on strength, as shown in Fig. 12(b). The significance of this data relates to commercial processing and different product forms, since a nonuniform distribution of deformation prior to aging can result in wide variances in strength within a product. Some product forms, such as com-

A complex precipitate structure is shown that includes Al_2Cu, Al_2CuLi, Al_3Li, and Al_3Zr.

Fig. 11. Bright-field transmission electron micrograph of Al-2.4Li-2.43Cu-0.18Zr alloy solution heat treated at 773 K and aged 24 h at 463 K. Courtesy of W.A. Cassada.

plex forgings and those formed by superplastic deformation, are not conducive to cold deformation prior to aging. Consequently, there is interest in examining alloying additions which may aid the nucleation of the strengthening precipitates similar to the effect of dislocations. Recent studies by Blackburn and Starke have shown that small additions of indium to 2090 can aid the nucleation of both T1 and theta prime, having an effect similar to a 3-to-4% stretch when compared with the indium-free alloy.

Magnesium additions to Al-Li-Cu alloys may provide a component of solid-solution strengthening, but in the quantities added to current commercial alloys magnesium coprecipitates with Al_3Li as S″ or S′ (Al_2CuMg). Since these metastable phases have large-coherency interfacial strains, their sizes and distributions are also sensitive to processing, as can be observed in Fig. 13. In the 8090 alloy the S′ phase replaces the T1 that is observed in 2090. However, in the 8091 alloy both

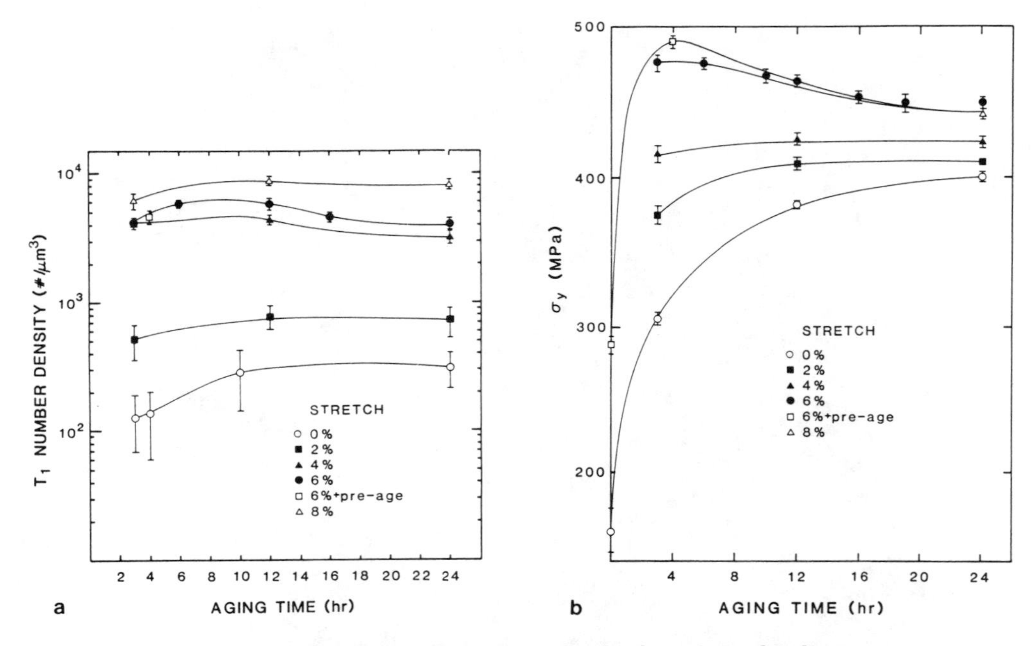

Fig. 12. (a) T1 number density as a function of aging time at 463 K for various degrees of stretch prior to aging and (b) corresponding yield strength vs. aging time for an Al-2.4Li-2.4Cu-0.18Zr alloy. Courtesy of W.A. Cassada.

T1 and S′ are present along with the Al_3Li precipitates. Zirconium is present in all the new Al-Li-X alloys and forms the metastable Al_3Zr dispersoid during solidification and/or during ingot preheating. This phase is very effective in inhibiting recrystallization and grain growth during solutionizing treatments. However, the new commercial alloys are also being produced in the recrystallized condition for thin-gage products in order to minimize anisotropy and maximize damage tolerance.

Unfortunately, grain-boundary precipitates form in addition to the beneficial strengthening precipitates in Al-Li-Cu and Al-Li-Cu-Mg alloys. These precipitates may form during quenching of thick products and during aging of all products. A variety of grain-boundary precipitates have been observed. In addition to the AlLi delta phase, the Al_6CuLi_3 (T2) has been observed as the dominant grain-boundary precipitate in both 2090 and 8090. The T2 precipitate exhibits icosahedral symmetry and has been observed to coarsen along the grain boundaries during aging, as shown in Fig. 14. Because of this precipitate morphology, and the fact that a very small volume fraction may cover a large amount of grain-boundary area, the T2 phase has a very detrimental effect on the fracture behavior of Al-Li-Cu-X alloys. Other grain-boundary precipitates—e.g., the R and S phases—have also been identified in age-hardened Al-Li-X alloys. Since the precipitation of the grain-boundary phases is competitive with the precipitation of the strength-

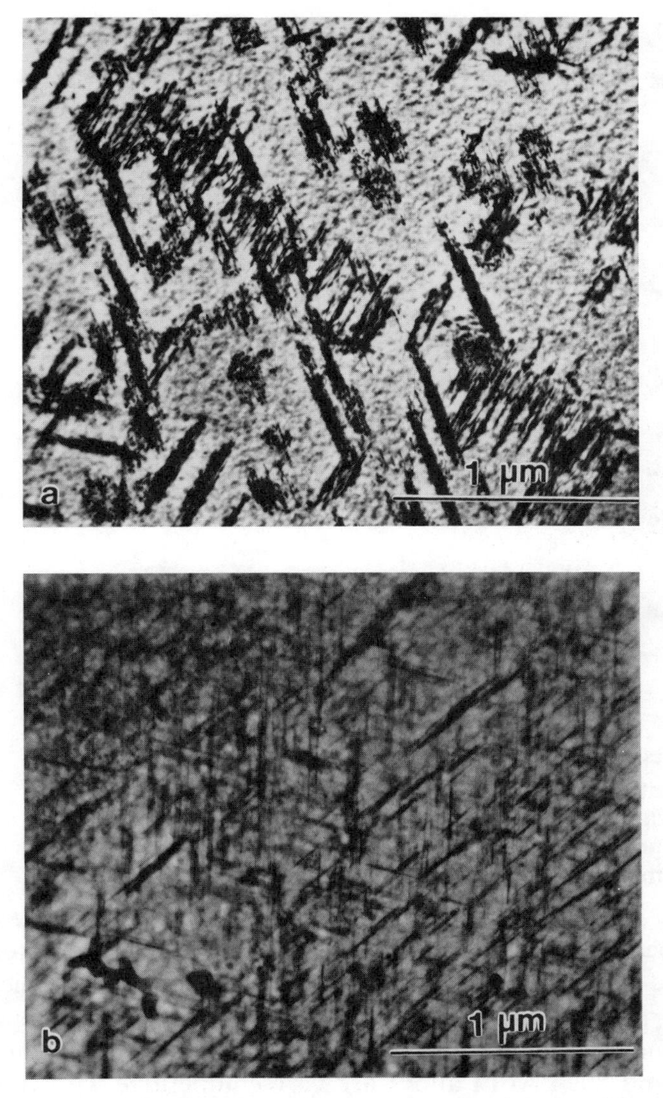

Fig. 13. Transmission electron micrographs showing the distribution of Al_2CuMg (S′) in an Al-2.7Li-1.5Cu-1Mg-0.15Zr alloy aged 310 h at 444 K and (a) not stretched before aging; (b) stretched 2% before aging

ening precipitates in the matrix during aging, they may be minimized by selective control of the aging time and temperature and amount of deformation prior to aging.

The elastic moduli of the commercial Al-Li-X alloys listed in Table 3 are shown in Fig. 15 as a function of the lithium content of the alloy, and mechanical-property data for the damage-tolerant variants are given in Table 4. Figure 16 compares the fatigue-crack-growth rates in two directions for alloy 8090 with the rates for both

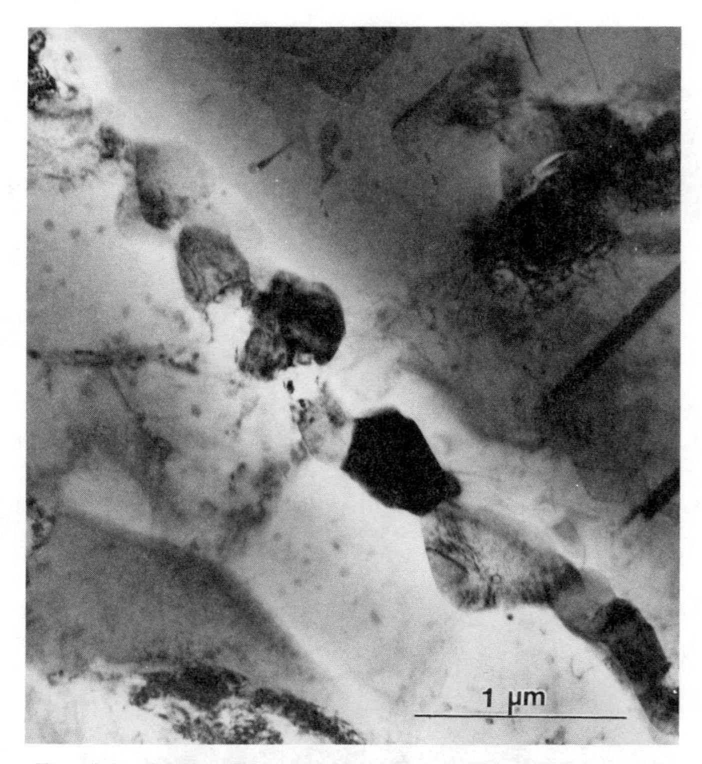

Fig. 14. Bright-field transmission electron micrograph of Al-2.4Li-2.4Cu-0.18Zr alloy aged 24 h at 573 K, showing icosahedral (T2) phase along a high-angle grain boundary and the T1 (Al_2CuLi) phase in the matrix. Courtesy of W.A. Cassada.

2024 and 2124. Figure 17 compares the fatigue-crack-growth rate of alloy 2090 tested under spectrum loading in both laboratory air and salt water with that of alloy 7075 tested in laboratory air. The crack-growth resistance of alloy 2090 is considerably better than that of alloy 7075, and 2090 is also less susceptible to corrosion fatigue. These data indicate that the properties currently being obtained on the new 2090 and 8090 Al-Li alloys are very competitive with those aluminum alloys now being used for aircraft construction, with the additional benefit of having significantly lower densities and higher elastic moduli.

There are two recent low-density development programs which utilize special manufacturing methods that appear to have special promise. One involves production of Al-Li-Be alloys by rapid solidification and the other involves production of a non-heat-treatable Al-Li-Mg alloy by mechanical alloying. As mentioned previously, the only two elemental additions to aluminum that have the effect of simultaneously decreasing density and increasing the elastic modulus by significant amounts are lithium and beryllium. Beryllium has limited solubility in aluminum (0.03 wt%), and this results in massive segregation of the Be-rich phase upon solidification under standard conditions, as shown in the upper optical micrograph

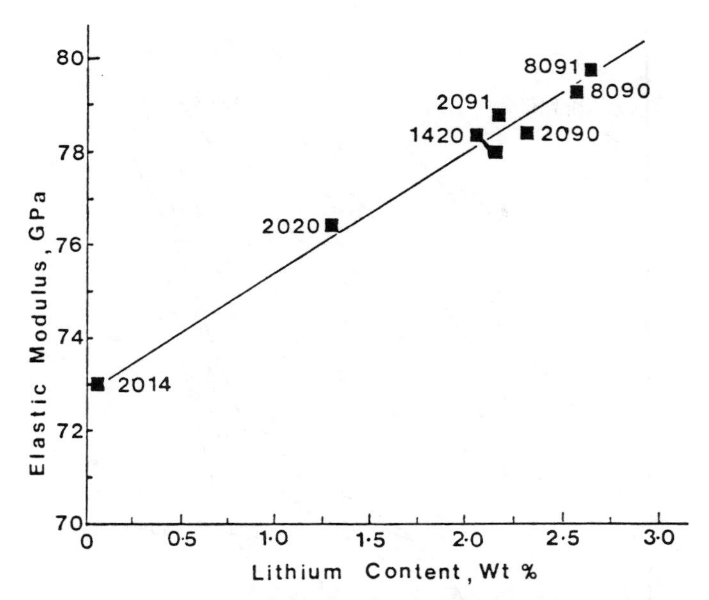

Fig. 15. Effect of lithium content on elastic modulus for Al-Li alloys. Courtesy of C. Peel.

Table 4. Plane-stress fracture toughness of damage-tolerant variants. From Peel.

Alloy	Treatment	0.2% PS, MPa	UTS, MPa	$K_{apparent}$, $MPa \cdot m^{1/2}$	K_c, $MPa \cdot m^{1/2}$	Grain structure
2090	T3	255	345	73	104	...
	T4	225	350	67	94	...
2090	T3 stabilized..	360	410	80	107	...
	T4 stabilized..	325	420	85	116	...
2091	T3	325	425	100	148	Recrystallized
8090	T3	285	365	81	126	Unrecrystallized
	T4	240	375	85	97	Unrecrystallized
8090	T4 stabilized..	345	425	77	102	Unrecrystallized
	T4 stabilized..	325	405	92	126	Recrystallized

of Fig. 18. The rapid-solidification method eliminates segregation of beryllium and greatly refines the microstructure of the alloy, as shown in the lower-micrograph in Fig. 18.

In aluminum alloys containing lithium and beryllium produced by rapid solidification, at least two phases are expected to precipitate. The Al_3Li phase should precipitate homogeneously and provide a significant contribution to strength, and the Be-rich phase should precipitate in the form of a dispersoid within the Al-Li matrix. The Be-rich phase will contribute to the mechanical properties of the alloy

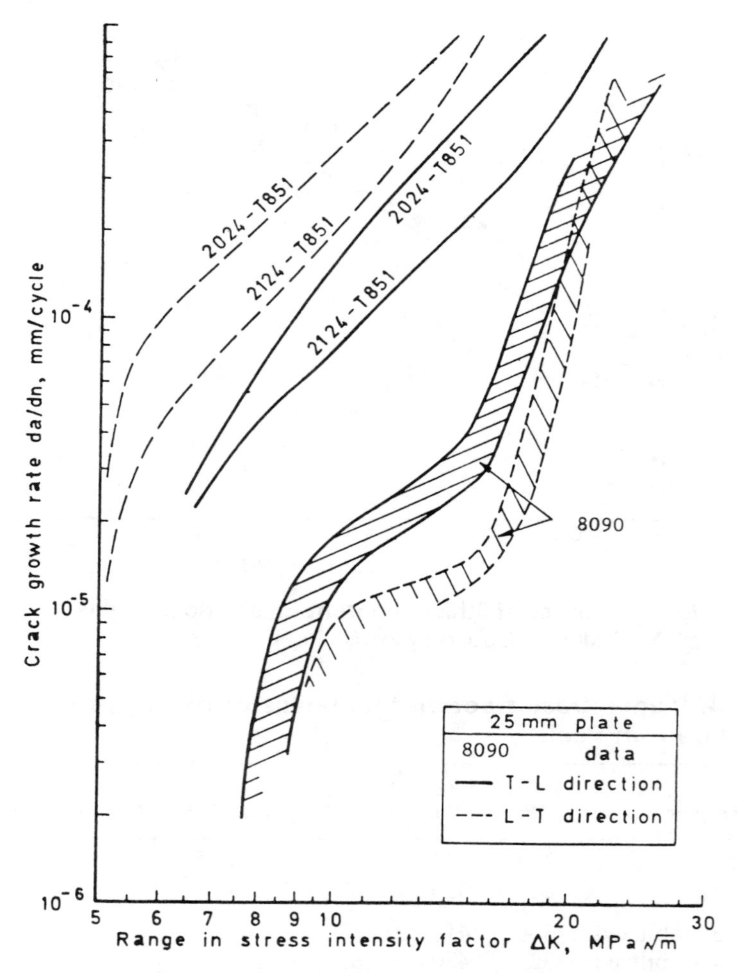

Fig. 16. Fatigue-crack-growth rate, da/dN, vs. stress-intensity range for two orientations of 8090 compared with similar orientations of 2024-T851 and 2124-T851. Courtesy of C. Peel.

in two different ways: (1) by adding a dispersion-strengthening component to the yield strength and (2) by acting to disperse dislocation glide, which will allow the Al_3Li to contribute more of its full potential to the alloy strength, and which will enhance the resistance to premature fracture. Figures 19(a) and (b) are transmission electron micrographs of an Al-3.6Li-9.8Be alloy showing the size and distribution of the Be-rich dispersoids and the Al_3Li phase in material that was prepared by rapid solidification. Table 5 compares the specific moduli and strengths of two Al-Li-Be alloys with those of aluminum alloy 7075. The structural weight savings that may be expected by using Al-Li-Be alloys are illustrated schematically in Fig. 20 using 7075 as a baseline. The Al-Li-Cu-X alloys are also shown for comparison.

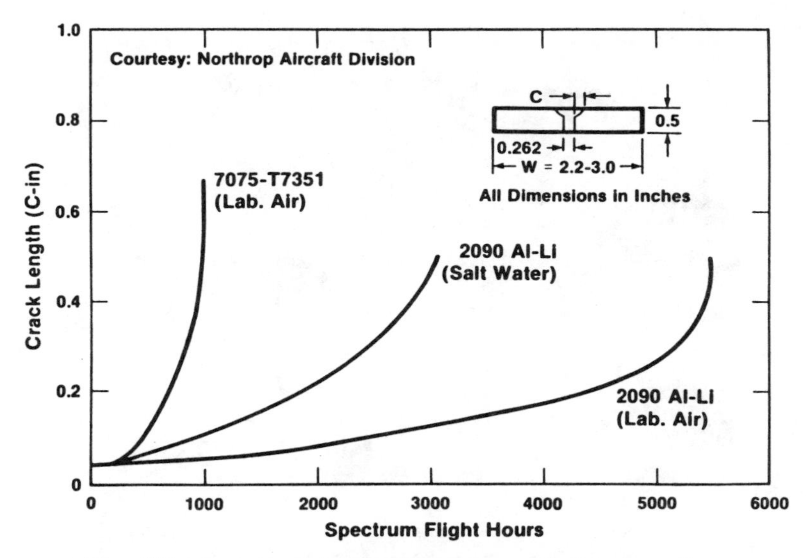

Fig. 17. A comparison of spectrum fatigue performance for alloy 2090 in salt water and laboratory air and alloy 7075-T7351 tested in laboratory air. Courtesy of Northrop Aircraft Division.

Table 5. Comparison of specific moduli and strengths of RSP Al-Li-Be with 7075-T6. From Lewis and Starke.

Alloy	E, kNM/g	TYS, NM/g	UTS, NM/g	ε, %	Density, Mg/m^3
Al-3.6Li-9.8Be ...	41.1	206	218	2.3	2.34
Al-2.5Li-11.0Be ..	38.6	177	207	5.57	2.42
Al-7075-T6.......	25.5	179	204	11.0	2.81

As mentioned earlier, deformation prior to aging is required in order to obtain the optimum distribution of the S′ and T1 precipitates necessary for maximizing mechanical properties. This is often difficult in certain product forms (e.g., forgings). However, mechanical alloying offers an alternative to ingot processing, and, because such products are strengthened by a very fine substructure that is stabilized by small carbides, such materials do not require quenching from high temperatures, a deformation step, or aging prior to use. Mechanical alloying is a high-energy milling, powder metallurgy process which involves repetitive plastic deformation, cold welding, and fracture of the powder particles of the initial raw materials followed by powder metallurgy consolidation.

A mechanically alloyed material designated IN-905XL, with a composition of Al-1.5Li-4Mg-0.4O-1.2C (wt%) has been developed for forged parts. This alloy is 8% lighter and 10% stiffer than 7075-T73. Properties in the transverse direction

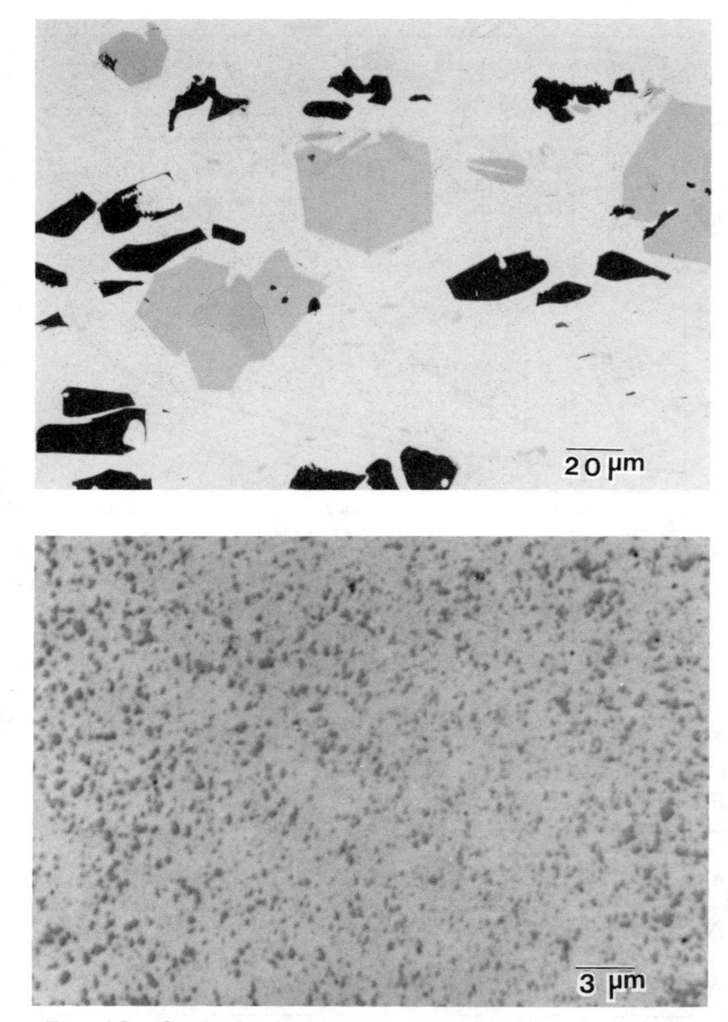

Fig. 18. Optical micrographs comparing the sizes of the Be phase in (above) an Al-Be alloy that was direct chill ingot cast and extruded and (below) an Al-Be alloy that was prepared by rapid solidification planar flow casting and extruded. Courtesy of R.E. Lewis.

are only slightly lower than those in the longitudinal direction. The corrosion resistance of the MA alloy is substantially better than those of competitive materials produced by ingot metallurgy methods. Properties of IN-905XL forgings are compared with those of 7075-T73 in Table 6.

High-Temperature Aluminum Alloys

The common aluminum alloys currently being used by the aircraft industry for high-temperature applications are based on the Al-Cu system and include 2219 (Al-

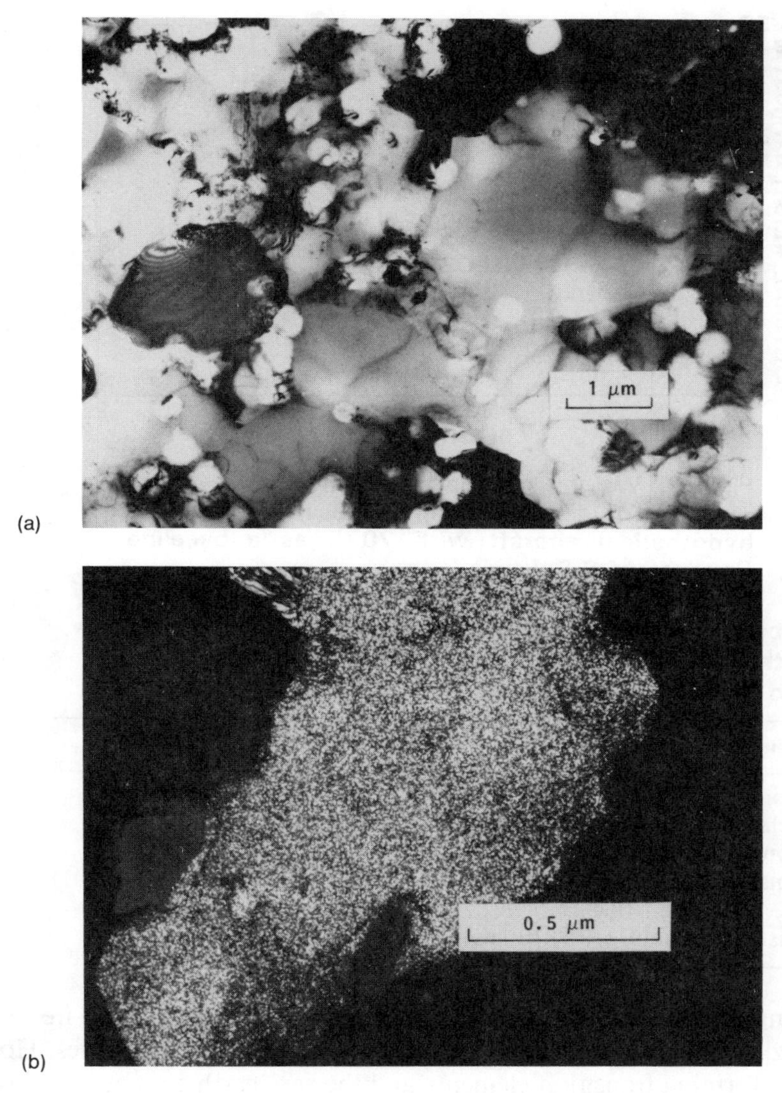

(a) Bright-field image showing the Be dispersoids. (b) Dark-field image showing Al_3Li.

Fig. 19. TEM's of an Al-3.6Li-9.8Be alloy in the peak aged condition. Courtesy of R.E. Lewis.

Cu-Mn) and 2618 (Al-Cu-Mg-Fe-Ni). The transition elements form intermetallic phases which are stable at elevated temperatures and hinder grain-boundary sliding. However, both alloys derive their strength from coherent and partially coherent intermediate precipitates which coarsen rapidly and transform to equilibrium precipitates at temperatures above 450 K. The rate of coarsening is controlled either by atom transfer across the precipitate/matrix interface or by volume or bulk diffusion. Low values of interfacial energy, solid solubility, and solute diffusivity are desirable for low coarsening rates.

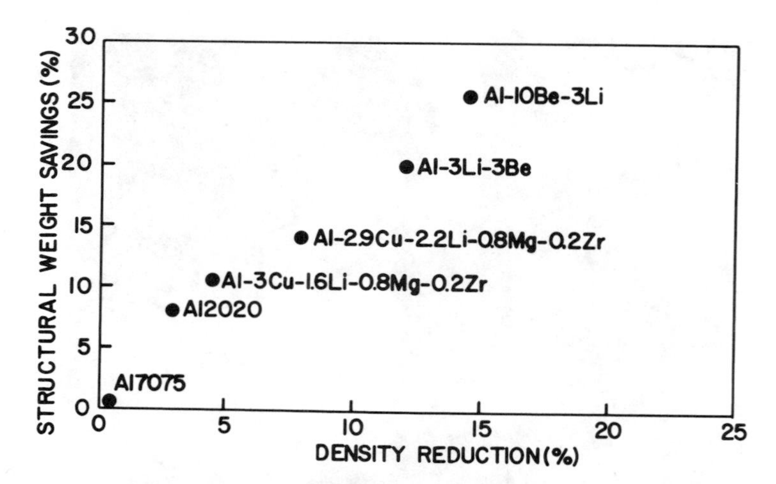

Fig. 20. Weight savings as a function of density reduction obtained by using various new alloys for a hypothetical aircraft with 7075 as a baseline. Courtesy of R.E. Lewis.

Table 6. Properties of IN 905XL forgings compared with alloy 7075-T73. From Schelling, Gilman, and Donachie.

Property	IN-905XL	7075-T73
Ultimate tensile strength, MPa (ksi)	517 (75)	503 (73)
Yield strength, 0.2%, MPa (ksi)	448 (65)	434 (63)
Elongation, %	9	13
Density, Mg/m^3 (lb/in.^{3})	258 (0.093)	2.80 (0.101)
K_{Ic}, MPa·m$^{1/2}$ (ksi·in.$^{1/2}$)	29.7 (27)	28.6 (26)
Elastic modulus, GPa (10^6 psi)	78 (11.3)	69 (10)

The alloying elements normally used for precipitation hardening—i.e., copper, magnesium, zinc, and lithium—have high solubilities and diffusion rates. However, the dispersoid-forming transition elements and the rare earth cerium have low solubilities and low diffusion rates. A high volume fraction of small dispersoids formed by these additions can give adequate high-temperature strength through Orowan strengthening and by stabilizing the substructure developed during processing. The dispersoids can also slow down all three contributions to steady-state creep: grain-boundary sliding, diffusional creep, and dislocation creep. Large additions of the transition elements can significantly increase the elastic modulus of aluminum, making the dispersion-hardened alloys especially attractive for stiffness-critical, high-temperature applications. Unfortunately, the limited solid solubility makes it impossible to obtain a fine dispersion of closely spaced incoherent particles using normal ingot casting procedures. The inherently low ingot solidification rates enhance the precipitation of coarse particles which lower ductility and fracture-toughness values. However, rapid solidification avoids the thermodynamic limitations normally imposed on aluminum–transition metal alloys, and very small intermetallics, as well as fine grains and a homogeneous distribution of alloying elements, may be developed using this process.

Recently there have been a number of programs directed toward developing high-temperature aluminum alloys that would be competitive with titanium on a density-compensated basis. Iron, molybdenum, vanadium, zirconium, silicon, and cerium have been chosen as alloying additions to aluminum because of their high liquid solubilities, limited solid solubilities, and low diffusivities. Both planar flow casting and atomized powder are being used in conjunction with powder metallurgy consolidation methods, and compositions have been selected which result in high volume fractions, 20 to 30% or more, of fine intermetallics. These intermetallics also have a high elastic modulus, and therefore also improve the stiffness of aluminum. The compositions of some of these intermetallics and their coarsening rates at 698 K are listed in Table 7. Figure 21 shows a band of strength-at-temperature performance for Al-Fe-Ce compared with various alloys processed using direct chill casting. At 232 °C the Al-Fe-Ce materials show strength capabilities about double that of I/M alloy 2219. In addition, the microstructures of these alloys remain relatively stable even after 100 h at 316 °C, thereby retaining room-temperature strength (Fig. 22).

In all practical alloys of the aluminum–transition metal systems, rapid solidification produces either cellular-dendritic, microeutectic, or icosahedral solidification structures. Planar flow casting has been used to produce a series of Al-Fe-V-Si alloys containing a very fine uniform distribution of small, nearly spherical $Al_{12}(Fe,V)_3Si$ (silicide) dispersoids (Fig. 23). The very low coarsening rates of these silicides (Table 7), and their ability to stabilize the substructure developed

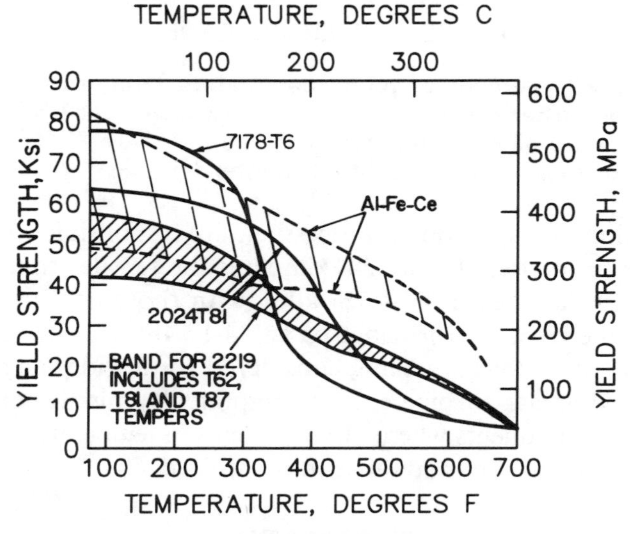

Fig. 21. Yield strength after 100 h at various temperatures for the new high-temperature RSP Al-Fe-Ce alloys compared with various conventional I/M aluminum alloys. Courtesy of Alcoa Technical Center.

Table 7. Coarsening rates at 698 K for various aluminum intermetallics. From Gilman, Skinner, Zedalis, and Raybould.

Intermetallic	Coarsening rate
$Al_{12}(Fe,V)_3Si$ (Fe:V ratio, 10; a = 1.260 nm)	8.4×10^{-27} m^3h^{-1}
$Al_{12}(Fe,V)_3Si$ (Fe:V ratio, 5; a = 1.260 nm)	2.9×10^{-26} m^3h^{-1}
Al_3Fe	2.2×10^{-23} m^3h^{-1}
$Al_3(Fe,Mo,V)$	1.7×10^{-23} m^3h^{-1}
Al_8Fe_4Ce	4.2×10^{-23} m^3h^{-1}

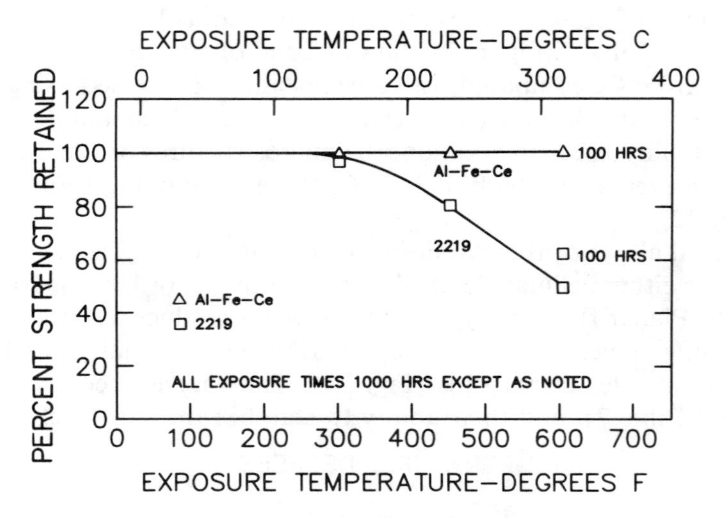

Fig. 22. Percent of room-temperature strength retained after elevated-temperature exposure for RSP Al-Fe-Ce alloys compared with the I/M alloy 2219. Courtesy of Alcoa Technical Center.

during consolidation and subsequent processing, are responsible for their superior high-temperature properties. Figure 24 compares the tensile strengths at various temperatures of an Al-8.5Fe-1.3V-1.7Si alloy with those of alloy 2014-T6, and Fig. 25 compares the creep performance of an Al-12.4Fe-1.2V-2.3Si alloy to alloy 2219-T851. These new alloys are being considered for a wide range of aerospace applications, including gas-turbine engine components, missile components, and airframe structural components where improvements in temperature capability, corrosion resistance, and elastic modulus can be utilized.

SUMMARY

Aluminum alloys have been the choice for structural components for aircraft since the 1930's. However, they have recently been subject to competition from carbon-

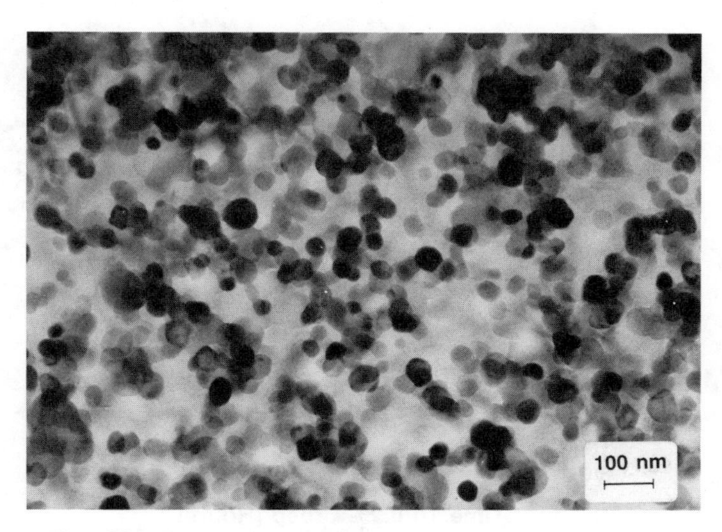

Fig. 23. Transmission electron micrograph of an as-extruded Al-8.5Fe-1.3V-1.7Si alloy prepared by Allied-Signal using rapid solidification and consolidation technology, showing the very fine dispersion of $Al_{12}(Fe,V)_3Si$ precipitates. Courtesy of P.S. Gilman.

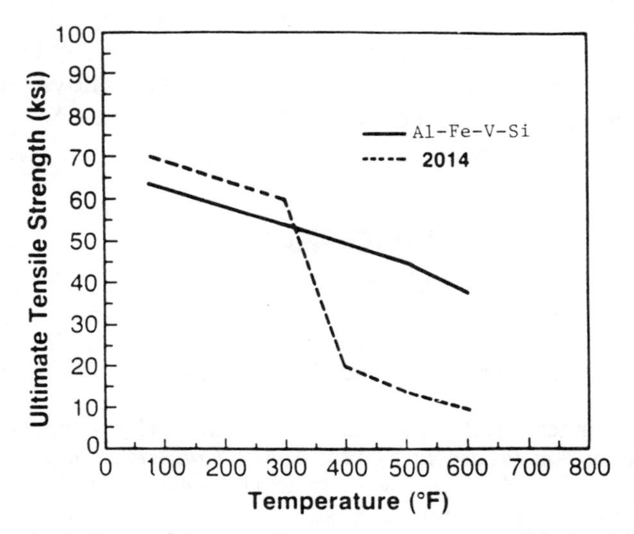

Fig. 24. Tensile strength vs. temperature for the Al-8.5Fe-1.3V-1.7Si alloy compared with 2014-T6. Courtesy of P.S. Gilman.

fiber-reinforced composites for low-density, high-stiffness applications and from titanium alloys for high-temperature applications. This competition has led to the development of new low-density Al-Li-X alloys that have better stiffness and damage tolerances than those of conventional aluminum materials. Although not competitive with the epoxy-matrix composites for all applications, they offer improvements over conventional aluminum alloys and cost advantages over com-

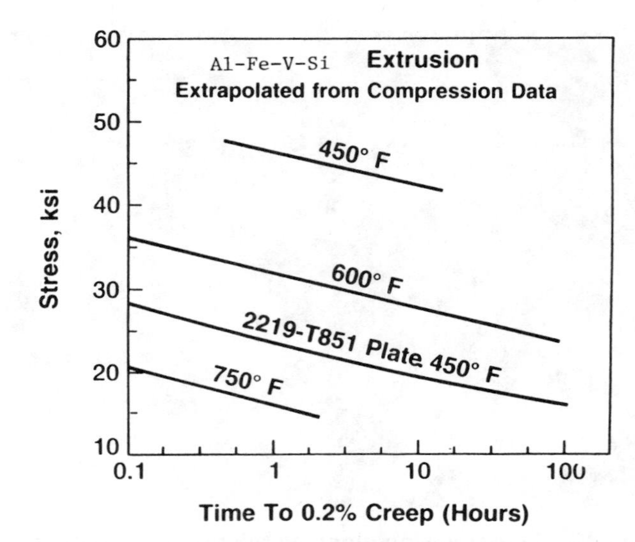

Fig. 25. Creep behavior of an RSP Al-12.4Fe-1.2V-2.3Si alloy at various temperatures compared with 2219-T851. Courtesy of P.S. Gilman.

posites because parts can be fabricated using conventional methods. The new powder alloys can provide significant improvements in high-temperature performance, compared with conventional aluminum alloys, and may one day be used to replace titanium alloys for applications up to 500 °C. These new alloys serve as an illustration of the use of physical metallurgy principles in the development of new materials for specific applications.

BIBLIOGRAPHY

Aluminum Standards and Data 1978 Metric SI, Aluminum Association, Washington

F.R. Billman and R.H. Graham, in *Proceedings of the Fourth Conference on Rapid Solidification, Processing, Principles, and Technologies*, Univ. of CA at Santa Barbara, 1987 (in press)

P.S. Gilman, D.J. Skinner, M. Zedalis, and D. Raybould, "High Temperature Aluminum Alloys: Applications," in *Proceedings of the Fourth Conference on Rapid Solidification, Processing, Principles, and Technologies*, 15-18 December, 1986, Univ. of CA at Santa Barbara, edited by R. Mehrabian (in press)

W.A. Cassada, G.J. Shiflet, and E.A. Starke, Jr., *Acta Met.*, Vol 34, 1986, p. 367

J.S. Ekvall, J.E. Rhodes, and G.G. Wald, in *Design of Fatigue and Fracture Resistant Structures*, ASTM STP 761, 1982, p. 328

G.T. Hahn and A.R. Rosenfield, *Met. Trans. A*, Vol 6A, 1975, p. 653

John E. Hatch, (Ed.), *Aluminum Properties and Physical Metallurgy*, American Society for Metals, Metals Park, Ohio, 1984

A. Kelly and R.B. Nicholson, in *Progress in Materials Science*, Vol 10, edited by Bruce Chalmers, Pergamon Press, New York, 1963, p. 151

R.E. Lewis and E.A. Starke, Jr., in *Mechanical Behavior of Rapidly Solidified Materials*, edited by Shankar M.L. Sastry and Bruce A. MacDonald, TMS-AIME, Warrendale, PA, 1986, p. 151

G.W. Lorimer, in *Precipitation Processes in Solids*, edited by K.C. Russell and H.I. Aaronson, the Metallurgical Society of AIME, New York, 1978, p. 87
L.F. Mondolfo, *Aluminum Alloys: Structures and Properties*, Butterworths, London, 1976
C.J. Peel, "Materials at Their Limits. Aluminum Alloys for Airframes. Limitations and Developments," Institute of Metals, Autumn Meeting, Birmingham, England, 1985
I.J. Polmear, *Light Metals: Metallurgy of the Light Metals*, Edward Arnold, Ltd., London, 1981
N. Ryum, in *Aluminum Alloys: Their Physical and Mechanical Properties*, Vol III, Engineering Materials Advisory Services Ltd., West Midlands, United Kingdom, 1986, p. 1511
B. Sarkar, M. Marek, and E.A. Starke, Jr., *Met. Trans. A*, Vol 12A, 1981, p. 1939
R.D. Schelleng, P.S. Gilman, and S.J. Donachie, "Aluminum-Magnesium-Lithium Forging Alloy Made by Mechanical Alloying," 1986, *Proceedings of Society for the Advancement of Material and Process Engineering Annual Meeting*, October, 1985, Kiamesa Lake, NY
Markus O. Speidel, *Met. Trans. A*, Vol 6A, 1975, p. 631
J.T. Staley, in ASTM STP 605, American Society for Testing and Materials, 1975, p. 71
E.A. Starke, Jr., *J. of Met.*, Vol 22, No. 1, 1970, p. 54
E.A. Starke, Jr., *Mater. Sci. Eng.*, Vol 29, 1977, p. 99
E.A. Starke, Jr., T.H. Sanders, Jr., and I.G. Palmer, *J. of Met.*, Vol 33, 1981, p. 24

8

Alloying Elements in Steel

H.W. PAXTON
U.S. Steel Professor
Carnegie-Mellon University

The alloying of iron is a very old subject, and it is also one where economics is the most crucial. The world-wide glut of steel and steel products means that any value-added concepts really have to be demonstrable before a consumer will choose steel over other metals or over other competitive materials such as polymers, concrete, or composites.

The next chapter, by Dr. Reed, discusses alloying in stainless steels, a fascinating subject in its own right. Here the alloy content is normally well over 10% and in many cases in the 20 to 30% range. The discussion here will cover effects of alloying elements which are rarely present in amounts over 5% and which frequently can cause measurable changes in properties when present in the parts per million range.

These low amounts of generally inexpensive alloying elements serve to keep down the cost of input materials, but it is also vital that, during the manufacture of finished products, the processing costs remain low and the yield remains high. Thus, to maintain appropriate high quality, it is clear that our understanding of each stage from metal production through solidification, hot and cold finishing, and subsequent heat treatment must be complete.

Fortunately, our predecessors have left us a rich supply of knowledge, often developed initially as the metallurgist's art and know-how. Increasingly, this knowledge has become better understood scientifically to the point that today's practitioners are confident that they can utilize automatic control on virtually any operation. Such control, of course, implies a quantitative understanding of the variables.

I think it is fair to say that we have come a long way from the time when the preferred quenchant for a master swordmaker was the urine of a red-headed boy.

While I would like to cover the full range of useful properties (electrical, mechanical, chemical, etc.) which can be controlled by alloying, I must restrict the following discussion to selected mechanical properties for both space and time considerations. Even with selection, there is more than enough material for this chapter.

I will begin with a discussion of modern steelmaking abilities, to illustrate that we can now start with a reproducible "pure" base on which we can impose the effects of alloying elements without concern for the unexpected interactions with other constituents.

I will follow with a brief discussion of important parameters which can be controlled during solidification, hot and cold rolling, and heat treatments, to give you a feel for processing.

Finally, I will develop in some detail the formal understanding of thermodynamic relations, factors influencing the kinetics of phase transformations (stable and metastable), and the success of various theories in predicting the resulting mechanical properties. Where relevant, I will include brief facets of history, so you can see how chronological developments were built on new insights.

NEW CHEMICAL COMPOSITIONS

The substitutional alloying elements which typically are present in the steels under discussion here are present in amounts of a few tenths to a few percent. Modern steelmaking, however, permits us to reduce and control the amounts of oxygen, hydrogen, nitrogen, carbon, sulfur, and phosphorus to levels unattainable a few years ago. This gives us two immediate advantages. Firstly, any deleterious effects which these elements have on properties are lessened because less is present. Secondly, and perhaps less obviously, the effective amounts of other elements which react with one or more of the six listed above are under better control. An example of this is aluminum, which has a primary purpose of controlling the oxygen in solution in liquid steel, but also controls grain size and shape through the formation of aluminum nitride particles which retard the motion of selected grain boundaries.

This is not the appropriate occasion to discuss the fine points of extractive metallurgy, but there is some very sophisticated high-temperature (1600 °C) chemistry involved. For example, the reduction of phosphorus to levels below 0.01% in a 200-ton heat involves first the reduction of silicon in the hot metal from the blast furnace below 0.2% and preferably below 0.1%. Not too long ago, this could easily have been as high as 0.8%. Recall if you will that the blast furnace contains a good deal of silica, which is readily reduced by coke. With these reduced amounts of silica in the vessel which oxidizes the pig iron to steel, the amount of silicate slag which is formed is also reduced, and can be made basic by moderate additions of lime, thereby favoring phosphorus transfer from metal to slag.

The other major process which is now relatively standard is treatment of the steel produced in a BOF, for example, in a separate vessel (ladle). Such treatment can involve, for example, removal of dissolved gases (H, N, O) by exposure to a vacuum or a stream of inert gas, reduction of carbon to very low levels (50ppm) by reaction to form CO under reduced partial pressures, or reduction of sulfur and control of sulfide shape by rare earths or, preferably, lime to make the process more economical. Use of "ladle metallurgy" can also involve changing the temperature of the liquid metal to permit preferred conditions for casting, usually continuously except for certain grades which segregate badly during solidification.

CONTROL OF PROCESS VARIABLES

Although the composition of the material is clearly a primary variable in determining properties, it will not be discussed further here. Rather, we will assume that the intended composition can be made within acceptable limits to the point where it leaves the tundish to enter the continuous casting mold. In practice, this frequently means that the stream has been protected from oxygen and nitrogen pickup, and that sufficient opportunity has been provided for any inclusions to rise out of the liquid and be absorbed in a suitable slag, rather than to be recirculated by turbulence and be trapped in the solidifying steel.

The adoption of continuous casting as a relatively routine process has revolutionized the industry in the last decade for almost all products, with the exception of very large pieces (heavy plate and forgings—e.g., rotors) and highly alloyed material. The ability which has been developed to cast long strings of heats with changes of composition between heats, and changes of cross section within a heat, has contributed greatly to yield and therefore to economy. The absence of composition variations along a cast slab removes a limitation which had plagued ingot casting for 200 years. The propensity for cracking of a cast slab, both superficial and internal, which was a real difficulty only a few years ago, is now so well under control that many slabs are subjected to immediate further hot processing without being cooled to room temperature for inspection, with corresponding improvements in energy consumption and productivity.

The casting process is in an exciting state at the moment. "Conventional casting" is being refined by mechanical design and by applying electromagnetic stirring at different points in the system. Serious attempts are also underway to cast material close to final size—in the 10-to-20-mm range, or even directly to 1 mm.

The next stage in processing, whether or not the metal has been cooled to room temperature, is to reheat to a temperature where the metal can be readily deformed. This deformation usually involves rolling, but, depending on the product, may involve, for example, forging, piercing, or extrusion. Until a few years ago, the purpose of this deformation was primarily to change the shape of the piece to that desired in the product, but, increasingly, the process is now controlled to achieve good combinations of properties—e.g., strength and toughness, formability at a given strength, etc., at the same time. This is achieved by control of the austenite grain size, especially during the later (i.e., cooler) stages of deformation, and, in particular, whether or not the grains are allowed to recrystallize before the eventual transformation. A small austenite grain size will lead to a fine structure in the transformation products, and thus to improved toughness.

Cooling from the austenite range is done with various degrees of care, depending on the properties desired. For parts where requirements are minimal, simple air cooling (normalizing) is often adequate. For development of good property combinations directly from hot rolling, the rate of cooling can now be closely controlled by laminar water jets through the ferrite formation temperature, and coiled into a slowly cooled large mass where the amount of carbon and nitrogen in solution, and the precipitation of other phases such as AlN or Nb(CN), can be controlled. At the other end of the spectrum, low-alloy high-strength plates which must possess simultaneously the required mechanical properties, adequate flatness, and good weldability are cooled rapidly in a controlled quench, sometimes with mechanical restraints to maintain shape.

While obtaining products with salable properties directly from the hot mill is clearly a desirable economic goal, and is practiced more and more, some requirements can be met only by further treatments involving deformation, heat treatment, or both. It is in these applications that alloying elements have a principal use, and discussion of our understanding of these effects—particularly on heat treatment—will be the theme of much of the rest of this chapter. To help calibrate the relative market sizes, it may be helpful to note that the "plain carbon" steels, which include aluminum-killed and microalloyed (containing small amounts of Nb, V, and/or Ti) steels, take about 90%, with "alloy" (up to about 5%) and "high-alloy" (5 to 50%) steels splitting the remainder in the ratio of about 9 to 1.

THERMODYNAMICS OF THE ALLOYS OF IRON

We are fortunate that, because of the technological importance of iron and its alloys, useful information on phase diagrams is readily available and that furthermore, because of some energetic peculiarities, a great deal of theoretical underpinning is also available which has been of special help in the all-important metastable regions where transformations take place.

Wever, in the early 1930's, made the first systematic study of the existing phase diagrams. He pointed out that, in binary diagrams, various elements either extended or reduced the temperature range over which the face-centered cubic austenite is stable (912 to 1394 °C). Elements of the first group are known as austenite stabilizers (such as nickel, manganese, cobalt, carbon, and nitrogen), and those of the second group are known as ferrite stabilizers (such as silicon, chromium, tungsten, molybdenum, vanadium, and niobium). Since binary alloys are not commonly used, the important fact to note is that the propensity to stabilize austenite or not carries over to ternary and higher alloys and is an important factor in selecting correct heat treatment temperatures. In some systems, the austenite field is closed in at higher concentrations by a compound. These situations are shown schematically in Fig. 1.

The reasons for these effects are clearly of great importance to our understanding

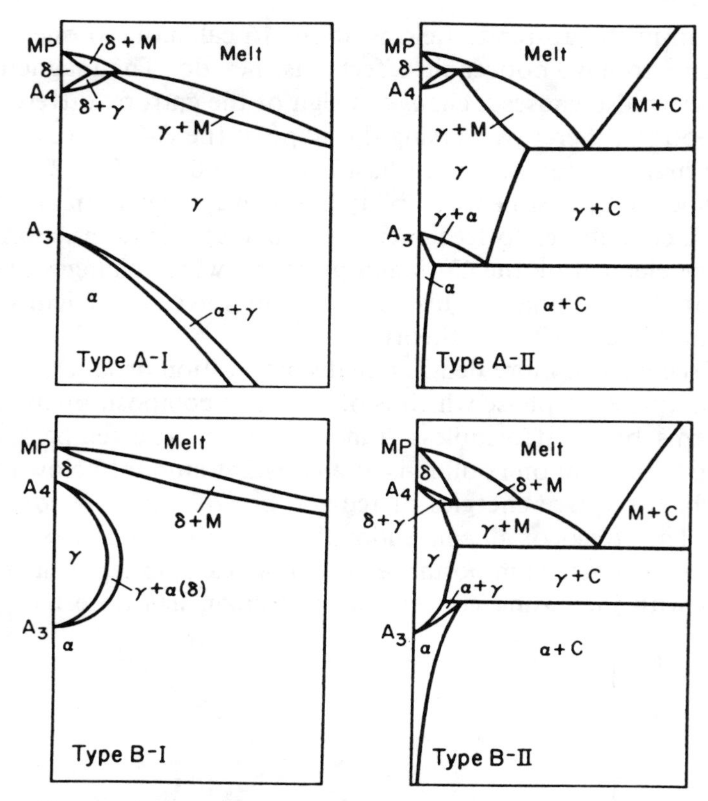

Fig. 1. Schematic representation of austenite stabilizers (type A) and ferrite stabilizers (type B) with and without an intermetallic compound C. After Wever.

of control by alloying elements. The significant breakthrough in this area, as were many others in the field of ferrous metallurgy and elsewhere, was originally made by the "man for all seasons," Clarence Zener, in a seminal paper on austenite decomposition in 1946 and was developed more completely a decade later.[1,2] Two observations are perhaps worth making on this. The first work was done at Watertown Arsenal, where Zener's interest had been stirred by John Herbert Hollomon. Secondly, the paper was received with more than the usual scorn by the metallurgical establishment, who were not about to be upstaged by a theoretical physicist. In those days, discussions of technical papers were published, and the discussion of this one ran to 12 pages, mostly later proved wrong.

Zener's basic idea grew from the observation that one would normally expect, on energetic grounds, that metals would be close packed at 0 K. Certainly this is not always true, since several metals are stably bcc down to absolute zero, but there are explanations for these exceptions. From our point of view, austenite would be stable at all temperatures if magnetic effects (ferromagnetism in bcc, antiferromagnetism in fcc) were absent. The enthalpy contribution from ferromagnetism in bcc iron is large relative to the energy involved in phase transformations and stabilizes this phase at low temperatures. The maximum difference in free energy between bcc and fcc iron in the austenite field is about 16 cal/mol, so one might expect alloying elements to have noticeable effects, as they do. The magnetic contribution to free energy also causes a change in sign of the entropy difference between the phases at 950 °C, thereby reversing the slope of the G-T curve and ultimately stabilizing the high-temperature bcc phase above 1400 °C (Fig. 2).

Alloying elements then influence stability in two ways: the normal effect of solid solution, which contributes to free energy through elastic interactions, etc.; and the effect of the element on the Curie temperature, which changes the "magnetic free energy" of the bcc phase. This second term solves a difficulty which had existed for some time in simpler theories.

As we shall see later in more detail, one decomposition product of austenite on cooling is "martensite," a phase which is of the same composition as the austenite and which forms by a diffusionless transformation. The temperature, M_s, at which this begins to form on cooling is at a constant amount below the temperature, T_0, at which the free energies of fcc and bcc *of the same composition* are equal (Fig. 3). The effects of all conventional alloys on this temperature have been studied, because of practical importance, and in all cases except silicon and cobalt T_0 goes down with increasing concentration. Silicon has no effect, and cobalt

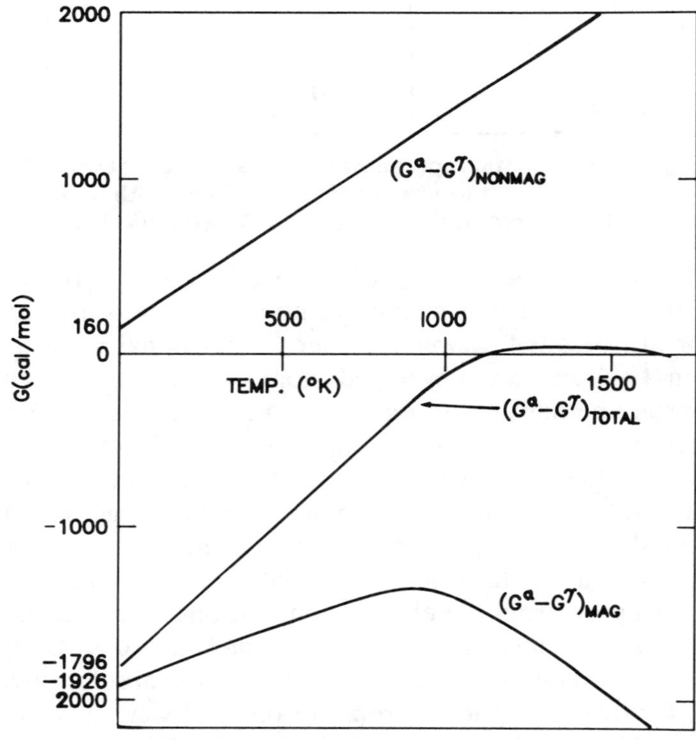

Fig. 2. Contributions to the free-energy difference between bcc and fcc iron

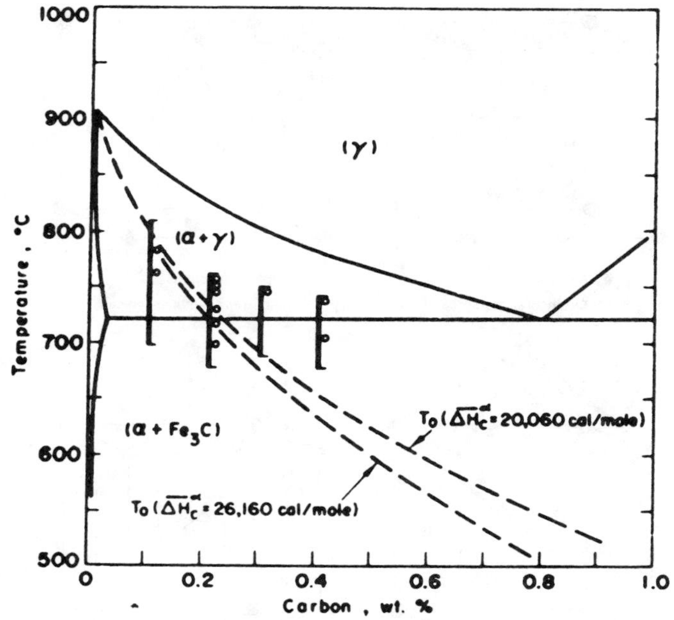

Two different values of the enthalpy of solution differential are shown.

Fig. 3. Schematic representation of C-T curve where free energies of bcc and fcc Fe-C alloys of the same composition are equal

increases M_s somewhat. Simple theory would predict that ferrite stabilizers would raise T_0 and hence M_s, and vice versa for austenite stabilizers. This does not occur, but the extra magnetic effects explain the results well.

Around 1967, these effects began to be introduced into quantitative calculations of iron-base phase diagrams by Hillert, Kaufman, Kubaschewski, and others, and the result today(3) is a very comprehensive methodology and set of data which are of special interest in calculating spinodal compositions in miscibility gaps. (Since the magnetic materials to which these often apply are not being discussed here, this will not be pursued.)

The thermodynamics of the interstitial solid solutions of carbon and nitrogen is another important area, especially as our range of alloys today relies heavily on the distributions of these elements between austenite and ferrite, and in compounds such as carbides, nitrides, and carbonitrides. In general, we can understand many of the effects by a simple hard-sphere model. We see immediately that the octahedral interstices in fcc are much larger than the tetrahedral, and that both of these are larger than either type of interstice in bcc (Fig. 4). The solubility of carbon or nitrogen in austenite is much larger than in ferrite, the heat of solution is less, and the boundaries of the fcc/bcc field as they leave 912 °C can be explained very nicely by statistical contributions to the entropy of mixing. Calculation of the phase

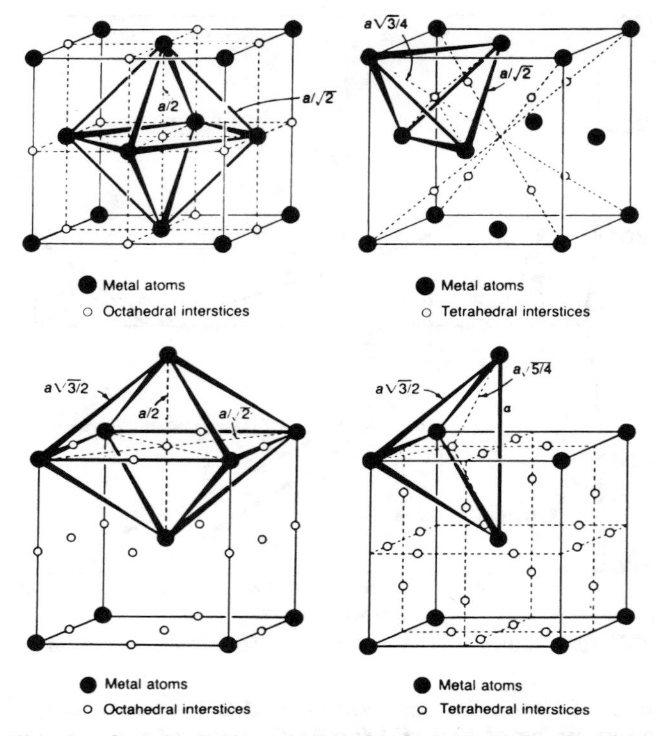

Fig. 4. Octahedral and tetrahedral sites in the fcc (above) and bcc (below) lattices

boundaries in the region where austenite is unstable is more difficult, but a number of attempts have been made. In order of increasing sophistication, these involve:

1. The assumption that the difference in heat of solution between fcc and bcc is independent of temperature and composition. Below Ac_1, this is not very satisfactory without a lot of corrections (Ac_1 is the lowest temperature at which austenite is stable).
2. Using activity determinations in austenite as a function of composition and extrapolating these below Ac_1.
3. Using statistical mechanics to calculate thermodynamic properties of austenite on various models. Some data are becoming available for checking the validity of models using Mössbauer and NMR techniques.(4)

Extension of these techniques to ternary and higher systems is obviously important, and, equally obviously, is not easy. There are two types of problems to evaluate. One is calculation of the isothermal section from a model involving the known properties of the binaries, and the second is extrapolation of the free energies calculated to metastable regions. Hillert and his group at KTH, Stockholm, have been active in this area. The first type of calculation has had more success than the second, and an example is shown in Fig. 5.

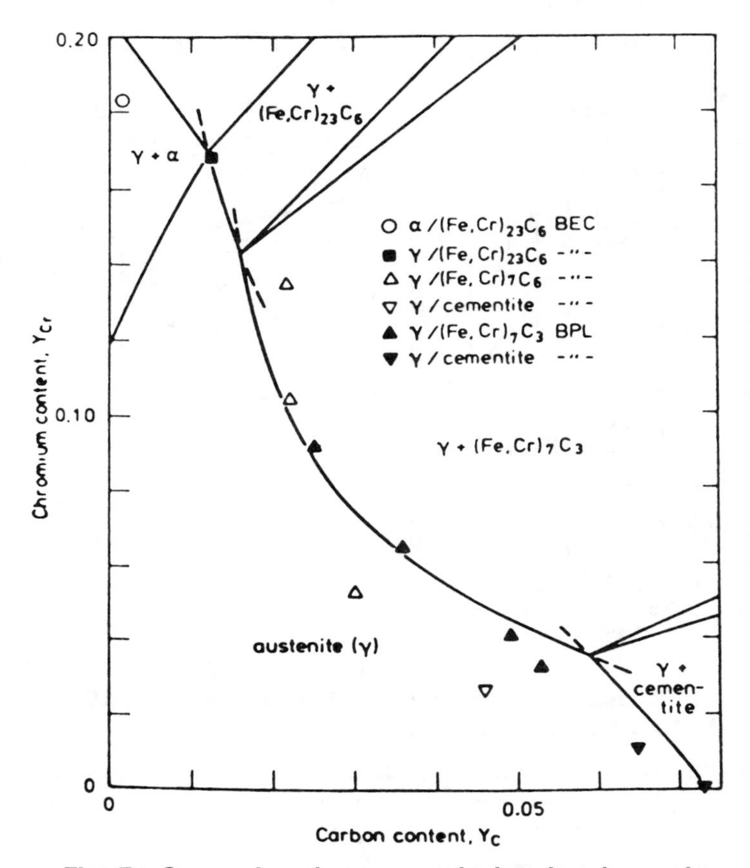

Fig. 5. Comparison between calculated and experimentally determined phase boundaries of the FeCrC system at 1273 °C

KINETICS OF AUSTENITE FORMATION AND TRANSFORMATION

The formation of austenite from a ferrite-carbide aggregate can occur by nucleation and growth or, in special circumstances, by a shear transformation which is the inverse of the martensitic reaction on cooling. These reactions are of scientific interest,[5] but up to now they have been of little practical concern and they will not be discussed further.

Decomposition of austenite is commonly represented by TTT (time-temperature-transformation) curves; a typical example is shown in Fig. 6. Understanding the factors which control these curves has been the subject of intense study by metallurgists for several decades. It is fair to say that while a semiquantitative understanding has been reached, and has enabled very adequate empirical practical tools to be developed, *a priori* calculations are not yet possible because of the difficulty of the problem.

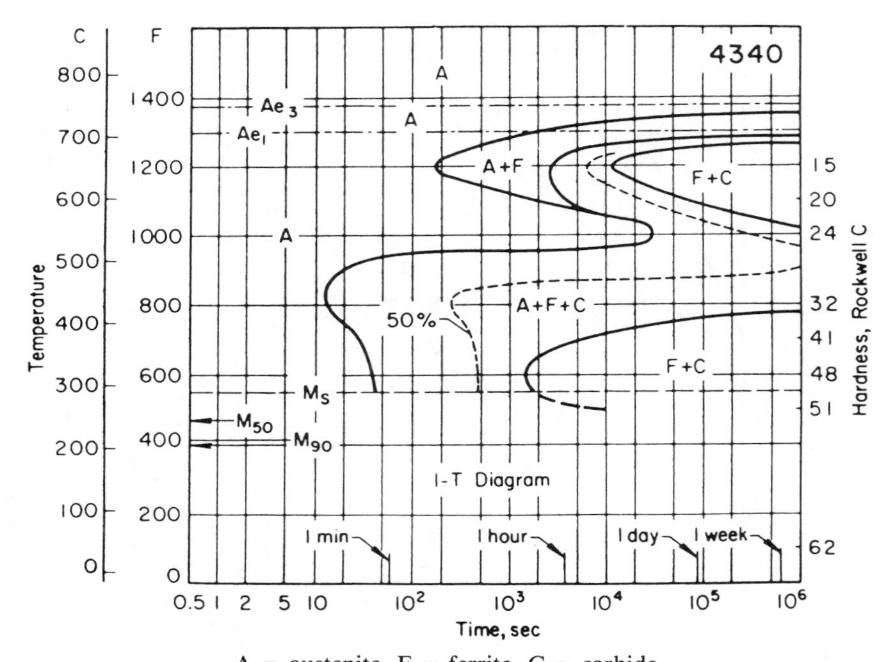

A = austenite, F = ferrite, C = carbide.

Fig. 6. Isothermal transformation curves for SAE 4340 after austenitizing at 850 °C

The transformation of austenite, especially in the steels discussed here, covers almost all of the classical types of mechanisms. Davenport and Bain, in their justifiably famous 1930 paper,[6] provided the technique for isolating these different types, and thus laid the foundation for quantitative studies. We note that they did not properly identify the characteristics of the martensite reaction, but perhaps this is not too surprising since the concepts of "nucleation and growth" dominated theoretical treatment at the time (Fig. 7). The precipitation of ferrite, cementite, and pearlite clearly occurs by nucleation and growth; kinetic studies have focused on the rate-controlling step—i.e., whether this occurs by the bulk diffusion of some solute, or whether some slower step at an interface is critical. Greninger and Troiano[7] gave an elegant demonstration of the "athermal" nature of martensite, and cleared up the confusion in which Davenport and Bain had unwittingly trapped themselves by the use of a dilatometric approach. In an athermal mechanism, the transformation begins at a very well-defined temperature (M_s) and continues on cooling. The undercooling below M_s controls the amount of transformation at that temperature, and at perhaps 200 °C below M_s the amount of martensite asymptotically approaches 100% (Fig. 8). This phase forms coherently with the austenite lattice, lattice shearing is involved, and generally a substantial volume change with associated stored energy occurs. If this volume change takes place at low temperatures (near or below room temperature), significant distortion or even cracking can occur—obviously a most unwelcome event since repair or rework is generally difficult or impossible. It is also an important consideration during welding, yet another topic we will not consider further.

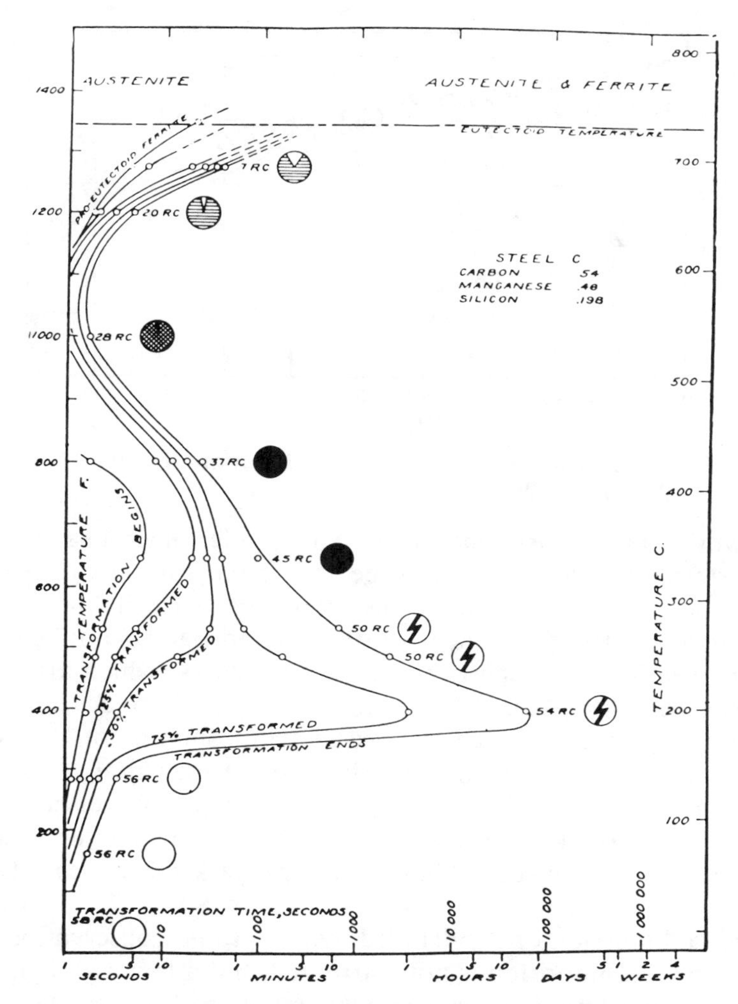

Fig. 7. One of Davenport and Bain's original "S-curves"

The intermediate transformation to bainite is more complex, since it may involve both diffusion and shear processes, leading to a ferrite-carbide aggregate which, with proper control, can have outstanding combinations of mechanical properties. In normal production operations, alloying elements are essential for delaying the formation of ferrite and pearlite so the workpiece can reach the temperature at which bainite can form. Recent developments have demonstrated that low-carbon bainites can be very attractive as engineering materials because they combine good mechanical properties with ease of welding.

While TTT curves offer a good basis for cataloging the effects of individual alloys, and for predicting microstructures associated with a specific transformation temperature, they are not directly useful in predicting the structures obtained during normal cooling of austenite. In this, transformation of local areas occurs over a range of temperatures rather than isothermally. We do not enter a com-

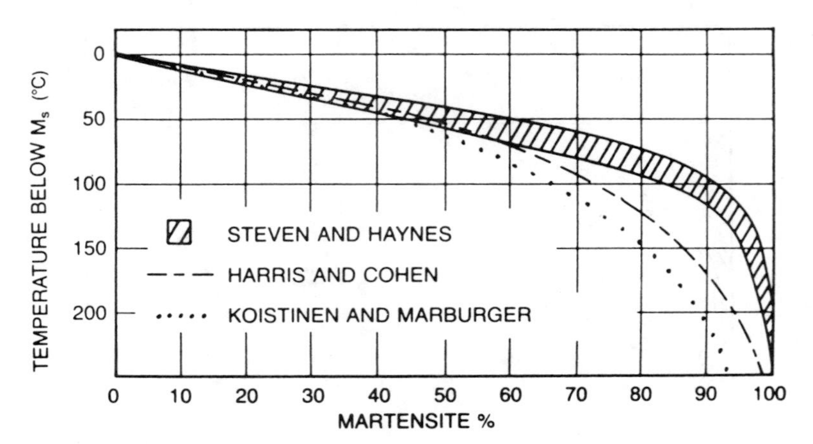

Fig. 8. Fraction of martensite as a function of undercooling below M_s

pletely new world, but adjustments have to be made. Since heat is abstracted through the external surface, we need to be able to measure or calculate rates of cooling at any point in the specimen. This subject of "hardenability" was developed by Grossman shortly after the work of Davenport and Bain; he was able to develop an engineering approach to the effects of alloys (which affect the TTT curves) in various cooling media. Many refinements have been developed over the years; these are described very well in a recent symposium report[8] and will not be further discussed here. The subject is critical in a wide variety of applications ranging from our inability to obtain totally satisfactory properties in a large rotor forging to our ability to control the residual stress pattern in a fatigue-sensitive part. Improvement in composition control has had a very marked beneficial effect in this area.

At this time, a discussion of the detailed metallographic observations and theories of austenite decomposition is not warranted, but a few highlights will help to convey the role of alloys, both individually and in combination. During service of most of the alloys discussed here, we have an aggregate of ferrite and carbide (usually, but not always, cementite) with perhaps small amounts of retained austenite—i.e., austenite which did not transform on cooling. In very few cases would we have untransformed martensite, because this is too brittle for normal service. Thus, in developing our understanding, we need to understand the rates of nucleation and growth from austenite and/or martensite, and how alloying elements influence these rates. To obtain high-strength materials, we need large numbers of small carbides distributed fairly uniformly in a ferrite matrix—that is, we need a high N (rate of nucleation) and a low g (rate of growth). This will not guarantee toughness, for which we need to add structural features which make nucleation and propagation of cracks or voids more difficult. Usually this is accomplished by keeping the austenite grain size small, and by minimizing the amount of nonmetallic inclusions.

If we consider the transformations of austenite which involve diffusion, there are several theoretical concepts which have been studied intensively but which are not yet resolved. The difficulty of evaluating the free-energy change available to drive a particular reaction at a particular temperature has been mentioned above for the equilibrium case. This difficulty is compounded when alloying elements are present, because unless the alloy partitions to equilibrium between matrix and precipitate, the maximum driving force will not be available (Fig. 9). Since alloying elements diffuse much more slowly than carbon at the temperatures of interest, this situation is common.

Aaronson, in one of his extensive studies of austenite decomposition,[9] offers an experimental evaluation of the partitioning of several elements during ferrite formation from austenite (Fig. 10). Near the A_3 temperature, partitioning occurs for manganese, nickel, and platinum, but at lower temperatures this rapidly becomes less and less until, at about 100 °C undercooling, it becomes effectively zero. For a number of other alloying elements, no partitioning seems to occur at all. In his calculations of growth rates expected with various assumptions, Aaronson notes that observed rates are lower, and often much lower, than the theoretical, and some additional interface drag must be present.

This retardation of ferrite formation during continuous cooling of austenite has important effects on hardenability, which may be defined as the ability to cause transformation at increasingly lower temperatures, where higher-strength mixtures of ferrite and carbide can form, either directly as bainite or by formation of martensite which can subsequently be reheated (tempered) to cause precipitation.

There are other effects of alloys which must also be considered in addition to hardenability. The M_s temperature is depressed approximately linearly by alloying additions *when these additions are dissolved in austenite*, and so once more we

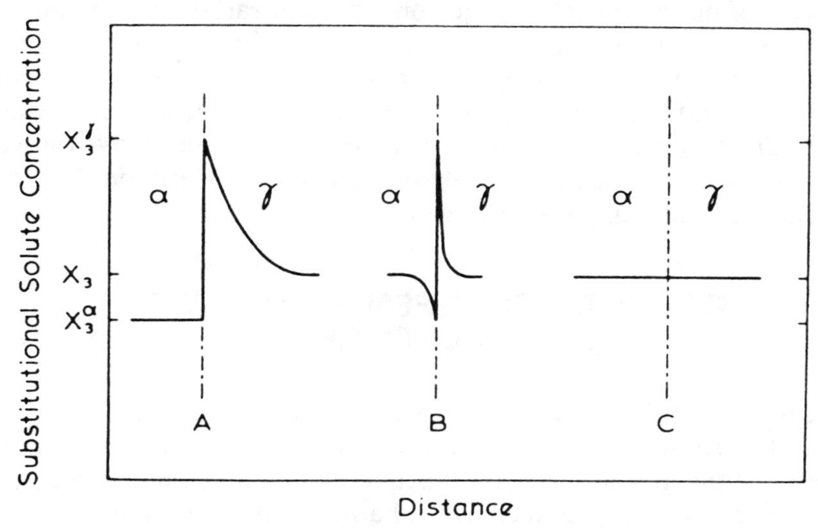

(A) Full equilibrium. (B) Local equilibrium. (C) Paraequilibrium.

Fig. 9. Schematic representation of alloying-element (X) distribution during ferrite growth

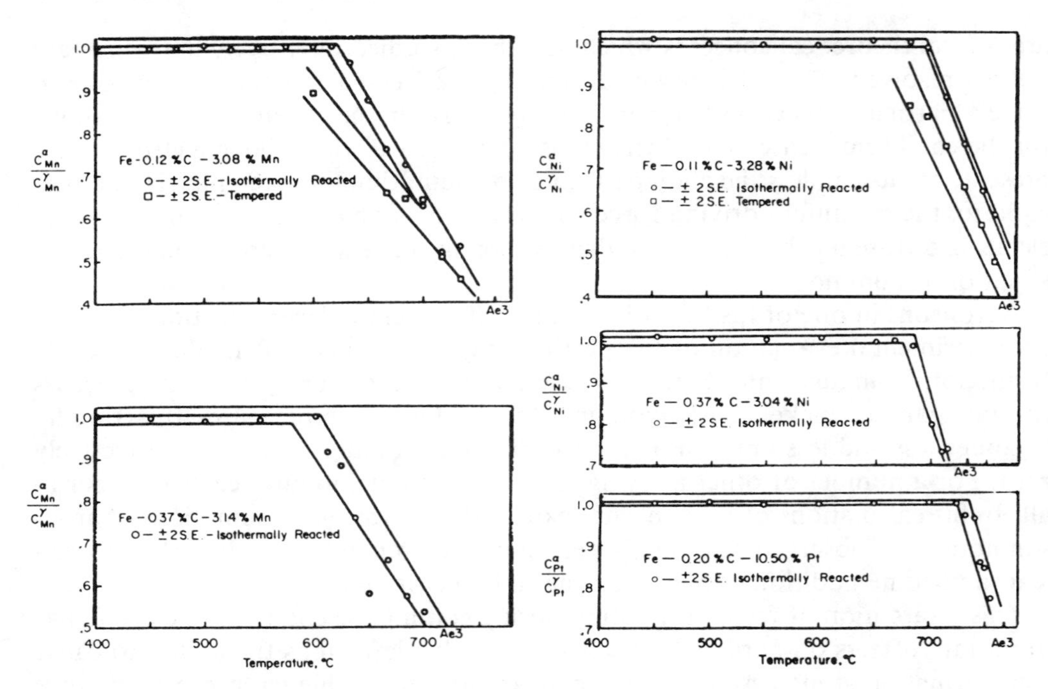

Fig. 10. Approach to "paraequilibrium" in several alloy steels during the isothermal formation of ferrite at the temperatures shown

must return to ternary and higher phase diagrams. Our goal generally is to keep M_s as high as possible, both to reduce the magnitude of the volume change during martensite formation (which depends on dissolved carbon content and on M_s itself) and to permit strain relief at as high a temperature as possible to prevent cracking. Alloys which are carbide formers may keep both carbon and alloying elements from contributing to hardenability and to subsequent changes in tempering behavior. Figure 11 illustrates for a series of iron-chromium alloys the various austenite compositions which can be obtained, and the effects on TTT curves, retained austenite, and tempering behavior.

RELATIONS BETWEEN STRUCTURE AND PROPERTIES

Although it is impossible in this review to give a comprehensive account of our knowledge in this subject, it may be helpful to offer a few examples to illustrate that conventional theories of mechanical behavior are obeyed in many cases, and that less well-understood topics such as the hardness of martensite are intriguing subjects in their own right.

As our understanding of mechanical properties has grown, we have learned to think of resistance to dislocation motion caused by various structural factors.

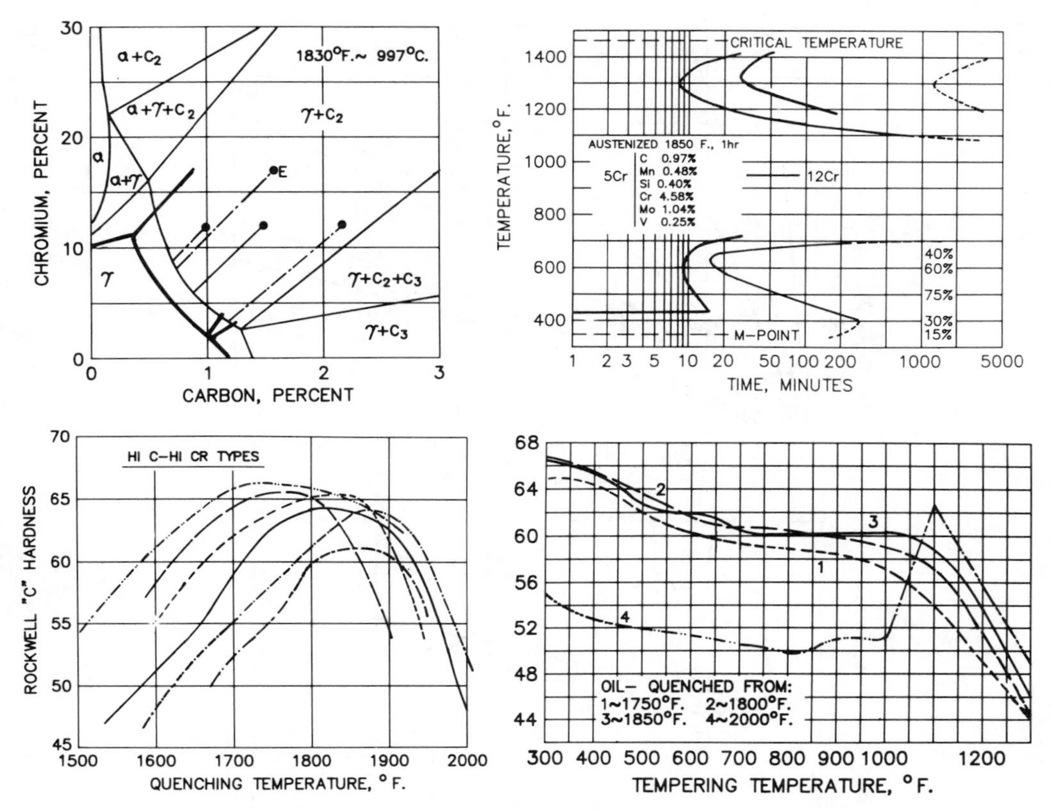

(Top left) Location of alloys on ternary diagrams at 1560 °F (heavier lines at lower left) and at 1830 °F. (Top right) Comparison of TTT curves for a 5Cr-1C steel and a 12Cr-1.5C steel. (Bottom left) Effect of austenitizing temperature on as-quenched hardness. (Bottom right) Effect of austenitizing temperature and quenching rate on subsequent tempering behavior.

Fig. 11. Effects of austenite composition in several FeCrC alloys

Examples are the lattice friction itself (important for iron), grain boundaries, precipitates (coherent or incoherent), other dislocations (produced by cold work or by transformation), twin boundaries, etc. Successful products arise from appropriate manipulation of the microstructure to create the *combination of properties* required in service.

A simple example of this comes from solid-solution hardening, a phenomenon long used in the generation of relatively inexpensive steels with strengths 20 to 50% above those of mild steels. To a first approximation, the misfit of solute atoms causes lattice strains proportional to the amount dissolved (Fig. 12) and provides strengthening through the lattice friction term. However, in doing so it also changes the impact transition temperature unfavorably, and this may preclude the choice of this type of strengthening. The exceptions are elements which, while producing strengthening, also markedly reduce the ferrite grain size. Pickering[10] has given

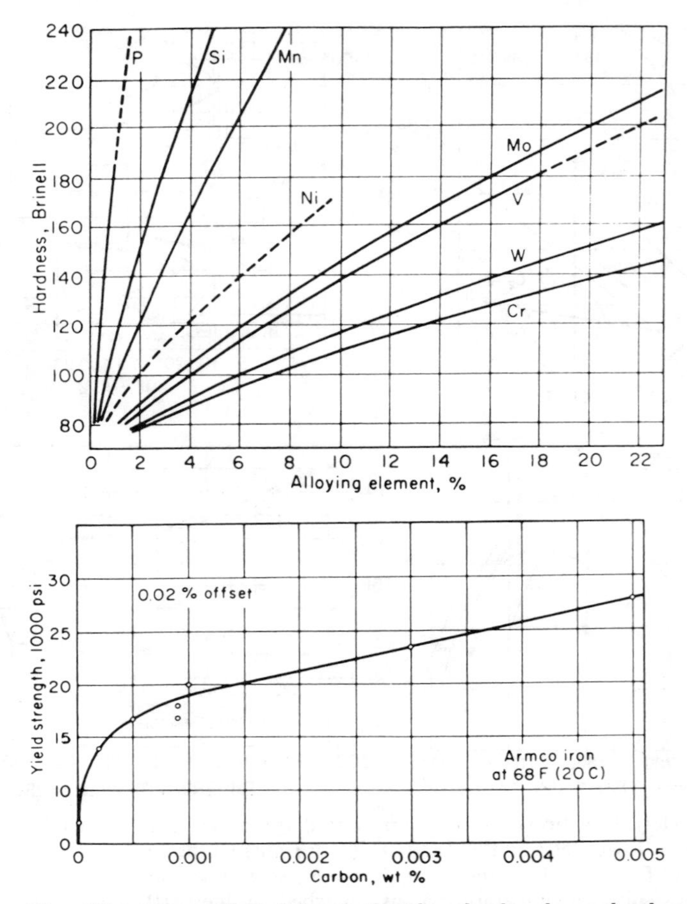

Fig. 12. (Above) Solid-solution hardening by substitutional alloys and (below) strengthening of ferrite by dissolved carbon.

us a clear way to represent this effect, as shown in Fig. 13. Each element has a vector in strength-toughness space. Phosphorus, for example, is a powerful strengthener but has very negative effects on toughness (hence the emphasis on dephosphorization earlier, although in some cases where toughness is not important, phosphorus is used as a strengthener). Aluminum also causes significant strain in the iron lattice, but because it is a powerful grain refiner, it is extremely beneficial to toughness. This comes about because AlN serves to restrain austenite grain growth, and these smaller grains offer much more area for ferrite nucleation and hence a smaller ferrite grain size. Manganese, with its intermediate position, offers modest strengthening, but its effect in refining ferrite grain size comes about because as an austenite stabilizer it reduces the ferrite transformation temperature and thereby increases the ratio of nucleation rate to growth rate.

An alternative way of achieving valuable combinations of strength and toughness was mentioned earlier in discussing process variables. When austenite is deformed in its stable range, it will recrystallize.[(11)] At high temperatures, this

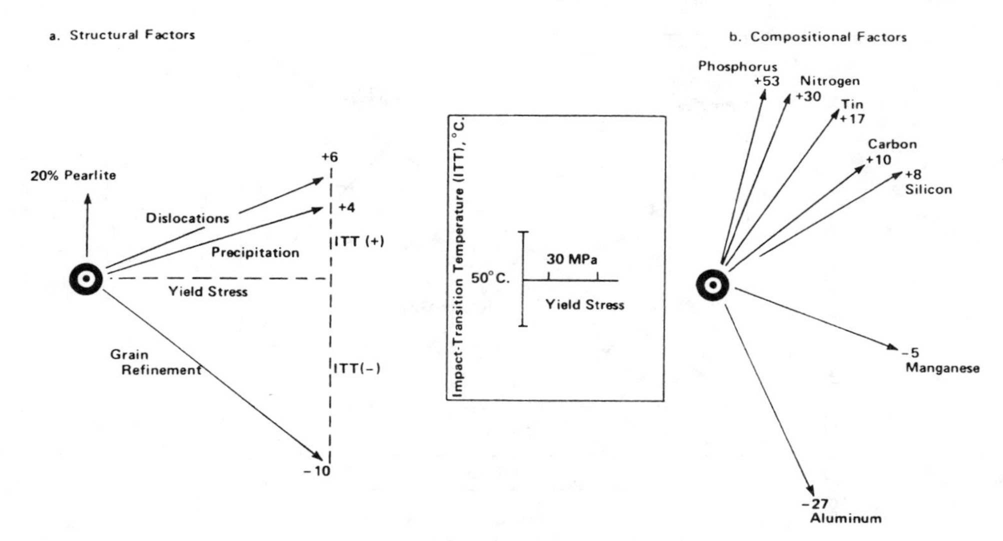

Fig. 13. Factors affecting yield strength and impact-transition temperature ratios indicating ΔITT per 15 MPa increase in yield strength

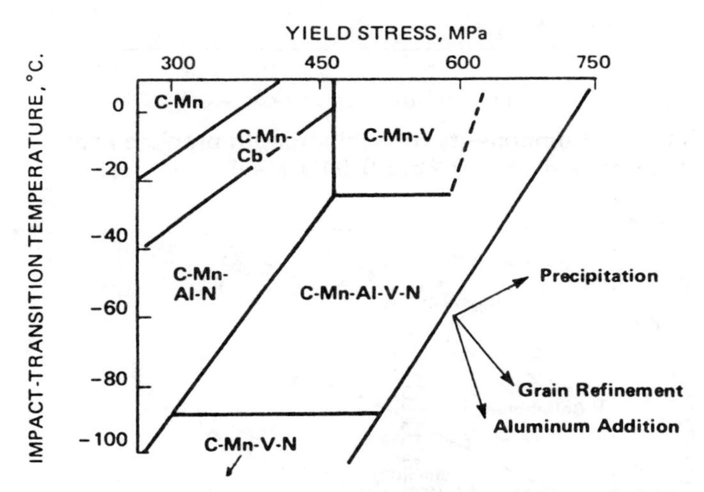

Fig. 14. Combinations of yield strength and ITT available in normalized high-strength, low-alloy steels

occurs very rapidly—in seconds at most. At intermediate points, the austenite does eventually recrystallize, but this can take several minutes, and on a modern continuous hot strip mill this time is not available unless the piece is deliberately held before entering the finishing stands. At still lower temperatures, the austenite does not recrystallize at all, and ferrite subsequently forms from pancake-shape deformed grains.

In many of the alloys to which this controlled rolling is applied, often called "microalloyed steels," the process can be further complicated by precipitation of the carbonitrides of niobium, vanadium, and titanium in the austenite. Occasion-

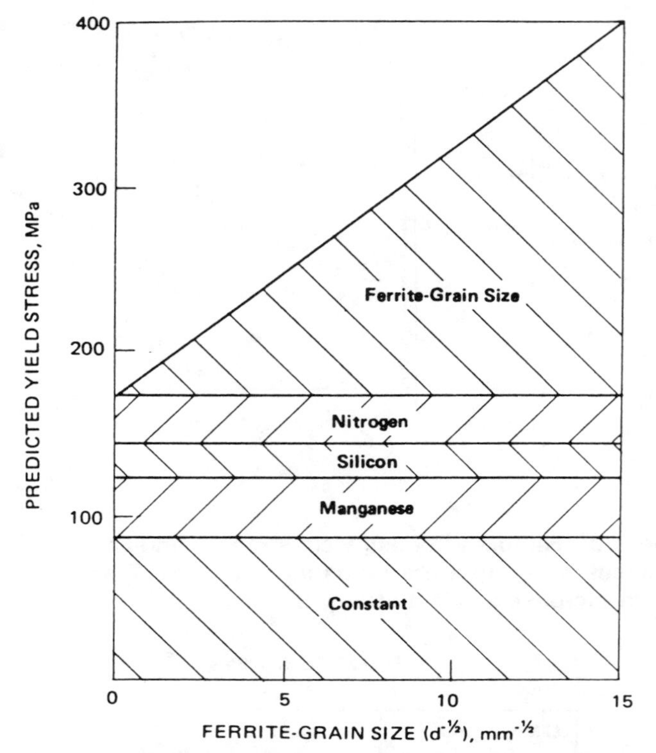

Fig. 15. Components of yield strength predicted for an air-cooled 1Mn-0.25Si-0.01N steel

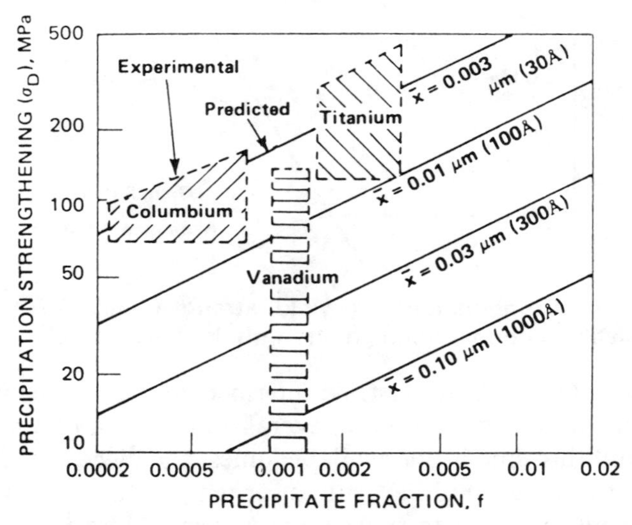

Fig. 16. Dependence of strengthening on precipitate size (x) and volume fraction compared with experimental observations for niobium, vanadium, and titanium

ally, hot rolling is continued below the Ac_3 temperature, and ferrite can form during the deformation and may be strain hardened (i.e., dislocated and textured) by the rolling. The ferrite grain size is fine, and the strength may be further increased by subsequent precipitation of carbonitrides in ferrite during cooling. A full discussion of this complex but important topic is not warranted here, but Fig. 14 will convey an idea of the properties which can be obtained in steels with various alloying elements.

In separating the various microstructural contributions to mechanical properties, we are particularly indebted to Irvine, Pickering, Gladman, and their colleagues. Building on the concepts of Hall and Petch for the effects of grain size,

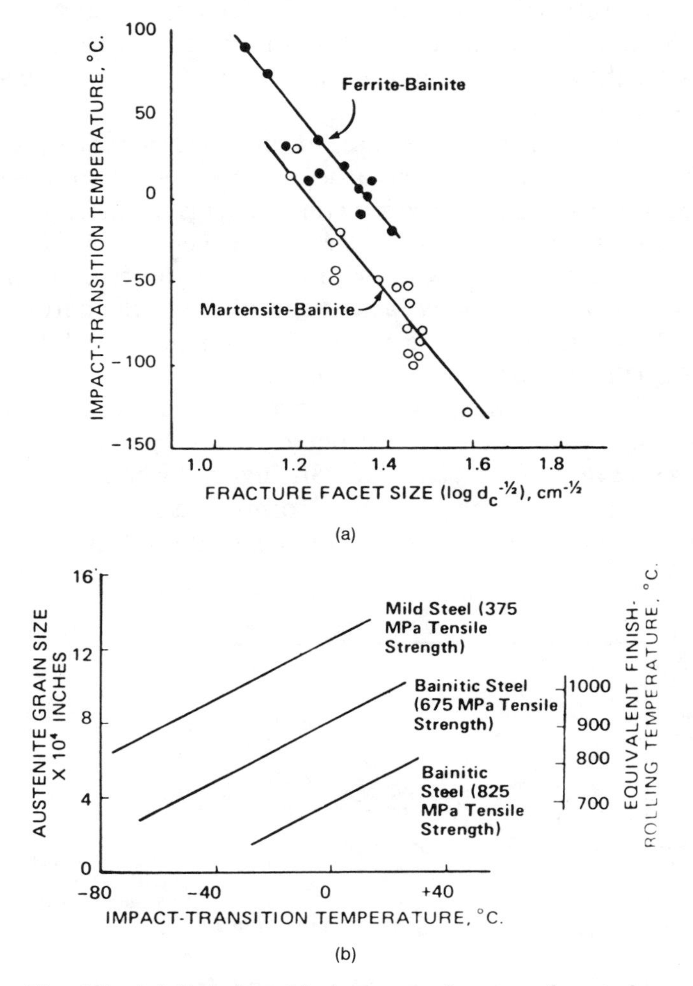

Fig. 17. (a) Relationship between fracture-facet size and ITT for martensite and bainite structures. The term d_c is the unit crack path or facet size. (b) Effect of finish-rolling temperature and austenite grain size on ITT of upper-bainite structures.

the Orowan model of precipitation hardening, and the individual solid-solution effects of the various elements (recalling that for carbon and nitrogen the amount *in solid solution in ferrite* is critical), they have been able to provide us with information sufficiently quantitative that design of compositions and processing requirements is greatly facilitated. Some examples of their work are shown in Fig. 15 and 16.

The same concepts can be applied to bainite, the structure formed at intermediate transformation temperatures especially if the proeutectoid ferrite transformation can be delayed by adding such elements as boron or molybdenum. Bainite is a complex microstructure with dislocations introduced because of the shear involved, large supersaturations of carbon in the ferrite, and a carbide dispersion which is quite different in the upper and lower parts of its formation range. The concepts discussed above once more prove useful; examples are shown in Fig. 17 and 18.

Although fully martensitic structures are, for practical purposes, never used, the structure itself is of intrinsic interest because of the extreme hardnesses which can be obtained, and because it is the basic structure which is tempered to obtain tough, high-strength materials. The hardness of martensite depends primarily on its carbon content (Fig. 19). The structure varies with the temperature of formation—i.e., with M_s, as shown in Fig. 20. Many factors contribute to the hardness, but these structural differences and the rearrangement of carbon atoms and dislocations during and after quenching appear to be very important.

The effects of alloying elements during tempering are profound. In plain carbon steels, a metastable carbide which is coherent with the martensite nucleates at low temperatures (below about 250 °C). In higher-carbon steels where some retained austenite may be present, this transforms to a ferrite-carbide aggregate between 200 and 300 °C. Finally, the transition carbide and the remaining lower-

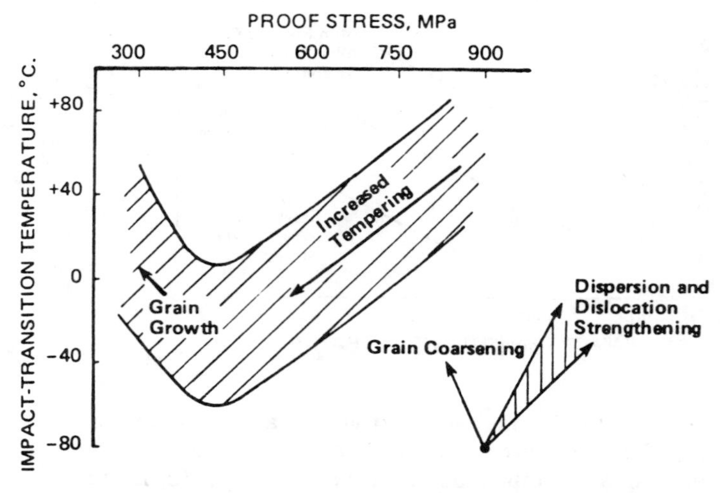

Fig. 18. Effect of tempering on the relation between proof stress and ITT for lower bainite

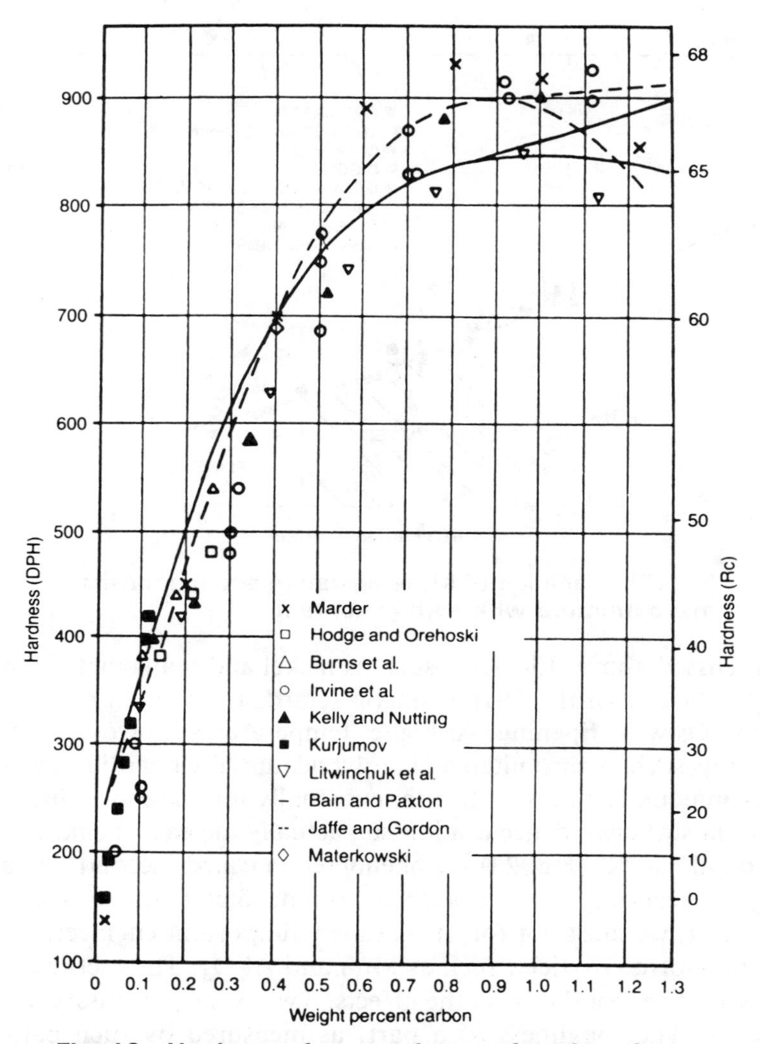

Fig. 19. Hardness of martensite as a function of carbon content

carbon martensite begin to form ferrite and cementite between 250 and 350 °C, with the cementite coarsening by classical Ostwald ripening as the temperature is raised further. The strength of the material falls off monotonically, the ductility increases, and the toughness generally increases, with some lower values in certain temperature ranges. These effects are shown in Fig. 21 and 22.

When alloys are present, the rate of softening is decreased, but the change in toughness with temperature is not significantly affected. This permits an immediate practical advantage: either an increased strength with the same toughness can be reached at the same tempering temperature, or the same strength with increased toughness can occur by increasing the tempering temperature somewhat. At lower temperatures, the reasons for these properties are much the same as in the normal-

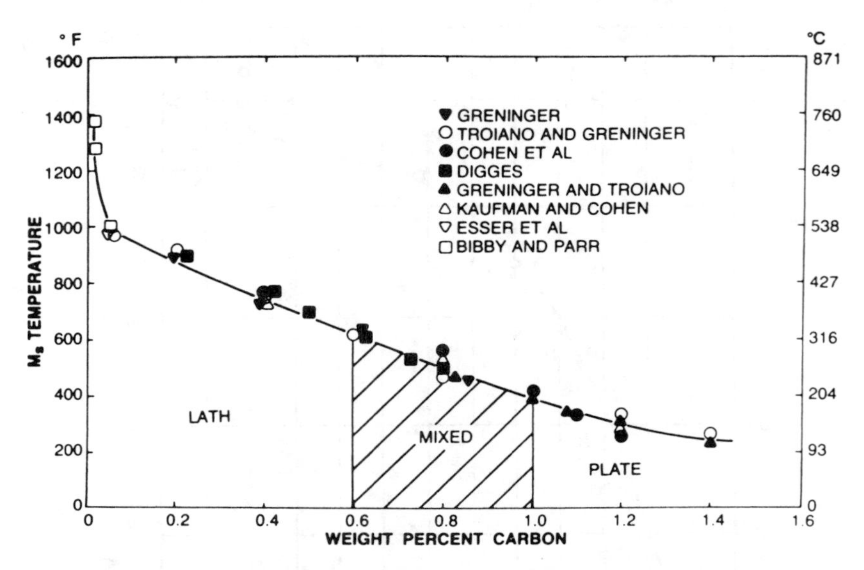

Fig. 20. Variation of M_s temperature and martensite microstructure with carbon content

ized steels discussed above. Elements such as nickel and manganese provide solid-solution strengthening of the ferrite; silicon retards the growth rate of cementite and thus slows Ostwald ripening. At higher temperatures, however, with carbide-forming elements such as chromium and molybdenum, their carbides, which are the stable phases in equilibrium with ferrite, eventually nucleate in a fine dispersion, and if present in sufficient concentration can actually increase the hardness at temperatures around 550 °C (Fig. 23) – a phenomenon called "secondary hardening."

In closing this section, which has largely been devoted to alloys dissolved in the phases of interest, we must not forget that many important engineering properties depend on undissolved particles such as MnS and Al_2O_3. The fact that the theory is more difficult does not diminish the effects. Two principal classes of properties are noteworthy. The toughness of a part, as measured by such parameters as Charpy shelf energy and crack-opening displacement (COD), is dependent on the volume fraction of inclusions which are either cracked or lose interface coherency during large plastic deformations. Anisotropy of properties depends also on the shape, composition, and distribution of inclusions, which, if sufficiently plastic, can be elongated very significantly during hot rolling.

One obvious solution is to reduce the sulfur and oxygen contents in the cast steel, thereby reducing the source of most of the inclusions, and this can now be done routinely where warranted. Changes in the composition of the sulfide inclusions to decrease their plasticity during hot rolling, leaving them basically spherical, can be (and is) done by additions of calcium, zirconium, or rare earths. Unless this is done, there is continuing danger of splitting during bending, lamellar tearing during welding, and lack of through-thickness ductility. An example of optimization of cerium additions to nearly equalize longitudinal and transverse shelf energies is shown in Fig. 24.

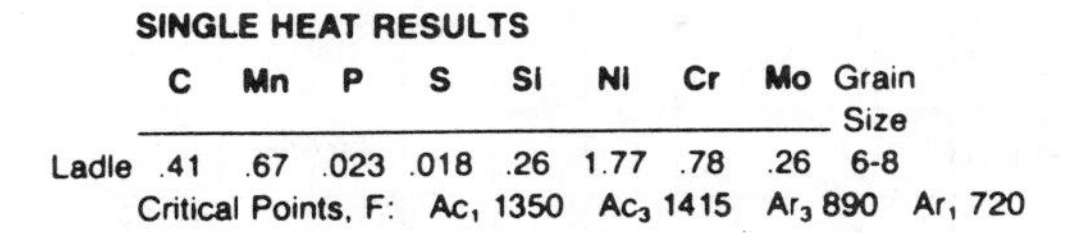

SINGLE HEAT RESULTS

	C	Mn	P	S	Si	Ni	Cr	Mo	Grain Size
Ladle	.41	.67	.023	.018	.26	1.77	.78	.26	6-8

Critical Points, F: Ac_1 1350 Ac_3 1415 Ar_3 890 Ar_1 720

Treatment: Normalized at 1600 F; reheated to 1475 F; quenched in agitated oil.
.530-in. Round Treated; .505-in. Round Tested. As-quenched HB 601.

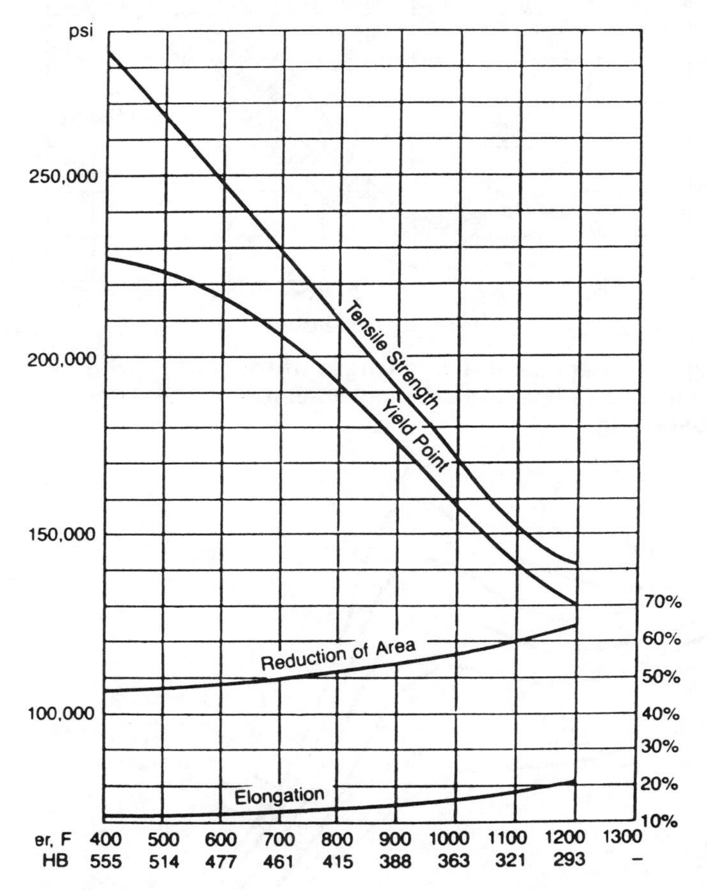

Fig. 21. Change of mechanical properties with tempering temperature for oil-quenched 4340 steel

SUMMARY

This is a difficult chapter to summarize, because I have approached a large subject by offering snapshots of understanding in a number of areas, some scientific and some more technological. I hope I have left you with the idea that we have in general the knowledge to obtain consistently the properties of which ferrous alloys are capable, and the processing abilities to create these materials in a cost-effective fashion. There is serious overcapacity in the world today, leading to some very strange commercial situations. A very important technical challenge—the challenge of being the low-cost producer—remains.

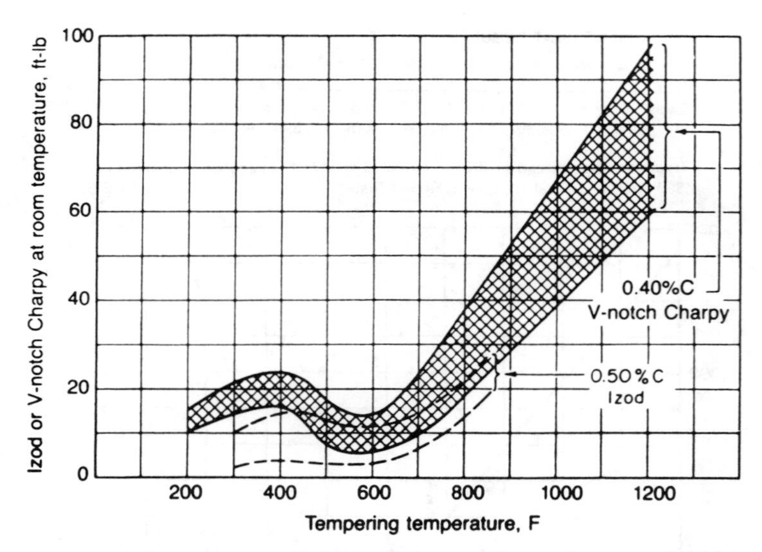

Fig. 22. Representative changes of toughness with tempering temperature in medium-carbon steels after quenching

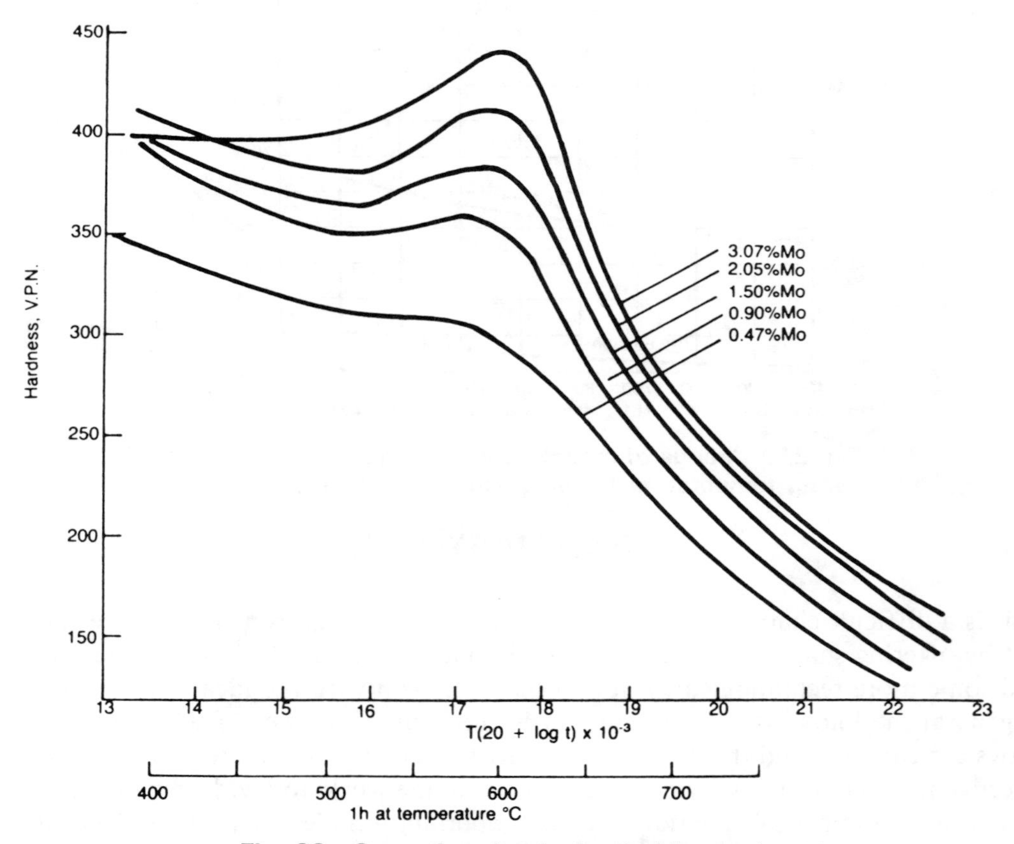

Fig. 23. Secondary hardening in Mo steels

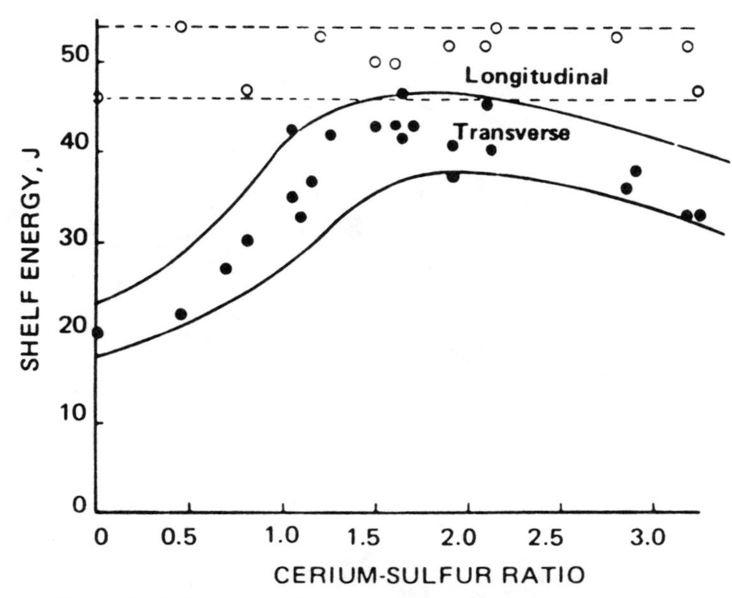

Fig. 24. Effect of cerium additions on longitudinal and transverse Charpy shelf energy

REFERENCES

1. C. Zener, *Trans. AIME*, Vol 167, 1946, p. 513
2. C. Zener, *Trans. AIME*, Vol 203, 1955, p. 619
3. M. Hillert, *Hardenability Concepts with Applications to Steel*, edited by D.V. Doane and J.S. Kirkaldy, TMS, Pittsburgh, 1978, p. 5
4. J.-M. Genin (private communication)
5. R.R. Judd and H.W. Paxton, *Trans. AIME*, Vol 242, 1968, p. 206
6. E.S. Davenport and E.C. Bain, *Trans. AIME*, Vol 90, 1930, p. 117
7. A.B. Greninger and A.R. Troiano, *Trans. ASM*, Vol 28, 1940, p. 537
8. D.V. Doane and J.S. Kirkaldy (ed.), *Hardenability Concepts With Applications to Steels*, TMS, Pittsburgh, 1978
9. H.I. Aaronson and H.A. Domian, *Trans. AIME*, Vol 236, 1966, p. 781
10. F.B. Pickering, "Microalloying 75," Union Carbide, New York, 1977, p. 9
11. R.A. Grange, *Fundamentals of Deformation Processing*, Syracuse University Press, 1962, p. 299

REFERENCES

9

Austenitic Stainless Steels With Emphasis on Strength at Low Temperatures

RICHARD P. REED
Fracture and Deformation Division
National Bureau of Standards

Rustproof steels, with increased passivity, were discovered at the turn of the century. These new steels, called stainless steels, are characterized by high chromium (>10%*) and low carbon (<0.3%). Research beginning in 1904 by Guillet,[1] followed by Portevin,[2] Giesen,[3] and Monnartz,[4] led to an understanding of the corrosion behavior of Fe-Cr-C alloys. During the same period, Guillet[5] and Giesen[3] published studies of Fe-Cr-Ni austenitic stainless steels. Clearly, the objective of these early studies was to develop a rustproof or corrosion-resistant alloy.

Since then, an enormous amount of research has been directed toward understanding and improving the properties of Fe-Cr-Ni alloys. From their initial use in cutlery, the applications of these steels have broadened considerably, along with their compositions. Only two criteria determine whether steels are classified as stainless (or rustproof or pitless): they must be iron-base alloys and they must contain about 11% Cr or more. This broad classification is divided into six principal subclasses:

1. *Ferritic steels* contain 11 to 30% Cr, have little or no nickel, and have low carbon. Ferritic structure is body-centered cubic (bcc) and ferromagnetic; it is usually obtained by slow cooling from the hot forming temperature.
2. *Martensitic steels* typically have high chromium and low carbon and nickel contents. Martensitic structure is bcc; its greater hardness and strength are obtained by quenching.
3. *Austenitic steels* typically contain 18% Cr and 8% Ni. The face-centered cubic (fcc) austenitic structure is stabilized by the nickel content.

*Weight percent is used, unless otherwise specified.

4. *High-manganese austenitic steels* typically contain 24% Mn, instead of 8% Ni, to stabilize the austenitic structure. Their development was driven by the relative scarcities and high costs of chromium and nickel.
5. *Duplex steels* contain varying amounts of nickel (4 to 8%) to adjust the relative amounts of ferrite and austenite and, thus, to achieve high strength (from the ferrite) and adequate toughness (from the austenite).
6. *Precipitation-hardened steels* contain aluminum, niobium, and titanium, which form carbides during aging at high temperatures. The carbides precipitate in either martensitic or austenitic structures, increasing the strength at the higher temperatures.

The ternary Fe-Cr-Ni and Fe-Cr-Mn diagrams in Fig. 1 show the chemical boundaries of these elements for the six types of stainless steels. Often these steels are strengthened by adding interstitial carbon or nitrogen or the solid-solution alloying element molybdenum.

The earliest book (1935) on stainless steels that has come to our attention is a series of technical papers on production and fabrication, properties, test techniques, and applications edited by Thum.(6) Then in 1949 Zapffe produced a simplified book(7) that emphasized the three structures (ferritic, martensitic, and austenitic) of stainless steels. This book also contains the fascinating history of the development and industrial manufacture of these steels. Two volumes devoted to stainless steel applications, production, corrosion resistance, properties, and alloying effects were written by Monypenny in 1951(8) and 1954.(9) In 1956, Keating(10) wrote a user-oriented book on Cr-Ni austenitic steels that focused on properties and fabrication techniques.

In 1976 the American Society for Metals (now ASM International) published *Source Book on Stainless Steels*,(11) which includes property, fabrication, corrosion, forging and heat treatment, phase-diagram, and other information central to the use of stainless steels. To date, the exhaustive handbook assembled in 1977 by Peckner and Bernstein(12) represents the most comprehensive review of these alloys; it documents stainless steel melting, fabrication, metallurgy, corrosion resistance, properties, and applications. Lula recently (1986) revised a general introductory book by Parr and Hanson(13) in which they discuss metallurgy, properties, corrosion, and fabrication of stainless steels. In the same year, he also wrote a careful review of the metallurgy, properties, and applications of high-manganese stainless steels.(14)

From the 1950's into the 1980's, several major international conferences have been held on stainless steels in general or on specific aspects of these alloys. Those of which we are aware are:

1. Symposium on σ-phase formation and verification, 1950, sponsored by the American Society for Testing and Materials(15)
2. Symposium on the residual element effects on properties, 1966, sponsored by the American Society for Testing and Materials(16)
3. Conference, Stainless Steels for the Fabrication and User, 1969, proceedings

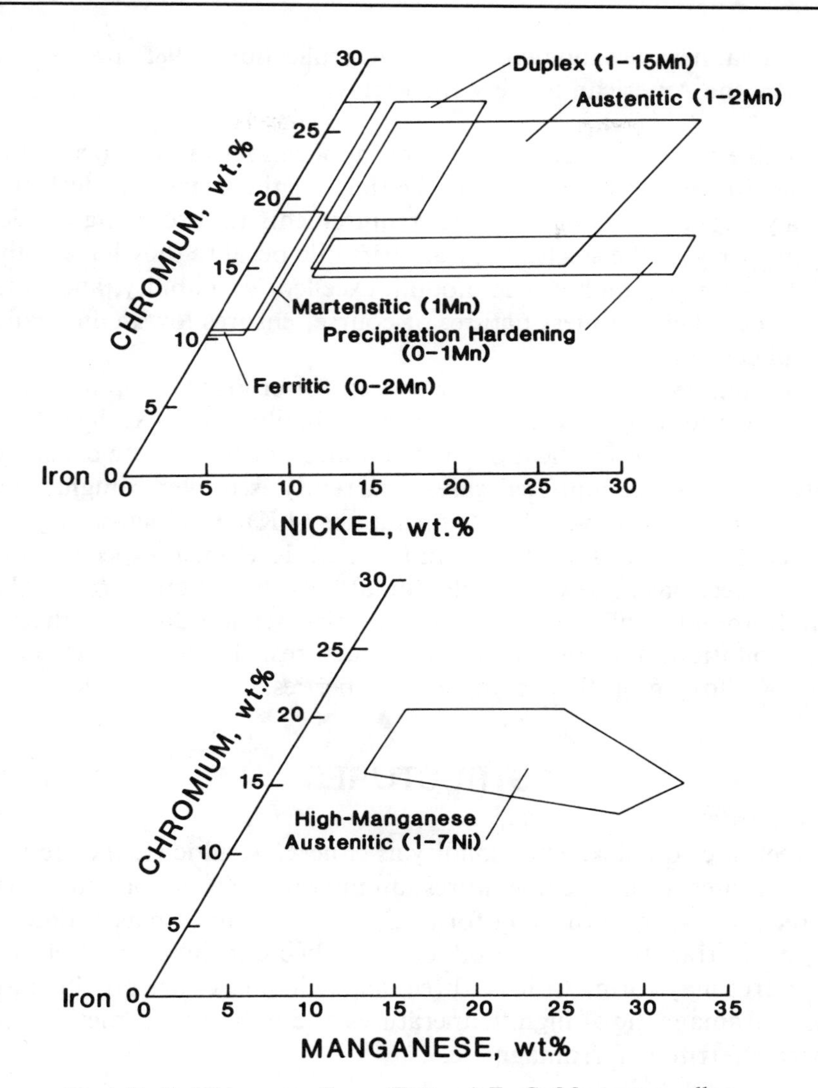

(Top) Fe-Cr-Ni ternary alloys. (Bottom) Fe-Cr-Mo ternary alloys.

Fig. 1. Composition ranges of various types of stainless steels

published by the Iron and Steel Institute, London.[17] At this conference, some of the first reports of dual-phase stainless steels were presented.

4. Conference on stainless steels in general, with emphasis on new applications, 1977, sponsored by Climax Molybdenum Company and held in London[18]
5. Symposium on stainless steel castings, 1980, sponsored by ASTM[19]
6. Conference on stainless steels with emphasis on properties and alloy development for cryogenic applications, 1982, sponsored by the International Cryogenic Materials Conference and held in Kobe, Japan[20]
7. Conference on stainless steels with emphasis on fabrication techniques, selec-

tion criteria, new developments, and new applications, 1985, proceedings published by the American Society for Metals.[21]

Thus, notable stainless steel development, applications, and associated research have been significant for about 50 years. During this time, stainless steels have been used in many applications with material requirements far exceeding the definitive rustproof property of the steels. They are versatile because they have high toughness, variable strength, high elastic moduli, excellent weldability, and other practical properties. Their stainless nature, of course, ensures low maintenance costs in most applications.

Austenitic stainless steels are usually chosen for cryogenic applications by virtue of their low thermal conductivity, good weldability, and excellent toughness. They are well suited for applications that require unusual safety considerations, such as storage of some liquefied gases. Currently, stronger, tougher austenitic steels are needed for use at very low temperatures (4 K). Nitrogen-strengthened Fe-Cr-Ni and Fe-Cr-Mn steels are being studied and developed, especially in Japan.

Subsequent sections review how alloying affects the general structural characteristics and properties of stainless steels. The first section discusses their solidification, precipitation, and transformation structures; the second section presents the effects of alloying on their cryogenic properties.

STRUCTURES

Stainless steels are complex. They solidify as either austenitic (fcc) or ferritic (bcc) structures. At intermediate temperatures, an intermetallic compound (σ-phase) or carbide precipitates, or both, may form. During cooling or plastic deformation, two martensitic transformation products with bcc and hexagonal close-packed (hcp) structures may form. At least three magnetic phases are possible: austenite is typically paramagnetic at high temperatures and antiferromagnetic at very low temperatures; ferrite is ferromagnetic.

Austenite

Austenite is naturally the dominant phase of austenitic stainless steels at ambient temperatures. Austenite is sometimes called γ phase and is always designated γ in phase diagrams. Solid-solution alloying elements, such as chromium, nickel, and manganese, assume normal lattice sites in the iron-base, fcc structure; there is no convincing evidence of ordering of these elements in the austenitic steels. Smaller atoms, such as carbon, nitrogen, phosphorus, and sulfur, are thought to be located interstitially in fcc octahedral lattice sites. For these elements (especially nitrogen), electronic bonding or short-range ordering may be present, but so far the evidence is only circumstantial. Austenite remains paramagnetic on cooling until about 50 K; then it usually becomes antiferromagnetic. The magnetic structures of alloys with high nickel contents may become mictomagnetic or spin-glass at low temperatures. The Néel temperatures, or temperatures of transition to mixed ferromagnetic states, are strongly dependent on alloying.

Ferrite

Ferrite may form during solidification or high-temperature treatment, depending on alloy content. It is occasionally present in small amounts (usually less than 10%) in alloys that are rapidly cooled, such as weldments. This bcc phase is conventionally called δ-ferrite when it forms at temperatures above the γ-loop or range and α-ferrite when it forms at lower temperatures. Most austenitic stainless steels have sufficient stabilizing elements, such as nickel, manganese, carbon, and nitrogen, to lower the temperature of the γ-loop or range sufficiently to prevent the diffusion-controlled $\gamma \rightarrow \alpha$ transformation. In wrought alloys, the retained δ-ferrite is typically transformed to austenite during forging or hot rolling. However, since the solubilities of many elements differ between the two phases, the solidification structure affects the location of these elements. This is especially significant for elements, such as phosphorus and sulfur, that are more soluble in ferrite; therefore, they are less prone to precipitate along grain boundaries during solidification when δ-ferrite is present. For this reason, δ-ferrite is thought to avert hot cracking in austenitic stainless steel weldments.

The relative amounts of austenite and ferrite that form during solidification are critically dependent on alloying. The liquidus and solidus surfaces of the Fe-Cr-Ni ternary have been presented by Speich[22] and by Schurmann and Brauckman.[23] The eutectic trough extends from near the Fe corner (76Fe-10Ni-14Cr) to a ternary eutectic (about 8Fe-43Ni-49Cr). Primary austenite solidifies if the composition is on the nickel side of the eutectic liquidus. Nearer the eutectic liquidus on the nickel side, limited ferrite forms, mainly in the dendritic cell boundaries. On the chromium-rich side of the liquidus line, primary ferrite solidifies.[24] Upon cooling, however, ferrite transforms to austenite in compositions typical of austenitic stainless steels. The ternary isothermal sections at 1350 and 1100 °C are shown in Fig. 2. At lower temperatures, the $\delta + \gamma$ region increases with attendant decreases in both δ and γ single-phase regions.

The tendency to transform from δ-ferrite to γ is illustrated by the phase diagrams at constant iron content but varying nickel and chromium contents (Fig. 3). Notice that the minimum nickel content for stable γ at 70% Fe is about 8% (22Cr) and at 60% Fe, about 14.5% (25.5Cr). In practice, these limits are adjusted by small additions of carbon, nitrogen, manganese, silicon, and, sometimes, molybdenum.

The binary Fe-Cr and Fe-Ni phase diagrams are shown in Fig. 4. Whereas addition of chromium encourages formation of ferrite, addition of nickel stabilizes the austenite. Indeed, in both binaries, there is an extensive solid-solution range (>85% alloy contents) at temperatures above 800 °C (below 800 °C, σ-phase forms in Fe-Cr alloys, and below 500 °C, high-nickel iron alloys tend to order). Binary iron-alloy phase diagrams for molybdenum, silicon, and manganese are reviewed by Novak.[25] Like chromium, molybdenum and silicon have extensive ferrite ranges; the γ-loop extends to only about 2% Mo and 2% Si. Austenite formation is

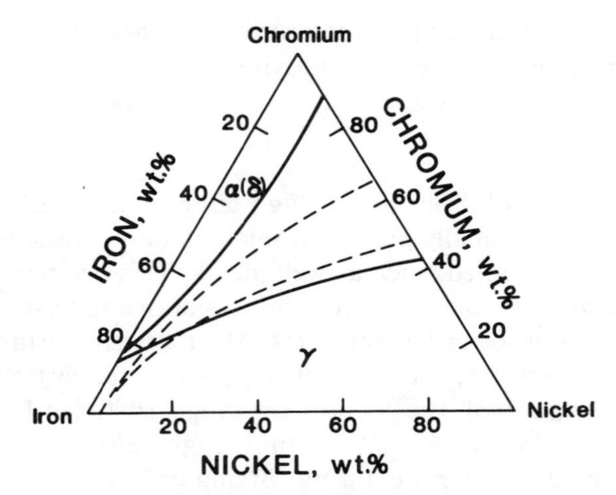

Fig. 2. Isothermal sections at 1350 °C (dashed lines) and 1100 °C (solid lines) for Fe-Cr-Ni ternary system

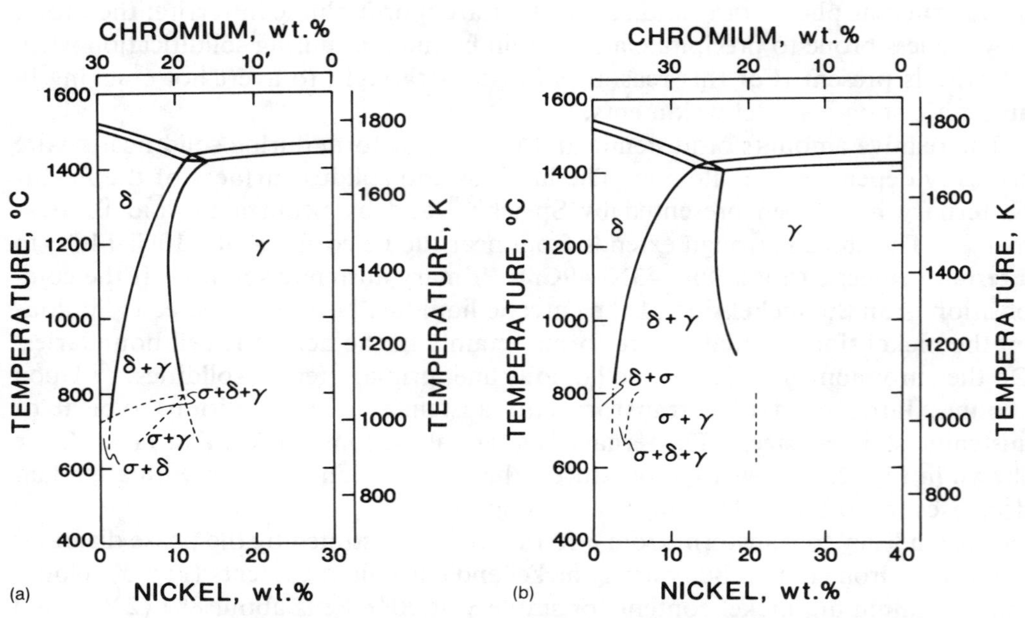

Fig. 3. Binary Fe-Cr_xNi_{30} (a) and Fe-Cr_yNi_{40-y} (b) phase diagrams depicting dependence on Cr and Ni at constant Fe contents of (a) 30 and (b) 40 wt%

encouraged by alloying iron with manganese; at 1000 °C, the γ-loop extends to 70% Mn.

For welding, the amount of δ-ferrite present is normally estimated from a Schaeffler[26] or DeLong[27] diagram. The ferrite portion of this type of diagram is shown in Fig. 5; the lines representing zero ferrite, or the demarcation between pure austenite (above) or small amounts of ferrite (below), are plotted. Tie lines depicting constant ferrite amounts have been developed empirically in terms of the nickel

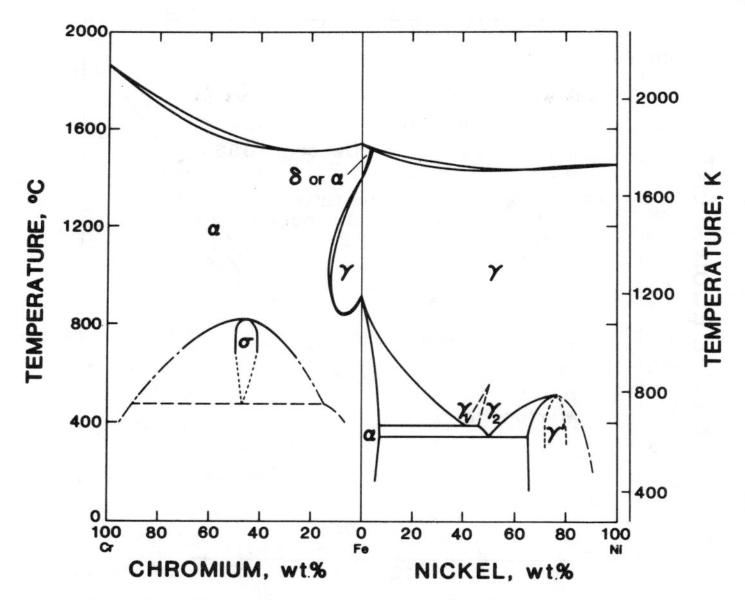

Fig. 4. Binary Fe-Cr and Fe-Ni phase diagrams

equivalent (austenite-forming tendency) and chromium equivalent (ferrite-forming tendency). These equivalents serve to identify the austenite-martensite and austenite-ferrite boundaries at ambient temperatures, when many alloying elements are present.

The proposed nickel- and chromium-equivalent equations are summarized in Table 1. Presumably, the Suutala formulations[28] are the most reliable equations because they are the most recent and were developed from the most data. Manganese additions of up to 2 to 4% have been found to promote austenite formation; greater amounts tend to promote ferrite formation.[29] Nitrogen additions result in greater austenite stability. Notice in Table 1 that the more recent Hammar[30] equivalents emphasize carbon more than nitrogen. Suutala concludes that at low (normal) contents, manganese additions stabilize austenite, but at higher contents (>5%), manganese additions enhance ferrite formation. Thus Hull's formulation,[29] which includes a negative Mn^2 term, provides the best representation of manganese effects.

Ferrite is ferromagnetic at temperatures below about 600 °C and paramagnetic at higher temperatures.

Martensite

The use of austenitic stainless steels is complicated by the metastability of the austenitic structures of most alloys. The metastability leads to martensitic transformation, which is a significant design consideration in applications requiring fracture-control planning, close dimensional tolerances, the absence of a ferromagnetic phase, and high toughness of weldment, heat-affected zone, and base metal. Upon cooling, under applied elastic stresses, or during plastic deformation, the

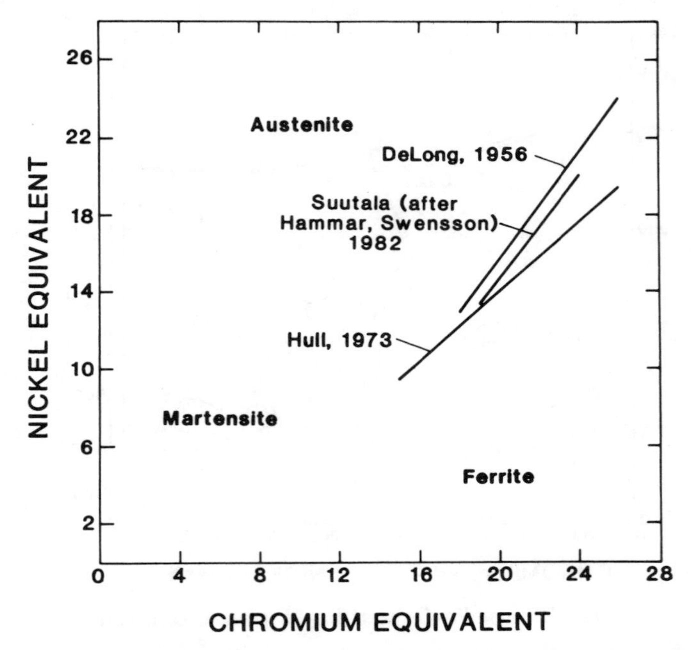

CHROMIUM EQUIVALENT

Fig. 5. Ferrite portion of Schaeffler diagram depicting lines that estimate the first solidification of ferrite during cooling to room temperature after welding

Table 1. Ferrite solidification equivalent coefficients

	Cr-equivalent coefficient (per wt %)					
Investigator (year)	Cr	Mo	Si	Nb	Ti	Al
DeLong *et al.* (1956)	1	1	1.5	0.5	...	...
Hull (1973)	1	1.21	0.48	0.14	2.2	2.48
Hammar, Svensson (1979)	1	1.37	1.5	2	3	...

	Ni-equivalent coefficient (per wt %)						
Investigator (year)	Ni	Mn	C	N	Cu	Co	Mn^2
DeLong *et al.* (1956)	1	0.5	30	30	...	...	...
Hull (1973)	1	0.11	24.5	18.4	0.44	0.41	−0.0086
Hammar, Svensson (1979)	1	0.31	22.5	14.2	1	...	...

austenitic phases of Fe-Cr-Ni alloys may transform to bcc (α') and hcp (ϵ) martensite phases. The alloy composition affects the chemical free-energy difference between the two phases.

Kaufman[31] derived free-energy relationships between γ and α' for the Fe-Cr-Ni ternary system. Reed[32] reviewed the Fe-Cr-Ni ternary free-energy relations between fcc and bcc and suggested a relationship for the free-energy differences between fcc and hcp structures based on the earlier work of Kaufman[31] on binary Fe-Cr and Fe-Ni alloys.

Other alloying elements, such as carbon, nitrogen, manganese, molybdenum, and silicon, contribute to austenitic stability in austenitic stainless steels. The complexity of the final structure has led to many studies for characterization of the effects of alloying on the stability of the austenitic structure with respect to α'-martensite.

Expressions relating the stability of the austenite and the temperatures of transformation during cooling (T_{ms}) or during deformation (T_{md}) have been developed empirically and are summarized in Table 2. Eichelman and Hull,[(33)] working with austenitic alloys with compositions of 10 to 18 Cr, 6 to 12 Ni, 0.6 to 5 Mn, 0.3 to 2.6 Si, 0.004 to 0.129 C, and 0.01 to 0.06 N, established that all these alloying elements stabilize the austenite and thus lower T_{ms}. Monkman *et al.*,[(34)] using a larger number of specimens with compositions of 5 to 13 Ni, 11 to 19 Cr, and 0.035 to 0.126 (C + N), produced a similar analysis and concluded that T_{ms} was linearly dependent on composition only to a first approximation. The dependence of T_{ms} on the carbon and nitrogen concentrations seemed to be influenced by the chromium and nickel concentrations.

Hull[(29)] studied the effects of nickel, chromium, manganese, carbon, nitrogen, silicon, and cobalt (aluminum, thorium, vanadium, and tungsten sometimes were added) on α'-martensite formation during low-temperature cooling-and-deformation experiments and on ferrite retention after cooling to room temperature from the melt. Hull's study confirmed that all the above elements suppress low-temperature α' formation during either cooling or deformation. To predict the effects of elemental additions on T_{ms}, Hull assumed that the nickel contribution was the average of the Monkman *et al.* and Eichelman and Hull results and then compared all the effects of alloying additions with that of nickel. The formulation of Andrews[(35)] applies to lower chromium and nickel concentrations, and thus higher T_{ms}; his results are included in Table 2 to indicate the disparity between the high- and low-temperature empirical results. In all studies, specimens of various compositions were prepared and cooled, and their T_{ms} values were measured.

Table 2. Temperature equivalent coefficients for calculation of T_{ms}

	Equivalent coefficient, K/wt %				
	Investigator (year)				
Element	Eichelman, Hull (1953)	Monkman, Cuff, Grant (1957)	Andrews (1965)	Hull (1973)	Self (1986)
Base	1578	1455	273	1755	794
Cr	−41.7	−36.7	−12.1	−47	−14.3
Ni	−61.1	−56.7	−17.7	−59	−17.5
Mn	−33.3	...	−30.4	−54	−28.9
Si	−27.8	...	...	−37	−37.6
C	−1670	−1460	...	−2390	−350
N	−1670	−1460	−423	−3720	...
Mo	...	...	−7.5	−56	−29.6
Other	...	...	...	−180 (Ti)	−1.19 (CrNi) +23.1 (Cr+Mo)C

Regression analyses were used to obtain the empirical dependence of T_{ms} on alloy concentration. Self[36] recently studied a wide range of compositions of weld deposits. For analyses, he assumed nonideal behavior, and with regression analysis, he derived a predictive equation containing interactive Cr-Ni and (Cr + Mo)-C terms (Table 2).

From Tables 1 and 2, it is apparent that the roles of chromium, molybdenum, and silicon are reversed in the stabilization of the bcc structure (ferrite and α'-martensite) between high and low temperatures. Binary Fe-Cr free-energy derivations[31,32] do not suggest this trend. Perhaps the explanation for the stabilizing effect at lower temperatures is the contribution of these elements to the increase in austenite flow strength. The effect of molybdenum on flow strength is discussed later in this chapter.

The analytic expressions of Schaeffler, Eichelmann and Hull, and Self are plotted in terms of nickel and chromium equivalents in Fig. 6. The less conservative analysis of Self implies that at room temperature, considerably less than 18% Cr is required to achieve the complete austenite stability of an 8Ni-equivalent alloy. Also, the nonlinear nature of the Self equivalent indicates the possibility of martensite control by means of alloy optimization rather than alloy "trade-offs," as offered by the linear trend lines.

Experiments have been conducted to assess the influence of cold work (either tensile or compressive) on T_{md}. Results are presented in Table 3. The experimental

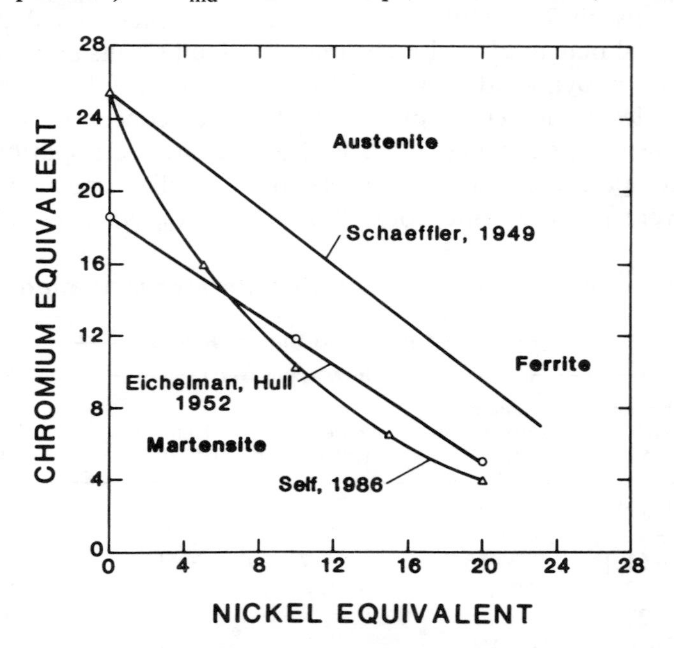

For constant amounts of 1 Mn, 0.03 C, and 0.5 Si.

Fig. 6. Martensite portion of Schaeffler diagram depicting lines that estimate the first formation of martensite during cooling to room temperature after welding.

Table 3. Temperature equivalents for calculation of T_{md}

Investigator (year)	Equivalent coefficients, K per wt %								
	Base	Cr	Ni	Mn	Si	C	N	Mo	Other
Angel (1954)	686	−14	−9.5	−8.1	−9.2	−46.2	−46.2	−18.5	...
Hull (1973)	1655	−23	−59	−41	−20	−777	−315	−24	−12 (Co)
Williams *et al.* (1976)	686	−6	−25	−16	+21	−222	−222	−11	...

definition of T_{md} varied in each study: Angel[37] used 50% α' at 30% tensile elongation; Williams *et al.*[38] used 2.5% α' at 45% compressive elongation; and Hull[29] used the minimum detection level (probably about 1% α') at 50% compressive elongation. There are subtle distinctions among the coefficients for T_{ms} and T_{md} calculations. After cooling, the chromium and nickel coefficients are nearly equal, manganese is higher, and carbon and nitrogen are higher by a factor of 50; after deformation, chromium and manganese are decidedly less than nickel, and carbon and nitrogen are higher by a factor of about 10. Careful research to delineate the dependence of composition on free energy, stacking-fault energy, and deformation parameters is needed to understand these distinctions.

Upon cooling or during deformation, fcc austenite may transform to two martensitic products. The hcp phase is associated with extended stacking faults and forms as thin sheets on (111) austenite planes. In most austenitic stainless steels, the α' product forms as laths. Breedis[39] reported a lath α' morphology for compositions ranging from Fe-19Cr-11Ni to Fe-10Cr-16Ni; at lower chromium and higher nickel concentrations, the α' morphology changed to surface martensite and then to a platelike structure (Fig. 7). From Reed,[40] the α' lathlike structure is

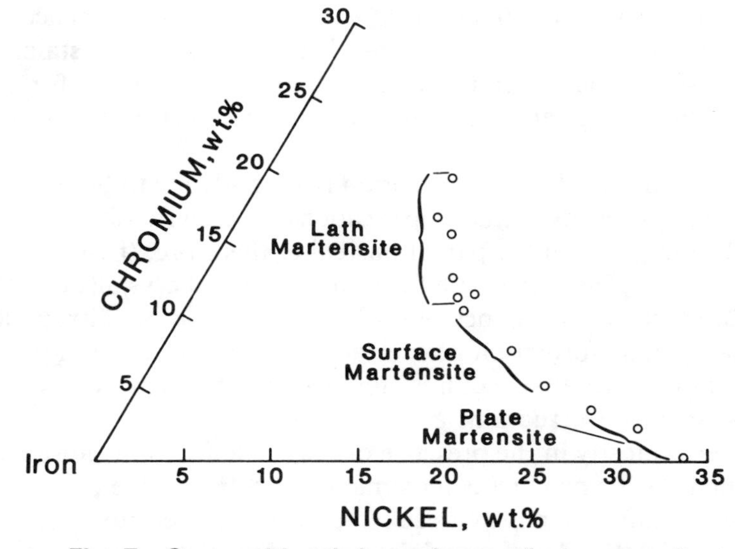

Fig. 7. Compositional dependence of α'-martensite morphology in Fe-Cr-Ni ternary alloys

parallel to $\langle 110 \rangle_\gamma$, with $(225)_\gamma$, $(112)_\gamma$, or both habit planes.[41] The laths are restricted within $\{111\}_\gamma$ bands, and usually three sets of habit planes form within the band.

There is a considerable amount of accommodation deformation in the austenite after transformation during cooling, particularly within the {111} bands containing the α' laths. Either ϵ-martensite or large amounts of stacking faults are present.

The orientation relationships among the γ-, ϵ-, and α'-phases[41] are:

$$(111)_\gamma \parallel (0001)_\epsilon \parallel (101)_{\alpha'}$$

$$[1\bar{1}0]_\gamma \parallel [1\bar{2}10]_\epsilon \parallel [11\bar{1}]_{\alpha'}$$

These relationships are apparently retained regardless of the manner of ϵ or α' formation.

Transformation during plastic deformation initially results in α'-martensite at intersections of active {111} deformation bands. The α' assumes the shape of laths along ⟨110⟩ common to the two active {111} systems. Evidence of dislocation pile-ups at α' laths has been observed.[32] Also, more than one lath forms at each transformation site, and, in contrast to the α'-phase that forms on cooling, all laths have the same habit-plane variant.

The internal defect structure of the α'-phase in Fe-Cr-Ni steels consists predominantly of dislocations. As chromium is replaced by nickel, Breedis[39] reports that the cellular, irregular distributions of dislocations that are typical of α' laths change to the planar, regular arrays that are typical of α' plates. Also, as nickel replaces chromium, the amount of ϵ transformation decreases, the sharpness of the hcp reflections decreases, and the fcc twin reflections become more diffuse.

Many metallurgists have been concerned with the role of the ϵ transformation: does ϵ act as a precursor transformation ($\gamma \rightarrow \epsilon \rightarrow \alpha'$), or is the ϵ an accommodation effect ($\gamma \rightarrow \alpha'$, $\gamma \rightarrow \epsilon$)? Both effects have been observed. The stainless steels with low stacking-fault energy tend to form ϵ, and the α' tends to form from ϵ. Higher-stacking-fault-energy alloys require transformation stresses for observable ϵ formation.

Several characteristics of the transformation from austenite to bcc-martensite in stainless steels cause problems in many applications. The martensitic bcc has a specific volume 1.7% larger than the parent austenite; therefore, these steels expand during transformation. The martensite forms as individual crystals, and the shear stress and volume expansion associated with transformation disrupt localized regions. For example, α' formation near or at the surface results in localized surface upheavals. For service that requires close tolerances, such as valves or bearings, these local surface fluctuations are disastrous.

Another concern, mostly in the presence of magnetic fields or pulsed currents, is that the bcc martensitic product is ferromagnetic, whereas the parent austenite is paramagnetic. A simple rule is that each 1% of bcc martensite results in a permeability increase of 0.01. Therefore, if time-dependent field changes corresponding to material changes on the order of 0.10 are significant, then alloy selection is important.

Stacking-Fault Energy

The stacking-fault energy is related to the free-energy difference between the fcc principal structure and the hcp structure. Lower stacking-fault energy leads to wider partial-dislocation separation (larger fault ribbons) and hence more planar dislocation slip structures and reduced cross slip.

Schramm and Reed[42] used x-ray peak-shift measurements, coupled with previous stacking-fault-energy measurements, to estimate the dependence of the stacking-fault energy on chemical composition. This dependence should correspond to the dependence of the $\gamma \rightarrow \epsilon$ transformation on composition, because a stacking fault represents a local planar area of ϵ. Rhodes and Thompson[43] suggested that the stacking-fault energy values of the Schramm and Reed analysis were too large, considering additional weak-beam electron-microscopy data. Weak-beam electron-microscopy measurements by Bampton, Jones, and Loretto[44] confirmed that the least-square analysis of Schramm and Reed produced stacking-fault energies that were too large. From measurements of individual nodes, they estimated a data spread of $\pm 25\%$. Brofman and Ansell[45] added carbon content to the factors affecting stacking-fault energy. Dependencies of stacking-fault energy on composition are given in Table 4.

Ledbetter and Austin[46] recently examined with x-rays the effects of additions of carbon and nitrogen ($C + N \leq 0.325$ wt %, $N \leq 0.21$ wt %) on the stacking-fault energy of an Fe-8Cr-10Ni alloy. They reported an increase of about 10% in the stacking-fault energy per at. % C + N. Earlier, Stoltz and VanderSande[47] determined with weak-beam microscopy that nitrogen in excess of about 0.24 wt % dramatically decreased the stacking-fault energy of an Fe-19Cr-7Ni-8Mn-0.03C alloy. Fujikuma *et al.*[48] measured stacking-fault probabilities for an Fe-18Cr-10Ni-8Mn alloy. They reported a minimum at about 0.15% N; a slight increase at lower nitrogen contents, and a rapid increase at higher nitrogen levels. Thus, all three sets of measurements appear to be consistent: the stacking-fault energy increased slightly with nitrogen contents up to about 0.20 wt %; it decreased substantially for higher nitrogen contents.

Carbides

The solid-solution solubility of the mobile interstitials carbon and nitrogen in austenitic alloys is very temperature dependent (Fig. 8).[49,50] At temperatures less

Table 4. Stacking-fault energy at room temperature

Investigator (year)	Energy equivalent, mJ/(m²·wt %)							
	Base	Cr	Ni	Mn	Si	C	Mo	Other
Dulieu, Nutting (1964)	0	...	0.5	1.4	...	3.4	0.1	3.6 (Ti), 3.2 (Cu), −0.55 (Co)
Schramm, Reed (1975)	−53	0.7	6.2	3.2	...	...	9.3	
Rhodes, Thompson (1977)	1.2	0.6	1.4	17.7	−4.7			
Brofman, Ansell (1978)	16.7	0.9	2.1	...	...	26	...	

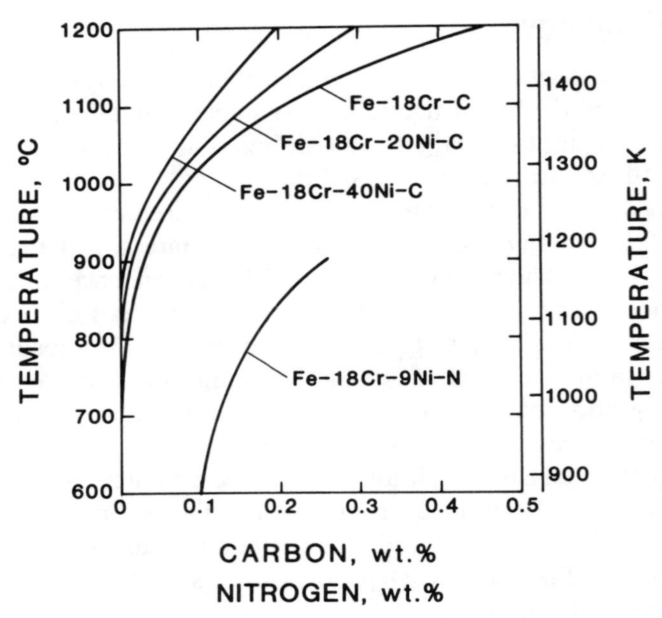

Fig. 8. Solubility limits for carbon[49] and nitrogen[50] as functions of temperature for various alloys

than about 950 °C, carbon begins to precipitate as a carbide in practically all austenitic steels, since the solubility limit is about 0.035. Using electron microscopy, the solubility of carbon in type 316 steel (Fe-19Cr-14Ni-2Mo) has been represented by the expression[51]:

$$\log [\mathrm{C\ (ppm)}] = 7.771 - 6272/T(\mathrm{K}) \qquad \text{(Eq 1)}$$

The solubility limits determined from Eq 1 are considerably less than those given in Fig. 8, which reflects the higher sensitivity of electron microscopy for detection of carbides. Carbides precipitate in the range 500 to 1000 °C; below 500 °C the atom mobility is too low, and above 1000 °C the solubility of carbon and nitrogen is sufficient. The most common carbide is $M_{23}C_6$, where M represents a metal, normally chromium. However, iron, molybdenum, titanium, and niobium also may assume the position of the metallic element, and nitrogen and boron may assume the position of carbon in the $M_{23}C_6$ structure.

The solubility limits of both carbon and nitrogen are reduced by the presence of nickel (see Fig. 8 for carbon). On the basis of binary Fe-Ni data,[52] Fig. 9 shows that this reduction is about 0.014% N from the addition of 20% Ni. The data of Sakamoto *et al.*,[53] obtained from observations of bubbling vacuum-induction melts, are quite consistent and represent the practical limits of maximum nitrogen contents (Fig. 9). These limits are particularly important to define the nitrogen-strengthening limitations that are discussed under "Strength" in the next section. Consistent with the binary alloy solubility limits, these data show that both chromium and manganese increase the ingot capacity for nitrogen.

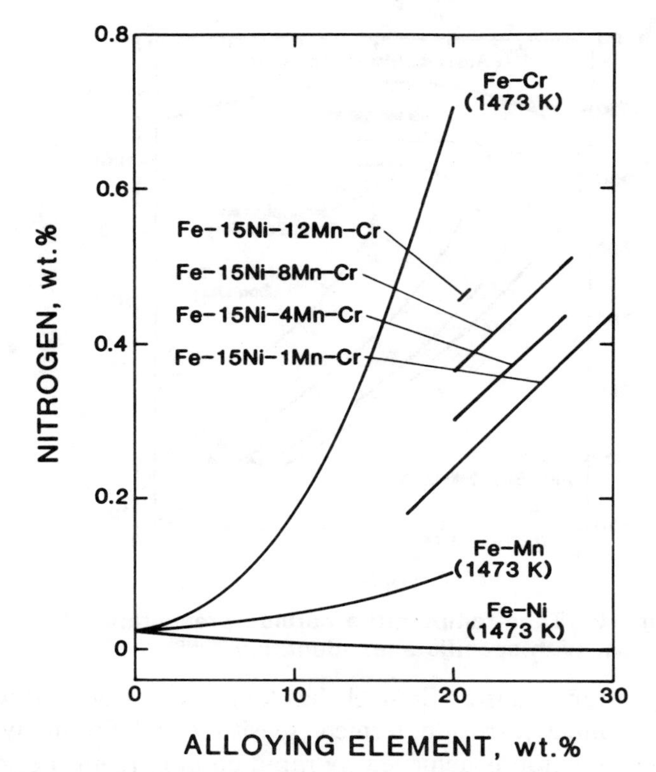

Fig. 9. Solubility limits for nitrogen in Fe-Cr, Fe-Ni, and Fe-Mn binary alloys at 1473 K[52] and in Fe-15Ni-Mn-Cr alloy liquids[53]

Carbides precipitate primarily at ferrite-austenite interfaces, followed by grain boundaries, noncoherent and coherent twin boundaries, and finally, within grains. This is depicted schematically in Fig. 10.[54,55] The temperature decreases and the time for nucleation increases as the precipitation site changes from high- to low-energy boundaries. Increased carbon content increases precipitation temperature and decreases time; increased nitrogen content has the reverse effect, suppressing carbide formation.[12]

The $M_{23}C_6$ carbide has a complex fcc structure with 92 metallic atoms and 24 smaller (C, N, or B) atoms and a lattice parameter of 10.61 to 10.64 Å.[12] But there is only minor mismatch between the {111} $M_{23}C_6$ and {111} austenite, estimated at 1.3% by Lewis and Hattersley.[56] Thus, the common plane of the interface between the two structures has been identified as {111}.[55,56] Alloy additions to the $M_{23}C_6$ structure, such as nitrogen, tend to decrease its lattice parameter and to increase the amount of interface mismatch. This should retard the growth kinetics.

Carbide precipitates normally have a sheetlike morphology at lower temperatures (480 to 730 °C) and dendritic shapes at intermediate temperatures (600 to 875 °C); particles form at the higher temperatures (>850 °C).[55] Higher temperatures, as well as longer aging times, promote coarsened structures.

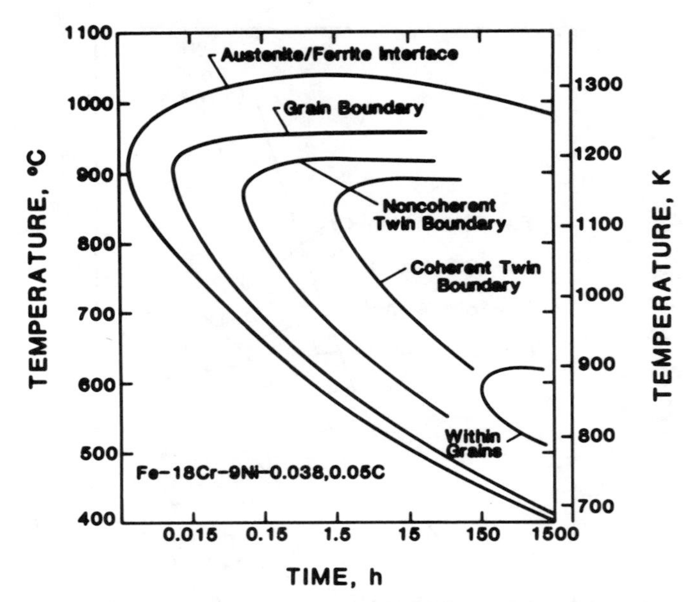

Fig. 10. Time-temperature carbide precipitation kinetics, with specific sites identified[54,55]

Carbide precipitation in austenitic steels has long been associated with increased sensitivity to intergranular corrosion, termed "sensitization." For many applications, suppression of sensitization is achieved by rapid cooling from the solution-treatment temperature. During welding, however, the heat-affected zone unavoidably enters the temperature range of carbide formation. To prevent severe sensitization during welding, two courses have been followed: (1) low-carbon (≤0.03% C) grades, designated by AISI with the "L" nomenclature, have been provided; and (2) alloys with titanium (AISI 321) and niobium (AISI 347) have been developed to achieve more random TiC or NbC precipitation within grains, rather than $M_{23}C_6$ precipitation at grain boundaries. These alloys are often called stabilized grades of stainless steel, since the carbon, taken out of solution, is unavailable for carbide formation and precipitation at grain boundaries.

More thorough reviews of carbide formation in austenitic steels are provided by Novak[25] and Lai.[57]

Sigma Phase

Notice the Fe-Cr part of the phase diagram in Fig. 4: at chromium contents greater than about 15%, the σ-phase may form when the alloy is exposed to temperatures between 400 and 800 °C. The σ-phase extends into the Fe-Cr-Ni ternary diagram, as indicated by Fig. 3. Alloys rich in chromium and lean in nickel tend toward σ-phase formation at lower temperatures. Maehara *et al.*[58] studied the effects of alloy additions on the amounts and aging times for σ-phase formation. They found that nickel decreases the amount of σ-phase formation, but increases the rate of formation in an Fe-25Cr-2.8Mo alloy. Additions of chromium and molybdenum increase both the aging kinetics and amounts of σ-phase in Fe-6.5Ni-2.8Mo and Fe-25Cr-6.5Ni alloys, respectively.

In the Fe-Cr system, the highest temperature of σ-phase formation is 821 °C (Fig. 4); nickel additions gradually increase this temperature (Fig. 3). The constant-iron vertical sections of the Fe-Cr-Ni ternary of Fig. 3 are consistent with the liquidus, solidus, 1100 °C, and 650 °C isotherms of Speich[22] and the 802 °C and 648 °C isotherms of Talbot and Furman.[59] Like carbides, the σ-phase forms preferentially at high-energy surfaces. Also, when ferrite is present, there is a strong tendency for σ-phase to form initially within the ferrite or at the ferrite side of the γ-δ boundary.

The σ-phase has a tetragonal structure, 30 atoms per unit cell, an a_0 lattice parameter range of 8.29 to 9.21 Å, and a c_0 lattice parameter range of 4.60 to 4.78 Å.[12] Sigma-phase alloy constituents do not have fixed stoichiometric ratios; they may range from B_4A to BA_4. Alloy additions, such as chromium, molybdenum, vanadium, and silicon, promote σ-phase formation.

For a more extensive review, see Novak[25] and Lai.[57]

PROPERTIES

Stress-Strain Characteristics

The AISI 300 series stainless steels used in cryogenic applications range from metastable to stable austenites. Alloys such as 304, containing 18% Cr and 8% Ni, are metastable, whereas 310, containing 26% Cr and 20% Ni, is stable with respect to martensitic transformation. The stress-strain behavior and temperature dependence of the flow strength of these alloys differ and depend on the austenite stability. In the less stable alloys, both ϵ- and α'-martensite form; in the slightly metastable alloys, such as 316, only α'-martensite forms; and in the stable alloys, neither ϵ- nor α'-martensite forms during deformation to fracture at any temperature.

At low temperatures, the tensile stress-strain characteristics of austenitic stainless steels depend on the stability of the austenitic structure. Consider the engineering stress-strain curves of a stable austenitic steel, alloy 310 (Fig. 11, top), and a metastable alloy 304 (Fig. 11, bottom) at low temperatures. As the temperature decreases, the yield strength of alloy 310 increases significantly, and the work-hardening rates, reflecting only dislocation interactions, remain relatively constant.

In a metastable austenitic, polycrystalline alloy at low temperatures, three distinct stages are present in the stress-strain curves. The contours of their stress-strain curves (Fig. 12) are remarkably similar to single-crystal shear stress–strain curves. Stage I represents the microstrain and early macrostrain behavior. The formation of α'-martensite is not thought to occur in this range; stacking-fault clusters or ϵ-martensite, or both, are most likely to complement dislocation interactions to reduce the rate of work hardening.

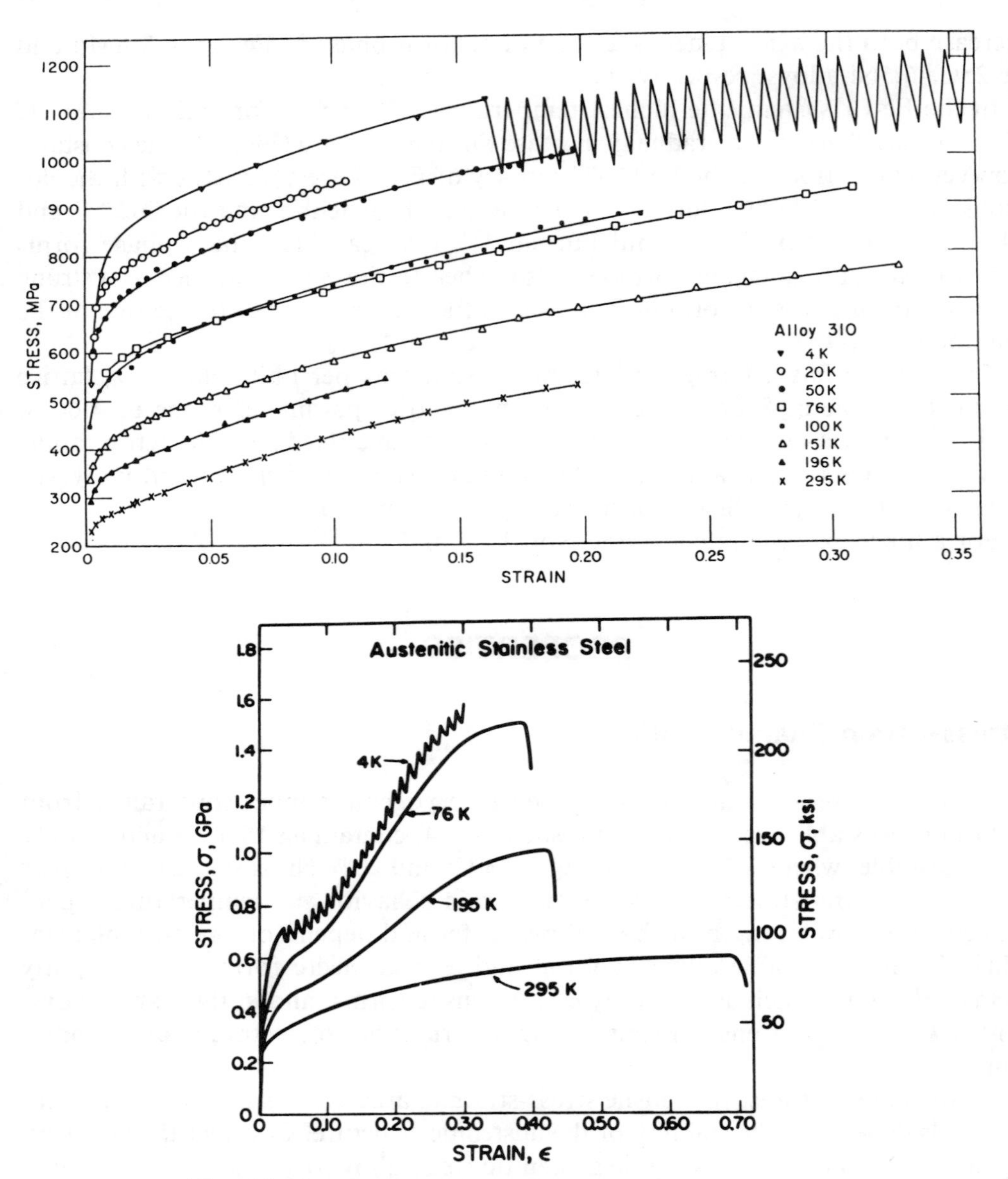

Fig. 11. Low-temperature stress-strain curves for (top) a stable austenitic alloy 310 (Fe-23Cr-20Ni) and (bottom) a metastable austenitic alloy 304 (Fe-19Cr-9Ni)

Stage II, the "easy glide" range, is associated with increasing ϵ-martensite formation and the formation of α'-martensite laths at cross-slip intersections. Suzuki *et al.*[60] proposed that such α' laths, with the long $\langle 110 \rangle$ direction representing the intersection of two active slip systems, act as windows to assist cross slip. Another possible explanation is that strain-induced ϵ-martensite (whose formation is prevalent in this stage) contributes a larger strain component (or, alternatively,

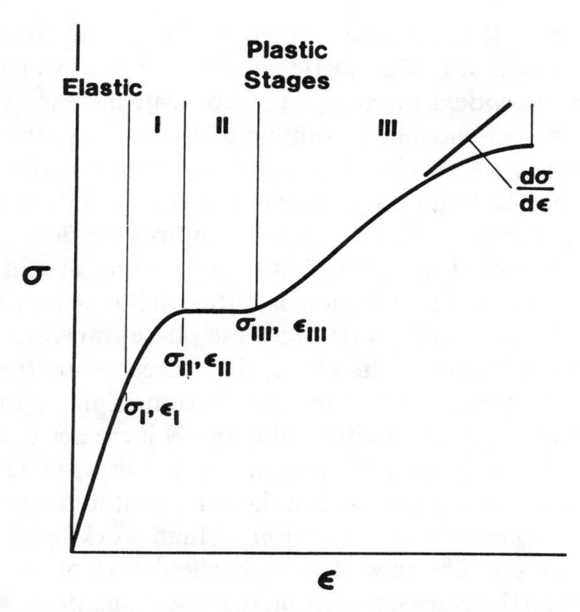

Fig. 12. Schematic depiction and characterization of three stages associated with the stress-strain curves for polycrystalline metastable austenitic stainless steels at low temperatures

exhibits lower work-hardening rates). Stage II is most prevalent in the temperature range 76 to 200 K. In this temperature region, the range of strain at which the stress remains relatively constant is large, usually from 0.04 to 0.10.

In stage III, the rate of work hardening increases and then becomes constant over a relatively large amount of plastic deformation (20 to 40%). The volume percentage of α'-martensite is linearly related to plastic deformation in stage III. The nature of the transition in the roles of α'-martensite formation—from possibly promoting easy glide in stage II to being associated with a linear, high rate of work hardening in stage III—is not clear. Perhaps in stage III essentially all active cross-slip sites have been transformed to α'-martensite, and subsequent α'-martensite formation occurs within active glide bands. Such α'-martensite formation would not be expected to promote stacking-fault glide. There is a hint from the x-ray data of Reed and Guntner[61] that the following sequence may occur: A maximum value of the volume concentration of ϵ-martensite occurs at about the transition from stage II to stage III. There is apparently no ϵ-martensite formation during stage III; the amount of ϵ-martensite decreases with strain, presumably transforming to α'-martensite.

Stages I and II decrease in significance in more stable austenites, but stage III is retained. For instance, AISI 316 exhibits neither stage I nor stage II, but does begin to transform to α'-martensite at strains of about 0.02 below 190 K. No ϵ-martensite was detected in 316LN alloys.

Alloying of austenitic stainless steels affects the stress and strain parameters of the three stages of low-temperature deformation. The effects of manganese alloying on these parameters at 4 K have been studied by Reed and Tobler.[62] The yield strength, σ_y (flow strength at 0.002 plastic strain), is within stage I. As shown in Fig. 13, σ_y is linearly dependent on manganese content, increasing with increasing manganese. The linear dependence is similar to that of σ_{II}, the stress at which stage II begins. The increase of σ_y with manganese content is considerable, about 33 MPa per wt % Mn. The addition of manganese to austenite solid solution does not alter the lattice parameter nor the shear modulus significantly (see later discussion), therefore traditional solid-solution strengthening should not result from manganese addition. But manganese increases the stacking-fault energy and suppresses α'- and ϵ-martensite transformation; these phase transformations influence the stress and strain parameters. Therefore, the increases in strength that occur when manganese is added are due to the contribution of manganese to austenite stability. Figure 13 shows that an addition of 0.1% N increases σ_y about 290 MPa. Nitrogen has the same strengthening effect in stable alloys, which implies that it has little effect on hcp or bcc stability within the composition range 0.1 to 0.2 wt %.

The stress denoting transition to the region of high work-hardening rates, σ_{III}, tracks very closely with σ_{II}. The rate of work hardening (from engineering stress-strain curves) of stage III decreases with increasing manganese and nitrogen, as shown in Fig. 14. The addition of nitrogen slightly affects the dependence of manganese content; nitrogen contributes to the reduction of $d\sigma/d\epsilon$ about ten times more strongly (on the basis of wt % addition) than does manganese. The dependence of the work-hardening rate in stage III on the percent α'-martensite per unit elongation is shown in Fig. 15. Clearly, there is a strong linear correlation, and the rate of work hardening can be regarded as a function of austenite stability with respect to α'-martensite formation, as determined by composition.

We have performed numerous correlations between the normalized amount of

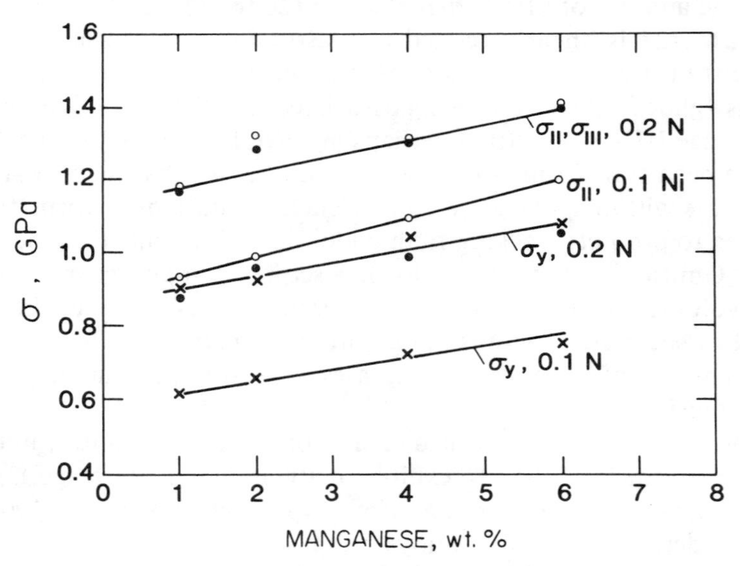

Fig. 13. Dependence of stages I and II strength parameters on Mn and N contents at 4 K

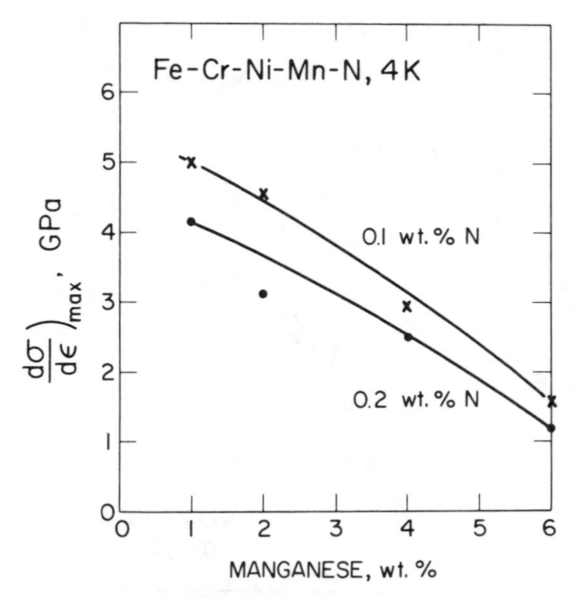

Fig. 14. Dependence of stage III work-hardening rate on Mn and N contents at 4 K

strain-induced α'-martensite at final elongation at 4 K and the alloy composition. Our best correlation for this normalized strain-induced martensite was obtained by using the Williams *et al.*[38] regression analysis for T_{md}. In Fig. 16 their expression is plotted versus percent α'/percent tensile elongation at 4 K for a series of modified 316LN alloys. Correlation is excellent for these alloys with a calculated T_{md} ranging from 120 to 240 K.

Therefore, the deformation characteristics of austenitic stainless steels can be significantly controlled by alloying. Sufficient alloy additions to stabilize the austenite result in typical solid-solution-strengthened, fcc stress-strain characteristics. The shape of the stress-strain curves of metastable austenites during strain-induced α'- and ϵ-martensite transformations is affected by alloying.

Strength

In addition to austenite stability, the strength of austenitic stainless steels is influenced by temperature, grain size, cold and hot working, and interstitial and solid-solution alloying. Temperature and grain-size effects have been discussed previously.[63-66] We focus here on the effects of alloy strengthening on stable austenitic steels at low temperatures.

At very low temperatures, the variation of tensile yield strength (σ_y) with nitrogen, carbon, molybdenum, manganese, and nickel additions to Fe-Cr-Ni austenitic steels has been recently and extensively studied. In Japan, a program to develop an alloy with σ_y of 1200 MPa and a toughness of 200 MPa·m$^{1/2}$has now been completed.[67-69] At NBS we have tested at low temperatures some of the new Japanese alloys, the European 316LN alloys, and selected laboratory heats from U.S.

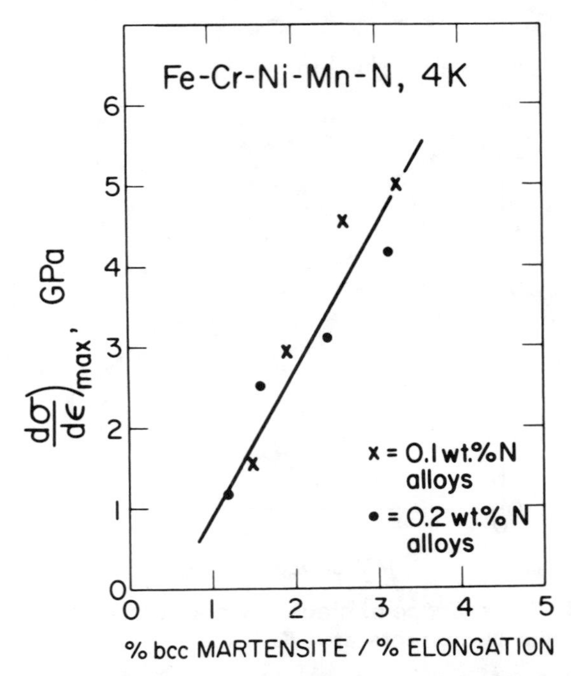

Fig. 15. Dependence of stage III work-hardening rate on the normalized amount of α'-martensite formed during uniform elongation at 4 K

suppliers in which nitrogen, carbon, molybdenum, manganese, and nickel alloy additions were systematically varied. A summary of these test results follows.

The interstitial additions substantially strengthen the alloys at low temperatures.(64,65,70-83) In Fig. 17, the effect of carbon plus nitrogen on σ_y at 4 K for an Fe-18Cr-10Ni steel is presented.(72) The break in the linear dependencies at about 0.15% (C + N) is related to martensitic transformation; at lower C + N contents, the strain-induced martensite reduces the lower flow strength, as discussed above under "Stress-Strain Characteristics." Regression analysis of 99 data points for alloy 316LN (and alloys with similar compositions) at 4 K indicated that σ_y increased 237 MPa per 0.1% N(77); an earlier analysis of 30 data points for 304LN-type alloys at 4 K found that σ_y increased 319 MPa per 0.1% N.(70) Strengthening from carbon is about half that from nitrogen at 4 K.(70) Thus, nitrogen additions strengthen considerably more than equivalent carbon additions at low temperatures. Alloying with nitrogen instead of carbon also reduces the tendency toward sensitization, as discussed under "Carbides." Furthermore, nitrogen is less expensive than carbon. For these reasons, nitrogen is now usually selected to increase the low-temperature strength of austenitic stainless steels. Many recent studies have measured the strengthening contributions from nitrogen for both Fe-Cr-Ni- and Fe-Cr-Mn-base austenitic structures at 4 K.(64,65,70-75,77-83) Studies of Reed and Simon(70,77) found a smaller standard deviation from regression analyses using the power of one to express the [N] dependence compared to 1/2 or 2/3 power dependence. Fundamental theories of solid-solution strengthening(84-86) predict depen-

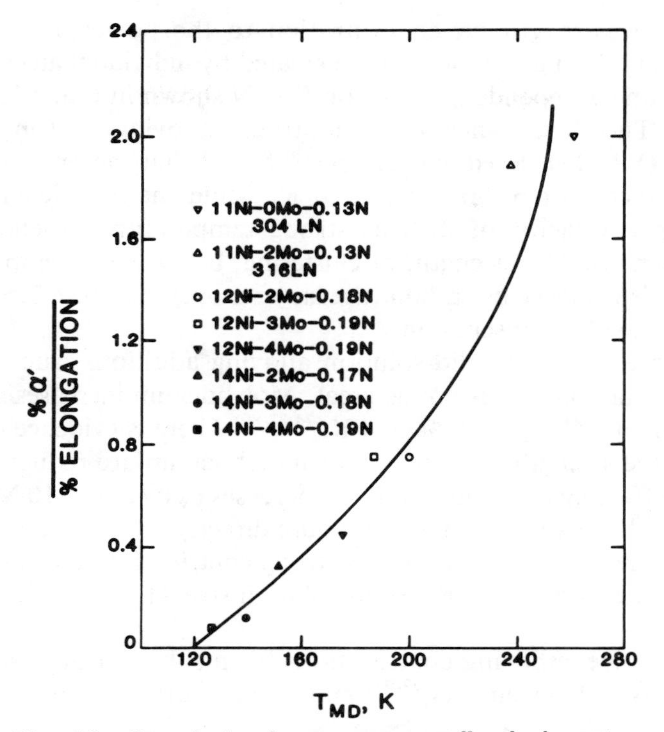

Fig. 16. Correlation between normalized α′-martensite formation in stage III with alloying, as characterized by the T_{md} calculations of Williams *et al.*[37]

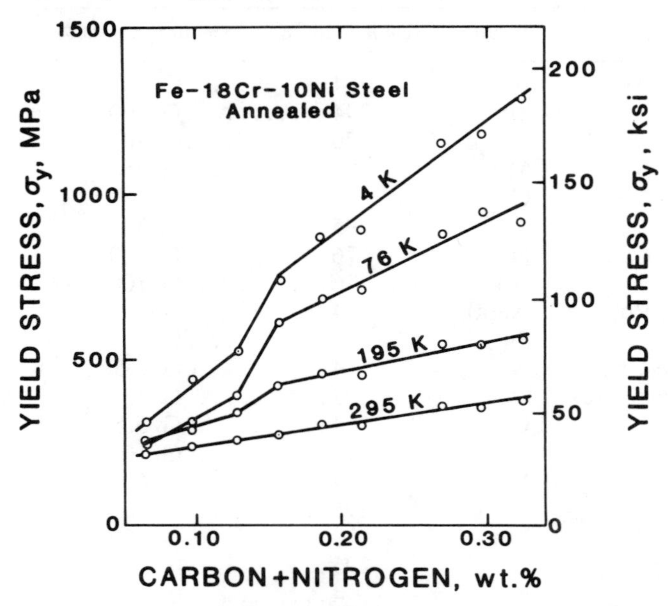

Fig. 17. Dependence of yield strength on C and N contents in Fe-18Cr-10Ni austenitic stainless steel at low temperatures

dence of the flow strength on concentration to the power 1/2 or 2/3 from dislocations cutting through the stress field created by individual atoms of this concentration. The linear dependence of σ_y on C + N shown in Fig. 17 also supports this assessment. This dependence of σ_y on nitrogen alloying is strongly dependent on temperature (Fig. 17). Reed and Simon[70] found that the nitrogen coefficient increased by a factor of 6 from 295 to 4 K, while the coefficient for carbon increased by only a factor of 2. This strong temperature dependence implies increased nitrogen bonding strength, or clustering, below room temperature, since the temperature dependence of carbon (the larger atom) should reflect the modulus and size effects (see later discussion).

The dependence of σ_y on solid-solution alloying additions, such as molybdenum, manganese, and nickel, is not as great. Molybdenum increases strength most effectively, about 50 MPa per wt % at 4 K.[74,87-89] There is evidence that at higher nickel contents the strengthening effects of molybdenum are reduced.[89] Manganese, when not affecting austenite stability, increases σ_y by about 10 MPa per wt % at 4 K.[53,73,75,89,90] Nickel does not contribute directly to solid-solution strengthening, but does affect strength through its strong contribution to austenite stability. The temperature dependence of the solid-solution strengthening additions has not been assessed.

In Table 5, the strengthening contributions from substitutional and interstitial alloying additions in fcc elements[92-97] and in austenitic stainless steels[62-91] are

Table 5. Strengthening contributions in fcc alloys

Solvent	Solute	Temperature, K	Amount, $\Delta\sigma$/100%
Ag	Al	4	G/30
	Sb	294	G/100
		76	G/30
Ni	Mo	295	G/45
	C	4	G/7
		295	G/30
Cu	Ni	76	G/120
	Au	76	G/100
Pb	Ag(a)	4	2G
	Sn	4	0
Fe-Cr-Ni (austenite)	N	4	G/2.5
		295	G/15
	C	4	G/5.5
		295	G/10
	Mo	4	G/60
		295	0
	Mn	4	G/200
		295	0
	Ni	4	0
		295	0

(a) Quenched from 300 °C.

listed in terms of a fraction of the shear modulus. This refers to the increase of σ_y from adding 100% (by atoms) of the solute, found by extrapolating from available data that usually range from 1 for interstitials to 5 to 20 at.%.

Adding carbon to nickel(95) or carbon to austenite produces roughly the same increases in strength at low temperatures. The addition of molybdenum to nickel or to Fe-Cr-Ni is also approximately equivalent. Galligan and Goldman(92) have achieved much stronger strengthening effects from alloying by quenching-in silver into probable split-interstitial sites in lead. In their work, strength increases are linearly proportional to silver content, not to a fractional power of it.

From the theory of solid-solution strengthening,(84-86) the shear strength, τ, resulting from a random array of single atoms interacting with mobile dislocations is:

$$\tau \simeq A[C_F^{2/3}][f_0^{4/3}] \qquad \text{(Eq 2)}$$

where C_F is the concentration of a solute species, and A is considered to be a constant for this discussion. The maximum interaction force is proportional to:

$$f_0^2 \sim \eta^2 + \alpha^2\delta^2 \qquad \text{(Eq 3)}$$

where the modulus misfit is

$$\delta = \frac{1}{G}\frac{dG}{dc} \qquad \text{(Eq 4)}$$

In Eq 4, G is the shear modulus, and c is the concentration. The size misfit is:

$$\delta = \frac{1}{a}\frac{da}{dc}$$

where a is the lattice parameter. The constant α is typically about 20.(84) Thus, the relative magnitudes of the strengthening contributions of various solutes may be estimated by comparing the modulus- and size-misfit parameters, assuming that other possible strengthening effects (e.g., electrostatic, stacking-fault interactions) are small.

Ledbetter and Austin(98) studied the elastic properties of austenitic stainless steel alloy series and measured their lattice parameters. Their measurements are summarized in Table 6, where E is Young's modulus, G is shear modulus, B is bulk modulus, ν is Poisson's ratio, and V is atomic volume. Lattice-parameter measurements at 4 K are planned in the near future.

At room temperature, both carbon and nitrogen show a large size misfit despite their small atom size [signified by the $V(X)/V(Fe)$ ratio]; obviously they must occupy interstitial lattice sites to achieve this increase. This size misfit is larger for nitrogen than for carbon although the atom size of nitrogen is less than that of carbon. The divergence of E and G trends from B and ν trends implies electronic bonding effects that alter shear modes differently than dilatational modes; this

Table 6. Effects of alloying on volume and elastic constants

Alloying element	$\frac{1}{a}\frac{da}{dc}$ 295 K	$\frac{1}{E}\frac{dE}{dc}$ 295 K	4 K	$\frac{1}{G}\frac{dG}{dc}$ 295 K	4 K	$\frac{1}{B}\frac{dB}{dc}$ 295 K	4 K	$\frac{1}{\nu}\frac{d\nu}{dc}$ 295 K	4 K	V(X)/V(Fe)
C	0.218	−0.78	0.61	−0.76	0.81	−0.93	−0.52	−0.11	−0.90	0.34
N	0.240									0.30
Ni.....	−0.003	0.07	...	0.09	...	−0.01	...	−0.06	...	0.93
Mn.....	0.022	−0.30	−0.38	−0.27	−0.37	−0.45	−0.48	−0.12	−0.08	1.04
Mo.....	0.150	−0.91	−0.14	−1.18	−0.38	0.79	1.30	1.22	1.13	1.32
Cr	0.075	...	...	...	...	...	...	...	...	1.02

divergence is found for nitrogen and carbon at 4 K and for molybdenum at both temperatures. Furthermore, the changes in sign of the Young's and shear moduli parameters of nitrogen and carbon on cooling from 295 to 4 K imply a low-temperature transition toward increased covalent bonding.

The maximum interaction forces, assuming $\alpha = 20$, for each solute are summarized in Table 7. The order of ratios of the interaction forces is correct when compared with the strengthening contribution listed in Table 5. However, the relative magnitudes of the interaction forces of carbon and nitrogen show greater disparity compared with the relative magnitudes of the solute contributions. The interstitial-element interaction forces are smaller than would be expected if one considered their strengthening contributions. This disparity undoubtedly will be increased at 4 K when the lattice-parameter measurements at 4 K are obtained: the ratio of strengthening contributions from interstitials increases relative to substitutional solid-solution elements at 4 K.

This leads to the suggestion that, particularly at low temperatures, the dislocation/interstitial interaction is distinct from that of the dislocation/solid solution interaction. The interaction-force ratios could be brought more in line with the strengthening contributions by increasing α, the parameter related to the curvature of the interaction potential of the solvent with the moving dislocations,(83) since the size-misfit parameters relate directly to the relative strengthening contributions for each type of atom. These disparities, combined with the linear dependence of strength on solute interstitial content, lead to the consideration of interstitial clustering.

Galligan and Goldman(92) measured strengthening factors of $2G$ at room tem-

Table 7. Maximum interaction force ratios(a) for solutes in austenitic stainless steel

T(K)	C	N	Ni	Mn	Mo
295	4.43	4.86	0.26	0.52	3.22

(a) $f_0^2 \sim \eta^2 + \alpha^2\delta^2$, where α is assumed to equal 20.

perature for presumably split interstitials of silver in lead, where silver has only slightly lower specific volume than lead. It is not difficult to envision that the much smaller carbon and nitrogen interstitials, in similar lattice positions, strengthen by factors of $G/10$ to $G/15$ at room temperature and that this strong, short-range interaction force is strongly temperature dependent.

SUMMARY

Austenitic stainless steel structures and strengths have been reviewed, with emphasis placed on low-temperature characteristics to include recent studies that have led to stronger alloys for cryogenic applications. The stress-strain characteristics of austenitic stainless steels depend on austenite stability. A region of low work hardening, following initial plastic deformation, is associated with hcp-martensite formation and bcc-martensite formation at cross-slip sites. The final work-hardening stage is associated with bcc-martensite formation. All stages are related to alloying effects on austenite stability.

The interstitials carbon and nitrogen have been shown to contribute more to strength than that predicted by solution-hardening dislocation theories. It is argued that carbon and nitrogen must act in clusters to increase solution hardening. Nitrogen has proven to be the most effective alloying element for increasing strength at low temperatures.

REFERENCES

1. L. Guillet, Chromium Steels, *Rev. Metall.*, Vol 1, 1904, p. 155
2. A.M. Portevin, Contribution to the Study of the Special Ternary Steels, Iron and Steel Institute, *Carnegie Sch. Mem.*, Vol 1, 1909, p. 230
3. W. Giesen, The Special Steels in Theory and Practice, Iron and Steel Institute, *Carnegie Sch. Mem.*, Vol 1, 1909
4. P. Monnartz, The Study of Iron-Chromium Alloys with Special Consideration of their Resistance to Acids, *Metallurgie*, Vol 8, 1911, p. 161
5. L. Guillet, Nickel-Chrome Steels, *Rev. Metall.*, Vol 3, 1906, p. 332
6. *The Book of Stainless Steels*, edited by E.E. Thum, American Society for Metals, Metals Park, OH, 1935, 787 pages
7. C.A. Zapffe, *Stainless Steels*, American Society for Metals, Metals Park, OH, 1949, 368 pages
8. J.H.G. Monypenny, *Stainless Iron and Steel*, Vol 1, Chapman and Hall, London, 1951, 524 pages
9. J.H.G. Monypenny, *Stainless Iron and Steel*, Vol 2, Chapman and Hall, London, 1954, 330 pages
10. F.H. Keating, *Chromium Nickel Austenitic Steels*, Butterworth Scientific, London, 1956, 138 pages
11. *Source Book on Stainless Steels*, American Society for Metals, Metals Park, OH, 1976, 408 pages
12. *Handbook of Stainless Steels*, edited by D. Peckner and I.M. Bernstein, McGraw-Hill, New York, 1977

13. J.G. Parr and A. Hanson (revised by R.A. Lula), *Stainless Steel*, American Society for Metals, Metals Park, OH, 1986, 173 pages
14. R.A. Lula, *Manganese Stainless Steels*, The Manganese Centre, Paris, 1986, 83 pages
15. *Symposium on the Nature, Occurrence, and Effects of Sigma Phase*, STP 110, American Society for Testing and Materials, Philadelphia, 1951, 181 pages
16. *Effects of Residual Elements on Properties of Austenitic Stainless Steels*, STP 418, American Society for Testing and Materials, Philadelphia, 1967, 136 pages
17. *Stainless Steels*, The Iron and Steel Institute, London, 1969, 214 pages
18. *Stainless Steel '77*, Climax Molybdenum Company, Climax, CO, 1977, 256 pages
19. *Stainless Steel Castings*, edited by V.G. Behal and A.S. Melilli, STP 756, American Society for Testing and Materials, Philadelphia, 1982, 444 pages
20. *Austenitic Steels at Low Temperatures*, edited by R.P. Reed and T. Horiuchi, Plenum Press, New York, 1983, 388 pages
21. *New Developments in Stainless Steel Technology Conference Proceedings*, edited by R.A. Lula, American Society for Metals, Metals Park, OH, 1985, 391 pages
22. G.R. Speich, in *Source Book on Stainless Steels*, American Society for Metals, Metals Park, OH, 1976, pp. 424-426
23. E. Schurmann and S. Brauckman, Untersuchungen über die Schmelzgleichgewichte in der Eisenecke des Dreistoffsystems Eisen-Chrom-Nickel, *Arch. Eisenhüttenw.*, Vol 48, 1977, pp. 3-8
24. J.A. Brooks, J.C. Williams, and A.W. Thompson, STEM Analysis of Primary Austenite Solidified Stainless Steel Welds, *Metall. Trans.*, Vol 14A, 1983, pp. 23-31
25. C.J. Novak, in *Handbook of Stainless Steels*, edited by D. Peckner and I.M. Bernstein, McGraw-Hill, New York, 1977, pp. 401-478
26. A. L. Schaeffler, Constitution Diagram for Stainless Steel Weld Metal, *Met. Prog.*, Vol 56, 1949, p. 680
27. W.T. DeLong, G.A. Ostrom, and E.R. Szumachowski, Measure and Calculation of Ferrite in Stainless Steel Weld Metal, *Weld. J.*, Vol 35, 1956, pp. 521s-528s
28. N. Suutala, Effects of Manganese and Nitrogen on the Solidification Mode in Austenitic Stainless Steel Welds, *Metall. Trans.*, Vol 13A, 1982, pp. 2121-2130
29. F.C. Hull, Delta Ferrite and Martensite Formation in Stainless Steels, *Weld. J.*, Vol 52, 1973, pp. 193s-203s
30. O. Hammar and U. Svensson, in *Solidification and Casting of Metals*, The Metals Society, London, 1979, pp. 401-410
31. L. Kaufman, "The Lattice Stability of the Transition Metals," in *Phase Stability in Metals and Alloys*, edited by P.S. Rudman, J. Stringer, and R.I. Jaffee, McGraw-Hill, New York, 1967, pp. 125-150
32. R.P. Reed, "Martensitic Phase Transformations," in *Materials at Low Temperatures*, edited by R.P. Reed and A.F. Clark, American Society for Metals, Metals Park, OH, 1983, pp. 295-341
33. G.H. Eichelman and F.C. Hull, The Effect of Composition on the Temperature of Spontaneous Transformation of Austenite to Martensite in 18-8 Type Stainless Steel, *Trans. Amer. Soc. Met.*, Vol 45, 1953, pp. 77-104
34. F.C. Monkman, F.B. Cuff, and N.J. Grant, Computation of M_s for Stainless Steels, *Met. Prog.*, Vol 71, 1957, pp. 94-96
35. K.W. Andrews, Empirical Formulae for the Calculation of Some Transformation Temperatures, *J. Iron Steel Inst.*, Vol 203, 1965, pp. 721-727
36. J.A. Self, "Effects of Compositions upon the Martensite Transformation Temperature of Austenitic Steel Welds," Center for Welding Research, MT-CWR-086-037, Colorado School of Mines, Golden, CO, 1986
37. T. Angel, Formation of Martensite in Austenitic Stainless Steels, *J. Iron Steel Inst.*, Vol 177, 1954, pp. 165-174
38. I. Williams, R.G. Williams, and R.C. Capellano, "Stability of Austenitic Stainless

Steels between 4 K and 373 K," in *Proceedings of the Sixth International Cryogenic Engineering Conference*, IPC Science and Technology Press, Guildford, Surrey, England, 1976, pp. 337-341
39. J.G. Breedis, Martensitic Transformation Iron-Chromium-Nickel Alloys, *Trans. AIME*, Vol 230, 1964, pp. 1583-1596
40. R.P. Reed, The Spontaneous Martensitic Transformations in 18% Cr, 8% Ni Steels, *Acta Metall.*, Vol 10, 1962, pp. 865-887
41. R.M. Kelly, The Martensite Transformation in Steels with Low Stacking Fault Energy, *Acta Metall.*, Vol 13, 1965, pp. 635-646
42. R.E. Schramm and R.P. Reed, Stacking Fault Energies of Seven Commercial Austenitic Stainless Steels, *Metall. Trans.*, Vol 6A, 1975, pp. 1345-1351
43. C.G. Rhodes and A.W. Thompson, The Composition Dependence of Stacking Fault Energy in Austenitic Stainless Steels, *Metall. Trans.*, Vol 8A, 1977, pp. 1901-1906
44. C.C. Bampton, L.P. Jones, and M.H. Loretto, Stacking Fault Energy Measurements in Some Austenitic Stainless Steels, *Acta Metall.*, Vol 26, 1978, pp. 39-51
45. P.J. Brofman and G.S. Ansell, On the Effect of Carbon on the Stacking Fault Energy of Austenitic Stainless Steels, *Metall. Trans.*, Vol 9A, 1978, pp. 879-880
46. H.M. Ledbetter and M.W. Austin, "Stacking-Fault Energies in 304-Type Stainless Steels: Effects of Interstitial Carbon and Nitrogen," in *Materials Studies for Magnetic Fusion Energy Applications at Low Temperatures—VIII*, edited by R.P. Reed, National Bureau of Standards, NBSIR 85-3025, Boulder, CO, 1985, pp. 271-294
47. R.E. Stoltz and J.B. VanderSande, "The Effect of Nitrogen on Stacking Fault Energy of Fe-Ni-Cr-Mn Steels," SAND 79-8735, Sandia Laboratories, Albuquerque, NM, 1979
48. M. Fujikuma, K. Takada, and K. Ishida, Effect of Manganese and Nitrogen on the Mechanical Properties of Fe18%Cr10%Ni Stainless Steels, *Trans. Iron Steel Inst. Jap.*, Vol 15, 1975, pp. 464-469
49. H. Tuma, M. Vyklicky, and K. Lobl, Activity and Solubility of C in Austenitic 18 Percent Cr-Ni Steels, *Arch. Eisenhüttenw.*, Vol 41, 1970, pp. 983-988
50. T. Masumoto and Y. Imai, Structural Diagrams and Tensile Properties of the 18 Percent Cr-Fe-Ni-N Quarternary System Alloys, *J. Jpn. Inst. Met.*, Vol 33, 1969, pp. 1364-1371
51. M. Deighton, Solubility of $M_{23}C_6$ in Type 316 Stainless Steel, *J. Iron Steel Inst.*, Vol 208, 1970, pp. 1012-1014
52. D. Kumar, A.D. King, and T. Bell, Mass Transfer of Nitrogen from N_2-H_2 Atmospheres into Fe-18Cr-Ni-Mn Alloys, *Met. Sci.*, Vol 17, 1983, pp. 32-40
53. T. Sakamoto, Y. Nakagawa, and I. Yamauchi, "Effect of Mn on the Cryogenic Properties of High Nitrogen Austenitic Stainless Steels," in *Advances in Cryogenic Engineering—Materials*, Vol 32, edited by R.P. Reed and A.F. Clark, Plenum Press, New York, 1986, pp. 65-71
54. S.J. Rosenberg and C.R. Irish, Solubility of Carbon in 18 Percent Chromium–10 Percent Nickel Austenite, *J. Res. Nat. Bur. Stand.*, Vol 48, 1952, pp. 40-48
55. R. Stickler and A. Vinckier, Morphology of Grain-Boundary Carbides and Its Influence on Intergranular Corrosion of 304 Stainless Steel, *Trans. Amer. Soc. Met.*, Vol 54, 1961, pp. 362-380
56. M.H. Lewis and B. Hattersley, Precipitation of $M_{23}C_6$ in Austenitic Steels, *Acta Metall.*, Vol 13, 1965, pp. 1159-1168
57. J.K.L. Lai, A Review of Precipitation Behavior in AISI Type 316 Stainless Steel, *Mater. Sci. Eng.*, Vol 61, 1983, pp. 101-109
58. Y. Maehara, Y. Ohmori, J. Murayama, N. Fujino, and T. Kunitake, Effects of Alloying Elements on σ Phase Precipitation in $\delta\gamma$ Duplex Phase Stainless Steels, *Met. Sci.*, Vol 17, 1983, pp. 541-547
59. A.M. Talbot and D.E. Furman, Sigma Formation and Its Effect on the Impact Prop-

erties of Iron-Nickel-Chromium Alloys, *Trans. Amer. Soc. Met.*, Vol 45, 1953, pp. 429-442
60. T. Suzuki, H. Kojima, K. Suzuki, T. Hashimoto, and M. Ichihara, An Experimental Study on the Martensite Nucleation and Growth in 18/8 Stainless Steel, *Acta Metall.*, Vol 25, 1977, pp. 1151-1162
61. R.P. Reed and C.J. Guntner, Stress-Induced Martensitic Transformations in 18Cr-8Ni Steel, *Trans. AIME*, Vol 230, 1964, pp. 1713-1720
62. R.P. Reed and R.L. Tobler, "Deformation of Metastable Austenitic Steels at Low Temperatures," in *Advances in Cryogenic Engineering—Materials*, Vol 28, edited by R.P. Reed and A.F. Clark, Plenum Press, New York, 1982, pp. 49-56
63. R.P. Reed and J.M. Arvidson, "The Temperature Dependence of the Tensile Yield Strength of Selected Austenitic Steels," in *Advances in Cryogenic Engineering—Materials*, Vol 30, edited by A.F. Clark and R.P. Reed, Plenum Press, New York, 1984, pp. 263-270
64. N.J. Simon and R.P. Reed, Strength and Toughness of AISI 304 and 316 at 4 K, *J. Nucl. Mater.*, Vol 141-143, 1986, pp. 44-48
65. R.P. Reed, N.J. Simon, P.T. Purtscher, and R.L. Tobler, "Alloy 316LN for Low Temperature Structures: A Summary of Tensile and Fracture Data," in *Proceedings, Eleventh International Cryogenic Engineering Conference*, edited by G. Klipping and I. Klipping, Butterworths, Guildford, Surrey, England, 1986, pp. 786-790
66. R.L. Tobler, R.P. Reed, and D.S. Burkhalter, "Temperature Dependence of Yielding in Austenitic Stainless Steels," in *Advances in Cryogenic Engineering—Materials*, Vol 26, edited by A.F. Clark and R.P. Reed, Plenum Press, New York, 1980, pp. 107-110
67. H. Nakajima, K. Yoshida, Y. Takahashi, E. Tada, M. Oshikiri, K. Koizumi, S. Shimamoto, R. Mira, M. Shimara, and S. Tone, "Development of the New Cryogenic Structural Material for Fusion Experimental Reactors," in *Advances in Cryogenic Engineering—Materials*, Vol 32, edited by R.P. Reed and A.F. Clark, Plenum Press, New York, 1984, pp. 219-226
68. K. Yoshida, N. Nakajima, K. Koizumi, M. Shimada, Y. Sanda, Y. Takahashi, E. Tada, H. Tsuji, and S. Shimamoto, "Development of Cryogenic Structural Materials for Tokamak Reactor," in *Austenitic Steels at Low Temperatures*, edited by R.P. Reed and T. Horiuchi, Plenum Press, New York, 1983, pp. 29-39
69. S. Shimamoto, H. Nakajima, K. Yoshida, and E. Tada, "Requirements for Structural Alloys for Superconducting Magnet Cases," in *Advances in Cryogenic Engineering—Materials*, Vol 32, edited by R.P. Reed and A.F. Clark, Plenum Press, New York, 1986, pp. 23-32
70. R.P. Reed and N.J. Simon, "Low Temperature Strengthening of Austenitic Stainless Steels with Nitrogen and Carbon," in *Advances in Cryogenic Engineering—Materials*, Vol 30, edited by A.F. Clark and R.P. Reed, Plenum Press, New York, 1984, pp. 127-136
71. K. Nohara, T. Kato, T. Sasaki, S. Suzuki, and A. Ejima, "Strengthening and Serrated Flow of High-Manganese Nonmagnetic Steel at Cryogenic Temperatures," in *Advances in Cryogenic Engineering—Materials*, Vol 30, edited by R.P. Reed and A.F. Clark, Plenum Press, New York, 1984, pp. 193-201
72. R.L. Tobler and R.P. Reed, "Tensile and Fracture Properties of Manganese-Modified AISI 304 Type Stainless Steel," in *Advances in Cryogenic Engineering—Materials*, Vol 28, edited by R.P. Reed and A.F. Clark, Plenum Press, New York, 1982, pp. 83-92
73. T. Horiuchi, R. Ogawa, M. Shimada, S. Tone, M. Yamaga, and Y. Kasamatsu, "Mechanical Properties of High Manganese Steels at Cryogenic Temperatures," in *Advances in Cryogenic Engineering—Materials*, Vol 28, edited by R.P. Reed and A.F. Clark, Plenum Press, New York, 1982, pp. 93-103
74. S. Yamamoto, N. Yamagami, and C. Ouchi, "Effect of Metallurgical Variables on

Strength and Toughness of Mn-Cr and Ni-Cr Stainless Steels at 4.2 K," in *Advances in Cryogenic Engineering—Materials*, Vol 32, edited by R.P. Reed and A.F. Clark, Plenum Press, New York, 1986, pp. 57-64

75. T. Horiuchi, R. Ogawa, and M. Shimada, "Cryogenic Fe-Mn Austenitic Steels," in *Advances in Cryogenic Engineering—Materials*, Vol 32, edited by R.P. Reed and A.F. Clark, Plenum Press, New York, 1986, pp. 33-42
76. R.L. Tobler, D.H. Beekman, and R.P. Reed, "Factors Influencing the Low Temperature Dependence of Yielding in AISI 316 Stainless Steels," in *Austenitic Steels at Low Temperatures*, edited by R.P. Reed and T. Horiuchi, Plenum Press, New York, 1983, pp. 135-157
77. N.J. Simon and R.P. Reed, "Design of 316LN-Type Alloys," in *Advances in Cryogenic Engineering—Materials*, Vol 34, edited by A.F. Clark and R.P. Reed, Plenum Press, New York, 1988, pp. 165-172
78. R.P. Reed, P.T. Purtscher, and K.A. Yushchenko, "Nickel and Nitrogen Alloying Effects on the Strength and Toughness of Austenitic Stainless at 4 K," in *Advances in Cryogenic Engineering—Materials*, Vol 32, edited by R.P. Reed and A.F. Clark, Plenum Press, New York, 1986, pp. 33-42
79. T. Sakamoto, Y. Nakagawa, and I. Yamauchi, "Effect of Mn on the Cryogenic Properties of High Nitrogen Austenitic Stainless Steels," in *Advances in Cryogenic Engineering—Materials*, Vol 32, edited by R.P. Reed and A.F. Clark, Plenum Press, New York, 1986, pp. 65-71
80. Y. Takahashi, K. Yoshida, M. Shimada, E. Tada, R. Miura, and S. Shimamoto, "Mechanical Evaluation of Nitrogen-Strengthened Stainless Steels at 4 K," in *Advances in Cryogenic Engineering—Materials*, Vol 28, edited by R.P. Reed and A.F. Clark, Plenum Press, New York, 1986, pp. 73-81
81. H. Masumoto, K. Suemune, H. Nakajima, and S. Shimamoto, "Development of High-Strength, High-Manganese Steels for Cryogenic Use," in *Advances in Cryogenic Engineering—Materials*, Vol 30, edited by A.F. Clark and R.P. Reed, Plenum Press, New York, 1984, pp. 169-176
82. K. Shibata, Y. Kobiti, Y. Kishimoto, and T. Fujita, "Mechanical Properties of High Yield Strength High Manganese Steels at Cryogenic Temperatures," in *Advances in Cryogenic Engineering—Materials*, Vol 30, edited by R.P. Reed and A.F. Clark, Plenum Press, New York, 1984, pp. 169-176
83. T. Sakamoto, Y. Nakagawa, I. Yamauchi, T. Zaizen, H. Nakajima, and S. Shimamoto, "Nitrogen-Containing 25Cr-13Ni Stainless Steel as a Cryogenic Structural Material," in *Advances in Cryogenic Engineering—Materials*, Vol 30, edited by A.F. Clark and R.P. Reed, Plenum Press, New York, 1984, pp. 145-152
84. P. Haasen, "Solution Hardening in f.c.c. Metals," in *Dislocations in Solids*, Chapter 15, edited by F.R.N. Nabarro, North-Holland, New York, 1976, pp. 155-189
85. F.R.N. Nabarro, The Theory of Solution Hardening, *Philos. Mag.*, Vol 35, 1977, pp. 613-622.
86. R. Labusch, Statistische Theorien der Mischkristallhartung, *Acta Metall.*, Vol 20, 1972, pp. 917-927
87. K. Suemune, K. Sugino, H. Matsumoto, H. Nakajima, and S. Shimamoto, "Improvement of Toughness of a High-Strength, High-Manganese Stainless Steel for Cryogenic Use," in *Advances in Cryogenic Engineering—Materials*, Vol 32, edited by R.P. Reed and A.F. Clark, Plenum Press, New York, 1986, pp. 51-56
88. K. Ishikawa, K. Hiraga, T. Ogata, and K. Nagai, "Low Temperature Properties of High-Manganese-Molybdenum Austenitic Iron Alloys," in *Austenitic Steels at Low Temperatures*, edited by R.P. Reed and T. Horiuchi, Plenum Press, New York, 1983, pp. 295-309
89. P.T. Purtscher, R.P. Walsh, and R.P. Reed, "Effect of Chemical Composition on the 4 K Mechanical Properties of 316LN-Type Alloys," in *Advances in Cryogenic*

Engineering—Materials, Vol 34, edited by A.F. Clark and R.P. Reed, Plenum Press, New York, 1987 (to be published)

90. R. Muira, H. Nakajima, Y. Takahashi, and K. Yoshida, "32Mn-7Cr Austenitic Steel for Cryogenic Applications," in *Advances in Cryogenic Engineering—Materials*, Vol 30, edited by A.F. Clark and R.P. Reed, Plenum Press, New York, 1984, pp. 245-252
91. K. Hiraga, K. Ishikawa, K. Nagai, and T. Ogata, "Mechanical Properties of Cold-Rolled and Aged Fe-Ni-Cr-Ti-Austenitic Alloys for Low Temperature Use," in *Advances in Cryogenic Engineering—Materials*, Vol 30, edited by R.P. Reed and A.F. Clark, Plenum Press, New York, 1984, pp. 203-210
92. J.M. Galligan and P.D. Goldman, "Metal Interstitial Solid-Solution Strengthening," in *Strength of Metals and Alloys, Proceedings of the Fifth International Conference*, Part II, Pergamon Press, New York, 1980, pp. 983-988
93. A.A. Hendrickson and M.E. Fine, Solid Solution Strengthening of Ag by Al, *Trans. AIME*, Vol 221, 1961, pp. 967-974
94. T.A. Bloom, U.F. Kocks, and P. Nash, Deformation Behavior of Ni-Mo Alloys, *Acta Metall.*, Vol 33, 1985, pp. 265-277
95. Y. Nakada and A.S. Keh, Solid-Solution Strengthening in Ni-C Alloys, *Metall. Trans.*, Vol 2, 1971, pp. 441-447
96. G.J. denOtter and A. Van den Beukel, Flow Stress and Activation Volume of Some Cold-Worked Copper-Based Solid Solutions, *Phys. Status Solidi(a)*, Vol 55, 1975, pp. 785-792
97. J.H. Tregilgas and J.M. Galligan, Hardening from Metal Interstitials in a Face-Centered Cubic Lattice—Pb-Ag, *Scr. Metall.*, Vol 9, 1975, pp. 1225-1227
98. H.M. Ledbetter and M.W. Austin (unpublished measurements), National Bureau of Standards, Boulder, CO, 1987

10

Introduction to Titanium Alloy Design

E.W. COLLINGS
Research Leader
Battelle Columbus Laboratories

1. CLASSIFICATION OF TITANIUM ALLOYS

Pure titanium undergoes an allotropic transformation from hcp (α) to bcc (β) as its temperature is raised through 882.5 °C.[1,2] Elements which when dissolved in titanium produce little change in the transformation temperature (e.g., tin) or cause it to increase (e.g., aluminum, oxygen) are known as "α stabilizers"; they are simple metals (SM) or the interstitial elements[1, p. 154] – generally nontransition elements. Alloying additions which decrease the phase-transformation temperature are referred to as "β stabilizers"; they are generally the transition metals (TM) and noble metals – i.e., metals which, like titanium, have unfilled or just-filled *d*-electron bands. In the alloys, of course, the single-phase-α and single-phase-β regions are not in contact as they are in pure titanium; they are instead separated by a two-phase $\alpha + \beta$ region whose width increases with increasing solute concentration. Based on these considerations, technical alloys of titanium are classified as "α," "β," and "$\alpha + \beta$."

1.1. Alpha Alloys

Unalloyed titanium and alloys of titanium with α stabilizers such as aluminum, gallium, and tin, either singly or in combination as in the commercial alloy Ti-5Al-2.5Sn or the experimental Ti-Al-Ga alloys,[3,4] are hcp at ordinary temperatures and as such are classified as α alloys. These alloys, according to Wood,[5] are characterized by satisfactory strength, toughness, creep resistance, and weldability. Furthermore, the absence of a ductile-brittle transformation, a property of the bcc structure, renders α alloys (typified by Ti-5Al-2.5Sn) suitable for cryogenic applications.[6]

Alpha-stabilizing solutes are those which, as a function of concentration, more or less elevate the temperature of the $(\alpha + \beta)/\alpha$ transus. Such solutes are generally nontransition metals (i.e., "simple metals," SM). An explanation of α stability based on electron-screening arguments proceeds as follows: When simple metals (e.g., aluminum) are dissolved in titanium, very few electrons appear at the Fermi level, most of them going to states within the lower part of the band. The titanium *d*-electrons tend to avoid the aluminum atoms, which thereby have the effect of diluting the titanium sublattice. The consequence of this is to emphasize any preexisting Ti-Ti bond directionality and thus to preserve the hcp structure characteristic of the titanium crystal. In general, when simple metals are added to titanium, the fields of titanium-like α stability are eventually terminated by intermetallic compounds, of composition Ti_3SM, which are also hexagonal in structure. The bond argument is consistent with the observation that α stabilizers are quite rapid solution strengtheners either in hcp solid solution or when added to bcc alloys.[7] The classification of α-phase alloys into systems whose phase diagrams exhibit (1) peritectic transformations or (2) peritectoid transformations, according to Molchanova's simplified scheme, is considered in Section 1.4.

1.2. Beta Alloys

As mentioned above, transition-metal (TM) solutes are stabilizers of the bcc phase. Thus all-β alloys generally contain large amounts of one or more of the so-called "β-isomorphous"-forming additions—vanadium, niobium, tantalum (group V TM's), and molybdenum (a group VI TM). The systematics of β stabilization in binary and multicomponent Ti-base alloys has been discussed in detail by Ageev and Petrova.[8] The archetypal binary β-stabilized Ti-base alloy, about which a great deal of physical and metallurgical information has been garnered over the years, is Ti-Mo. For a useful overview of the mechanical properties and aging characteristics of a pair of typical β alloys, Ti-15Mo-5Zr and Ti-15Mo-5Zr-3Al, the work of Nishimura *et al.*[9] is recommended. There are several important commercial β alloys; one which has been attracting considerable attention recently is Ti-11.5Mo-6Zr-4.5Sn (β-III).[10-13] Beta alloys, according to Wood,[5] are extremely formable. They are, however, prone to ductile-brittle transformation[14] and, along with other bcc-phase alloys, are unsuitable for low-temperature applications.[6]

The transition-metal block of the periodic table may be regarded as commencing with group III, scandium, yttrium, and lanthanum (or perhaps more precisely, lutetium). In this scheme, the alkaline-earth metals, calcium, strontium, and barium, may be regarded as "pretransition metals," and the noble metals, copper, silver, and gold, as "post-transition metals." As indicated in most periodic charts of the elements, the structures of the transition metals all change from hcp to bcc as *e/a* increases from 4 through 6. It is possible that stabilization of the bcc structure can be justified within the framework of a screening model in terms of which a high conduction-electron concentration, which enhances the screening of ion cores, may favor a symmetrical, hence cubic, structure. Thus an increase in electron density (as in the groups V and VI elements), which tends to symmetrize the screening, increases the stability of the bcc structure. Symmetrization may also be accomplished through lattice vibrations; thus, all six of the groups III and IV elements transform to the bcc structure at high temperatures (as compared with their Debye temperatures). With regard to alloys, the addition of transition elements to titanium increases the electron density and consequently stabilizes the bcc or β structure. Thus, as a general rule, the transition elements are β stabilizers. The systematics of β stabilization by transition elements has been discussed in detail by Ageev and Petrova,[8] according to whom: (1) the β-stabilizing action of TM solutes is greater the "farther" they are from titanium in the periodic table; and (2) for the retention of the metastable-β solid solution during quenching, the β stabilizer has to provide for an electron/atom ratio of at least 4.2.

According to Zener,[15] and subsequently Fisher,[16,17] who has considered the problem of bcc stability in considerable detail, the magnitude of the elastic shear modulus $C' = (C_{11} - C_{12})/2$ is a useful parameter for ranking the stabilities of bcc transition metals and alloys. The variation of C' with conventional electron/atom ratio is plotted in Fig. 1, which shows that the alloying of group-IV ele-

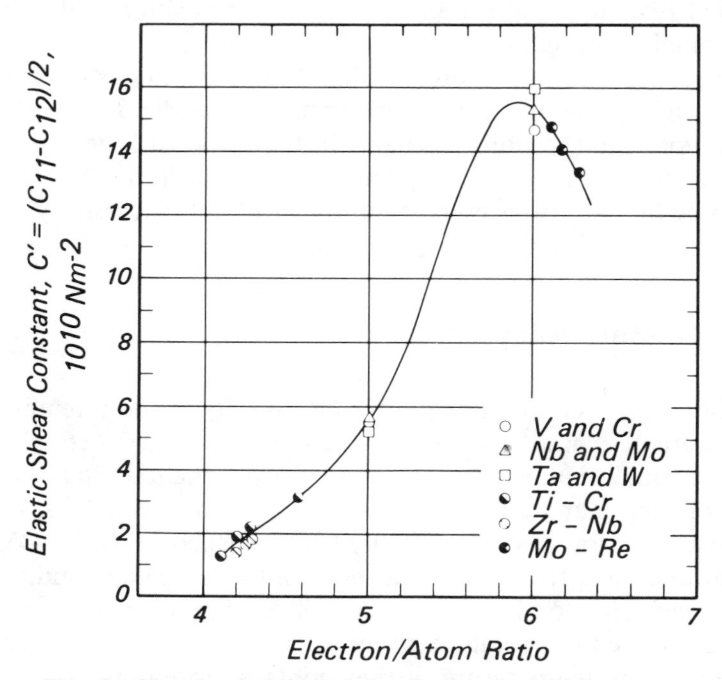

Fig. 1. Elastic shear modulus for bcc transition metals and some of their binary alloys as a function of electron/atom ratio[18]

ments with other elements to the "right" of them in the periodic table increases the bcc stability, which rises to a maximum near $e/a = 6$ for the elements chromium, molybdenum, and tungsten. On the other hand, with decreasing e/a, the vanishing of C' for $e/a = 4.1$ corresponds to the compositional threshold for martensitic transformation—to be discussed below. As a result of the alloying of titanium with transition elements of higher group number, the continuous increase of bcc stability manifests itself as a lowering of the $\beta/(\alpha + \beta)$ transus temperature.

As indicated below, within the context of β stabilization two subclasses of phase diagrams exist—the "β isomorphous" and the "β eutectoid," depending on whether or not a solid-solution/compound eutectoid exists at a sufficiently elevated temperature. It is instructive in the present context to consider a group of simplified, compositionally truncated, binary Ti-TM equilibrium phase diagrams, arranged according to the positions that the solute elements occupy in the TM block of the periodic table. Some representative diagrams selected from such a postulated arrangement are presented in Fig. 2. In order to focus attention on the alloys of most interest, the limiting composition (in at. %) in each group (except group IV itself) has been selected such that $e/a \leq 5.0$. In so doing it has been assumed that the numbers of $s + d$ valence electrons belonging to the elements in the columns headed by Fe, Co, Ni, and Cu are 8, 9, 10, and 11, respectively. An alternative way of deriving a reduced composition scale for intercomparison purposes, and one that would focus attention on alloy chemistry rather than electron density, might have been to normalize composition (i.e., stretch the composition scale) to that of the first β-eutectoidal intermetallic compound.

Interesting systematics to be noted in Fig. 2 are that: (1) as the solute element moves to the "right," the phase diagram changes from the β-isomorphous to the β-eutectoidal type; and (2) along the row Mn-Fe-Co-Ni-Cu, the eutectoid temperature increases monotonically. Extrapolating this trend to the "left" suggests that Ti-V can also be thought of as eutectoidal, but with an inaccessibly low eutectoid temperature.

1.3. Alpha-Plus-Beta Alloys

The $\alpha + \beta$ alloys are such that at equilibrium, usually at room temperature, they support a mixture of α and β phases. Although many binary β-stabilized alloys in thermodynamic equilibrium are two-phase, in practice the $\alpha + \beta$ alloys usually contain mixtures of both α and β stabilizers. The simplest of such alloys, and one upon which the most attention has undoubtedly been lavished, is Ti-6Al-4V. Although this particular alloy is difficult to form, even in the annealed condition,[6] $\alpha + \beta$ alloys generally exhibit good fabricability as well as high room-temperature strength and moderate elevated-temperature strength. They may contain between 10 and 50% β phase at room temperature; if they contain more than 20%, they are not weldable. The properties of $\alpha + \beta$ alloys can be controlled by heat treatment, which is used to adjust the microstructural and precipitational states of the β component.

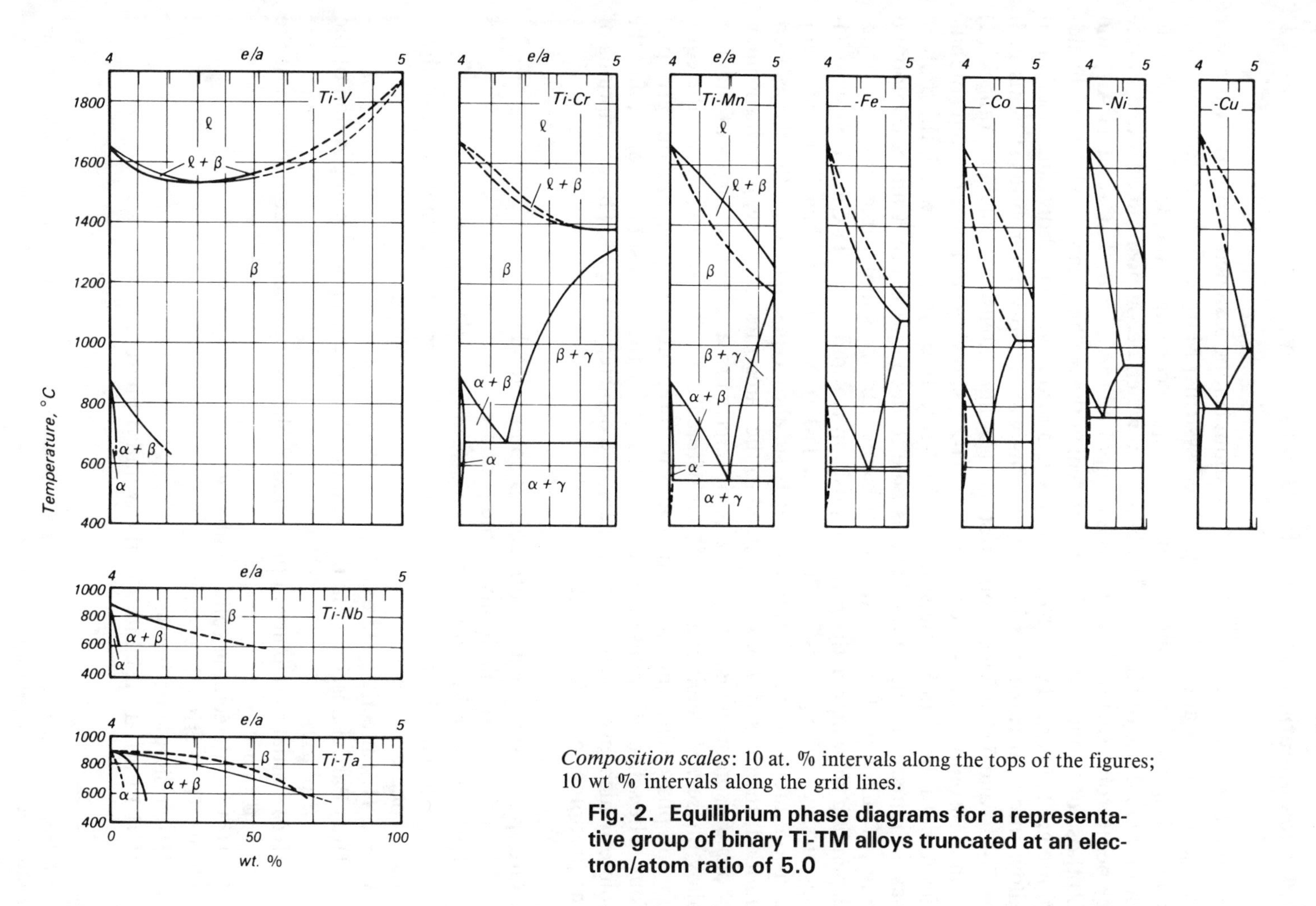

Composition scales: 10 at. % intervals along the tops of the figures; 10 wt % intervals along the grid lines.

Fig. 2. Equilibrium phase diagrams for a representative group of binary Ti-TM alloys truncated at an electron/atom ratio of 5.0

1.4. Classification Schemes for Binary Alloys

All authors agree that the alloys of titanium can be assigned to one of two major categories* — α-stabilized or β-stabilized systems. Margolin[19] has recommended subdividing the former into two more groups according to the degree of α stabilization: (1) those of "limited α stability," in which decomposition of α takes place by peritectoid reaction into β plus a compound (e.g., Ti-B, Ti-C, and Ti-Al); and (2) those of "complete α stability," in which the α phase can coexist with the liquid (e.g., Ti-O and Ti-N). Margolin has also recommended subdividing β-stabilized alloys into four categories in the following way: (1) β-isomorphous systems such as Ti-Mo and Ti-Ta which show restricted α- and extensive β-solubility ranges; (2) β-and-α-isomorphous systems, such as Ti-Zr, showing complete mutual solubilities in both the α and β phases; and (3) β-eutectoid systems in which the β phase has a limited solubility range and is able to decompose into α and a compound (e.g., Ti-Cr and Ti-Cu) — this class being further subdivisible into two more depending on whether the β decomposition is rapid (e.g., Ti-Cu, Ti-Ni, and Ti-Sn) or sluggish (e.g., Ti-Cr, Ti-Mn, and Ti-Fe). Kornilov[20] has discussed a subdivision into what he refers to as four basic alloy types. Although the classes were untitled, their descriptions conformed to the categories: α phase, β-and-α isomorphous, β isomorphous, and β eutectoid, referred to above. In a survey of titanium alloy phases, Molchanova[1, p. xiv] has offered a detailed subdivision of the equilibrium phase diagrams into two groups of three subcategories, and one group of four. The schematic phase diagrams which typify these ten subcategories, and the solutes which give rise to each of them, are depicted in Fig. 3. The descriptions of the groupings are as follows:

Group I: Systems with continuous β solid solubility
- I(a) Complete miscibility in the α phase
- I(b) Partial miscibility in the α phase
- I(c) Partial miscibility in the α phase and eutectoid decomposition of the β phase

Group II: Eutectic systems
- II(a) Partial miscibility in the α and β phases; eutectoid decomposition of the β phase
- II(b) Partial miscibility in the α and β phases; peritectic decomposition of the β phase
- II(c) No detectable solid solubility

Group III: Peritectic systems
- III(a) Simple peritectic
- III(b) Partial miscibility in the α and β phases

*As indicated above, technical alloys are, of course, classified as α, β, and $\alpha + \beta$ according to their microstructural states when placed in service.

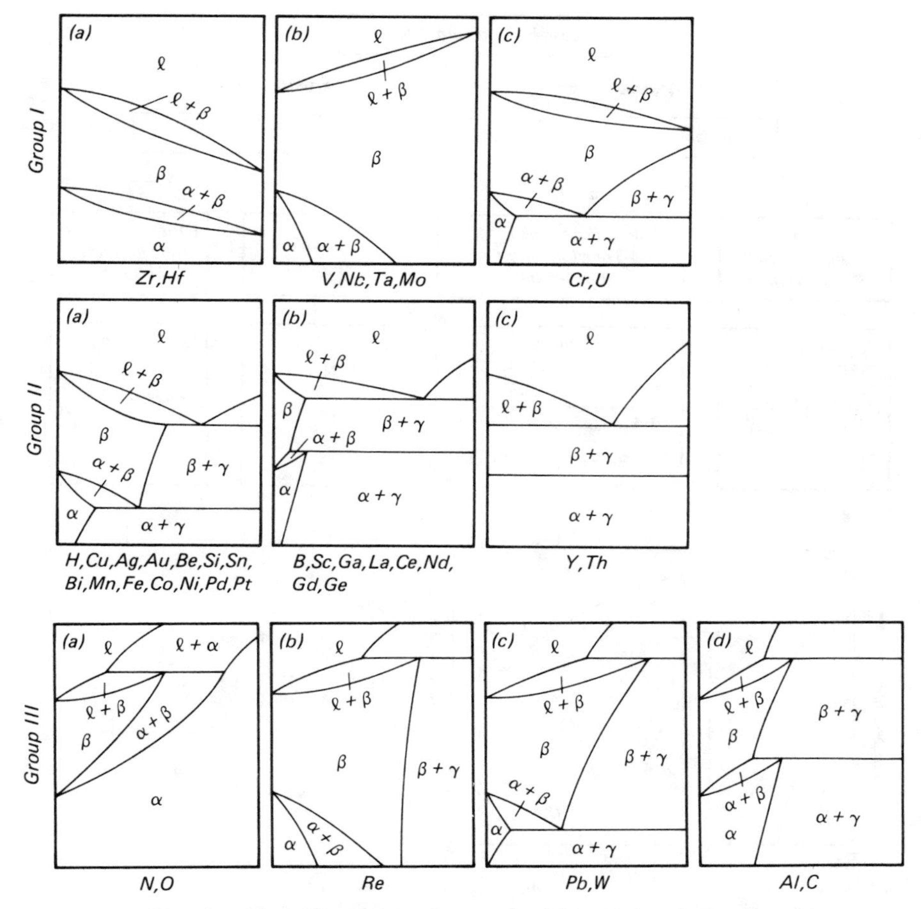

Fig. 3. Classification scheme for binary titanium-alloy phase diagrams[1, p. xiv]

III(c) Partial miscibility in the α and β phases; eutectoid decomposition of the β phase

III(d) Partial miscibility in the α and β phases; peritectic dissociations of the β phase

In the same book, Molchanova[1, p. 154] has also offered a simpler subdivision into the four categories depicted in Fig. 4.

1.5. Classification of Technical Multicomponent Alloys

1.5.1. Classification Schemes. Technical multicomponent alloys are generally composed of mixtures of α and β stabilizers (cf. Section 1.3) depending on the ratio of which they may be classified broadly as "α," "β," or "$\alpha + \beta$." And within the last category are the subclasses "near-α" and "near-β," referring to alloys whose

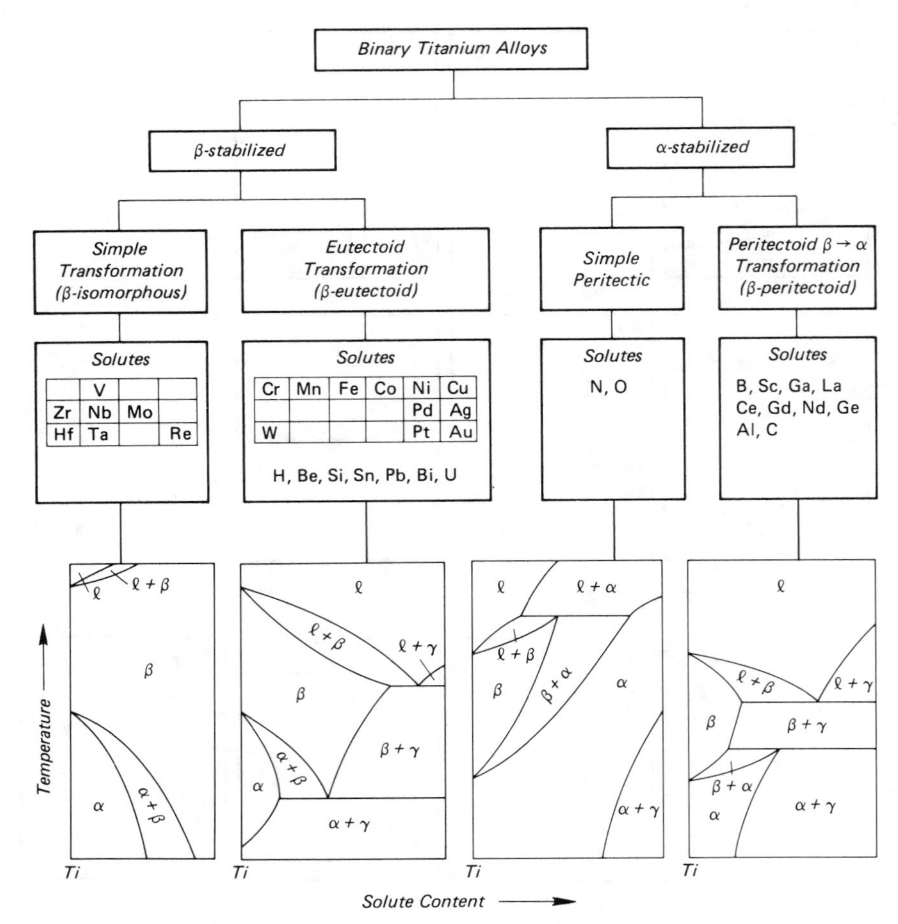

"α" and "β" are hcp and bcc solid-solution alloys, respectively, and "γ" represents an intermetallic compound.

Fig. 4. Classification scheme for binary titanium-alloy phase diagrams — an alternative to the scheme in Fig. 3[1, p. 154]

compositions place them near the $\alpha/(\alpha + \beta)$ or $(\alpha + \beta)/\beta$ phase boundaries, respectively. A list of U.S. alloys subdivided into these categories is presented in Table 1. The compositions of some other U.S. alloys and of some technical British and Soviet alloys are listed in Tables 2, 3, and 4, respectively. These alloys may also be sorted into microstructural classifications with the aid of a scheme to be discussed in the following subsection.

In an alternative attempt at alloy classification, Nishimura *et al.*[23] have mapped (Fig. 5) the locations of a series of U.S. technical alloys along the abscissa of a "β-isomorphous" (see Fig. 4) binary-alloy phase diagram. Presented in this way, it is clear that position along the abscissa is controlled by the relative abundances of the alloy's α- and β-stabilizing components. This ratio can be most conveniently expressed in terms of α-stabilizing and β-stabilizing "equivalencies."

Table 1. Classification of U.S. technical multicomponent alloys[5,21]

Composition, wt %		Classification
Ti-5Al-2.5Sn		α
Ti-8Al-1Mo-1V	Near-α	α + β
Ti-6Al-2Sn-4Zr-2Mo	Near-α	α + β
Ti-6Al-4V		α + β
Ti-6Al-2Sn-6V		α + β
Ti-3Al-2.5V		α + β
Ti-6Al-2Sn-4Zr-6Mo	Near-β	α + β
Ti-5Al-2Sn-2Zr-4Cr-4Mo	Near-β	α + β
Ti-3Al-10V-2Fe	Near-β	α + β
Ti-13V-11Cr-3Al		β
Ti-15V-3Cr-3Al-3Sn		β
Ti-4Mo-8V-6Cr-4Zr-3Al		β
Ti-8Mo-8V-2Fe-3Al		β
Ti-11.5Mo-6Zr-4.5Sn		β

Table 2. Commercial and semicommercial U.S. titanium alloys[22] (supplement to Table 1)

Composition, wt %	Classification
Ti-0.8Ni-0.3Mo	α and near-α
Ti-6Al-2Nb-1Ta-0.8Mo	α and near-α
Ti-2.25Al-11Sn-5Zr-1Mo	α and near-α
Ti-5Al-5Sn-2Zr-2Mo	α and near-α
Ti-7Al-4Mo	α + β
Ti-4.5Al-5Mo-1.5Cr	α + β
Ti-6Al-2Sn-2Zr-2Mo-2Cr	α + β
Ti-8Mn	Metastable β

Table 3. British technical commercial alloys[22]

Designation	Composition, wt %
IMI 318 ...	Ti-6Al-4V
IMI 550 ...	Ti-4Al-4Mo-2Sn-0.5Si
IMI 679 ...	Ti-11Sn-1Mo-5Zr-2.25Al-0.25Si
IMI 680 ...	Ti-11Sn-4Mo-2.25Al-0.25Si
IMI 685 ...	Ti-6Al-5Zr-0.5Mo-0.3Si
IMI 829 ...	Ti-5.5Al-3.5Sn-3Zr-1Nb-0.3Mo-0.3Si
IMI 834 ...	Ti-5.5Al-4Sn-4Zr-1Nb-0.3Mo-0.5Si

Table 4. Soviet titanium alloys(a)[22]

Designation	Composition, wt %
TG-00	99.7 Ti
TG-2	99.2 Ti
VT1-1 VT1D-1 VT1-2	Commercial unalloyed grades
VT-2	Ti-1.6Al-2.5Cr
VT-3	Ti-4.6Al-2.5Cr
VT-3-1	Ti-4.6Al-2Cr-1.7Mo-0.5Fe
VT-4	Ti-4.6Al-1.5Mn
VT-5	Ti-4.5Al
VT-5-1	Ti-5Al-2.5Sn
VT-6	Ti-6Al-4V
VT-8	Ti-6Al-3Mo
VT-14	Ti-4Al-3Mo-1V
VT-15	Ti-3Al-6.5Mo-11Cr
VT-16	Ti-2Al-7Mo
OT-4	Ti-3Al-1.5Mn
OT-4-1	Ti-1.7Al-1.4Mn
48-OT3	Ti-4Al-0.1Si-0.1Fe-0.005B
IRM-1	Ti-4Al-4Nb
IRM-2	Ti-4Al-4Nb-0.1Re
IRM-3	Ti-4Al-3.5Mo
IRM-4	Ti-3.5Al-3.5Mo-0.1Re
AT-2-1	Ti-Zr-(Mo or Nb or V)
AT-2-2	Ti-Zr-(Mo or Nb or V)
AT-2-4	Ti-Zr-(Mo or Nb or V)
AT-3	Ti-3Al-0.7Cr-0.4Fe-0.3Si-0.01B
AT-4	Ti-4Al-0.6Cr-0.23Fe-0.4Si-0.01B
AT-6	Ti-6Al-0.6Cr-0.4Fe-0.3Si-0.01B
AT-8	Ti-7Al-0.6Cr-0.2Fe-0.3Si-0.01B

(a) See also Table 18.

1.5.2. Alpha-Stabilizing and Beta-Stabilizing Equivalencies. The prototypical α-stabilizing and β-stabilizing additions to titanium are aluminum and molybdenum, respectively. Accordingly it is useful to be able to classify a multicomponent titanium-base alloy in terms of its equivalent aluminum and molybdenum contents.

(a) *Equivalent Aluminum Content*: In that elements such as aluminum and oxygen elevate the $(\alpha + \beta)/\alpha$ transus when alloyed into titanium, they are regarded as strong stabilizers of the α phase. Tin is also an α stabilizer, although not a strong one. Since zirconium has the effect of lowering the temperature of the $(\alpha + \beta)/\alpha$ transus at a very low rate, it may from that standpoint be regarded as a neutral addition. On the other hand, zirconium occupies the same column of the periodic

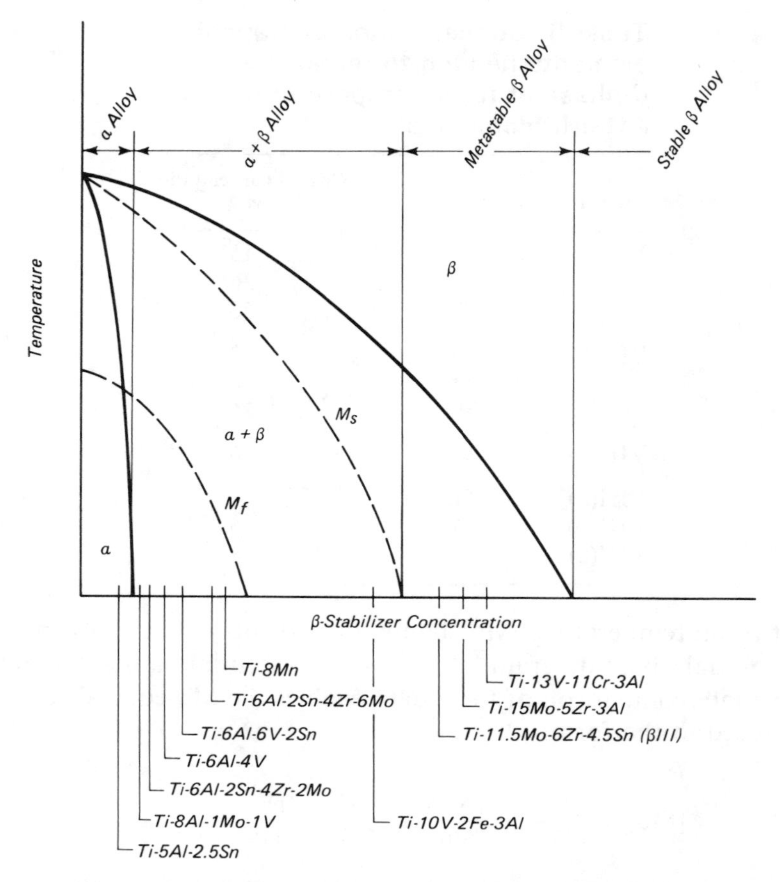

Fig. 5. Compositions of U.S. technical alloys mapped onto a pseudobinary β-isomorphous phase diagram[23]

table as titanium (viz. group IV). As a consequence of this chemical similarity, zirconium may substitute for titanium in a multicomponent alloy and thereby add weight to its α-stabilizing component. For this reason it may be regarded as an α stabilizer.

Aluminum is the canonical α-stabilizing addition against which other such additions may be compared. According to Rosenberg,[24] the equivalent aluminum content of an alloy containing aluminum, zirconium, tin, and oxygen is

$$[\text{Al}]_{\text{eq}} = [\text{Al}] + \frac{[\text{Zr}]}{6} + \frac{[\text{Sn}]}{3} + 10[\text{O}] \qquad \text{(Eq 1)}$$

where [x] indicates the concentration of element "x" in weight percent.

(b) *Equivalent Molybdenum Content*: The β-stabilizing strength of transition-element additions to titanium can be gaged by the rates at which they lower the martensite transus and hence the degree to which they permit the retention of the

Table 5. Concentration of transition elements needed to retain the β-phase at room temperature. After Molchanova[1, p. 158]

Group number	Element	Critical concentration, wt %
V	V	15
	Nb	36
	Ta	50
VI	Cr	8
	Mo	10
	W	25
VII	Mn	6
VIII(a)	Fe	4
VIII(b)	Co	6
VIII(c)	Ni	8

β phase at room temperature. Molchanova has displayed this information in the form of a partial phase diagram[1, p. 158] whose essential data are summarized in Table 5. An intercomparison of these data enables the Mo-equivalency of an alloy to be expressed in the form:

$$[\mathrm{Mo}]_{\mathrm{eq}} = [\mathrm{Mo}] + \frac{[\mathrm{Ta}]}{5} + \frac{[\mathrm{Nb}]}{3.6} + \frac{[\mathrm{W}]}{2.5} + \frac{[\mathrm{V}]}{1.5} + 1.25[\mathrm{Cr}] + 1.25[\mathrm{Ni}] + 1.7[\mathrm{Mn}] + 1.7[\mathrm{Co}] + 2.5[\mathrm{Fe}] \qquad \text{(Eq 2)}$$

Transformation of a number of multicomponent titanium-base alloys into their Al- and Mo-equivalent formats provides a rationalization for their placement into one or another of the previously discussed phase-stability classifications (Table 6).

2. EQUILIBRIUM PHASES IN BINARY TITANIUM ALLOYS

The equilibrium phase diagrams of numerous Ti-base binary alloys have been presented and discussed by McQuillan and McQuillan,[25] Imgram *et al.*,[26] and Zwicker,[2] and the properties of several important systems have been reviewed by Jaffee[27] and Margolin and Nielson.[19] The most comprehensive compendium of binary phase diagrams has of course been provided by Molchanova.[1]

In what follows, two important alloy systems will be briefly reviewed. They are: Ti-Al, the basis of technical α-Ti alloys; and Ti-Mo, a β-isomorphous alloy and the basis of several technical β-Ti alloys.

Table 6. Aluminum and molybdenum equivalencies of a series of U.S. titanium alloys

Alloy classification and composition, wt %	Aluminum equivalency, wt %				Molybdenum equivalency, wt %									
	[Al]	[Zr]/6	[Sn]/3	$[Al]_{eq}$	[Mo]	[Ta]/5	[Nb]/3.6	[V]/1.5	1.25[Cr]	1.25[Ni]	1.7[Mn]	1.7[Co]	2.5[Fe]	$[Mo]_{eq}$
Alpha and Near-Alpha Alloys														
Ti-0.8Ni-0.3Mo	...	...	...	...	0.3	...	...	...	...	1.0	...	...	...	**1.3**
Ti-5Al-2.5Sn	5.0	...	0.8	**5.8**	...	...	...	...	...	...	...	...	...	...
Ti-8Al-1Mo-1V	8.0	...	...	**8.0**	1.0	...	...	0.7	...	...	...	...	...	**1.7**
Ti-6Al-2Sn-4Zr-2Mo-0.1Si	6.0	0.7	0.7	**7.4**	2.0	...	...	...	...	...	...	...	...	**2.0**
Ti-6Al-2Nb-1Ta-0.8Mo	6.0	...	...	**6.0**	0.8	0.2	0.6	...	...	...	...	...	...	**1.6**
Ti-2.25Al-11Sn-5Zr-1Mo	2.3	0.8	3.7	**6.8**	1.0	...	...	...	...	...	...	...	...	**1.0**
Ti-5Al-5Sn-2Zr-2Mo	5.0	0.3	1.7	**7.0**	2.0	...	...	...	...	...	...	...	...	**2.0**
Alpha-Beta Alloys														
Ti-6Al-4V	6.0	...	...	**6.0**	...	...	...	2.7	...	...	...	...	...	**2.7**
Ti-6Al-6V-2Sn	6.0	...	0.7	**6.7**	...	...	...	4.0	...	...	...	...	...	**4.0**
Ti-7Al-4Mo	7.0	...	...	**7.0**	4.0	...	...	...	...	...	...	...	...	**4.0**
Ti-4.5Al-5Mo-1.5Cr	4.5	...	...	**4.5**	5.0	...	...	...	1.9	...	...	...	...	**6.9**
Ti-6Al-2Sn-4Zr-6Mo	6.0	0.7	0.7	**7.4**	6.0	...	...	...	...	...	...	...	...	**6.0**
Ti-5Al-2Sn-2Zr-4Mo-4Cr	5.0	0.3	0.7	**6.0**	4.0	...	...	...	5.0	...	...	...	...	**9.0**
Ti-6Al-2Sn-2Zr-2Mo-2Cr	6.0	0.3	0.7	**7.0**	2.0	...	...	...	2.5	...	...	...	...	**4.5**
Ti-3Al-2.5V	3.0	...	...	**3.0**	...	...	...	1.7	...	...	...	...	...	**1.7**
Ti-10V-2Fe-3Al	3.0	...	...	**3.0**	...	...	...	6.7	...	...	...	...	5.0	**11.7**
Beta Alloys (Metastable)														
Ti-8Mn	...	...	...	...	...	...	...	...	...	...	13.6	...	...	**13.6**
Ti-11.5Mo-6Zr-4.5Sn	...	1.0	0.4	**1.4**	11.5	...	...	...	...	...	...	...	...	**11.5**
Ti-15V-3Cr-3Al-3Sn	3.0	...	1.0	**4.0**	...	...	...	10.0	3.8	...	...	...	...	**13.8**
Ti-13V-11Cr-3Al	3.0	...	...	**3.0**	...	...	...	8.7	13.8	...	...	...	...	**22.5**
Ti-8Mo-8V-2Fe-3Al	3.0	...	...	**3.0**	8.0	...	...	5.3	...	...	...	...	5.0	**18.3**
Ti-3Al-8V-6Cr-4Mo-4Zr	3.0	0.7	...	**3.7**	4.0	...	...	5.3	7.5	...	...	...	...	**16.8**

2.1. The Typical Alpha Alloy, Ti-Al

Equilibrium phase diagrams for Ti-Al, a representative α-stabilized titanium-base binary alloy, are to be found in McQuillan and McQuillan,[25, p. 174] Molchanova[1, p. 137] (see Fig. 6), and Zwicker.[2, p. 147] Research articles in which descriptions of portions of the equilibrium diagram have been discussed are listed in Table 7. Two of the most interesting and important regions of the diagram surround the ordered intermetallic compounds Ti_3Al and TiAl. Short-range and long-range order in Ti-Al alloys and the occurrence of the long-range-ordered α_2 phase in the vicinity of the compound Ti_3Al were discussed by Blackburn,[38] who offered an equilibrium partial phase diagram for the composition range 0 to 30 at. % (Fig. 7). Confirmatory evidence for the existences and ranges of the disordered and ordered phases in Ti-Al (20 ~ 30 at. %) is to be found in the results of the magnetic susceptibility measurements of Collings *et al.*[45] Gehlen[46] has investigated the crystallography of Ti_3Al, which was found to possess the DO_{19} structure—a unit cell composed of four regular hcp cells apparently supported by covalent-like directional bonds connecting the aluminum and titanium atoms. Solution strengthening in Ti-Al alloys in terms of localized Ti-Al bonds, leading at sufficiently high solute concentrations to the above-mentioned long-range-ordered structure and electronic effects related to it, have been discussed by Collings and coworkers.[45,47,48] Some physical properties of Ti-Al alloys attributable to the occurrence of brittle intermetallic compounds at compositions near 25 at. % (Ti_3Al), 50 at. % (TiAl), and ~70 at. % (~$TiAl_3$) have been discussed, with particular reference to

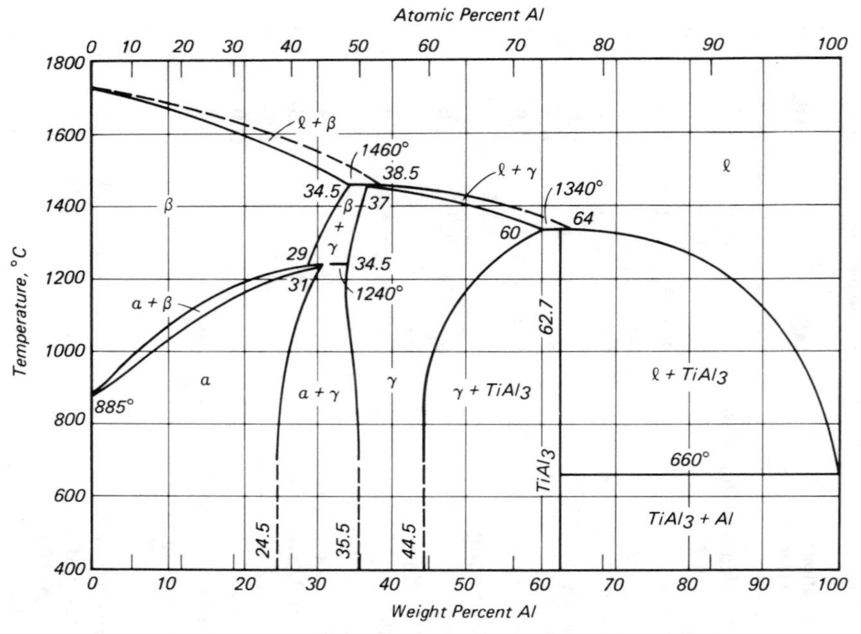

Fig. 6. A Ti-Al phase diagram[1, p. 137] (but see Table 7)

Table 7. List of investigations directed toward a determination of the Ti-Al equilibrium phase diagram

Aluminum concentration range, at. %	Temperature range for equilibrated solid alloys, °C	Principal and auxiliary techniques described	Reference
0 to 64	750 to 1100	Optical metallography; with x-ray diffraction and thermal analysis	28
0 to 75	700 to 1400	Optical metallography; with x-ray diffraction and Vickers hardness	29
0 to 75	700 to 1200	Optical metallography; with x-ray diffraction, Vickers hardness, thermal analysis, and centrifugal bend tests	30
5 to 49	550 to 1050	Electrical resistivity, magnetic susceptibility; with optical metallography and x-ray diffraction	31
0 to 63	450 to 1350	Electrical resistivity; with optical metallography and x-ray diffraction	32
0 to 48	800 to 1450	Optical metallography and x-ray diffraction	33
5 to 38	400 to 1100	Magnetic susceptibility	34
0 to 38	550 to 1200	Optical metallography, electrical resistivity, and x-ray diffraction	35
5 to 43	550 to 1200	Electrical resistivity and Vickers hardness; with thermal analysis, dilatometry, and x-ray diffraction	36
7 to 35	550 to 1100	Optical metallography; with electron microscopy, x-ray diffraction, differential thermal analysis, electrical resistivity, and dilatometry	37
5 to 25	500 to 1100	Electron microscopy	38
27 to 45	1025 to 1225	Electron microscopy	39
7 to 19	200 to 900	Electrical resistivity; with electron microscopy	40
0 to 33	500 to 1100	Differential thermal analysis, x-ray diffractometry, electrical resistivity, and hardness; with optical metallography	41
30 to 57	900 to 1365	Magnetic susceptibility; with optical metallography	42
0 to 30	625 to 1100	Magnetic susceptibility; with electron microscopy	43
7 to 44	450 to 1150	Differential thermal analysis; with electron microscopy	44

electronic bonding and its relationship to intermetallic compound formation, by Collings.[48]

The wide discrepancies that exist among the numerous Ti-Al phase diagrams presently in existence are evident in Zwicker's collection of six equilibrium diagrams.[2, p. 147] In order to shed further light on the position of a particularly important feature, the $(\alpha_2 + \gamma)/\gamma$ phase boundary, Collings[42] employed magnetic susceptibility techniques (augmented by optical metallography) in order to develop

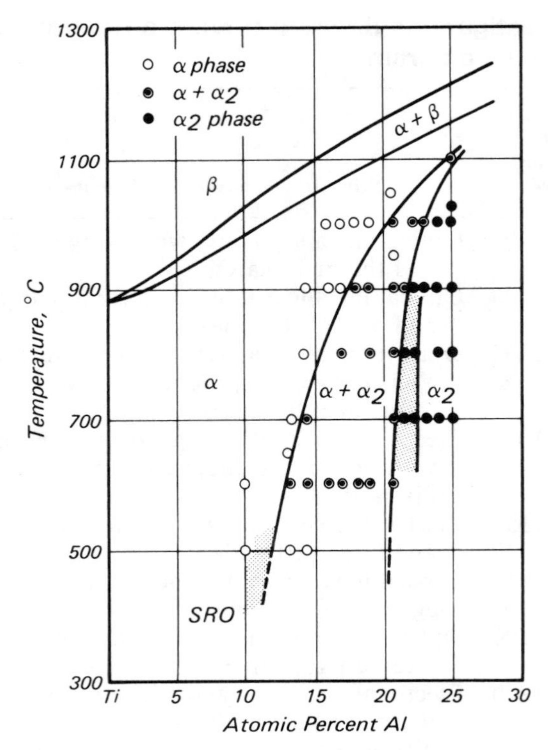

Fig. 7. Partial Ti-Al equilibrium phase diagram for the range 0 to 25 at. % Al[38]

an equilibrium partial diagram for Ti-Al (30-57 at. %) within the temperature range 900 ~ 1300 °C (Fig. 8). Swartzendruber *et al.*[43] also used magnetic susceptibility as a technique for studying a portion of the Ti-Al phase diagram. The most recent investigation of the Ti-Al system was by Shull *et al.*,[44] who employed differential thermal analysis, assisted by transmission electron microscopy, to develop a phase diagram for the composition range 7 to 44 at. % Al (Fig. 9).

2.2. The Typical Beta Alloy, Ti-Mo

Equilibrium phase diagrams for Ti-Mo have been developed by Craighead and coworkers (1950),* Hansen and coworkers (1951),* Duwez (1951),* Molchanova,[1, pp. 27-32] Terauchi *et al.*,[49] and, most recently, Hayman.[50] The diagram according to Hansen *et al.*[51, p. 977] is reproduced in Fig. 10. The alloys used in developing this diagram had been homogenized for 20 to 40 h at 1250 °C prior to being annealed at eight temperatures between 855 and 600 °C for times ranging from 90 to 650 h. Diagrams such as Fig. 10 are not usually continued below 600 °C

*See references in Molchanova.[1, p. 32]

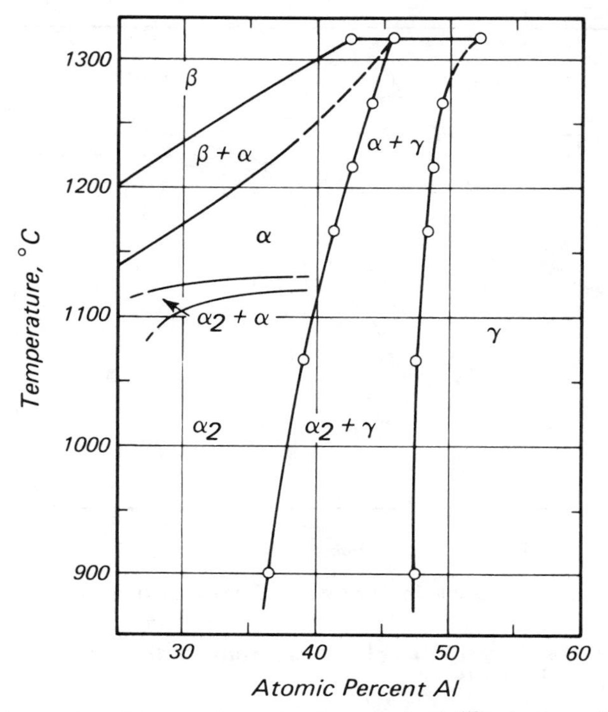

The data points (o) were determined magnetically[42]; the boundaries of the $\beta + \alpha$ and $\alpha_2 + \alpha$ fields were earlier established by Blackburn.[39]

Fig. 8. Partial Ti-Al phase diagram for the range 25 to 57 at. % Al

owing to the difficulties which are always encountered in attempts to attain thermodynamic equilibrium in reactive alloys when the diffusion rates are low. According to Molchanova,[1, pp. 27-32] the β phase is stable at all temperatures in alloys containing more than 16 at. % Mo (28 wt %). The $\beta/(\alpha + \beta)$ phase boundary is almost linear and intersects the 650 °C line at 14 at. % Mo (24 wt %). The maximum solubility of molybdenum at 600 °C in α-Ti-Mo alloys was stated to be only about 0.4 at. % (0.8 wt %), with the cautionary note that there was some uncertainty associated with that number.

In numerous studies of quenched nonequilibrium β-phase titanium alloys and the effects of aging on them as they proceed toward thermodynamic equilibrium, it has been noted that within the equilibrium $\alpha + \beta$ field, and bordering the meta-equilibrium $\omega + \beta$ zone, the aging of quenched β-stabilized alloys can result in a separation of the β phase into a solute-rich β matrix and a solute-lean β' precipitate. The β'/β interfaces,[52] or the interiors of the β' precipitates themselves,[53] are the sites of α-phase precipitation during further aging. Clearly the phase-separated $\beta' + \beta$ is a nonequilibrium condition, and as such is discussed in Section 3.3. A double-bcc phase can, however, exist as an *equilibrium* two-phase state in some alloy systems. Referred to as "β-phase immiscibility," it occurs, for exam-

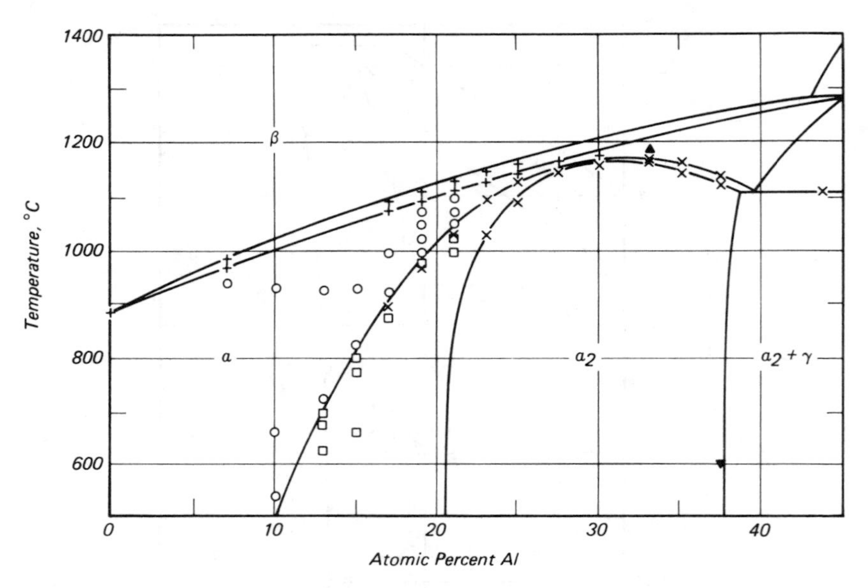

The squares represent single-phase α; the circles indicate two-phase $\alpha + \alpha_2$.

Fig. 9. Partial Ti-Al phase diagram for the range 0 to 45 at. % Al[44]

ple, in Zr-Nb[54] and related systems such as Ti-Zr-Nb.[55] The purpose of this digression into the existences of phase-separated and phase-immiscible double-bcc phases is to provide a suitable context for introducing the results of some studies of the Ti-Mo system by Terauchi and colleagues.[49] From optical observations, electrical resistivity measurements, x-ray diffractometry, lattice-parameter measurement, and transmission electron microscopy (TEM), those authors have deduced the existence of a pair of bcc phases—referred to as $\beta_1 + \beta_2$—occupying an area of the equilibrium phase diagram lying outside the $\alpha + \beta$ field (Fig. 11). Strong confirmation of the validity of Terauchi's diagram has recently been presented by Hayman.[50]

2.3. Departures From Equilibrium in Beta Alloys

Phase diagrams are usually developed by examining at room temperature the phases that are retained after quenching from an elevated-temperature anneal, reliance being placed on (*a*) the attainment of thermodynamic equilibrium at the elevated temperature of interest and (*b*) the retention of the elevated-temperature phase during the quench. In studies of quenched microstructures, an important but not always accomplished goal is the control and quantification of the quench rate. If the quench is too slow, diffusional processes intervene to obscure the result. When the primary aim is to study microstructure (rather than the production of material for physical- or mechanical-property testing) and the highest possible

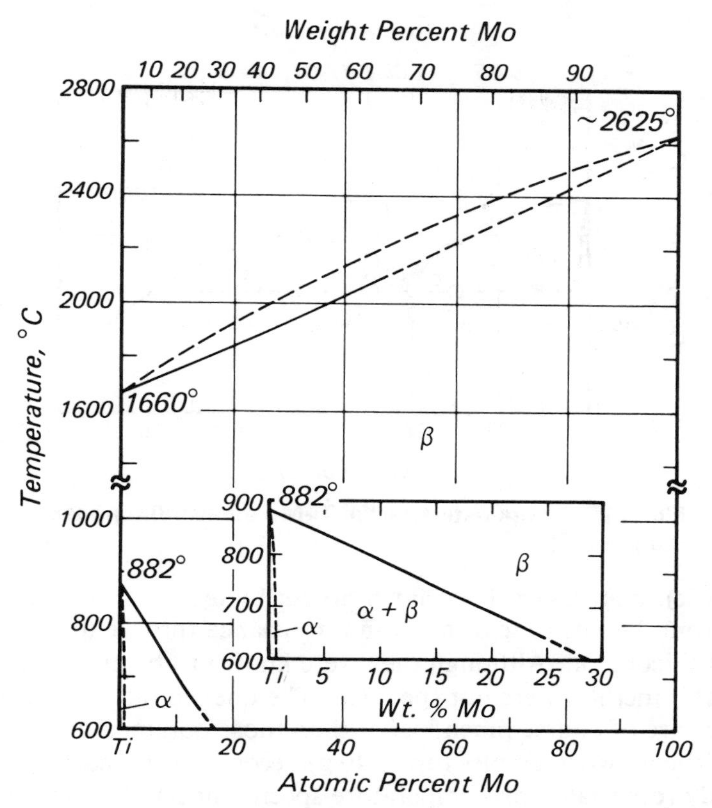

Fig. 10. The Ti-Mo equilibrium phase diagram[51, p. 977]

quench rates are mandatory, thin foils are generally heated in a controlled environment and subjected to *in situ** gas or liquid quenching. For example in Hickman's study of Ti-TM alloys, rolled strips self-heated under high vacuum to 1250 °C by the passage of direct current were quenched by admitting helium gas to a pressure of 0.1 atm; sample temperature was then restored, and the current switched off.[56,57] Such techniques are generally capable of quench rates of 50 $°C \cdot s^{-1}$ to 2×10^4 $°C \cdot s^{-1}$, and in Hickman's case about 10^3 $°C \cdot s^{-1}$ was claimed. Helium is about three times as effective a quench medium as argon under the same conditions.

Brown, Jepson, and Heavens[58] heated indirectly, under vacuum, specimens varying in thickness from 0.05 to 5.1 mm (0.002 to 0.20 in.); they then applied a 150 to 700 torr head-pressure of argon to suppress the boiling of the iced water or refrigerated calcium chloride solution subsequently admitted to quench the sample. In this way quench rates of 2.5×10^4 to 2×10^5 $°C \cdot s^{-1}$ were achieved.[59] Balcerzak and Sass[60] attached rolled specimens (0.05 to 0.08 mm, or 0.002 to

*In which the quenching medium is introduced into the furnace space containing a fixed sample.

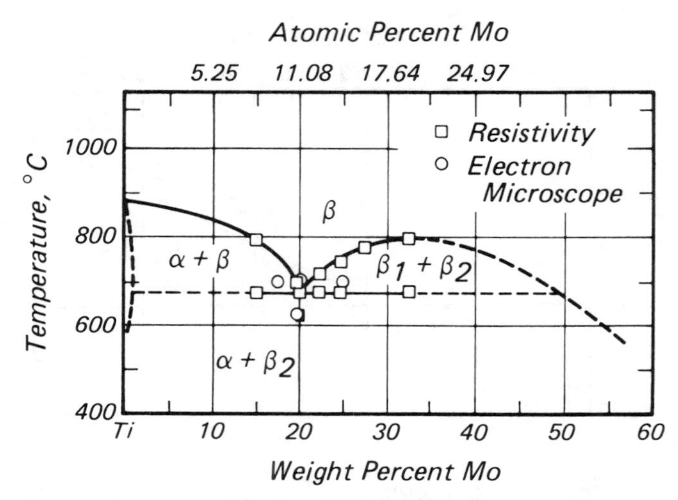

Fig. 11. A suggested partial Ti-Mo equilibrium phase diagram[49]

0.003 in., thick) to an Inconel specimen holder by means of which they could be transferred from the hot zone of a vacuum resistance furnace to a waiting pool of water-cooled silicone oil. Although they were known to be high, the quench rates achieved by this method were not specified. The quench rates achievable by all of these methods are of course much higher than those obtained during the ice-brine quenching of the massive samples (up to 40 g) needed for mechanical- and physical-property study (especially low-temperature specific heat). Accordingly, some discrepancies must be expected between the microstructural results obtained from thin foils and those derived from the quenching of bulk specimens.

The measured M_s temperature for a given alloy composition is itself a function of quench rate. In Ti-Nb(5 at. %), for example, Jepson *et al.*[59] noted that the M_s temperature decreased from 760 to 710 °C as the cooling rate increased from 10^{-3} to 10 $°C \cdot s^{-1}$, but that once a critical cooling rate of 32 $°C \cdot s^{-1}$ was exceeded, M_s was independent of the cooling rate. The critical threshold itself was a function of alloy composition and decreased from 200 to ~0.4 $°C \cdot s^{-1}$ as the niobium content increased from 0 to 15 at. %.

In Ti-TM alloys, as with other systems, the quench rates necessary to achieve structural transformation while preserving compositional homogeneity are strongly constitution dependent. Thus, whereas bulk dilute alloys of titanium with early transition elements can be water quenched without evidencing serious decomposition, the same is not true of alloys such as Ti-Fe, -Ni, and -Co, whose anomalous physical properties could be partially interpreted in terms of compositional, hence structural, segregation. The pronounced differences between the properties of the quenched dilute Ti-V, -Nb, etc. alloys and those of Ti-Fe, -Co, and -Ni can be simply explained in terms of differences among the solute tracer diffusion coefficients in β-Ti at 1000 °C. As shown in Fig. 12, the diffusion coefficients of vanadium, niobium, and molybdenum are less than 1.3×10^{-9} $cm^2 \cdot s^{-1}$, while those of the Fe-group elements are 60×10^{-9} $cm^2 \cdot s^{-1}$. The extreme examples are cobalt

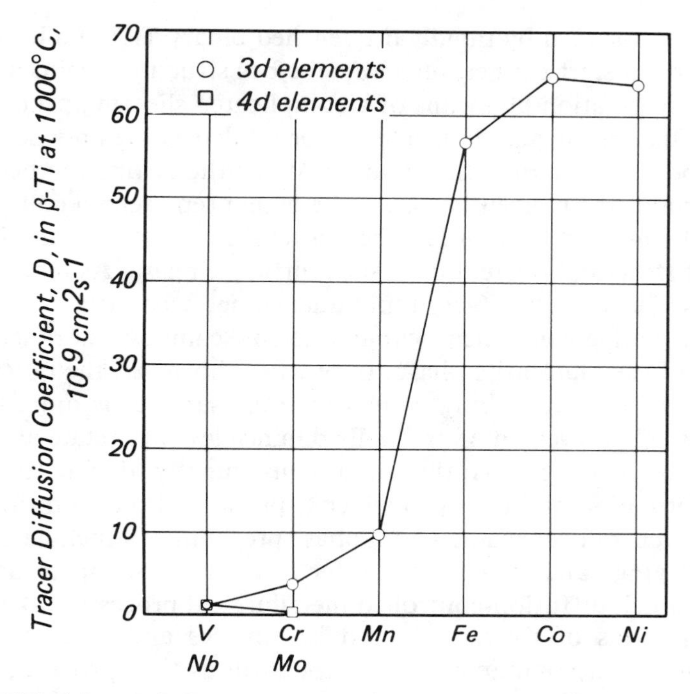

Diffusivity controls the rate at which thermodynamic equilibrium is attained at a given temperature; note that the diffusivities of iron, cobalt, and nickel are almost two orders of magnitude higher than those of vanadium, niobium, and molybdenum.

Fig. 12. Tracer diffusion coefficient for transition metals in β-Ti at 1000 °C computed from frequency-factor and activation-energy data of Zwicker.[2, p. 174]

on one hand and molybdenum on the other; their diffusion coefficients (β-Ti, 1000 °C) are in the ratio 200:1. This must be taken into consideration in comparing the properties of the two classes of titanium-base alloys, and in selecting a quenching technique.

3. NONEQUILIBRIUM PHASES IN BINARY TITANIUM ALLOYS AND THE APPROACH TO EQUILIBRIUM – AGING

Equilibrium phase diagrams of the type discussed in the previous section are usually developed by *deducing the initial states* of alloys which have been quenched to room temperature. The nonequilibrium phases to be considered herein represent the *final* states of such quenching processes.

The structure of α-stabilized alloys quenched from the β field is martensitic. When quenched from below the $(\alpha + \beta)/\alpha$ transus the structures found are of course simply the frozen-in untransformed results of equilibration at the prequenched temperature.

The structures assumed by rapidly β-quenched binary Ti-TM alloys are mapped in Fig. 13. Below a start temperature, M_s, the bcc structure begins a spontaneous allotropic transformation by means of a complicated shearing process to a structure known as martensite and designated α' or α'' depending on whether the transformation product is hcp or orthorhombic. When the distinction between α' and α'' is unimportant, the martensites are to be herein represented collectively by the notation α^m. Being of second order, the martensitic transformation is anticipated by a regime of structural fluctuations called diffuse ω phase. As represented in Fig. 13, the ω phase, as a result of very rapid quenching, exists as a crystalline precipitate plus a fluctuating component within a narrow composition range overlapping the boundary of the martensite phase. In practice, however, the range over which it occurs during brine quenching of macroscopic samples is quite broad and is depicted in Fig. 13 as a region of gradually diminishing precipitate abundance. The free energy of α^m is lower than that of ω; consequently, during the partial martensitic transformation of an alloy in which ω phase is also able to form, the martensite needles generally consume any ω-phase precipitates which lie in their paths.

The terms "aging" and "tempering" refer to moderate-temperature heat treatments during which diffusion-controlled metallurgical processes take place within macroscopic periods of time, measured in minutes and hours. The aging of metastable alloys is accompanied by precipitation as they proceed, generally by

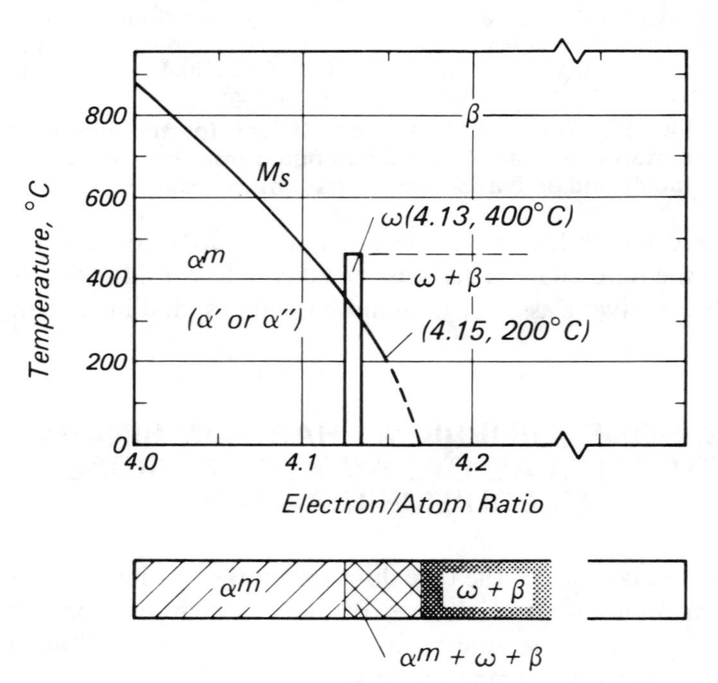

Both "quenched data" and "aged data" are included—see also Tables 8 (α^m phases) and 10 (ω phase).

Fig. 13. Schematic representation of the occurrences of the martensitic phases α' and α'' (i.e., α^m collectively) and the ω phase in Ti-TM alloys

means of a nucleation-and-growth mechanism, toward thermodynamic equilibrium. The equilibrium phases considered in the previous section are of course achieved by the prolonged heating of previously metastable structures. Although heat treatment or annealing at any temperature will permit a metastable alloy to approach more closely its state of thermodynamic equilibrium, the term "aging" is generally understood to imply a heat treatment in the low- to moderate-temperature range.

The aging of α-phase alloys, and in particular the long-range-ordering of the DO_{19} α_2-phase Ti_3SM-base structure, is not treated here. What will be considered, though, with reference to binary Ti-TM alloys, are: (1) the transformation under aging conditions of the α' and α'' martensitic phases; (2) aging in the $\omega + \beta$-phase field and precipitation of the isothermal ω phase; and (3) precipitation out of the β-phase field, adjacent to the $\omega + \beta$ region, of a solute-lean bcc phase designated β'—the so-called "phase-separation" reaction.

3.1. The Martensitic Phases

The word "martensite," named for Professor A. Martens, was originally adopted by metallurgists to define the acicular structure in quenched carbon steel that was responsible for its outstanding hardness.[61] The occurrence and structures of martensites in numerous other alloy systems have been described by Cohen[61] and by Bilby and Christian.[62] Detailed discussions of martensitic transformations, but with particular reference to titanium alloys, have been offered by the McQuillans,[25, Ch. 9] Margolin and Nielsen,[19] Hammond and Kelly,[63] and Otte.[64] Unalloyed titanium transforms martensitically from bcc to hcp during cooling through its $\beta \rightarrow \alpha$ allotropic transformation temperature, 882.5 °C. In alloys of titanium, the equilibrium α and β fields are separated by a two-phase, $\alpha + \beta$, region and the $\beta \rightarrow \alpha^m$ transformation temperature,* M_s, is composition dependent. In α-stabilized alloys, typified by Ti-Al, M_s may lie a little below the $(\alpha + \beta)/\alpha$ transus[59]; in the β-stabilized alloys it always lies within the $\alpha + \beta$ field. For some recent information on martensitic transformations in the β-isomorphous systems Ti-V, Ti-Nb, and Ti-Mo, and the manner in which the transformed structures revert to β or decompose on aging, the papers of Davis, Flower, and West[65-67] should be consulted.

3.1.1. Quenched Martensite

(a) *Occurrence*: The optical microstructures of a series of β-quenched Ti-TM alloys are shown in Fig. 14. The five alloys Ti-Mo, Ti-V, Ti-Nb, Ti-Mn, and Ti-Fe have comparable *e/a* ratios, had similar masses, and were ice-brine quenched under similar conditions. Noticeable in the photomicrographs of these alloys, which are arranged in ascending order of solute diffusion coefficient in β-Ti (Fig. 12), is what appears to be a gradual transition from a diffusionless to a diffusion-

*The symbol α^m is used herein as a shorthand notation for the product of martensitic transformation, whether it be α' or α''. M_s refers to the start of the transformation during cooling; M_f usually designates its finish.

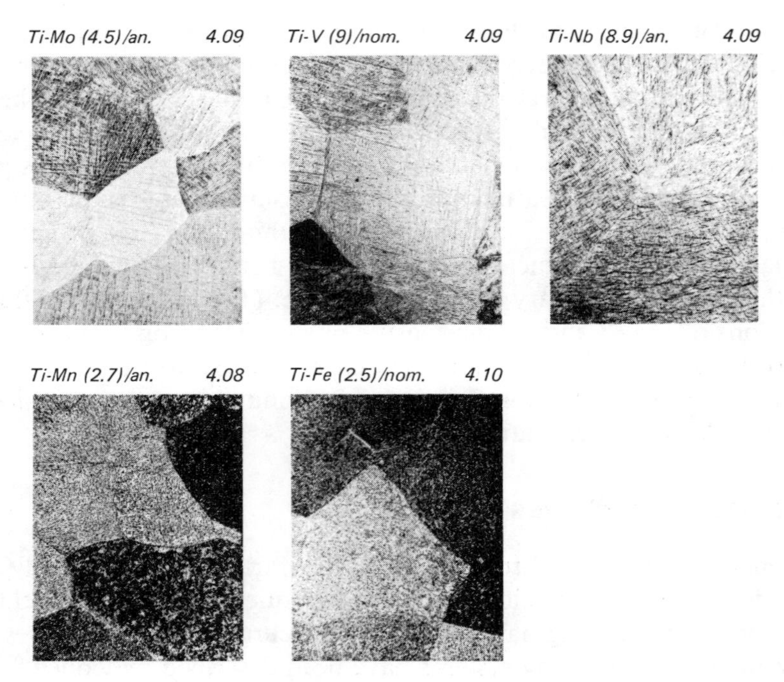

Indicated are the analyzed (an.) or nominal (nom.) solute concentrations and the conventional electron/atom ratio. Magnifications of original 11.5-by-9-cm micrographs, 50×.

Fig. 14. Optical micrographs of five low-concentration Ti-TM alloys after quenching into iced brine from the β phase, arranged in ascending order of solute atomic diffusion coefficient in β-Ti (see Fig. 12)

influenced as-quenched structure on proceeding from Ti-Mo to Ti-Fe. These results suggest that the atomic diffusion coefficient must be introduced as a scaling factor when estimating the effects of quench rate on martensitically transformable Ti-TM alloys.

In Table 8 are listed the $M_{s,200°C}$ compositions of eight Ti-TM alloys bounded constitutionally by Ti-Fe and Ti-Nb, together with the corresponding conventional electron/atom ratios. Quite remarkable is the fact that all the e/a ratios, except for those of Ti-Co and Ti-Ni (which may be exceptional cases), lie within ± 0.03 of a common value, 4.15, suggesting that the martensitic transformation in Ti-TM alloys is of common origin and related to electronic factors.

(b) *Morphology*: When conditions are particularly favorable, transformation from β to α^m takes place completely, on a large scale, and with considerable structural coherence. The result is the so-called "massive martensite" (otherwise known as packet, or lath, martensite), which consists of large irregular zones on the scale of 50 to 100 μm, subdivided into parallel arrays of fine platelets less than 1 μm across (Fig. 15). In massive martensite, the lack of retained β phase prevents direct determination of the habit plane. With increasing solute concentration, the coherence between the platelets, which would otherwise make up a massive colony, is

Table 8. Compositions of the $M_{s,200°C}$ intercepts expressed in terms of conventional electron/atom ratio

Solute group number, *GN*	Solute element	Concentration, *c*, corresponding to M_s at 200 °C, at. %(a)	Conventional *e/a* based on group number(b)	
	V	13.3	4.13	
V	Nb	20.5	4.21	
	Ta	19.1	4.19	
	Cr	6.0	4.12	
VI	Mo	6.7	4.13	
	W	8.2	4.16	
VII	Mn	5.0	4.15	
	Fe	3.3	4.13	
VIII	Co	6.0	4.24	4.30(c)
	Ni	7.6	4.30	4.46(c)
			Mean(d) 4.15 ± 0.03	

(a) After Zwicker.[2, p. 174]
(b) Calculated according to: $e/a = 4 + \Delta GN\, c/100$, where $\Delta GN = GN_{solute} - GN_{Ti} = GN_{solute} - 4$.
(c) Based on number of valence ($s + d$) electrons.
(d) Excluding cobalt and nickel.

lost. The result of this is a partially disordered array of individual platelets referred to as "acicular martensite" (Fig. 16). Further increase in solute concentration prevents a complete transformation from taking place, and β phase trapped between the platelets of the acicular martensite enables direct habit-plane determination to be accomplished.

Not far removed from the β-plus-acicular-martensite quenched structure is the Widmanstätten arrangement consisting of groups of α-phase needles lying with their long axes parallel to the {110} planes of the parent retained β (Fig. 17). Widmanstätten $\alpha + \beta$ (or Widmanstätten α, if the focus is primarily on α-phase precipitation), which is characteristic of dilute Ti-TM alloys or near-α $\alpha + \beta$ alloys (such as Ti-6Al-4V) appropriately cooled, is usually treated as a product of α-phase nucleation and growth. It is introduced into this section on nonequilibrium phases since, when solute diffusion coefficients are sufficiently large, either intrinsically so (large frequency factor, D_0) or if the temperature is high, diffusional processes compete with diffusionless martensitic transformation during the quenching of β-Ti alloys.[59,67,68]

3.1.2. Deformation Martensite. Metastable-bcc* Ti-TM alloys will also transform under the application of mechanical stress. Although some confusion has arisen in the past over the structure of the deformation product, the situation has been adequately clarified by Williams.[53] The β-quenched solute-lean but untrans-

*Alloys β-quenched into the $\alpha + \beta$ field but without intersecting M_s.

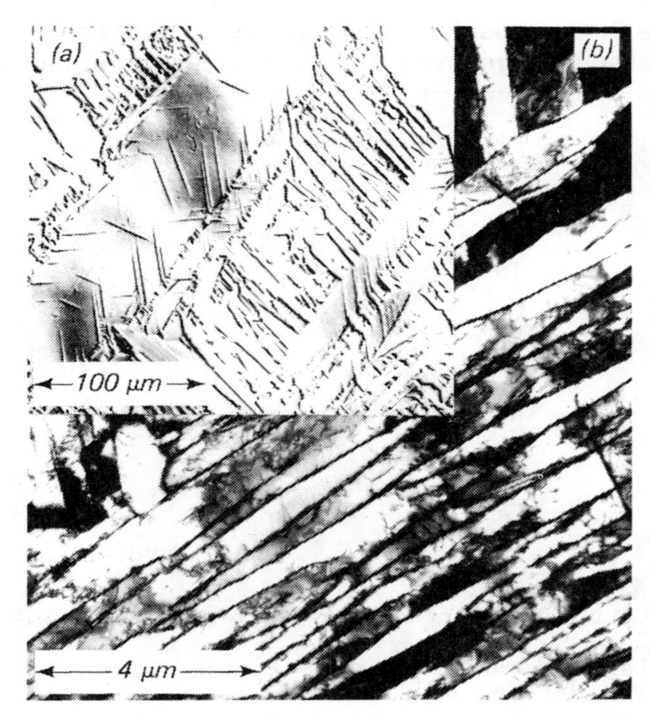

Specimen: Ti-1.78Cu quenched from 900 °C. (a) Optical micrograph showing large colonies. (b) Electron micrograph showing individual plates within the colonies. Micrographs courtesy of J.C. Williams, Carnegie-Mellon University.

Fig. 15. An example of "massive martensite"[53]

formed Ti-TM alloys are characterized by low shear moduli. In such an alloy, which may be represented in the phase diagram by a point close to M_s, the application of stress will trigger a transformation to "deformation-induced" or "stress-induced" martensite. Farther away from the M_s transus the bcc lattice responds directly to the influence of an applied stress by twinning. This, as pointed out by Suzuki and Wuttig,[69] can in the present context be regarded as a special case of martensitic transformation to a crystallographically equivalent structure. Indistinguishable optically from deformation martensite, mechanical twins are easily identified by diffractometry—their structure, of course, being identical to that of the parent lattice. Stress-induced martensitic transformation and twinning in Ti-Mo alloys have been discussed by Oka and Taniguchi.[70] Deformation-induced orthorhombic martensite produced by compressively deforming a β-quenched Ti-Mo (5 at. %) alloy is depicted in Fig. 18.

3.1.3. Reversion and Aging of Martensites. Numerous authors have investigated the thermal stabilities and modes of decomposition of tempered (aged) martensites. Using a thermal-arrest technique, Jepson *et al.*[59] have determined both the M_s temperature and the temperature, β_s, at which the *reverse* transformation commences. The results are given in Table 9. If M_s and β_s are fairly close, the tem-

Specimen: Ti-12V quenched from 900 °C. (a) Optical micrograph. (b) Electron micrograph showing lenticular-shape plates, some of which are internally twinned. Micrographs courtesy of J.C. Williams, Carnegie-Mellon University.

Fig. 16. An example of "acicular martensite"[53]

Table 9. Phase-transformation and equilibrium-phase-boundary temperatures for Ti-Nb alloys

Niobium content, at. %	Transformation temperatures(a), °C — M_s (Ref 59)	M_s (Ref 71)	β_s (Ref 59)	$\beta/(\alpha + \beta)$ boundary, °C (Ref 59)	(Ref 72)
0	855	855	...	885	885
2½	753	...	...	...	...
5	720	760	...	760	810
7½	619	...	646	...	...
9	567	...	592	...	...
10	560	600	540	650	765
11	517	...	530	...	...
12½	455	500	455	620	740
15	385	400	387	585	725
17½	300	...	317	545	705

(a) M_s = Martensite start ($\beta \rightarrow \alpha'$) temperature. β_s = Martensite reversion ($\alpha' \rightarrow \beta$) temperature.

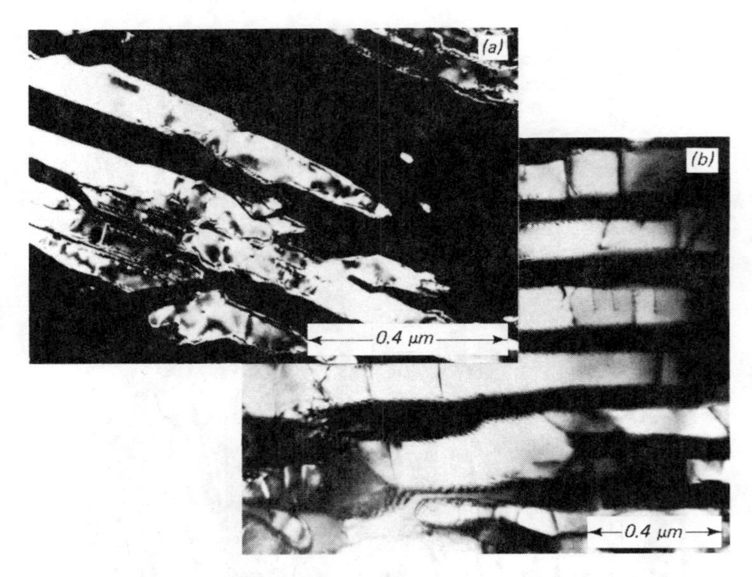

(a) Dark-field micrograph showing the absence of internal structure in the platelets. (b) Bright-field micrograph showing the α platelets separated by the β matrix. Micrographs courtesy of J.C. Williams, Carnegie-Mellon University.

Fig. 17. An example of Widmanstätten $\alpha + \beta$ phase

Specimen: Ti-Mo(5 at. %) quenched from the β phase and deformed 23% by compression. Magnification of original 9-by-9-cm micrograph, 50×.

Fig. 18. An example of deformation martensite and/or twinning

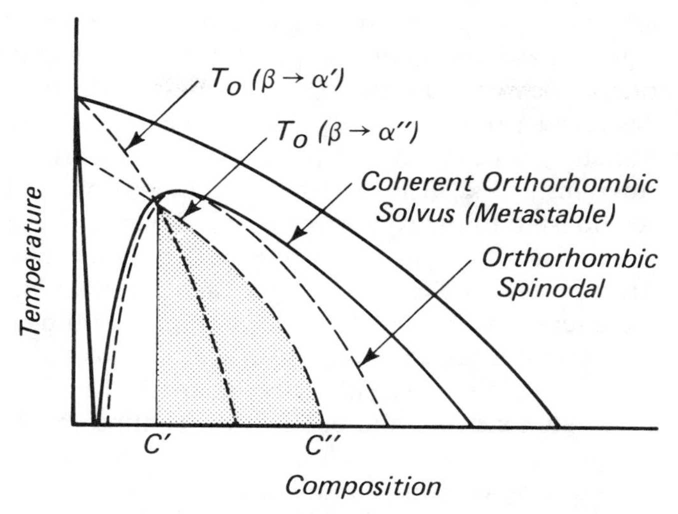

Fig. 19. Schematic representation of phase relationships in β-isomorphous Ti-TM alloys (i.e., Ti-V, Ti-Nb, and Ti-Mo) for purpose of illustrating modes of martensitic transformation and spinodal decomposition of martensites[67; see also 66]

perature T_0, corresponding to equality of the β and α^m free energies, can be taken as their mean. Other workers, notably Hatt and Rivlin,[73] Baker and Sutton,[74] Hickman,[57] and Flower *et al.*,[66,67] have considered the manner in which the α^m decomposes and the natures of the products, which may be β, $\omega + \beta$, or $\alpha + \beta$, depending on the conditions. The situation can be loosely, but instructively, summarized with the aid of Fig. 19, a heuristic nonequilibrium/equilibrium phase diagram illustrating the relative positions of the $M_s^{\alpha'}$ and $M_s^{\alpha''}$ transi (after Flower, Davis, and West[67]) and the isothermal $\omega + \beta$ phase region. Several classes of decomposition take place, depending on the composition of the martensite and the temperature. In the Ti-Nb α' martensites ($\lesssim$11 wt %, 6 at. %), for example, the equilibrium β phase was found to nucleate heterogeneously. The α'' martensites within a certain composition range are able to decompose spinodally into α''_{lean} + α''_{rich}[66,67] (Fig. 19). This process, which has been observed to take place in Ti-Nb (14-20 wt %, 8-11 at. % Nb),[67] begins during the quench, continues on further aging, and eventually proceeds to $\alpha + \beta$. Alloys sufficiently high in niobium content and aged at moderate temperatures revert to metastable β, which then decomposes into $\omega + \beta$. The process has been noted in water-quenched α''-Ti-Nb (20 at. %) aged at 330 °C,[74] water-quenched α''-Ti-Nb(20.7 at. %) aged at 335 °C,[73] Ti-Nb(22 at. %) aged above 200 °C,[57] and Ti-Nb(25 at. %) aged above 150 °C,[57] both of the latter alloys having been previously gas quenched.

3.2. The Omega Phases

In discussing the occurrence of ω phase in titanium alloys, Fig. 13 makes a useful starting point. As represented therein, a narrow region exists in which ω phase appears *athermally* during very rapid quenching from the β phase. Over a broader composition range ω phase will occur as a precipitation product of β decomposition during moderate-temperature ($\lesssim$400 °C) *isothermal* aging; alloys *rapidly quenched* into this broader region are host to a "diffuse ω phase," so-called because of the existence of straight or curvilinear lines of diffuse intensity in selected-area electron diffractograms.(52,75,76) The occurrence, composition, and structure of ω phase, and its relationship to the competing α and β phases, have been reviewed by Hickman(77) and Sass,(76) who made reference to important earlier work, including the often-quoted studies of Silcock(78) and Bagariatskii *et al.*(79)

During moderate-temperature ($\lesssim$400 °C) aging for several days, a metastable equilibrium $\omega + \beta$ state is attained, analysis of which yields the compositions of the ω (solute-lean) and β (solute-rich) phase boundaries. The ω-phase regime does not extend to zero solute concentration, but is tightly confined by the nearby lower-energy martensitic-phase field. As a result of the limited composition range of the athermal ω phase, good agreement is obtained between the "saturation composition of ω phase after aging at 400 °C"(57) and the "solute concentrations which yield ω phase on quenching",(79) as shown in Table 10. The latter is of course a difficult quantity to determine accurately, since away from the narrow zone of athermal ω phase, rapidly quenched alloys support the diffuse ω phase, and more slowly quenched larger samples will contain isothermal ω. Table 10 gives the conventionally calculated e/a ratios for ω phase, separated into two listings accord-

Table 10. Compositions of the $\omega/(\omega + \beta)$ data points expressed in terms of conventional electron/atom ratio

Solute group number	Solute element	Saturation composition of ω phase (at. %) after aging at ~400 °C (Ref 57)	Solute concentration (at. %) in Ti alloys for which ω phase is formed on quenching (Ref 79)	Conventional e/a based on group number (see Table 8): "Aged data"	"Quenched data"
V	V	13.8 ± 0.3	...	4.14	...
			13	...	4.13
	Nb	~9 ± 2	...	4.09	...
			18	...	4.18
VI	Cr	6.5 ± 0.2	...	4.13	...
			7	...	4.14
	Mo	4.3 ± 0.4	...	4.09	...
			4.5	...	4.09
	W	... ...	7.5	...	4.15
VII	Mn	5.1 ± 0.2	...	4.15	...
			5.5	...	4.16
	Re	... ...	4.5	...	4.14
VIII	Fe	4.3 ± 0.2	...	4.17	...
			3	...	4.12
				Mean 4.13 ± 0.03	Mean 4.14 ± 0.03

ing to the source and the method of data acquisition. It is remarkable to note that in each case the critical *e/a* ratios for the eight alloys listed are constant [within 20% in terms of $\Delta(e/a)$], and that the two independent mean values (which differ by <0.01) agree within experimental scatter. Secondly, a comparison of Tables 8 and 10, which yield, respectively, $\langle e/a \rangle_{\alpha^m} = 4.15 \pm 0.03$ and (from an overall mean) $\langle e/a \rangle_{\omega} = 4.13 \pm 0.03$, emphasizes that athermal ω phase occurs at the threshold of martensitic transformation and suggests that the ω and α^m transformations are interrelated through a common electronic mechanism. The results also reconfirm the validity of Fig. 13.

The results of a literature study of collected transformation data for nine Ti-TM alloys, conducted by Luke, Taggart, and Polonis[81] (Table 11), independently of that of Bagariatskii *et al.*[79] and prior to that of Hickman,[57] and indexed in terms of an *e/a* ratio based on Pauling valence[82] rather than group number, led to a similar conclusion.

3.2.1. Athermal Omega Phase. The presence in quenched Ti-TM alloys of athermal ω-phase particles, too small (~20 to 40 Å) to be detected optically, can be unequivocally diagnosed using TEM and selected-area electron diffractometry (SAD). By way of example, their occurrence in ice-brine-quenched Ti-Mo(5 at. %), which is just on the edge of the quenched martensitic regime, has been confirmed using these techniques, the results of which are shown in Fig. 20.

The occurrence of athermal ω phase in such close proximity to the martensitic phase boundary is a phenomenon of general validity and of fundamental importance in the theory of the ω transformation (to be considered later in more detail). Although from the standpoint of the crystallographer the absence of a habit-plane description precludes the defining of the ω-phase transformation as martensitic,[53] both transformations, according to Suzuki and Wuttig[69] and Clapp,[84] possess

Table 11. Compositional limits of the $\omega + \beta$ phase in quenched alloys of titanium with other transition elements[80,81]

Alloy		Solute concentration range, at. %		Pauling *e/a*-ratio range	
Solvent	Solute	Minimum	Maximum	Minimum	Maximum
Ti	V	14	21	4.14	4.21
Ti	Nb	12	20	4.12	4.20
Ti	Cr	...	11	...	4.20
Ti	Mo	6	10	4.11	4.18
Ti	Fe	4	10	4.07	4.18
Ti	Ru	5	10	4.09	4.18
Ti	Ni	7	...	4.12	...
Ti-V (2.5 at. %)	Ru	5	...	4.13	...
Ti-V (5 at. %)	Ru	3.5	7.5	4.13	4.20
Ti-V (7.5 at. %)	Ru	2.5	6	4.13	4.20
Ti-V (10 at. %)	Ru	1.5	4	4.13	4.18
Ti-V (15 at. %)	Ru	...	2.5	...	4.20

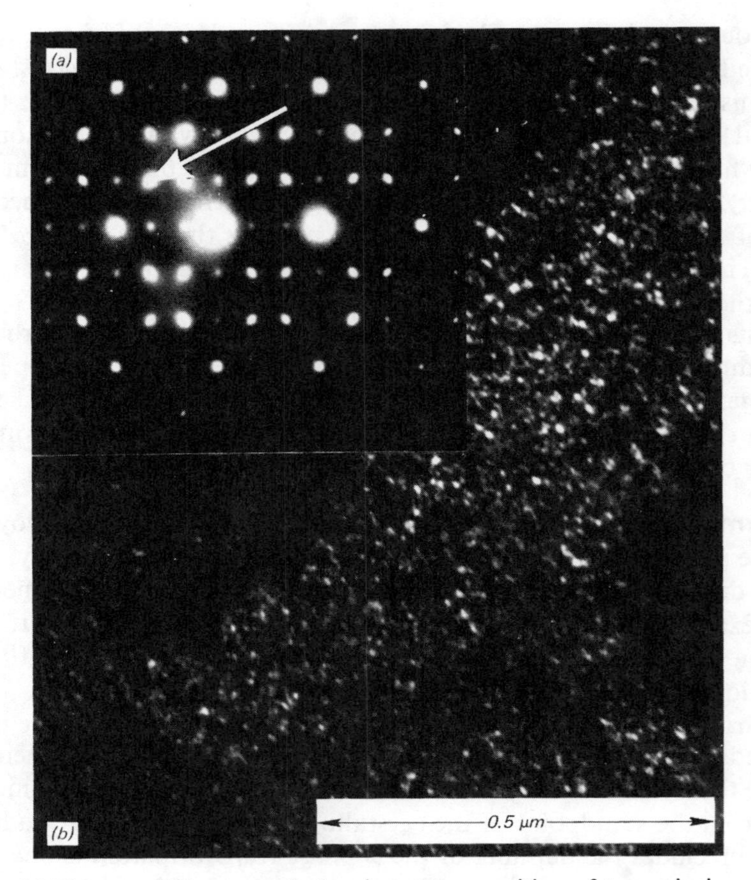

(a) Electron diffractograph showing a superposition of two principal spot patterns: a rectangular arrangement of round spots originating from the β matrix, and a group of elongated spots originating from the ω phase. (b) Dark-field electron micrograph of the ω-phase precipitate originating from the diffraction spot indicated by the arrow in the diffractograph.

Fig. 20. Transmission electron micrograph of Ti-Mo (5 at. %) quenched into iced brine from the β field[83]

a common ingredient in the form of a soft-phonon instability. Arguments supporting this view are developed below.

(a) *Linear-Fault Mechanism for Athermal* ω: Athermal ω phase may be regarded as being developed within the bcc lattice by applying to pairs of adjacent $(110)_\beta$ planes equal and opposite shears, in the $\langle 111\rangle_\beta$ direction, through distances about equal to 1/6 of the separation of the $(111)_\beta$ planes. The arrows in Fig. 21 indicate the planes, or rows, of atoms involved and the directions of the shears required. The coherent β/ω interface at the boundary $(110)_\beta$ plane is an important feature. If z is the separation of the $(111)_\beta$ planes $(=a\sqrt{3}/2$, where a is the bcc lattice parameter), then: (1) displacements of the A and B atoms by the amounts $\pm(1/6)z$ lead to the hexagonal structure in Fig. 21 proposed originally by Silcock[78] as a

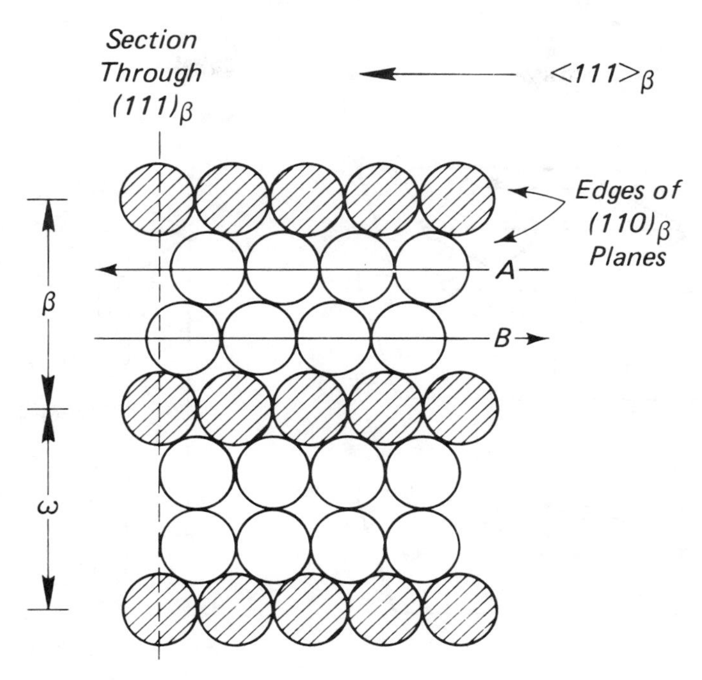

$(111)_\beta$ section (in plane of page) through bcc crystal depicting the transformation to ω phase as a displacement of adjacent $(110)_\beta$ planes (say, planes A and B).

Fig. 21. "Linear-fault" model for ω-phase transformation[79]

result of measurements on a low-solute-content ω phase; and (2) displacements of $\pm 0.15z$ yield the trigonal structure originally proposed by Bagariatskii[79] and now understood to be characteristic of higher-solute-content ω phases.

Assuming touching hard spheres of constant diameter (ϕ) in the simple geometrical model of Fig. 21, the separation of the $(110)_\beta$ planes, originally $2\sqrt{2}\ \phi$, shrinks to $(1 + \sqrt{3})\phi$ on transformation to ω phase, a 3% contraction which justifies the tendency for ω to form in titanium and zirconium under pressure.[85] Adoption of this conventional crystallographic approach enabled most of the gross features of quenched ω phase, as determined by TEM and electron diffraction,[60,86-88] to be interpreted. Figure 21 suggests a ⟨111⟩ texture to the ω precipitation. Indeed, the model proposed by Sass *et al.*[60,87,88] was based on ⟨111⟩ rows of particles, 10 to 15 Å in diameter and 15 to 25 Å apart. The way in which the displacements depicted in Fig. 21 lead to the collapsing together of selected $(111)_\beta$ planes, and hence to the ω transformation, is depicted in Fig. 22.

According to the static-particle model, athermal ω phase, whose electron diffractograms (Fig. 23) are characterized by sharp spots and straight "lines of intensity," was made up of clusters of such rows, while the broad reflections and the either straight or curved lines of intensity ("diffuse streaking") of "diffuse ω" were supposed to originate from either individual rows of particles, or isolated particles, respectively.

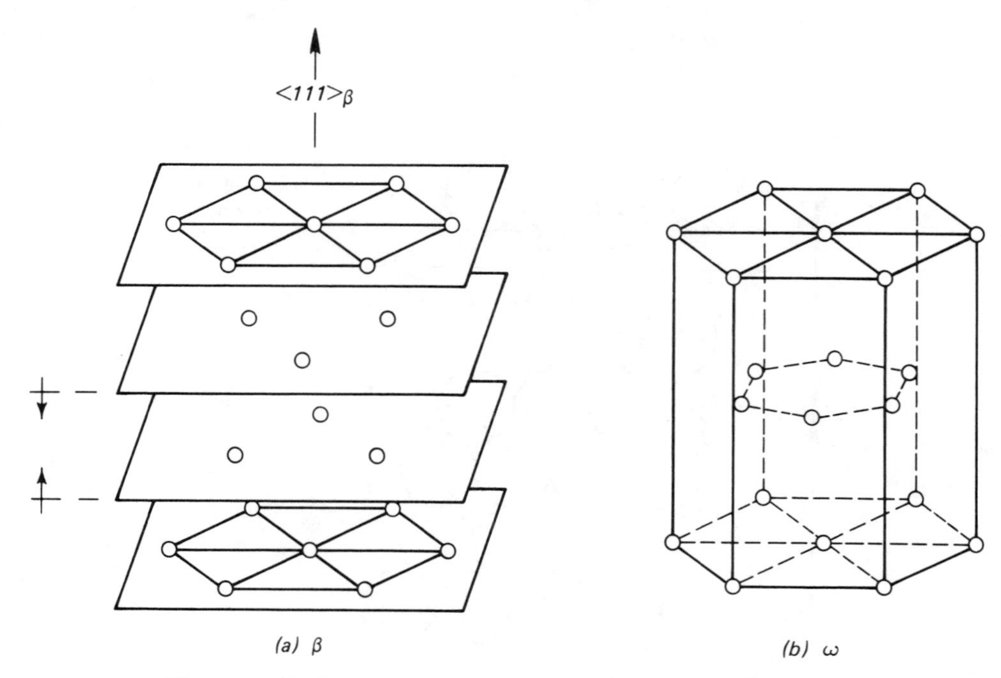

The same displacement process as that depicted in Fig. 21 is represented here, but viewed from a different direction.

Fig. 22. "Linear-fault" model for ω-phase transformation

But the reversible nature of ω precipitation under temperature cycling between room temperature and 100 K calls for more than a static crystallographic interpretation. Figure 23 compares the composition dependence of the electron diffractograms of Ti-Nb with the temperature dependence of those of Ti-Mo. It seems that the effect of lowering temperature is similar to that of lowering solute concentration, in that in both cases curvilinear lines of diffuse intensity become straight and well defined. The figure serves to demonstrate, moreover, the reason for the uncertainties and arguments which have been associated with the assignment of compositional limits for athermal-ω-phase formation. As pointed out by Williams(53): (1) since the diffuse streaking tends to coincide with the positions of the ω-particle reflections when they are present, there is no sharp line of demarcation separating the regions of athermal and diffuse ω; and (2) the reversibility of the ω makes the specification of temperature particularly important, especially when relating structure to low-temperature physical properties such as the superconducting transition temperature. The soft-phonon mechanistic model of the ω-phase effect, originating with the work of de Fontaine(90) and developed more fully in the paper by de Fontaine, Paton, and Williams,(89) provided a satisfactory rationalization, in lattice-dynamical terms, for both the temperature- and composition-dependences of the athermal and diffuse ω phases.

(b) *Phonon Mechanism for Athermal ω*: The dynamical equivalent of the linear-fault crystallographic model centers about the proposed existence of a longitudinal phonon propagating in the ⟨111⟩ direction. The wavelength necessary to achieve

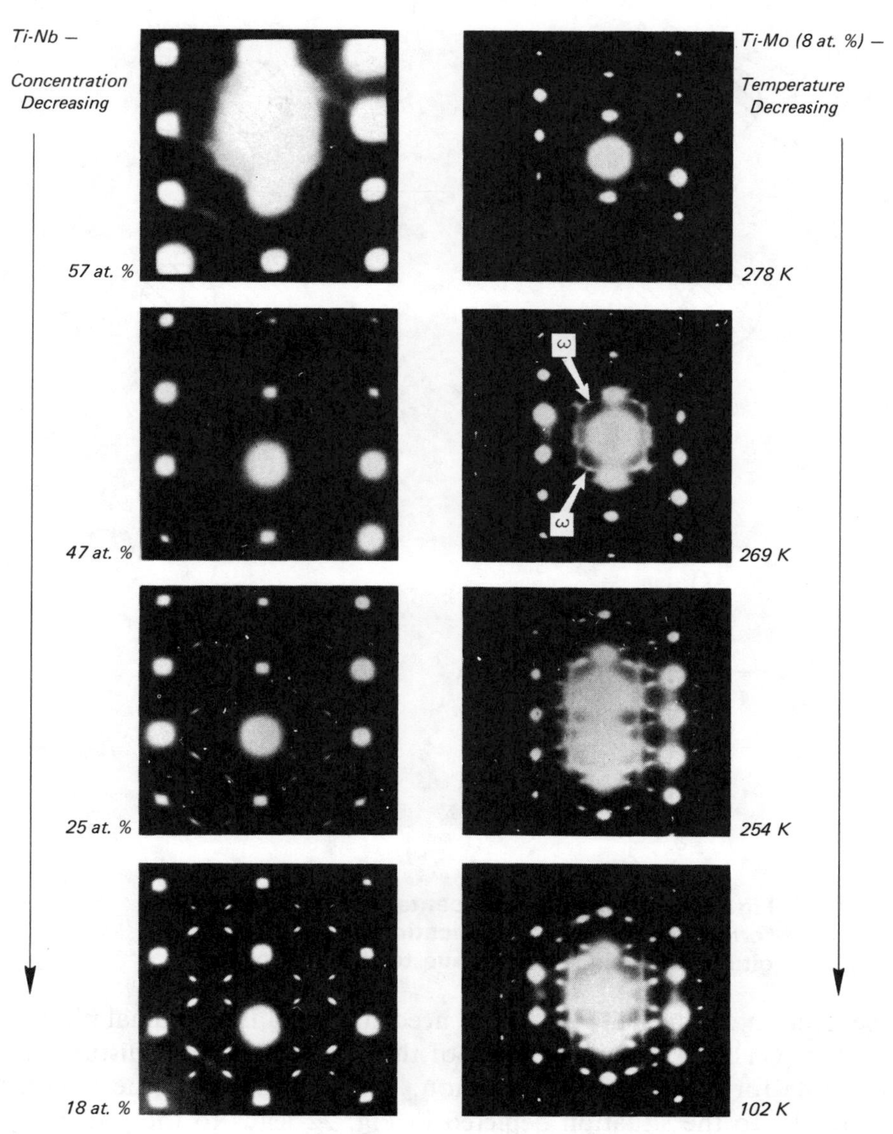

Left side: As-quenched Ti-Nb alloys in the (110) reciprocal-lattice section[60,75,76]; photographs courtesy of S.L. Sass, Cornell University. *Right side*: As-quenched Ti-Mo (8 at. %) in the (131) reciprocal-lattice section[89]; photographs courtesy of J.C. Williams, Carnegie-Mellon University.

Fig. 23. Changes from diffuse to sharp ω reflections from quenched Ti-TM alloys in response to either a decrease in solute content or a decrease in temperature

(with the aid of anharmonicity) the necessary displacements of the *A* and *B* atoms (Fig. 21) is illustrated in Fig. 24, which represents a unit cell of the earlier figure. Clearly, if the shaded atoms are to remain unmoved, while *A* and *B* are to be shifted in opposite directions as shown, a longitudinal wave of wavelength equal

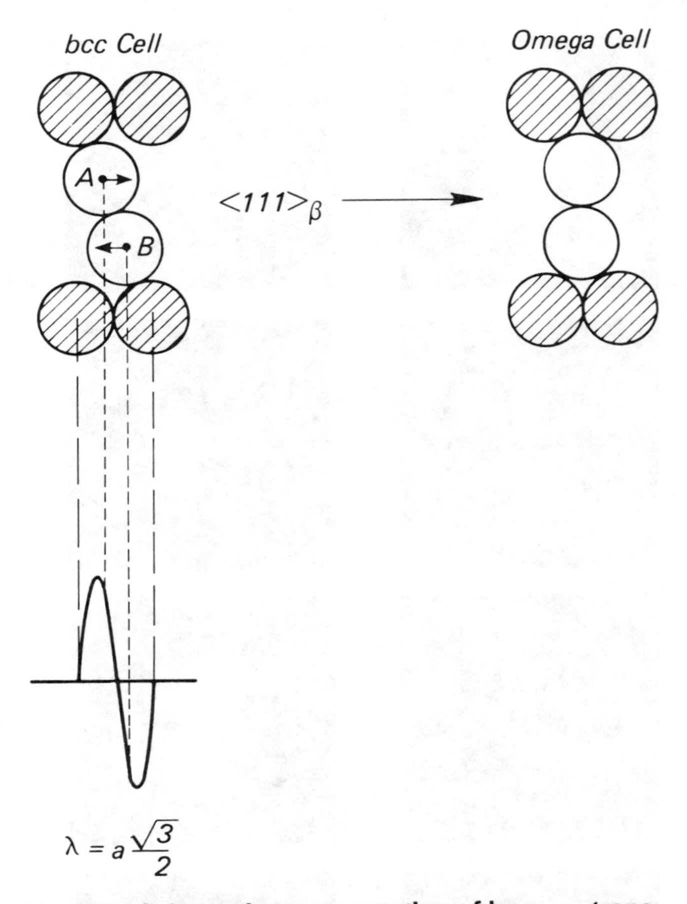

Fig. 24. Schematic representation of bcc → ω transformation induced by application of a 2/3 ⟨111⟩ longitudinal displacement wave to the bcc lattice[76]

to the separation of the $(111)_\beta$ planes is needed, viz., a longitudinal phonon with wave-vector $\frac{2}{3}\langle 111\rangle$. It is the instability of the bcc lattice to this disturbance that is responsible for the athermal transition. An application of the longitudinal-phonon model to the situation depicted in Fig. 22 leads to the possibility of a continuum of incomplete-ω states intermediate between the original β and the completely transformed ideal-ω phases. This is illustrated in Fig. 25.

3.2.2. Isothermal Omega Phase

(a) *Occurrence*: Athermal ω phase has been shown to occur as a crystalline precipitate within a narrow composition range in quenched Ti-TM alloys. Then, provided that the temperature is below about 400 °C, after prolonged aging a metastable $\omega + \beta$ state is attained, characterized at a given temperature by a fixed volume-fraction and composition of the ω and β end-points.[57] It has been determined that the *aged* product bears the same crystallographic relationship to the parent lattice as does the *athermal* ω phase.[91] After sufficiently long aging times at

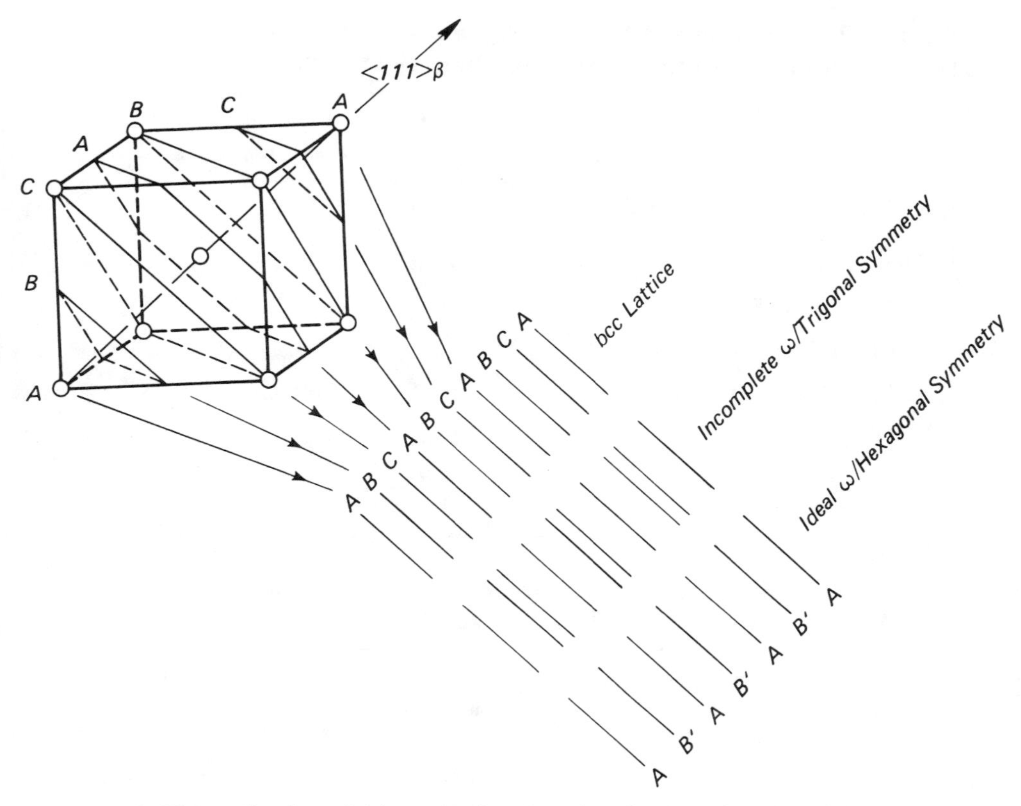

The application of this mechanism (see also Fig. 24) raises the possibility of an incomplete ω phase.

Fig. 25. Representation of the bcc → ω transformation in terms of the displacement-wave mechanism

450 and 500 °C, α-phase precipitation can be expected, as indicated in Table 12. The relative sluggishness of the decomposition process in Ti-Mo compared with that in Ti-Fe is a reflection of the great difference in the diffusion coefficients of these two solute atoms (see Fig. 12).

The compositions of the meta-equilibrium isothermal ω phases have already been given in Tables 10 and 11 within the context of a discussion of the limits of β-phase stability. With the aid of corresponding data for the β phase,[56,57] such as that presented in Table 13, a semiquantitative metastable equilibrium phase diagram for ω + β may be assembled. Figure 26 is such a diagram for Ti-Nb.

(b) *Morphology*: The isothermal ω particles assume one of two types of morphology—cubic or ellipsoidal—depending on the linear lattice misfit $(V_\omega - V_\beta)/3V_\beta$, where V represents the unit-cell volume divided by the number of atoms per unit cell.[57] If this is large (1 to 3%), as in Ti-V, Ti-Cr, Ti-Mn, and Ti-Fe,[77] minimization of elastic strains in the cubic matrix dictates a cubic morphology (Fig. 27). If the misfit is small (<0.5%), as in Ti-Mo and Ti-Nb,[57,77] the morphology is dominated by surface-energy considerations, leading to the ellipsoidal particle shape depicted in Fig. 28. The influence of misfit on ω-particle morphology has

Table 12. Time needed for the appearance of α-phase precipitation during aging of quenched Ti-TM alloys

Alloy	Aging time, h, at temperature of: 400 °C	450 °C	500 °C	Reference
Ti-V (15, 19 at. %)	20 to 30	<20	...	56
Ti-V (19 at. %)	...	...	<4	92
Ti-V (25 at. %)	20 to 30	(No ω)	...	56
Ti-Cr (9.3 at. %)	50	...	...	57
Ti-Mn (6.7 at. %)	68	...	...	57
Ti-Fe (6.0 at. %)	150	12	(No ω)	57
Ti-Nb (22 at. %)	72	...	...	73
Ti-Mo (8 at. %)	...	320	50	57
Ti-Mo (10 at. %)	...	150	(No ω)	57

Table 13. Niobium contents of the ω and β phases in aged metastable-equilibrium Ti-Nb alloys[57]

Average niobium concentration, at. %	Aging time(a), h	Volume fraction of ω phase	Nb concentration, at. %: ω phase	β phase
22	10	0.36	6-11	29 ± 1
	30	0.34	5-10	30 ± 1
	50	0.33	6-11	31 ± 1
25	10	0.25	7.5-12	30 ± 1
	24	0.26	7.5-12	30 ± 1

(a) Aging temperature, 450 °C.

been graphically demonstrated by Williams *et al.*[93] in experiments in which the addition of 5.5 at. % Zr to Ti-V(20 at. %) resulted in a decrease in the misfit from 1.5-2.0% to ~0.25% and caused the precipitate shape to change from cubic to ellipsoidal.

(c) *Influence of Aging on Alloy Properties*: Phenomenological studies of the effects of aging on the properties of ω + β Ti-TM alloys were conducted more than 30 years ago by Frost *et al.*[94] and Brotzen *et al.*[95] Their results, along with numerous others, have already been thoroughly reviewed by McQuillan,[96, pp. 51-57] and Margolin and Nielsen.[19]

It was the hardening and embrittling properties of ω phase that originally led to its discovery[97]; subsequently it was found that the hardness, already characteristic of quenched ω + β-phase samples, increased with aging time.[94,95] During aging, the tensile and yield strengths increased and the ductility (elongation at fracture) decreased. Upon overaging, during which the ω phase dissolves and is replaced somehow by α phase, the ductility is restored. Most of the common tran-

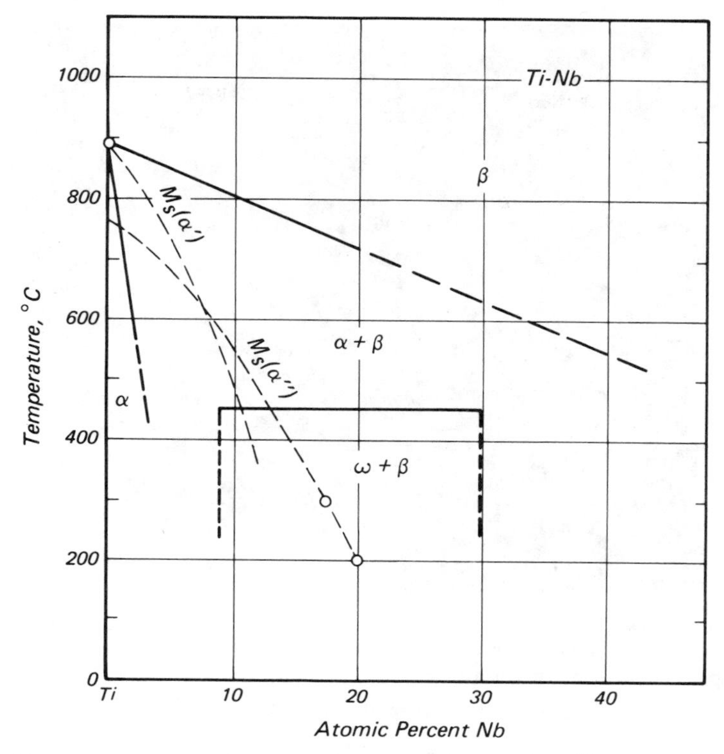

Data sources: α and β transi from Molchanova[1, p. 20]; M_s transi from Jepson *et al.*[59] and Flower *et al.*[67]; ω + β phase data from Hickman.[57]

Fig. 26. Locations of the α- and β-equilibrium transi, the $M_s(\alpha')$ and $M_s(\alpha'')$ transi, and the regimes of occurrence of athermal and isothermal ω phases in Ti-Nb

sition elements (niobium is an exception) decrease the lattice parameter of the bcc phase; thus, during the isothermal aging of ω + β, the lattice generally shrinks as the β component becomes enriched in solute. If the aging temperature is increased and the alloy overaged to α + β, the lattice may expand to accompany a readjustment of the volume fraction of β phase to its equilibrium value.[19,95]

3.3. Beta-Phase Separation

When the temperature is too high[52,93] or the alloys too concentrated[91] to support ω-phase precipitation, a solute-lean bcc precipitate, designated β′, separates out. The relationship in temperature/composition space between the metastable ω + β- and β′ + β-phase fields is indicated schematically in Fig. 29, which itself is based on Fig. 26 and some suggestions by Williams *et al.*[93] As with ω + β, the β′ + β mixed phase is metastable; but unlike ω, which can be generated by a displacement wave in a virtual crystal if need be, the β′ precipitate stems from the chemical differences between the solute and solvent atoms. Thus, the β → β′ + β

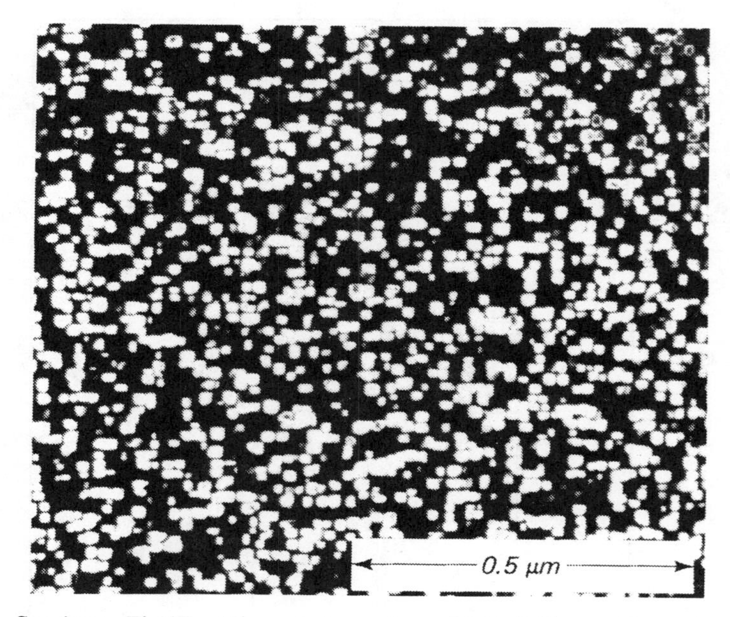

Specimen: Ti-10Fe. Micrograph courtesy of J.C. Williams, Carnegie-Mellon University; copyright 1973, Plenum Publishing Corp., reprinted with permission.

Fig. 27. An example of cubic ω phase

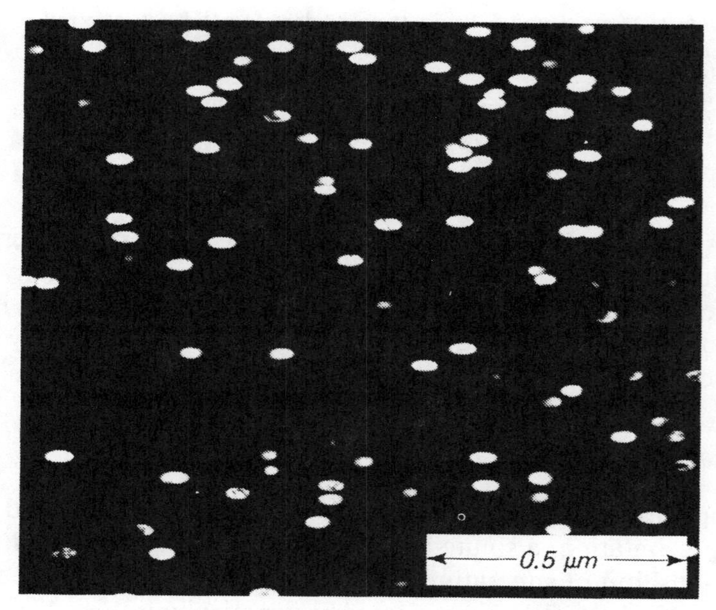

Specimen: Ti-11.5Mo-6Zr-4.5Sn. Micrograph courtesy of J.C. Williams, Carnegie-Mellon University; copyright 1973, Plenum Publishing Corp., reprinted with permission.

Fig. 28. An example of ellipsoidal ω phase

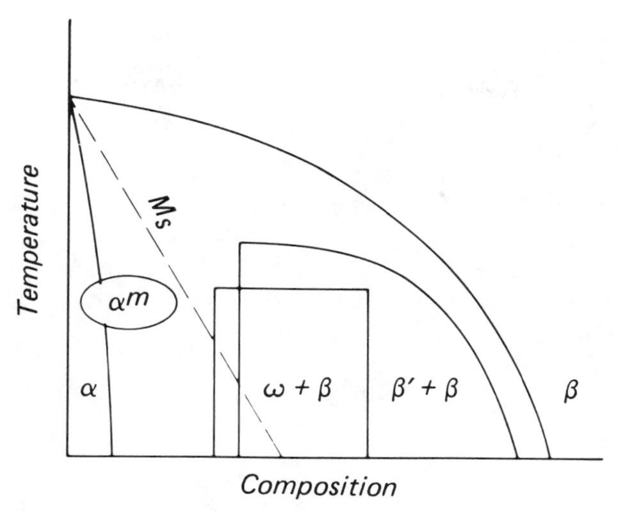

Fig. 29. Schematic representation of the locations of the two metastable phases ω + β and β′ + β within the equilibrium α + β-phase field in a typical Ti-TM alloy[93]

"phase-separation" reaction, as it is called, is a clustering reaction characteristic of alloy systems which show positive heats of mixing[98] or equivalent manifestations of a tendency for the alloying constituents to unmix. Partly as a consequence of the time-temperature conditions for β' formation and ω-phase precipitation, the former (or at least the precursor stages of it) was for a time confused with ω phase. This is no longer the case, the status of β' as a metastable bcc precipitate now being well established.[53,91]

The $\beta \rightarrow \beta' + \beta$ reaction has been studied in considerable detail using Ti-Cr[52,99] and Ti-Mo[98,100-102] as model systems, although its appearance in more complicated alloys (e.g., Ti_{72}-Zr_8-V_{20}) is well documented.[53,91] Typical examples of β' precipitate morphology are given in Fig. 30.

4. MULTICOMPONENT EQUILIBRIUM AND NONEQUILIBRIUM TITANIUM ALLOYS

The properties of multicomponent alloys of titanium can be regarded as deriving from a combination of the properties of two binary-alloy prototypes: the α-stabilized Ti-Al, and the β-stabilized Ti-Mo. Ti-Al provides a model for solid-solution strengthening; it is, moreover, a direct precursor of technical ternary-α alloys such as Ti-5Al-2.5Sn. Ti-Mo (and other β-isomorphous alloys such as Ti-V and Ti-Nb) provide models for both metastable and stable β alloys; the foregoing discussions of their properties aid in the understanding of those of the β components of $\alpha + \beta$ alloys as well as those of the technical metastable-β alloys themselves—Ti-11.5Mo-6Zr-4.5Sn (so-called "β-III"), for example.

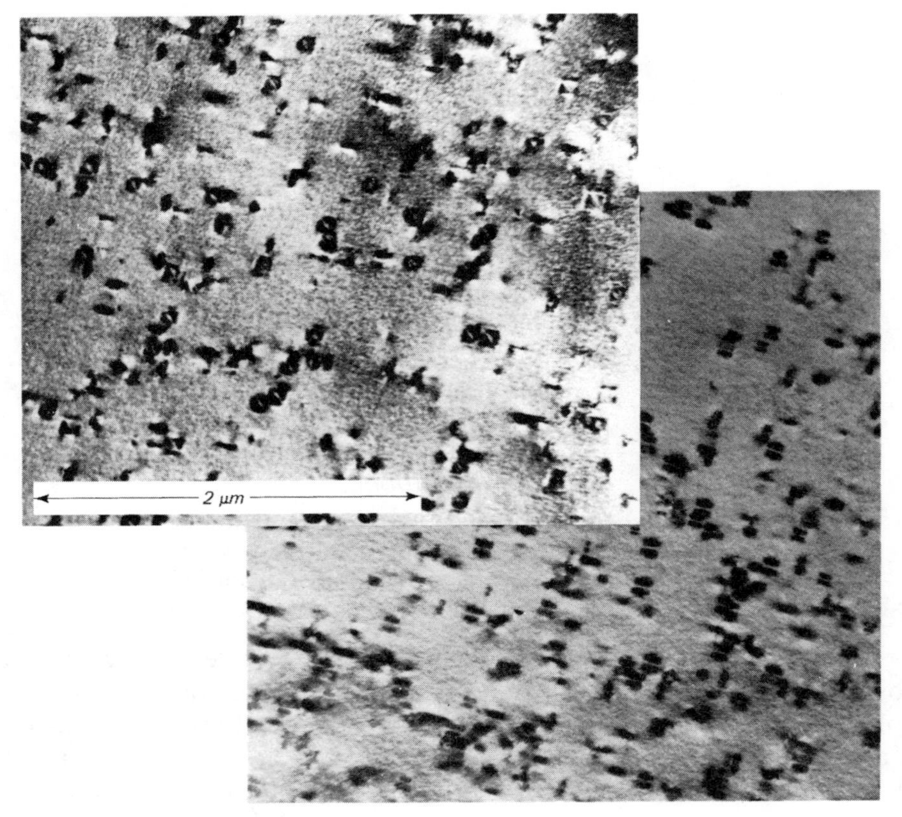

Two orientations of the same sample are represented. *Specimen*: Ti_{72}-V_{20}-Zr_8. Micrographs courtesy of J.C. Williams, Carnegie-Mellon University.

Fig. 30. An example of β' precipitation in a β matrix[53]

An example of the effect of alloying with both α *and* β stabilizers is to be found in the well-studied Ti-Al-V system. In Ti-Al itself, the two-phase $\alpha + \beta$ region is both *narrow* and *high in temperature*. With Ti-6Al (10 at. % Al), for example, the $\beta/(\alpha + \beta)$ and $(\alpha + \beta)/\alpha$ transformations take place at ~1010 °C and 970 °C, respectively. The introduction of vanadium at constant aluminum concentration, although it has a comparatively small influence on the position of the $\beta/(\alpha + \beta)$ transus, produces a rapid decrease in the $(\alpha + \beta)/\alpha$ transus.

4.1. Equilibrium Phases

4.1.1. The Near-Alpha Alloy Ti-6Al-4V. The alloy Ti-6Al-4V could be regarded as being derived from unalloyed titanium by (1) the addition of aluminum to produce solution strengthening and raise what becomes the $\beta/(\alpha + \beta)$ transus, and (2) the addition of vanadium to lower the $(\alpha + \beta)/\alpha$ transus. These effects are

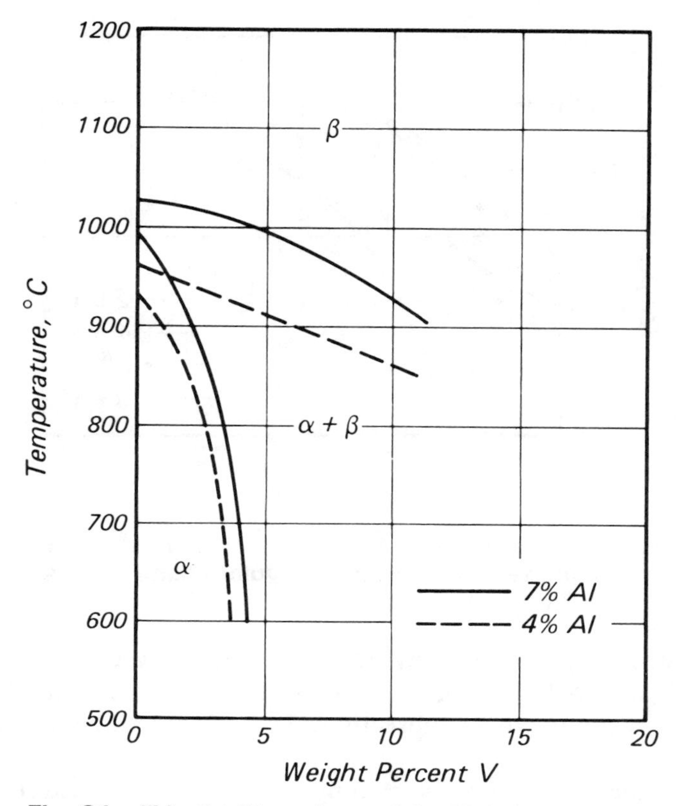

Fig. 31. "Vertical" sections of the Ti-Al-V vs. T equilibrium-phase solid (a right triangular prism) at 4 and 7 wt % Al[103]

shown in Fig. 31, which plots transformation temperature versus vanadium concentration for two fixed levels of aluminum, and Fig. 32, which performs a complementary function in terms of a continuous variation of the aluminum concentration at four fixed levels vanadium. The corresponding equilibrium ternary phase diagrams for Ti-Al-V are given in Fig. 33. The equilibrium states of Ti-6Al-4V are indicated by special points on that figure.

A wide range of processing techniques are applied to Ti-6Al-4V to produce numerous kinds of mill products exhibiting a wide range of microstructures. An example of the manner in which thermal processing can control the microstructure of Ti-6Al-4V is depicted schematically in Fig. 34.

4.1.2. The Near-Alpha Alloy Ti-6Al-2Sn-4Zr-2Mo. The alloy Ti-6Al-2Sn-4Zr-2Mo (Ti-6242) was developed in order to extend the previously existing upper-temperature limit of titanium-alloy service. The development philosophy seems to have been to replace the "4V" of Ti-6Al-4V with "2Mo" (the latter being a stronger β stabilizer than V) and to insert tin and zirconium, which are neutral in this regard (see Section 1.5). Tin is well known as a solution strengthener; zirconium con-

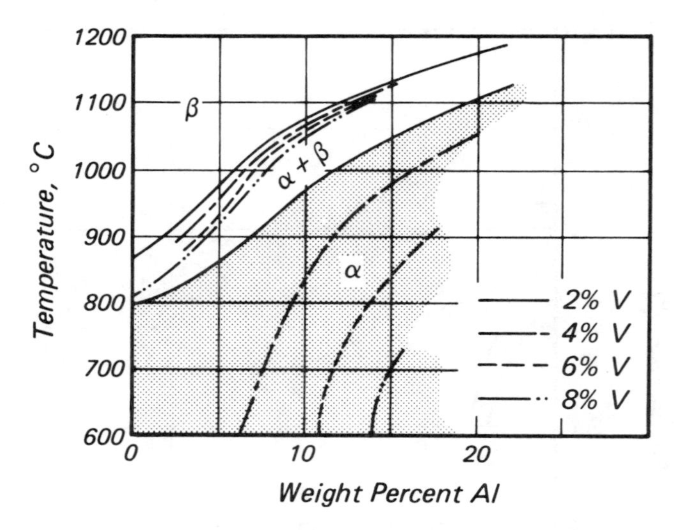

Fig. 32. "Vertical" sections of the Ti-Al-V vs. T equilibrium-phase solid (a right triangular prism) at 2, 4, 6, and 8 wt % V[103]

tributed little strengthening but may have been included to improve the stability of the α phase (see Eq 1) and/or to assist with homogenization.[105] As with the alloy described in the previous subsection, Ti-6242 functions as a near-α $\alpha + \beta$ alloy.

Although the microstructures assumed by this alloy are characteristic of its class, and are generally quite similar to those exhibited by its ternary prototype, Ti-6Al-4V, a wide range of variants can be generated by adjusting the thermochemical-process parameters. In this manner a set of eight microstructural types have been generated by Chen and Coyne as part of an investigation of the influence of microstructural variation on the mechanical properties.[106] Since the mechanical properties, particularly at elevated temperatures, are very sensitive to the microstructure, a great deal of attention has been given recently to the study of their interrelationships.[106] The two basic microstructures are depicted in Fig. 35. The $\beta/(\alpha + \beta)$ transus is at about 990 °C.[5, p. 1-6:72-2] On cooling from above this temperature, the transformed-β structure varies from martensitic (which is, of course, a "non-equilibrium phase"—see below) to Widmanstätten $\alpha + \beta$ as the cooling rate decreases from water-quench speeds to those characteristic of air cooling. The structure preserved after an anneal below the $\beta/(\alpha + \beta)$ transus consists of globular α (the original α phase*) plus a transformed version (usually Widmanstätten) of the original β. Fig. 36 shows a set of three typical microstructures selected to represent the effects of air cooling from anneals at temperatures of 900 °C and 980 °C. The highest annealing temperature yields the largest volume-fraction of

*Also referred to as "primary-α."

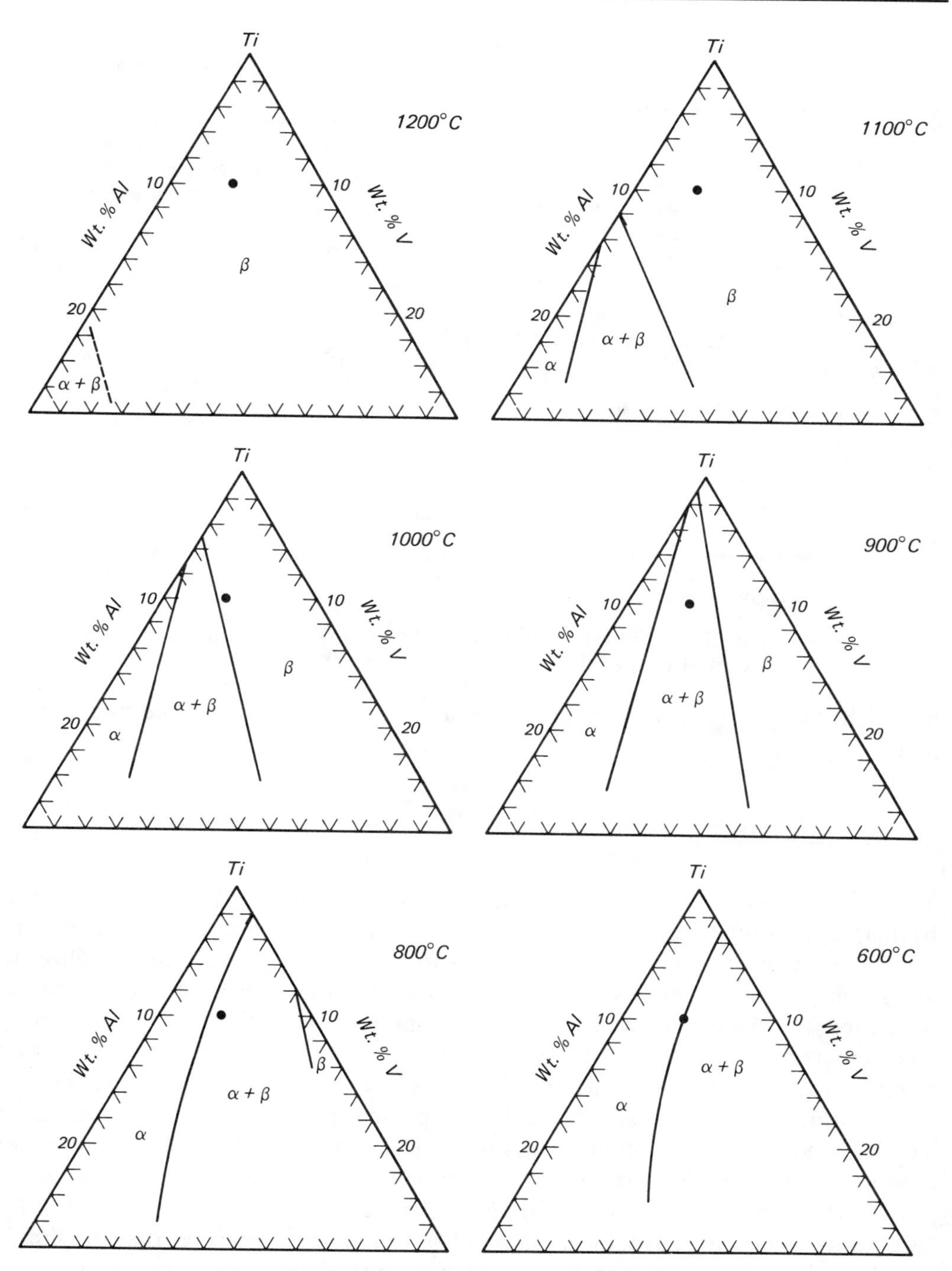

The solid circle (•) represents the alloy Ti-6Al-4V.

Fig. 33. "Horizontal" (or isothermal) sections of the Ti-Al-V vs. T equilibrium-phase prism at the temperatures indicated[103]

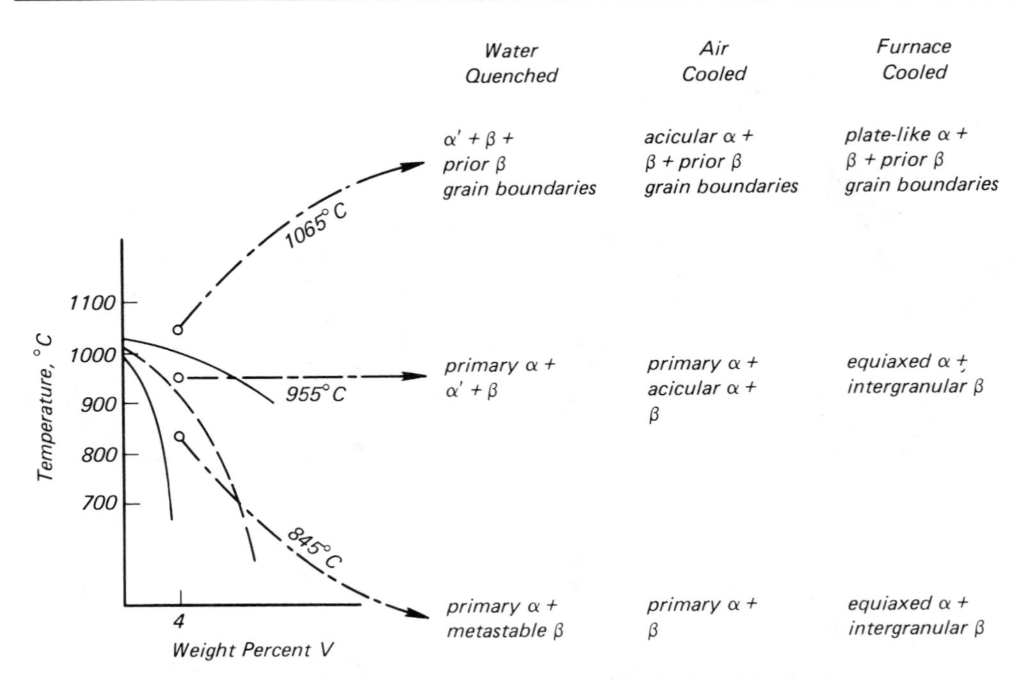

Fig. 34. Microstructural control of Ti-6Al-4V through thermal processing. After Donachie.[104, p. 43]

β, which, being the most unstable at ordinary temperatures, yields the finest transformed structure.

4.2. Nonequilibrium Phases

The multicomponent β alloys have been discussed above with particular reference to their equilibrium properties, which are of course achieved only after aging. The β alloys result from the inclusion in their formulations of sufficient β stabilizer to drop their M_s temperatures to well below room temperature. The properties of quenched multicomponent β alloys are comparable to those of their binary counterparts—e.g., Ti-12Mo in the case of β-III.[107] It has long been known that the presence of solutes such as aluminum have a retarding influence on the $\beta \rightarrow \omega$ reaction.[27, p. 141] Thus, although the presence of iron and aluminum in Ti-10V-2Fe-3Al and of zirconium and tin in Ti-11.5Mo-6Zr-4.5Sn (β-III) does not completely prevent the formation of ω phase during quenching from above the $\beta/(\alpha + \beta)$ transus,[108,109] the fact that the ω reflections in the latter alloy are diffuse rather than sharp, as they are in Ti-11.6Mo itself,[110] indicates that the simple metals do inhibit the transformation. Since the diffusionless athermal ω transformation is a response to bcc lattice instability, this inhibiting effect by SM additions in general[93] is presumably a result of their increasing the stiffness of the bcc lattice.

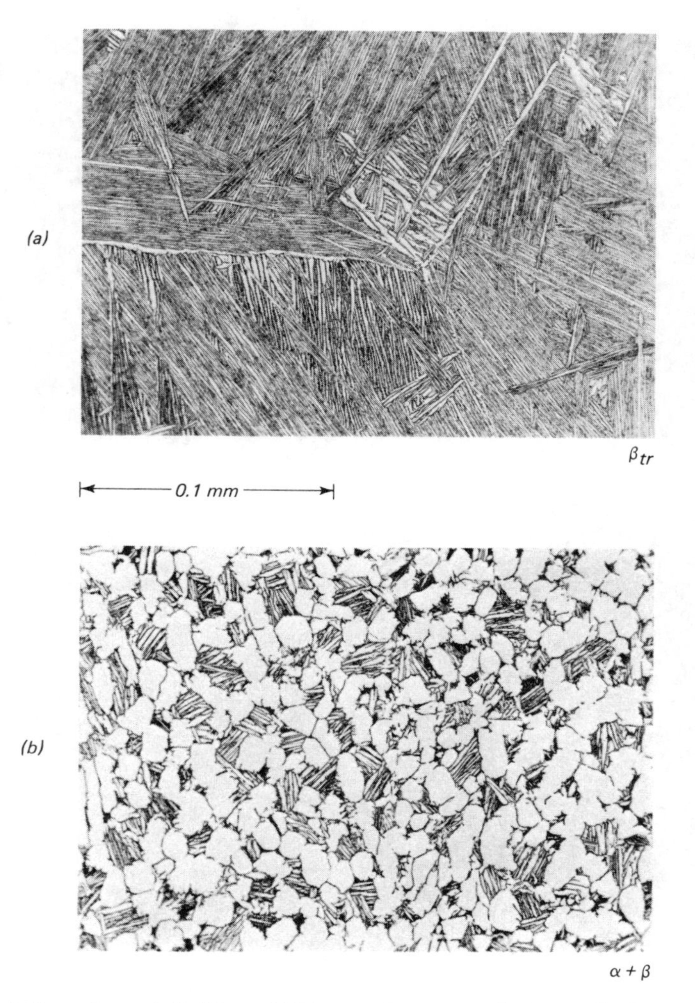

(a) Transformed β [β_{tr} or Widmanstätten $\alpha + \beta$, $(\alpha + \beta)_w$] as a result of a heat treatment consisting of 2 h at 1024 °C, air cool. (b) $\alpha + \beta$, as a result of 2 h at 968 °C, air cool. Micrographs courtesy of S.L. Semiatin, Battelle.

Fig. 35. Optical micrographs of Ti-6Al-2Sn-4Zr-2Mo-0.1Si (Ti-6242-Si) in two characteristic metallurgical conditions

4.2.1. Nonequilibrium Ti-10V-2Fe-3Al. The commercial alloy Ti-10V-2Fe-3Al (i.e., "Ti-10-2-3"), whether or not it is "near-β"[111] or "β",[108] is certainly a modified Ti-V alloy just on the threshold of β stability. Its e/a ratio of 4.11_4* identifies it with Ti-V(11 at. %), whose β phase is not fully retained on quenching.[1, p. 14; 112] According to Toran and Biederman,[111] the M_s temperature of

*$(e/a) = 4.00 + (0.0954 \times 1) + (0.0186 \times 4) - (0.0559 \times 1) = 4.11_4$.

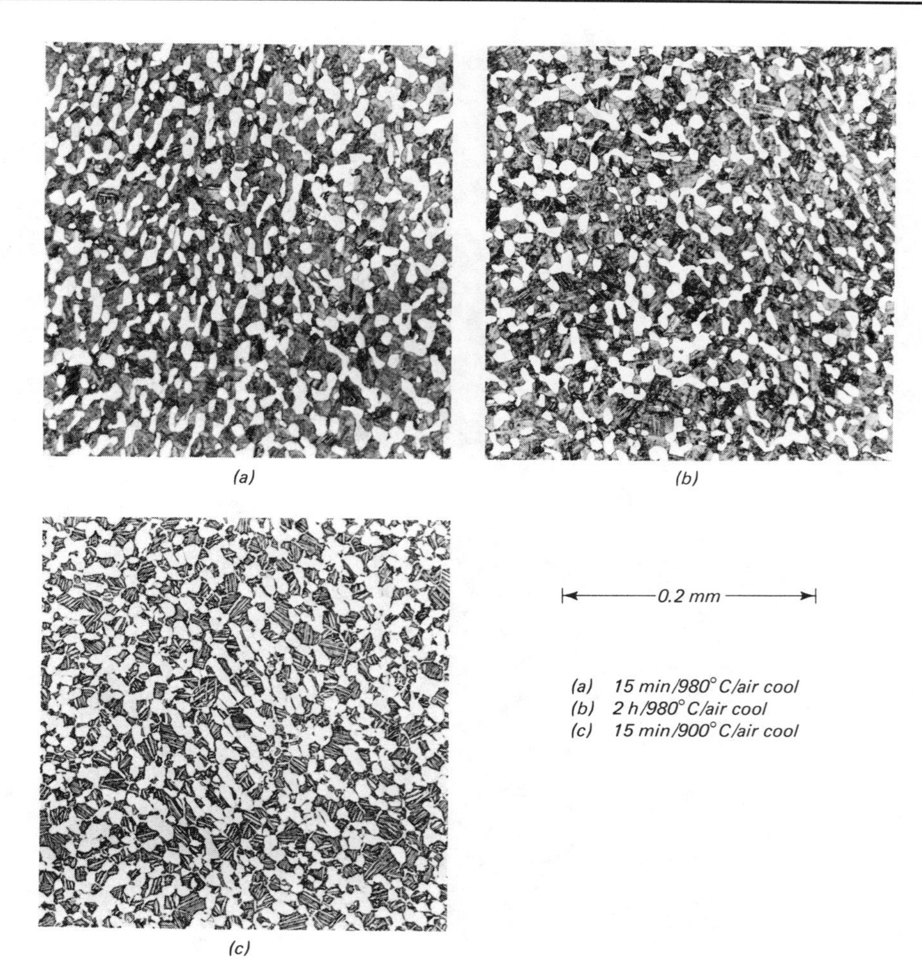

Micrographs courtesy of S.L. Semiatin, Battelle

Fig. 36. Microstructures in $\alpha + \beta$-phase Ti-6Al-2Sn-4Zr-2Mo-0.1Si (see Fig. 35b) after further exposure to the heat treatments specified

Ti-1023 is 555 °C, and quenching from above the β transus (788 °C) yields both martensite and ellipsoidal ω-phase particles. On the other hand, Duerig, Middleton, Terlinde, and Williams[108] claimed that quenching into water, presumably at room temperature, yielded β plus athermal-ω phase. If, following Duerig *et al.*, room temperature lies between M_d (the transus for deformation martensite) and M_s, it is possible that mechanical stress, if it takes place during quenching, could be responsible for a partial martensitic transformation.

In any case, whether quenching stress or deliberately applied post-quench mechanical stress is responsible for a martensitic transformation in this alloy, it is interesting to note that the transformed structure and the ω phase are coexistent. Thus, as Duerig *et al.*[108] have noted, the transformation to deformation α'' leaves the ω particles intact; of course, the ω phase is equally durable during mechanical twinning. In other systems it has been noted that martensitic transformation obliterates the ω phase. The energetics of its stability in this case have been discussed by Duerig *et al.*[108]

4.2.2. Nonequilibrium Ti-11.5Mo-6Zr-4.5Sn. The commercial alloy Ti-11.5Mo-6Zr-4.5Sn (or "β-III") is generally used in the $\alpha + \beta$ aged condition since athermal ω in the quenched alloy, and isothermal ω in an alloy aged in service, are precipitation hardeners which, although they increase the flow stress, reduce the ductility (i.e., they embrittle the alloy). Consequently, there has been little inducement to investigate the fundamental properties of as-quenched β-III. One such study has, however, been undertaken by Rack, Kalish, and Fike, the results of whose x-ray, TEM, and other observations of quenched and deformation-transformed material have been reported in considerable detail.[109] Quenching of β-III from above the β transus (755 °C) resulted in a bcc lattice and a finely dispersed ω-phase precipitate with a mean particle size of $\lesssim$30 Å. No trace of martensitic transformation was observed either as a result of quenching into water, or after subsequent quenching into liquid nitrogen, indicating that M_s was below 77 K. Foils being thinned for TEM observation underwent the usual spontaneous transformation which, however, in this case did not obliterate the ω-phase particles. The athermal ω particles could not be distinctly imaged in the electron microscope, making it impossible to specify their shape.[109] Likewise, selected-area electron diffraction performed on the as-quenched alloys revealed the streaked reflections characteristic of diffuse ω. The occurrence of sharp and diffuse, respectively, athermal-ω reflections and their implication with respect to ω-phase precipitation have been considered by Sass and colleagues, some of whose results have been summarized in the foregoing discussions. The morphology of β-III's isothermal ω, which has been photographed many times in aged samples,[e.g., 13,107] is distinctly ellipsoidal and similar to that present in binary Ti-Mo (cf. Fig. 28).

4.3. Aging of Technical Multicomponent Alloys

The aging effects encountered in commercial multicomponent alloys, although often complicated by the mixture of phases present in the starting materials depending on their compositions or thermal histories, can be interpreted in terms of the aging properties of binary alloys. The precipitational effects likely to be encountered during the aging of technical titanium-base alloys are illustrated below in three case studies.

4.3.1. Aging of Ti-6Al-4V. The effect of heat treatment on the properties of Ti-6Al-4V, as catalogued by McQuillan[96, p. 72] and Wood,[5, p. 1-4:72-11] can be appreciated to some extent with the aid of the equilibrium phase diagrams of Fig. 31 to 33. The aging behavior will of course depend on the initial state of the alloy. For example, depending on the temperature at which it was held prior to being quenched, the alloy could consist of either: (1) all martensite, (2) primary α plus martensite, or (3) primary α plus retained β (see McQuillan[96]; see also Fig. 34).

In case the initial condition is martensitic: (1) the α' recovers toward an α possessing the appropriate equilibrium composition; and (2) the boundaries and internal structures (see Section 3.1.1) of the plates provide heterogeneous nucleation sites for the precipitation of β.[53] The vanadium content of the β and the aluminum content of the α naturally play important roles in the precipitation process. With regard to the aluminum component, the precipitation of α_2 can contribute to the strengthening of the alloy.

4.3.2. Aging of Ti-6Al-2Sn-4Zr-2Mo. Alloys such as Ti-6Al-2Sn-4Zr-2Mo provide enormous opportunities for studying the effects of thermomechanical processing on microstructure. Again, the aged properties will depend on the combination of initial heat treatment and mechanical deformation to which the material has been subjected. In actual practice, for this alloy the thermomechanical processing will be an integral part of a forging operation. Properties sought are a combination of high strength and fracture toughness and good creep resistance. The creep properties of Ti-6242(Si) alloys can be maximized in the following manner[106]: (1) for $\alpha + \beta$ forgings, by a β solution treatment* followed by aging for 8 h at 593 °C and air cooling; (2) for β forgings, the solution heat treatment should be conducted in the $\alpha + \beta$ field. The microstructural goal of these processing operations is to achieve a fine Widmanstätten structure and to avoid the occurrence of both martensitic α' and globular α, both of which seem to degrade the creep resistance and the fracture toughness.[106]

4.3.3. Aging of Ti-11.5Mo-6Zr-4.5Sn (β-III). A useful insight into the aging/mechanical-property characteristics of β-Ti alloys in general can be acquired from a study of the properties of the alloy sequence Ti-15Mo, Ti-15Mo-5Zr.[9,113] Aging of these alloys at temperatures below about 400 °C resulted in pronounced ω-phase-induced hardening. In Ti-15Mo-5Zr, the isochronal curves of tensile strength versus temperature and elongation versus temperature were practically mirror images of each other about a suitably selected horizontal axis; at 400 °C, the tensile strengths rose to maxima while the elongations dropped to zero.[9] In order to develop α-phase precipitation in Ti-15Mo-5Zr, the aging temperature had to be raised above 400 °C. The exchange of ω phase for α phase resulted in a recovery of the ductility. Nishimura *et al.*[9] found that a satisfactory combination of strength and ductility could be acquired by developing a suitable mixture of ω-phase and α-phase precipitates.

The metallurgical behavior exhibited by β-III is similar to that found in Ti-12Mo[107] which may, therefore, be regarded as its binary prototype. The aging of Ti-Mo($\leq$21 at. %) for several hours at temperatures below 400 °C results in the nucleation and growth of isothermal ω-phase precipitates. In particular, during aging of Ti-Mo(8 at. %) for 150 h at that temperature, Hickman[57] found that 68 vol % of the alloy had become ω phase with a molybdenum concentration of 4.7 at. %, while the composition of the β matrix was 14.5 at. % Mo (25.4 wt %).

*The β transus is at 990 °C and M_s is at 800 °C.

The presence of ω phase severely embrittles any titanium-base alloy in which it occurs; for this reason an early proclaimed disadvantage of β-III was its unsuitability for service at temperatures between about 200 and 425 °C, a very important temperature range within which any high-strength titanium alloy would be expected to perform reliably. It was, however, a simple matter to devise a stabilization heat treatment to avoid this difficulty. In doing so, Froes *et al.*[10] developed the pseudobinary equilibrium and meta-equilibrium diagram for β-III (Mo variation) depicted in Fig. 37. The effect of heat treatment on β-III, as discussed by Froes *et al.*,[10] Boyer *et al.*,[107] and Williams *et al.*,[13] can be appreciated with the aid of such a diagram.

The stabilization treatment suggested by Froes *et al.*[10] was aging for 8 h in the temperature range 550 ~ 600 °C to produce about 40% α phase and a β matrix too rich in molybdenum to be susceptible to ω-phase embrittlement during subsequent service ($\gtrsim$8,000 h) at 200 ~ 425 °C. The molybdenum content of the β-III matrix was 17.2 wt % (9.4 at. %) after the stabilization heat treatment. Polonis and colleagues (e.g., Luhman *et al.*[114]; see also Chandrasekaran *et al.*[115]) have conducted extensive investigations of the ω-reversion effect with particular reference to the Ti-Cr system in which the kinetics of the $\omega \xrightarrow{500°C} \beta$ reaction are quite rapid. While extending these studies to β-III, the strengths and ductilities of the variously thermomechanically processed alloy were correlated with the observed microstructures.[107] It was hoped that the success previously obtained with Ti-Cr[116] would be repeated in β-III. In the former case, the $\omega + \beta \xrightarrow{500°C} \beta' + \beta$

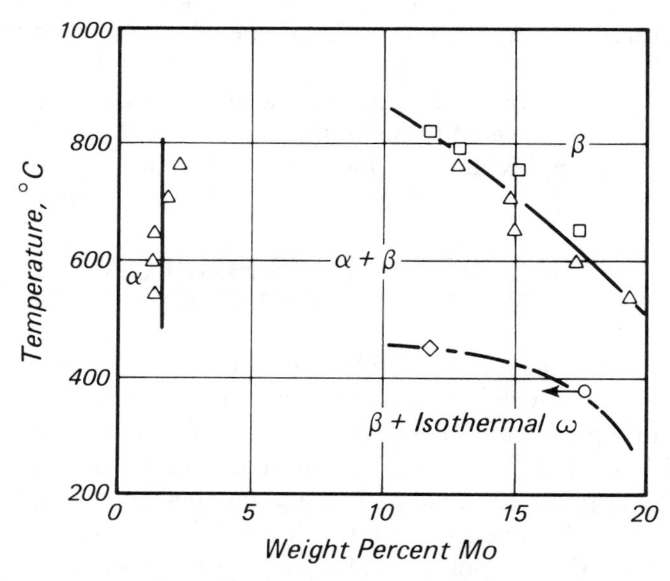

Data sources: (△) partitioning data; (○) behavior of "17.3Mo-β-III" upon aging at 370 °C (700 °F); (◇) upper limit of isothermal ω in regular β-III; (□) β transus for 0.28%O β-III-type alloys.

Fig. 37. Pseudobinary phase diagram for (Ti-6Zr-4.2Sn)-XMo for values of X (wt % Mo) of up to about 19[10]

reversion was complete after 5 min; the strength of the original ω-strengthened aged alloy was preserved and the ductility was increased. But with β-III, after a similar treatment, the ω phase was only partially reverted. Some of the strengthening arising from the presence of ω phase was retained, and some ductility, as a result of the increased volume fraction of β phase, was acquired. But it was estimated that if complete reversion had been achieved, the strength would have been no greater than that of the solution-treated-and-quenched alloy. Further interesting results were obtained by repeating the aging and reversion heat treatments on alloys which had been deformed after solution heat treatment. In these cases, the α phase which formed at dislocation sites during $\omega + \beta$ aging played a strengthening role at all stages of the heat treatment.

The experiments of Froes *et al.*[10] and Polonis *et al.*[107] were subsequently extended in detail by Williams *et al.*,[13] who studied the microstructural and mechanical responses of β-III to: (1) variation of solution treatment temperature, (2) variation of $\alpha + \beta$ aging temperature, (3) duplex aging (ω reversion), and (4) cold work administered prior to various heat treatments. Some of the results can be summarized with reference, again, to Fig. 37: (1) in alloys $\alpha + \beta$ aged for 8 h at 590 °C after solution treatment (ST) and quenching, a change of ST temperature from 790 °C [i.e., above $\beta/(\alpha + \beta)$], in which case the α precipitation was uniformly distributed, to 720 °C [i.e., below $\beta/(\alpha + \beta)$], which was responsible for a very irregular α-phase precipitate, had no significant effect on the mechanical properties; (2) aging at a temperature lower than 590 °C, in particular 8 h at 480 °C, resulted in a closely spaced array of finer α-phase particles and an increase in strength. Also studied were the properties of quenched-plus-aged (8 h at 510 °C) and quenched-plus-duplex-aged (50 h at 370 °C + 8 h at 510 °C) alloys. In the latter case, partial ω reversion, followed in time by α-phase precipitation at the remaining ω-phase sites, seemed to be taking place. As compared with the single aging treatment, the duplex aging, which included a preliminary $\omega + \beta$ aging, engendered a deterioration in the ductility with no improvement in strength.

5. TITANIUM ALLOYS FOR LOW-TEMPERATURE SERVICE

Cryogenic applications of titanium alloys include rotors for superconducting generators, and components with aerospace applications such as reuseable upper-stage spacecraft and high-pressure rocket engines (for craft such as the Space Shuttle, the Aerospace Plane, and "space tugs"). The reuseable rocket engine is a particularly challenging application. Fuel pumps and engines operate at high pressures and must be small and lightweight; high turbine and fuel-pump-impeller tip speeds place high stresses on the components. The search for suitable alloys for cryogenic service is, as usual, the search for materials which under service conditions (in this case, at temperatures near 4.2 K) have adequate strength, ductility, and fracture toughness. The advantages of titanium alloys are high strength, high specific strength (strength/weight), and low thermal conductivity. Disadvantages are lower fracture toughness than fcc ferrous materials under the same conditions.

5.1. Alloy Phase Selection for Low-Temperature Service

A summary of the mechanical properties of several representative α-phase, $\alpha + \beta$-phase, and β-phase alloys is given in Table 14.

The immediate choice for an alloy intended for use at cryogenic temperatures would be an all-α alloy, since the bcc structure at low temperatures generally undergoes a ductile-to-brittle transition (bcc steel is a classical example). To improve the low-temperature ductility of α-phase alloys, the interstitial level is reduced, typically, to below about 0.1 wt % per element, giving rise to the ELI* grade of alloy. But the reduction in interstitial level is accompanied by a reduction in strength.

With the quest for strength as their driving force, designers went from unalloyed titanium to α-Ti alloys such as Ti-5Al-2.5Sn(ELI); likewise, a recent trend has been to introduce other alloys having even higher strengths, and suitably modified for acceptable toughness, into cryogenic service.[117,118] With this in mind, Ti-6Al-4V(ELI) has been the subject of recent investigations. Recognizing that Ti-6Al-4V is an $\alpha + \beta$ alloy, it is natural to inquire into the role played by the β-phase component. This will be dealt with subsequently in this section.

Due to their poor low-temperature ductility, bcc-phase alloys have never been seriously considered as structural materials for use at cryogenic temperatures. However, one such alloy, Ti-50Nb, necessarily finds widespread use in superconducting machinery. The low-temperature deformation of this alloy is considered below.

Seemingly to demonstrate a nonapplicability of a β-Ti alloy to cryogenic service, Salmon[6] has presented the properties of Ti-13V-11Cr-3Al within the context of those of more cryogenically useful materials (see Table 14). Alloy Ti-13V-11Cr-3Al tends to embrittle below 170 K. The solution-treated (ST)-plus-annealed alloy has an elongation of 5% at 20 K; the elongation of the ST-and-aged material (a two-phase mixture of α and $TiCr_2$) is 0.2% at 77 K, and evidently is nonexistent at liquid-helium temperatures (Table 14[6]).

5.1.1. Unalloyed Titanium and Dilute Alloys of Titanium

(a) *Addition of Transition-Metal Solutes*: The transition metals (TM)—i.e., the β stabilizers—are not rapid strengtheners of titanium. Accordingly, the strength-versus-temperature curves for various dilute α-phase Ti-TM alloys are not strong functions of solute concentration (Fig. 38).

Small levels of some β stabilizers are known to be detrimental to the extreme low-temperature properties.[121] For this reason, iron and manganese, common contaminants of commercial titanium must be removed from material intended for cryogenic service. The levels of these elements are adequately low in the ELI grades[121]—e.g., Fe < 0.25 wt % in Ti-5Al-2.5Sn(ELI).

(b) *Addition of Yttrium*: Yttrium in small amounts has been added to titanium alloys to reduce grain size and improve forgeability. The mechanical properties of titanium alloys are known to have been deleteriously affected through segregation

*Extra-low interstitial content.

Table 14. Tensile strengths at low temperatures for several commercial titanium-base alloys[6]

Alloy	Condition	Test temperature, K	Ultimate strength, 10^8 N·m^{-2}	Yield strength, 10^8 N·m^{-2}	Elongation, %
Ti-5Al-2.5Sn (5-2.5)	Annealed(a), normal interstitial	295	8.8	8.6	16
		200	11.0	10.6	14
		77	14.0	13.7	12
		20	16.9	16.8	5.1
	Annealed, extra-low interstitial (ELI)	295	7.6	7.1	17
		200	9.2	8.6	16
		77	12.6	11.9	17
		20	15.4	14.4	15
Ti-8Al-1Mo-1V (8-1-1)	Annealed(b)	295	10.3	9.7	16
		200	11.9	11.3	14
		77	15.6	14.4	13
		20	17.5	16.2	2.4
	Duplex annealed(c)	295	10.2	9.5	15
		200	11.2	10.3	15
		77	14.9	13.4	22
		20	16.9	16.1	1.2
Ti-6Al-4V (6-4)	Annealed(d), normal interstitial	295	9.9	8.9	12
		200	11.6	10.7	11
		77	15.3	14.3	11
		20	17.9	17.3	2.4
	Solution treated(e) and aged(f), normal interstitial	295	12.2	11.3	8
		200	13.2	12.8	6
		77	17.6	17.0	5
		20	20.4	19.9	0.7
	Annealed, ELI	295	9.9	9.3	12
		200	11.5	10.9	12
		77	15.1	14.6	10
		20	18.2	17.9	2.9
	Solution treated and aged, ELI	295	11.2	10.6	9
		200	13.2	13.2	7
		77	17.2	16.7	5
		20	19.6	19.6	1.0
Ti-13V-11Cr-3Al (13-11-3)	Annealed or solution treated(g)	295	9.7	9.4	19
		200	12.5	12.2	12
		77	19.5	18.9	2.1
		20	22.6	...	0.5
	Solution treated and aged(h)	295	13.6	12.4	7
		200	15.7	14.7	2.1
		77	16.5	...	0.2
		20	...	...	...

(a) 15 min to 4 h at 707 to 867 °C, air cool. (b) 8 h at 787 °C, furnace cool. (c) 8 h at 787 °C, furnace cool, plus 15 min at 787 °C, air cool. (d) 30 min to 4 h at 707 to 817 °C, air or furnace cool. (e) 847 to 957 °C. (f) 1 to 10 h at 482 to 597 °C. (g) 10 to 30 min at 757 to 787 °C. (h) 20 to 100 h at 427 to 507 °C.

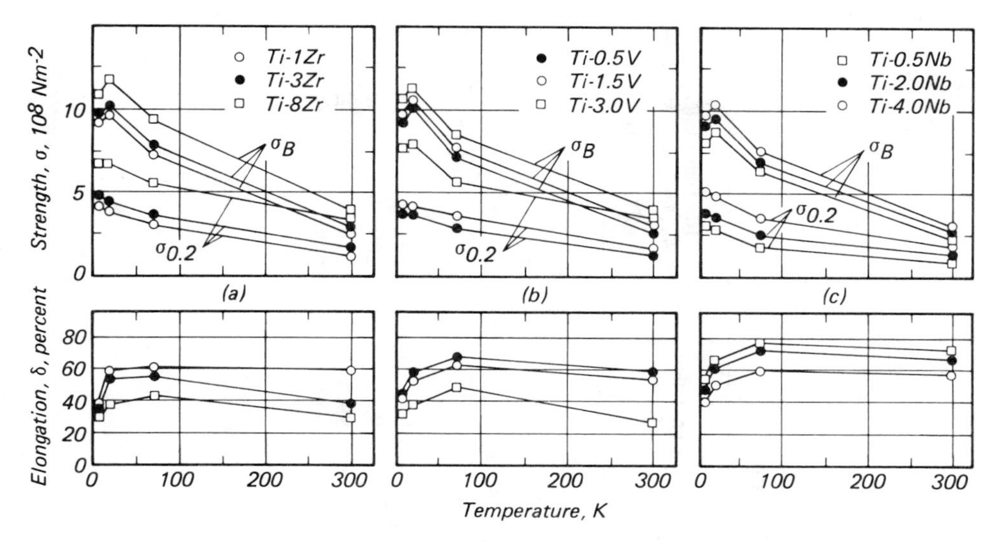

Fig. 38. Yield strength (0.2%) and tensile strength, with associated elongations at fracture, of low-concentration Ti-Zr, Ti-V, and Ti-Nb alloys as functions of test temperature[119,120,132]

of yttrium, the possibility of whose presence must be considered a potential problem.[121]

(c) *Addition of Interstitial-Element Solutes*: The properties of unalloyed titanium at low temperatures have been considered by numerous workers.[6,120,122-126] Their studies had to do with the influence of the interstitial elements carbon, nitrogen, and oxygen on the plastic mechanical properties. Interstitial elements, of which oxygen is an example, are potent strengtheners of titanium (Table 15). Some commercial grades, although exhibiting modest ductility down to 20 K, lose most of it upon cooling into the liquid-helium range.[6] A reduction in the interstitial level is needed to restore low-temperature ductility. A penalty one pays for this is a reduction in strength[120,124] (Fig. 39). To compensate for this, it then becomes necessary to turn to solution-strengthened alloys of titanium, bearing in mind that steps must always be taken to preserve adequate fracture toughness.[117,118]

5.1.2. Beta-Phase Titanium Alloys. The bcc structure is not generally considered suitable for use at low temperatures since the most frequently used structural alloys undergo ductile-to-brittle transitions at sufficiently low temperatures. The effect is particularly pronounced in alloys of iron and chromium, but the fact that a ductile-to-brittle transition is not a property of sodium (a simple metal) or niobium (a transition metal) demonstrates that the phenomenon is *not necessarily* characteristic of the bcc lattice. Beta titanium alloys are metastable (unless extraordinarily concentrated) and hence would not be expected to fail by conventional brittle cleavage, as is the case with bcc ferrous alloys. Nevertheless, as a comparison of Fig. 38 and 40 indicates, Ti-50Nb, a typical all-β alloy, has a very much lower ductil-

Table 15. Solution hardening of titanium by interstitial elements and α stabilizers

Alloying addition	Concentration range, at. %	Condition	Law(a)	Slope, b, $kg \cdot mm^{-2} \cdot at. \%^{-1/2}$ or $kg \cdot mm^{-2} \cdot at. \%^{-1}$	Intercept, a, $kg \cdot mm^{-2}$	Correlation coefficient, %	Hardening rate, $\partial H_V/\partial c$, $kg \cdot mm^{-2} \cdot at. \%^{-1}$ At 0.1 at. %	Hardening rate, $\partial H_V/\partial c$, $kg \cdot mm^{-2} \cdot at. \%^{-1}$ At 1.0 at. %
Simple-Metal Additions								
Al	0-10	100 h/850 °C/IBQ	c	15	102	99.6	...	15
Ga	0-5	As-cast	c	24	108	99	...	24
Sn	0-7	As-cast	c	24	112	99	...	24
Interstitial-Element Additions								
B	0-0.2	120 h/800 °C/IBQ	$c^{1/2}$	218	110	92	344	...
C	0-0.5	120 h/800 °C/IBQ	$c^{1/2}$	170	104	99.9	269	...
N	0-5	120 h/800 °C/IBQ	$c^{1/2}$	239	98	99.8	378	...
O	0-3	120 h/800 °C/IBQ	$c^{1/2}$	194	100	99.9	307	...

(a) Data fitted to either $H_V = a + bc$ or $H_V = a + bc^{1/2}$.

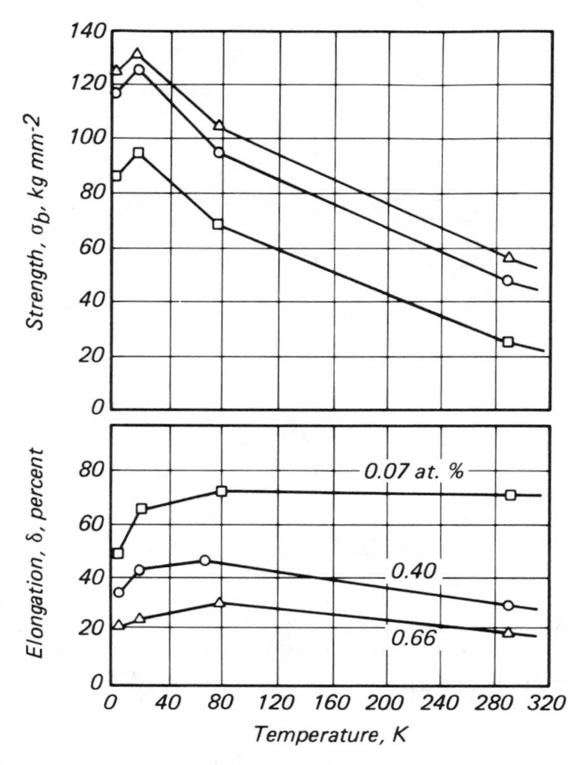

In this case, O_{eq} was given by: [O] + [H] + 2/3[C] + 2[N], all in at. %.

Fig. 39. Ultimate tensile strength and elongation as functions of test temperature for unalloyed titanium containing various levels of interstitial impurity expressed in terms of equivalent-oxygen content, O_{eq}.[120,124]

ity than α-phase titanium. By way of further example, the elongation of Ti-13V-11Cr-3Al [β-phase solution-treated (10 to 30 min at 760 to 790 °C)] at 20 K is 0.5%,[6] which may be compared with 14.7% for α-phase annealed (15 to 240 min at 700 to 870 °C) Ti-5Al-2.5Sn(ELI).

Deformation mechanisms in α- and β-phase alloys are discussed below.

5.1.3. Alpha-Plus-Beta-Phase Titanium Alloys. Available in the literature is detailed information on the low-temperature properties of three representative $\alpha + \beta$ alloys: the "near-α" alloys Ti-8Al-1Mo-1V[6] and Ti-6Al-3Nb-2Zr,[129] and the $\alpha + \beta$ alloy Ti-6Al-4V.[6,117,118] The low-temperature ductilities of these three materials are listed in Table 16. Evidently it is possible to exert considerable control over the low-temperature ductility through appropriate variation (by heat treatment) of the two-phase microstructure. Detailed electron-microscope studies of low-temperature dislocation pile-ups at acicular β-phase precipitates in Ti-6Al-3Nb-2Zr were undertaken by Lavrentev *et al.*[129] Nagai *et al.*[117,118] studied the

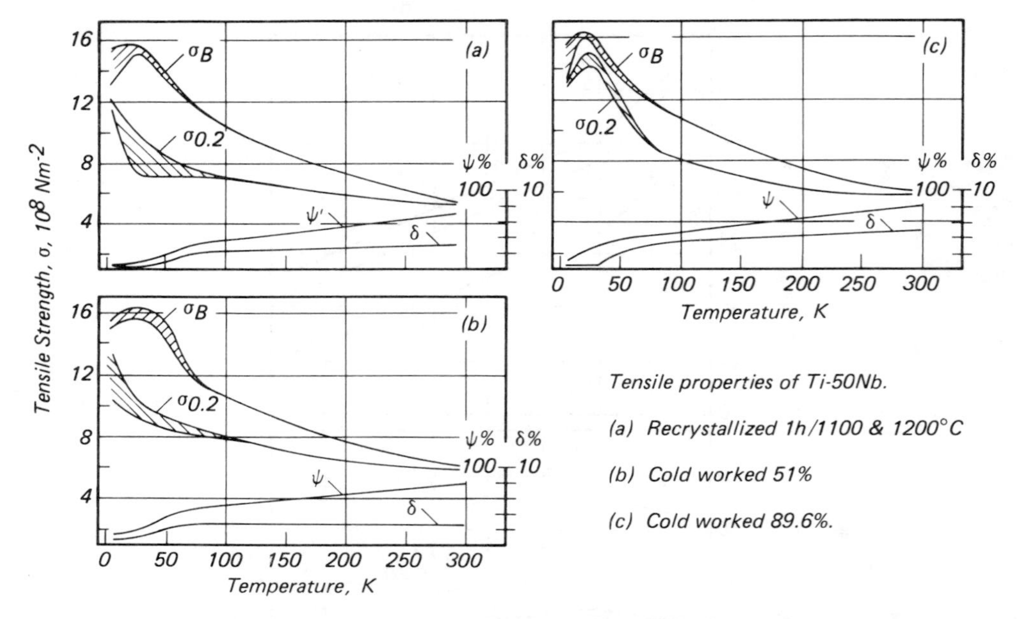

Fig. 40. Tensile properties of Ti-50Nb in various starting conditions as functions of test temperature (room temperature, 77 K, 30 K, and 4.2 K)[127,128]

Table 16. Low-temperature ductility of some $\alpha + \beta$ titanium alloys

Composition	Condition(a)	Elongation(b)	Reference
Ti-8Al-1Mo-1V	8 h at 790 °C + FC + 15 min at 790 °C + AC	1.2% at 20 K	6
Ti-6Al-3Nb-2Zr	1 h at 800 °C + AC	4 to 5% at 4.2 K	129
Ti-6Al-4V	1 h at 1050 °C(c) + AC	4% at 4 K (ELI) 1.5% at 4 K (normal)	118
Ti-6Al-4V	1/2 to 4 h at 710 to 820 °C	2.9% at 20 K (ELI) 2.4% at 20 K (normal)	6

(a) FC = furnace cool; AC = air cool. (b) ELI = extra-low interstitial content. (c) β anneal.

influence of the optical microstructure on the plastic tensile properties and fracture toughness of variously heat treated samples of Ti-6Al-4V.

5.1.4. Alpha-Phase Titanium Alloys. The α-Ti alloy generally selected for low-temperature service is Ti-5Al-2.5Sn(ELI). It possesses a considerably lower yield strength than, say, Ti-6Al-4V(ELI) (14.4×10^8 N·m^{-2} compared with 17.9×10^8 N·m^{-2}, for annealed alloys; see Table 14), but the absence of β phase renders it considerably more ductile (15% elongation at 20 K compared with 2.9%, again for annealed alloys; see Table 14). The fracture toughness of Ti-5Al-2.5Sn is generally thought of as being greater than that of Ti-6Al-4V.[6] However, recent studies have shown that the toughness of Ti-6Al-4V can be doubled in response to β annealing followed by suitably slow cooling[117,118]; in such a metallurgical condition, and for some applications, it would then be preferable to an all-α alloy.

5.1.5. Summary

(a) *Unalloyed Titanium*: All commercial grades of unalloyed titanium exhibit moderately good ductility at temperatures down to about 20 K. Their elongations at fracture actually increase as the temperature is decreased from 300 K, and pass through broad maxima (of about 40 to 50%) at about 77 K before descending rapidly as the temperature approaches 4.2 K. In some samples the elongation, δ_B, becomes negligibly small at liquid-helium temperatures. Cold rolling increases the yield and ultimate strengths but at the expense of ductility, as usual. The effects of interstitial elements on plastic properties have been considered in great detail by Conrad and coworkers (e.g., Conrad[124]). It has been pointed out that the solutes carbon, nitrogen, and oxygen, which bond in a covalent-like manner to the surrounding titanium atoms, have pronounced influences of the strength of otherwise unalloyed titanium at temperatures below about one-half of the melting point.

(b) *The α Alloy Ti-5Al-2.5Sn*: The ductility of Ti-5Al-2.5Sn(ELI) is fairly independent of temperature between room temperature and 20 K, δ_B remaining at about 16% throughout that range. The ductility of the normal-interstitial grade is considerably lower; in fact δ_B decreases monotonically below room temperature, dropping to 12% at 77 K and to only 5% at 20 K.

(c) *The $\alpha + \beta$ Alloy Ti-6Al-4V*: The ductility of annealed Ti-6Al-4V is fairly independent of temperature between room temperature and 77 K. Below 77 K, it decreases rapidly as the temperature continues to fall toward 20 K. The ductility of the annealed alloy is twice as great as that of the solution-treated-and-aged material—e.g., $\delta_{B,77K} = 11.4\%$ as compared with 4.9% (at normal interstitial levels). Reducing the interstitial content influences the tensile properties only marginally but improves the fracture toughness by 130% at room temperature, and by 40% at 20 K.

(d) *The Near-α $\alpha + \beta$ Alloy Ti-8Al-1Mo-1V*: The near-α $\alpha + \beta$ alloy Ti-8Al-1Mo-1V, although originally developed for high-temperature applications, can be used reliably down to moderate subambient temperatures in either the single-annealed (SA, "mill-annealed," 8 h at 790 °C + furnace cool) or duplex annealed (DA, mill annealed + 15 min at 790 °C + air cool) condition. The room-temperature ductilities of SA and DA alloys are similar ($\delta_B \cong 15\%$), but upon cooling, that of the SA alloy decreases, while that of the DA alloy increases before passing through a maximum ($\delta_B \cong 22\%$) at about 77 K and dropping to low values at 20 K ($\delta_B \cong 1\%$).

(e) *The β Alloy Ti-13V-11Cr-3Al*: As a β-Ti alloy, Ti-13V-11Cr-3Al would be expected to possess poor low-temperature ductility. Indeed it does, the elongation at fracture of the solution-treated-and-aged (STA) material becoming insignificantly small below about 100 K; at 77 K, $\delta_B = 0.2\%$. Some improvement results if the aging stage (i.e., 20 to 100 h at 430 to 500 °C) of the STA heat treatment is omitted, in which case $\delta_{B,77K}$ becomes about 2%.

5.2. Physical Metallurgy and Low-Temperature Strength

5.2.1. Influence of Interstitial Content. Strengthening of titanium alloys by interstitial elements has been referred to above and also in Section 5.1.1(c). The interstitial solutes carbon, nitrogen, and oxygen are relatively immobile below about 300 °C and provide stable solution strengthening below that temperature. The solution-strengthening potencies of boron, carbon, nitrogen, and oxygen are compared with those of the α stabilizers aluminum, gallium, and tin in Table 15. There it can be seen that the hardening rates produced by the interstitials are more than an order of magnitude greater than those produced by the α stabilizers, which are themselves potent strengtheners. Table 15 also shows that the hardening potency of the interstitials increases in the sequence C < O < B < N. These results agree qualitatively with those of Conrad *et al.*,[122,124] who showed that the measured Gibbs free energies of activation associated with thermally activated plastic flow of dilute Ti-interstitial alloys had the values 1.50, 1.64, and 1.73 eV, respectively, for Ti-C, Ti-O, and Ti-N alloys.

5.2.2. Influence of Interstitial Content and Grain Size. The influence of interstitial content and grain size on the low-temperature tensile properties of titanium alloys has been considered by Conrad.[124] A set of typical results is given in Fig. 41. There it can be seen that at low temperatures (from 400 down to 77 K) the true fracture stress, ϵ_F, is generally less dependent on temperature and grain size than at higher temperatures. At temperatures above 400 K, the effect of interstitial content depends on grain size. Conrad[124] also confirmed that an increase in the interstitial content leads to a decrease in ductility, with the rate of decrease in elongation with concentration increasing in the sequence C < O < N.

5.2.3. Influence of Heat Treatment (Alloy Phase Morphology). The strength and temperature of an $\alpha + \beta$ titanium alloy can be adjusted by controlling the volume-fractions and morphologies of the α- and β-phase components through suitable heat treatments. Nagai *et al.*[117,118] have shown how the structure of Ti-6Al-4V varies in response to β annealing (1 h at 1050 °C in this case) followed by variable-time cooling (in 48 s to 2.8 h) from 1050 to 550 °C followed by water quenching to room temperature (Fig. 42).

The results of mechanical-property testing are given in Fig. 43, in which it can be seen that slow cooling ($\gtrsim$162 s) following β annealing brings about excellent fracture toughness and good fracture strain, with only a slight loss of strength at 4 K. The slow-cooled Widmanstätten structure is characterized by a "basket-weave" arrangement of "packets" of α-phase plates. It turns out that a crack, although it will propagate in a straight line within such a packet, becomes deflected or arrested at the boundary between packets. This inhibition of crack propagation seems to

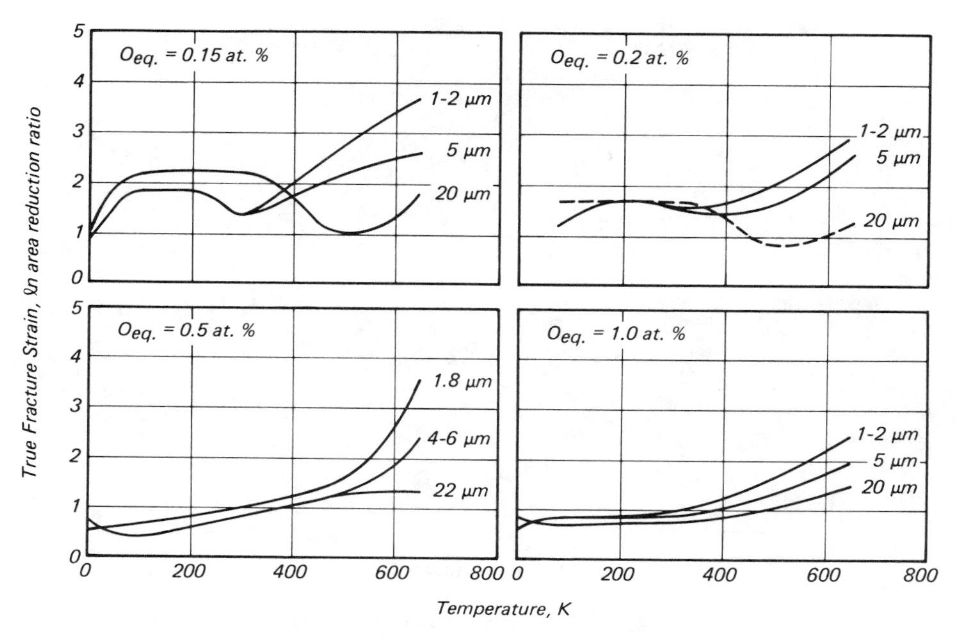

Fig. 41. Effects of temperature, interstitial level (expressed in terms of an equivalent-oxygen concentration, O_{eq}), and grain size on the true fracture strain of unalloyed titanium wires[124]

be responsible for the increased fracture toughness exhibited by the slow-cooled material.

5.3. Low-Temperature Deformation Modes

At low temperatures, titanium alloys generally exhibit serrated yielding in their stress-strain curves. This is true of both α-phase and near-α-phase alloys (see Fig. 44 for commercial-purity titanium[124] and Ti-6Al-3Nb-2Zr[129]) as well as of metastable β-phase alloys (see Fig. 45). The mechanisms of serrated yielding are different in the two cases: the α-phase alloys seem to undergo conventional adiabatic-heating oscillations under tensile loading at low temperatures; in the β-phase alloys the serrated yielding is clearly a manifestation of lattice instability.

5.3.1. Serrated Yielding in Alpha-Titanium Alloys. Serrated yielding may be regarded as a quite general low-temperature mechanical effect which, when it occurs, arises from the following combination of factors: (*a*) adiabatic heating; (*b*) a strong negative temperature coefficient of the yield stress; and (*c*) the small specific heat of a metal at low temperatures. Moskalenko *et al.*,[119,120,132] who studied unalloyed titanium (iodide and commercial), Ti-2.4Zr-1.2Mo, Ti-Al(1.5, 3, 5.5 wt %), Ti-Zr(1, 3, 8 wt %), Ti-V(0.5, 1.5, 3 wt %), and Ti-Nb(0.5, 2, 4 wt %), and Conrad,[124] who studied commercial titanium in considerable detail,

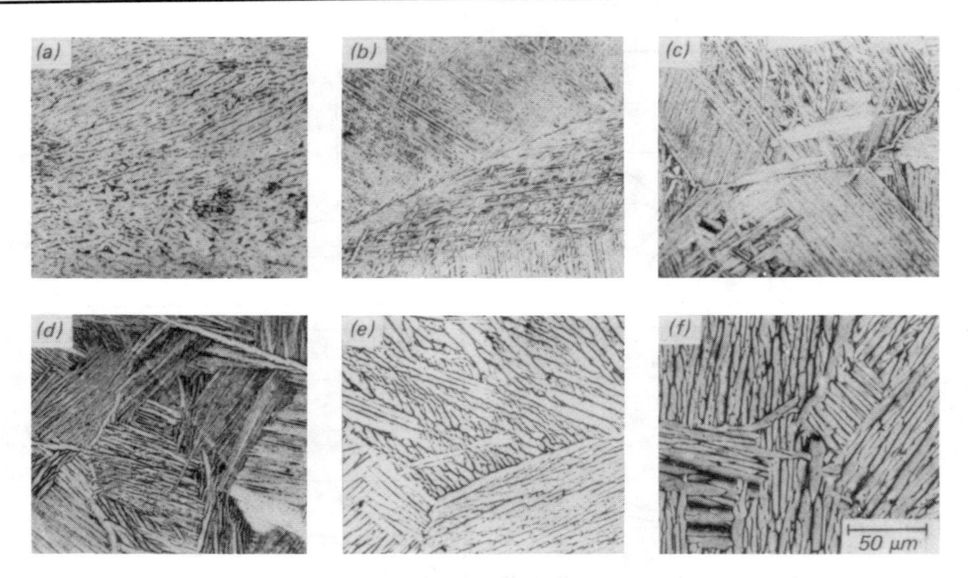

Conditions: (a) as-received (a mill-annealed fine lamellar $\alpha + \beta$ structure (cf. Fig. 60); (b) β-annealed 3.6 ks at 1050 °C plus cooled in 48 s to 550 °C and water quenched; (c) cooled in 162 s; (d) cooled in 360 s; (e) cooled in 3600 s; (f) furnace cooled in 10.2 ks. Micrographs courtesy of K. Nagai, National Research Institute for Metals, Tsukuba.

Fig. 42. Optical microstructure of normal-interstitial-level (as distinct from ELI-grade) Ti-6Al-4V as function of heat treatment[117,118]

agreed that the adiabatic heating was due to dislocation avalanching (which follows in response to the sequence: localized plastic flow/dislocation pile-up/stress increase). As mentioned above, serrated yielding is a frequently observed phenomenon in low-temperature testing. Since the discovery of the effect in aluminum and its alloys by Basinsky,[133] it has been known to be *not necessarily* associated with either twinning or martensitic transformation. But in the commercial titanium studied by Conrad, and in the dilute titanium alloys studied by Moskalenko and Pupsova[119] (see Fig. 46), the serrated yielding seemed to be invariably accompanied by twinning, which, according to Conrad, appears to assist in the nucleation of the dislocation avalanche.

5.3.2. Serrated Yielding in Beta-Titanium Alloys. Serrated yielding (Fig. 45) and twinning (Fig. 47) have both been observed in Ti-50Nb, a representative metastable-β titanium alloy. However, the mechanism in such systems has been identified as an "anomalous plastic property"[135, pp. 100-108] related to thermoelasticity and pseudoelasticity (Fig. 48) and traceable to the instability of the β-Ti lattice.[136,137] The starting point for a discussion of these properties is Fig. 49, which shows the measured M_s for Ti-Nb extrapolated into the liquid-helium temperature range. This figure suggests that $\beta \rightleftharpoons \alpha''$ transformation will take place whenever the extrapolated line is crossed—as a function of either composition or

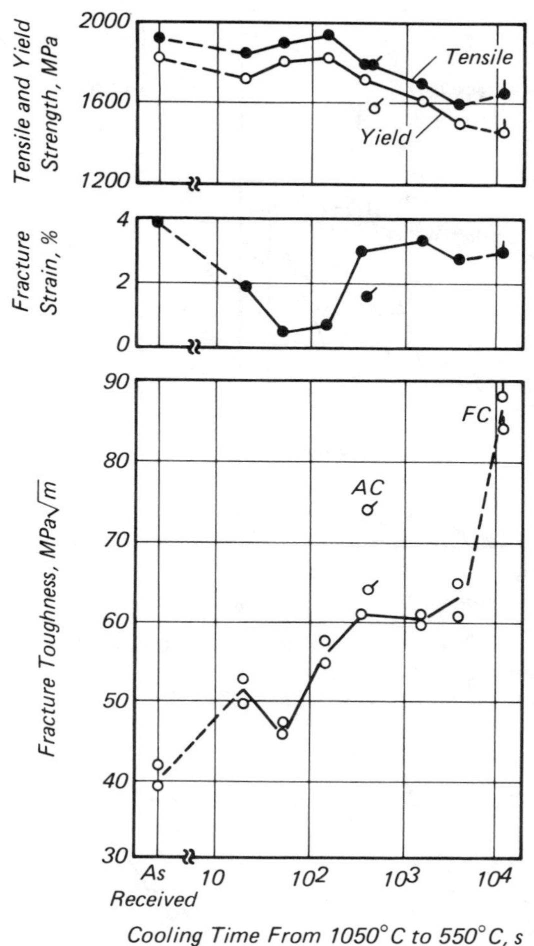

Fig. 43. Mechanical properties of normal-interstitial-level (as distinct from ELI-grade) Ti-6Al-4V in the as-received (mill-annealed) condition (see Fig. 42a) and as a function of cooling time from 1050 to 550 °C following a β anneal for 3.6 ks at 1050 °C[117,118]

temperature. It turns out, however, that the transformation does not take place spontaneously. For example, a physical-property measurement sensitive to martensite has shown that no fresh martensite is produced when a water-quenched alloy is cooled to 4 K for measurement[140] (Fig. 50). Some extra energy, in the form of lattice strain energy, for example, is needed to initiate the transformation. Thus the transformation will take place under either of the following two circumstances: (1) the crystal is prestrained and lowered in temperature through the extrapolated M_s[136,137]; or (2) the crystal is lowered in temperature below M_s and then strained, as in a low-temperature tensile test. The effects in both cases are reversible (unless, as is usually the case in option 2, the sample is strained to fracture), thereby pro-

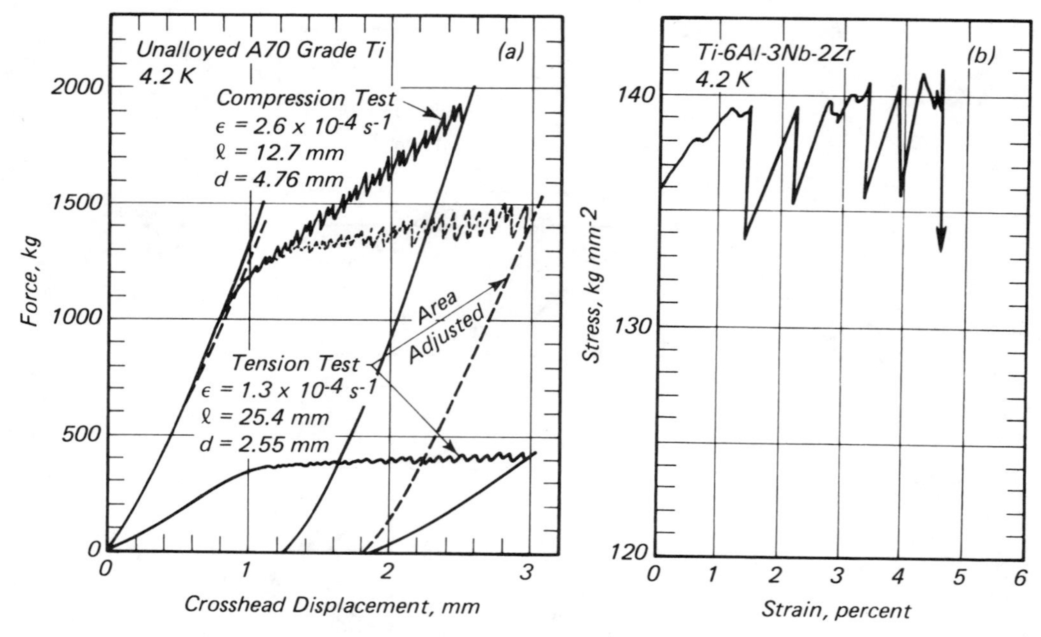

Fig. 44. Occurrence of serrated yielding during tensile testing at 4.2 K of (a) α-Ti[130; see also 124] and (b) the near-α alloy Ti-6Al-3Nb-2Zr[129]

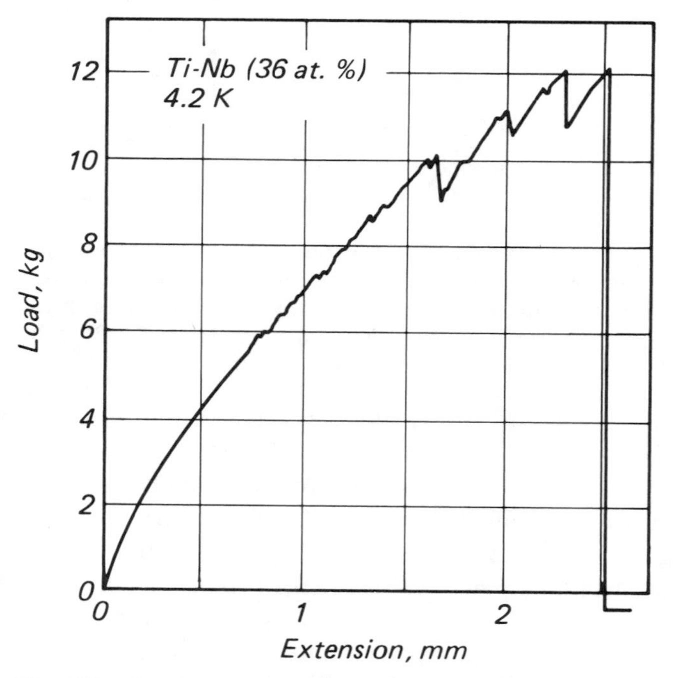

Fig. 45. Occurrence of serrated yielding during tensile testing at 4.2 K of the β alloy Ti-Nb(36 at. %)[131]

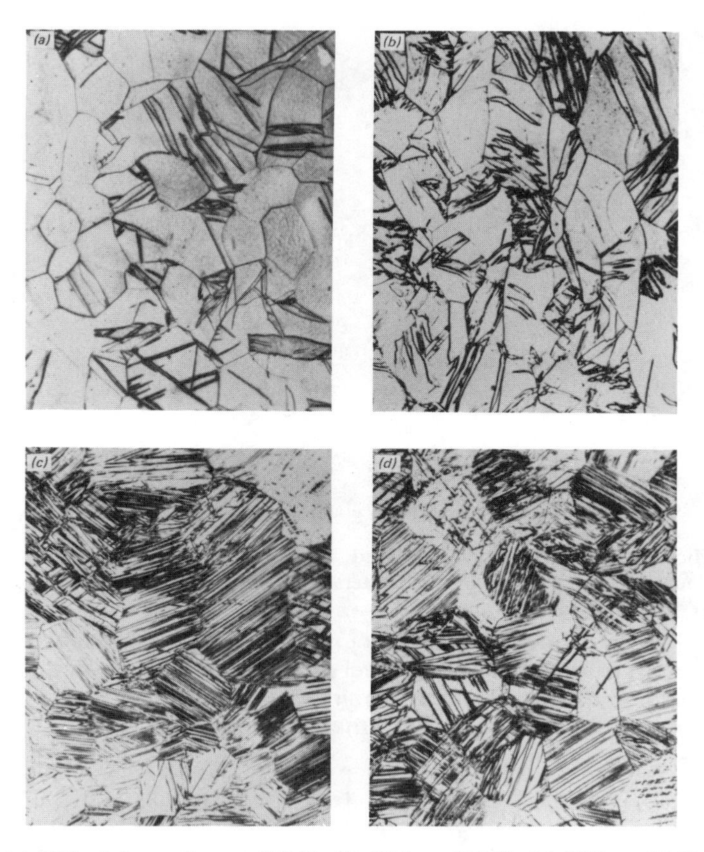

(a) 12% deformation at 293 K; (b) 25% at 293 K; (c) 12% at 77 K; (d) 13% at 4.2 K. Micrographs courtesy of B.I. Verkin, Physicotechnical Institute of Low Temperatures, Kharkov.

Fig. 46. Microstructure of Ti-0.5Nb in response to tensile deformation at various temperatures[119]

viding examples, respectively, of: (1) thermoelasticity and (2) pseudoelasticity (see Fig. 48).

Evidence for reversible martensitic transformation at zero applied stress (thermoelasticity) in room-temperature-strained Ti-Nb alloys has been provided by Hochstuhl, who demonstrated the existence of *negative*, reversible, thermal-expansion components in Ti-Nb(21 at. %) and other alloys.[136,137] With *decreasing* temperature the sample's length *increased* monotonically as its temperature was lowered from room temperature into the liquid-helium range (the formation of athermal-ω would have resulted in enhanced lattice contraction; see Section 3.2.1a). The effect, which was reversible, did not occur in the as-quenched samples, presumably because they lacked the strain energy needed to trigger the transformation.

Evidence for pseudoelasticity under tensile-test conditions has been provided by Koch and Easton.[see 135, p. 107] In most cases the tensile test is continued through the serrated-yielding region to sample failure (Fig. 45), which, when it takes place,

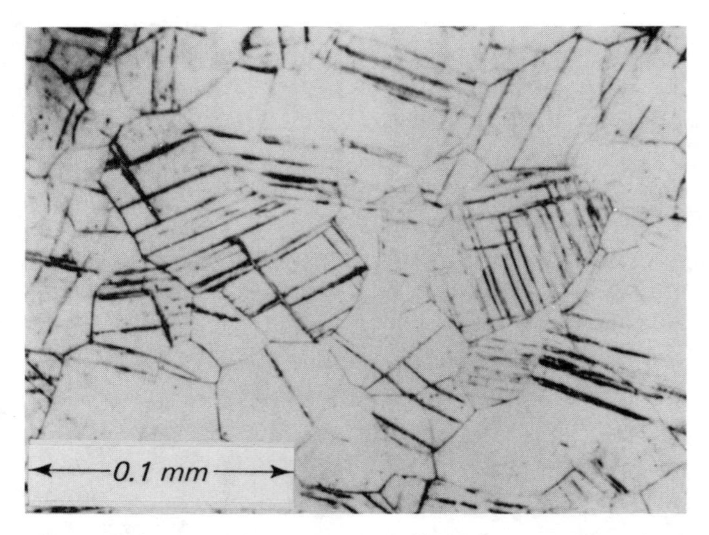

Micrograph courtesy of D.T. Read, National Bureau of Standards, Boulder, CO, copyright 1978, Butterworth and Co. (Publishers) Ltd. (reproduced with permission).

Fig. 47. Subsurface micrograph (original magnification, 600×) of Ti-45Nb (recrystallized at 800 °C and water quenched) after straining to fracture at 4 K, revealing a twinlike deformation structure[134]

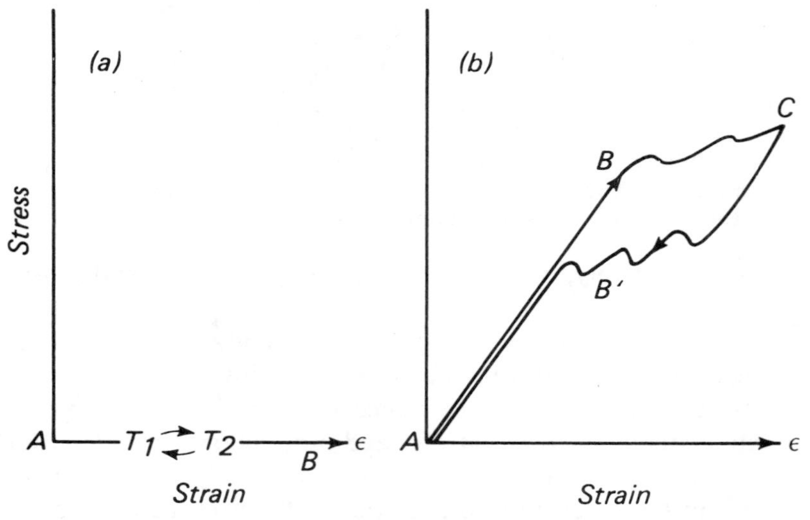

Fig. 48. Schematic representations of (a) thermoelasticity and (b) pseudoelasticity

is of the shear-rupture type (Fig. 51). If tensile unloading is commenced before the sample ruptures, evidence for reverse transformation, in the form of acoustic emission, is obtained as the stress is released.[131]

Just as was the case for α-Ti alloys, the β alloy's tensile strength drops as it enters the serrated-yielding regime (cf. Fig. 38 and 40).

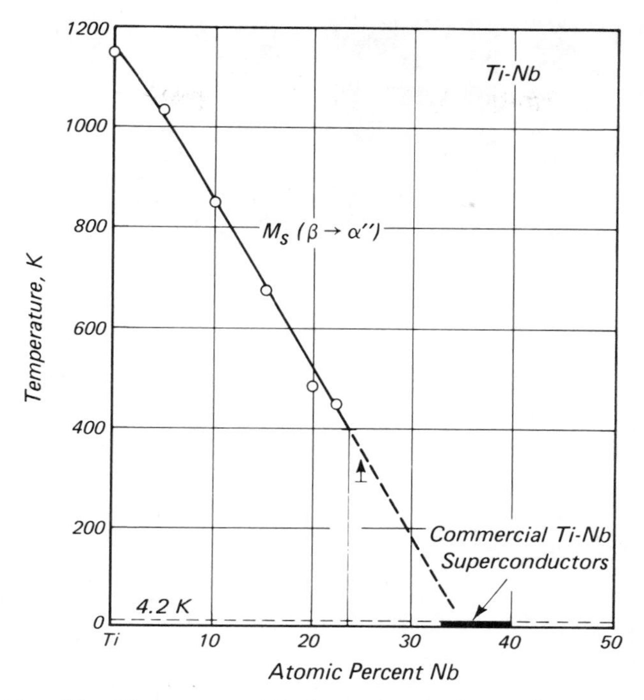

Fig. 49. The martensitic $\beta \rightarrow \alpha''$ transformation curve for Ti-Nb based on the results of Duwez,[71] Brown *et al.*,[138] and Baker,[139] and its extrapolation by Koch and Easton[131]

6. EVOLUTION OF CONVENTIONAL (INGOT-METALLURGY) HIGH-TEMPERATURE TITANIUM ALLOYS

As pointed out in Section 5, the β alloys are unsuited for low-temperature applications. But whereas the α alloys perform satisfactorily at low temperatures where solid-solution strengthening is important, their mechanical properties decrease rapidly with increasing temperature. As the temperature increases above room temperature, since the strengths of all the α alloys, such as Ti-5Al-2.5Sn (Fig. 52a), continue to decrease rapidly, they must be abandoned in the intermediate-temperature range in favor of alloys such as those depicted in Fig. 52(b), in which the strengthening mechanisms are "microstructural" in nature.

The α-phase alloys may be suitable for elevated-temperature service if they are (a) thermally stable in the temperature range of interest and (b) further strengthened or reinforced with precipitates, particles, or second phases.

6.1. Analytical Design of Conventional High-Temperature Titanium Alloys

The properties of technical titanium alloys can be generally understood in terms of the prototypical α- and β-stabilized binary alloys, Ti-Al and Ti-Mo,[141] respec-

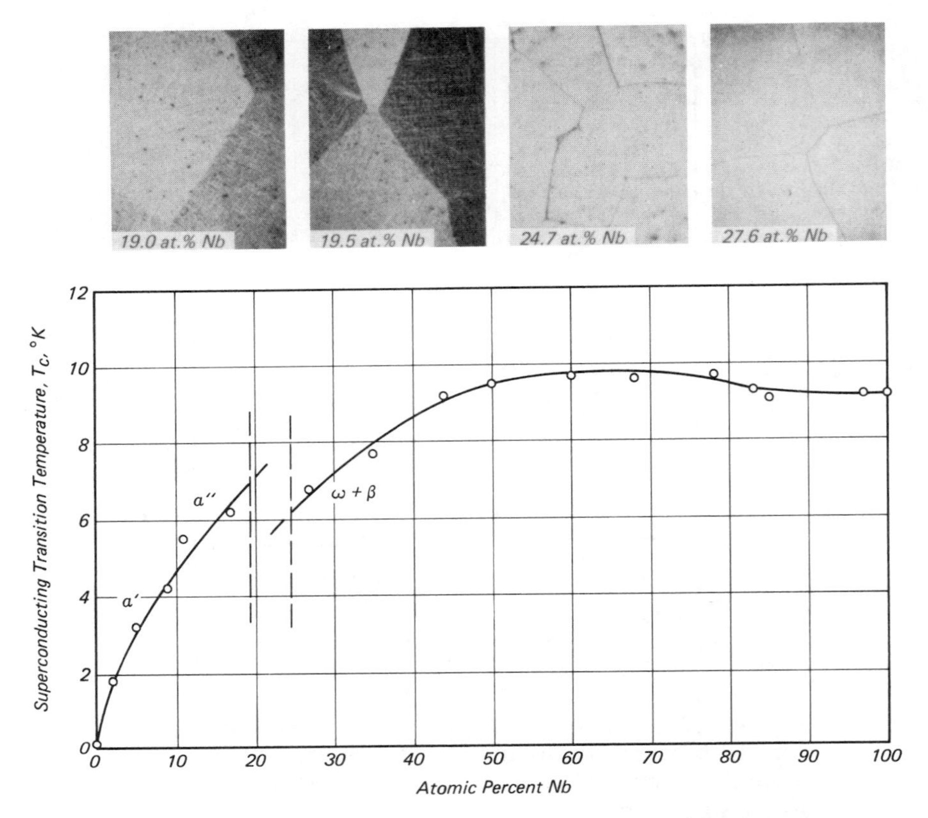

The martensitic regime terminates at the same composition in each case; evidently no further transformation of the quenched alloys takes place on cooling them into the liquid-He range.

Fig. 50. Optical microstructures of quenched Ti-Nb alloys referred to the composition dependence of the superconducting transition temperature

tively. The high-temperature metallurgical stability of titanium alloys (ignoring environmental or chemical effects) is limited by $\beta \rightleftharpoons \alpha + \beta$ transformation, which in unalloyed titanium occurs at 883 °C. Thus, although titanium melts at 1670 °C, its service temperature, based on metallurgical stability, is governed by the so-called "homologous temperature"—i.e., the temperature adjusted to $\beta \rightleftharpoons \alpha + \beta$ transformation rather than to melting. It would therefore seem that the addition of α stabilizers, of which aluminum is an excellent example, would not only provide potent solution strengthening, but would also increase the alloy's "heat resistance" by raising its transformation temperature (cf. Collings[141, p. 56]). This prospect encouraged the early development of technical α alloys. Working on the assumption that if a modest amount of α stabilizer is good, even more would be better, early developers produced alloys such as Ti-5Al-5Sn-5Zr only to find that they became embrittled by α_2-phase precipitation during elevated-temperature service. After exploring the boundary of the single-phase-α solid solution, Rosenberg[24]

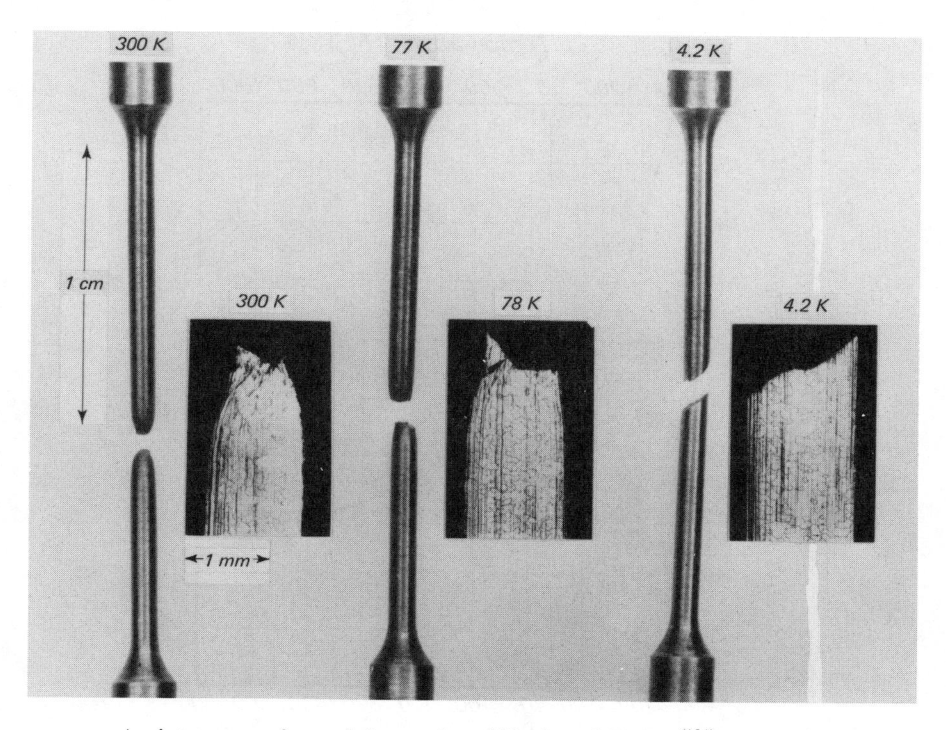

An intercomparison of the results of Koch and Easton[131] on Ti-Nb (38.5 at. %) (tensile bars depicted) and Albert and Pfeiffer[127] on Ti-Nb(34 at. %) (fracture ends depicted) demonstrates the reproducibility of the fracture behavior at each of the test temperatures. Photographs courtesy of D.S. Easton, Oak Ridge National Laboratory, and H. Hillmann, Vacuumschmeltze GmbH.

Fig. 51. Tensile specimens and fracture ends of samples after testing to fracture at the temperatures indicated

recommended that the total "equivalent-Al" content of an alloy should be kept below 9 wt %, such that for an alloy containing aluminum, zirconium, tin, and oxygen, for example:

$$[\mathrm{Al}]_{\mathrm{eq}} = [\mathrm{Al}] + \frac{[\mathrm{Zr}]}{6} + \frac{[\mathrm{Sn}]}{3} + 10[\mathrm{O}] \lesssim 9 \qquad \text{(Eq 3)}$$

where [x] is the concentration of element x in weight percent. Based on more recent work we may now include gallium in terms of [Ga]/2. Alloys with $[\mathrm{Al}]_{\mathrm{eq}} > 9$ are the now-obsolete "super-alpha" alloys.

An alternative approach to the development of titanium-base alloys for high-temperature service is to "skip over" the β transus and to operate in the β-phase field. This approach led to the development of the β-Ti alloys typified by the U.S. alloy "β-III" of which the binary prototype is Ti-Mo (see Section 4.2). The guid-

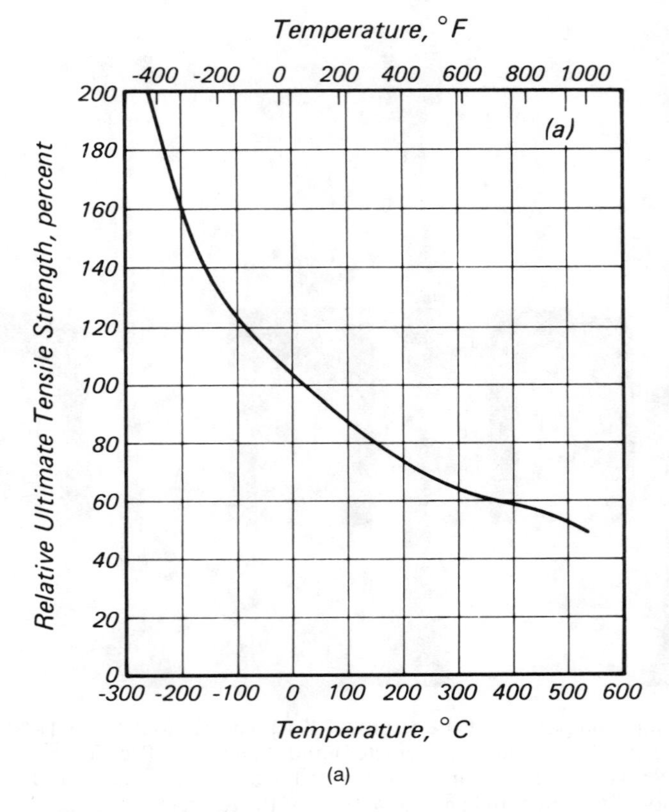

(a)

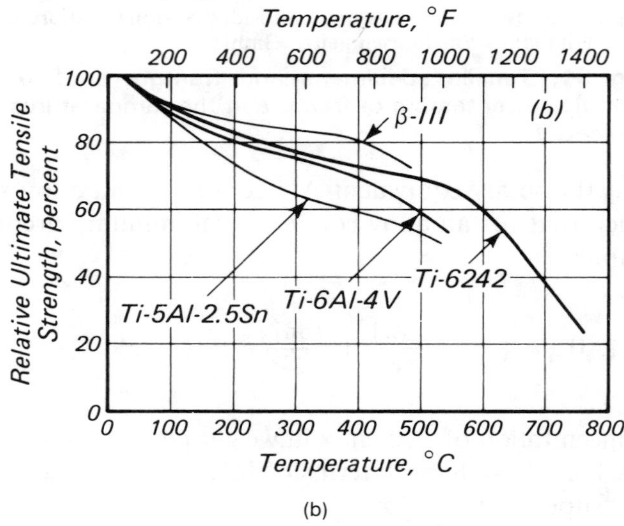

(b)

Fig. 52. Temperature dependence of relative ultimate tensile strengths for (a) annealed Ti-5Al-2.5Sn sheet, and (b) the three commercial alloys Ti-6Al-4V, Ti-6242, and β-III, in comparison with Ti-5Al-2.5Sn[5]

ing philosophy behind this approach can be appreciated by visualizing the typical β-isomorphous phase diagram whose transformation temperatures drop rapidly with solute concentration (see Sections 1.2 and 1.5). The β alloys tend to be metastable (see Section 4.3.3), but if suitably processed they can be operated successfully at moderately high temperatures.

An important advantage that β-Ti alloys, such as β-III, have over the α-Ti alloys is that their strengths tend to decrease much less rapidly with increasing temperature. In attempts to combine the better temperature-dependence characteristic of the β-Ti alloys with the metallurgical stability of the α-Ti alloys, and for other reasons associated with the details of thermomechanical processing, the two-phase $\alpha + \beta$ alloys were developed. Fine adjustments in the compositions of the $\alpha + \beta$ alloys over the years, in both the U.S. and the U.K., have resulted in alloys suitable for service at temperatures up to about 590 °C (IMI 834; see Table 3). An important ingredient of IMI 834 is the 0.5 wt % Si. In the U.S., experiments with the alloy Ti-6Al-2Sn-4Zr-2Mo with varying silicon contents yielded a maximum in creep resistance at 0.1 wt % Si (Fig. 53), resulting in a limiting silicon-level specification in the U.S. of 0.1 to 0.2 wt %. But comparable experiments on other alloys have demonstrated a continuous increase in creep resistance with increasing silicon content up to the limit of solubility.[162] Accordingly, alloys in the U.K. have tended to include more silicon than their U.S. counterparts. The U.K. designs have also tended to demonstrate more boldness in the choice of minor alloying elements than their U.S. counterparts.

The development chronology of conventional ingot-metallurgy (IM) heat-resistant alloys is depicted in Fig. 54, in which the service temperature is plotted as a function of year of introduction. Their relative heat resistances, as gaged by the Larson-Miller parameter, are plotted in Fig. 55. The Larson-Miller parameter (a measure of "time at temperature" for a given strain) is given by $T(C + \log t)$, where T is temperature in degrees Rankine (°F + 460), t is time in hours, and C is a constant (about 20 for a large number of alloys).

The compositions of the most heat resistant of the alloys depicted in the figures are listed in Table 17. The philosophy guiding the design of these alloys can be appreciated with reference to that table: (1) First of all we notice that the alloys have almost the same $[Al]_{eq}$ (except IMI 550) which, moreover, has a mean value of 7.3, only 1.7 units less than the limiting value of 9 referred to in Eq 3. The inclusion of 1700 wt ppm of oxygen in an $[Al]_{eq} = 7.3$ alloy would raise it to the limit, and thus for all practical purposes the alloys could be said to be operating at the limit of α stability. (2) The level of β stabilizer is very low—sufficient to confer some microstructural strengthening, but not enough to engender metallurgical instability. It is notable that the level of heat resistance goes up as the β-stabilizer content decreases; in proceeding from Ti-6242 to IMI 834, the molybdenum content steadily decreases. As pointed out by Hoch *et al.*,[3] the β-stabilizing strengths of the alloying additions niobium, vanadium, and molybdenum decrease in the sequence Mo > V > Nb; thus, to provide β-phase microstructural strengthening with the least possible lowering of the β transus, it is best to substitute niobium for some

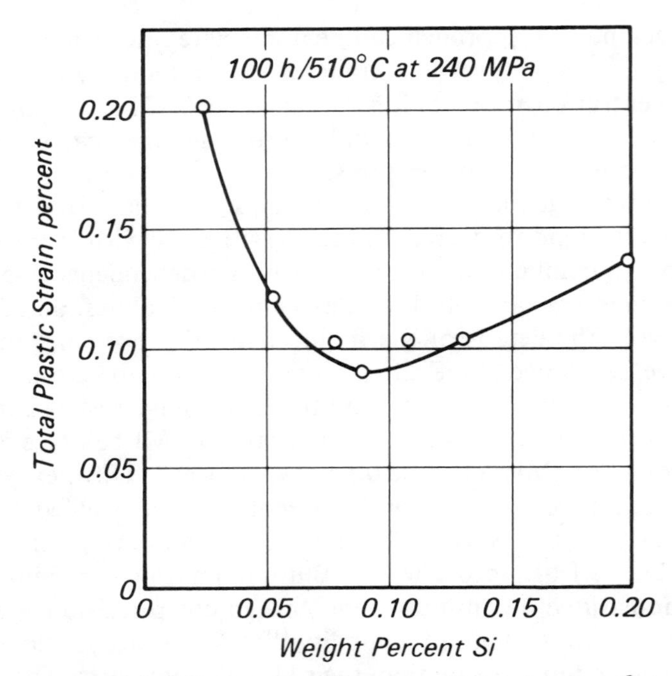

Fig. 53. Influence of silicon on the creep performance of Ti-6262. After Seagle *et al.* (see Blenkinsop[142]).

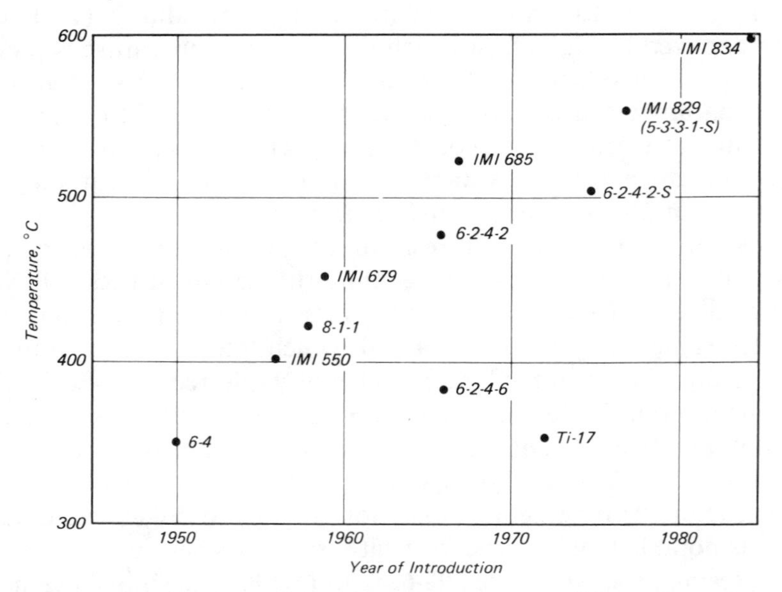

Year of introduction and operating temperature capability under optimum stress conditions. The suffix "S" indicates that a small amount of silicon (usually 0.25 wt %) has been added to the basic formulation.[142]

Fig. 54. High-temperature titanium-base alloys for aircraft engine applications

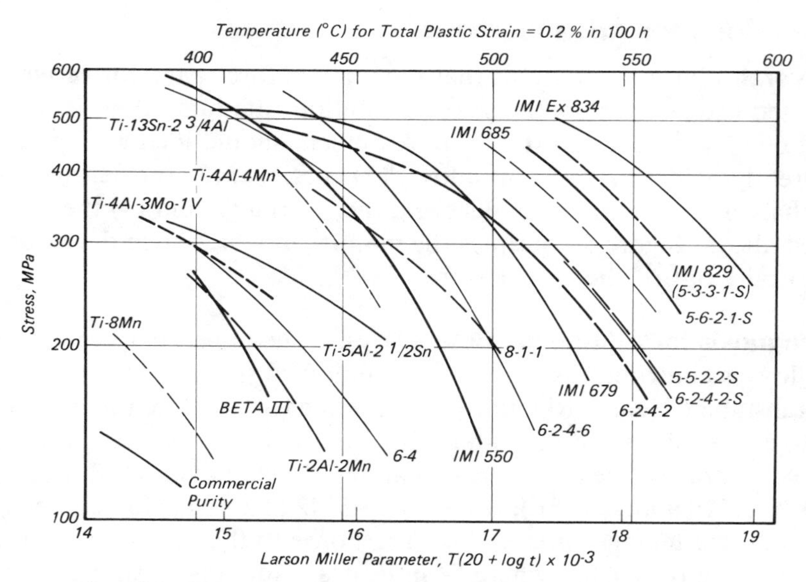

Fig. 55. Elevated-temperature creep performances of the alloys referred to in the previous figure (except Ti-17, viz. Ti-5Al-2Sn-2Zr-4Mo-4Cr, which was developed for applications up to 350 °C) and some others[142]

Table 17. Evolutionary design(a) of multicomponent $\alpha + \beta$ titanium alloys(b)

	α				β			
Alloy	Al	Sn	Zr	$[Al]_{eq}$	Nb	V	Mo	Si
IMI 834	5.5	4	4	7.5	1	...	0.3	0.5
IMI 829	5.5	3.5	3	7.2	1	...	0.3	0.3
IMI 685	6	...	5	6.8	...	...	0.5	0.25
6242-Si	6	2	4	7.4	...	...	2	0.2
6242	6	2	4	7.4	...	...	2	...
IMI 679	2¼	11	5	6.8	...	...	1	0.2
811	8	...	...	8	...	1	1	...
IMI 550	4	2	...	4.7	...	...	4	0.5

(a) Analysis: (1) Maintain $[Al]_{eq}$; (2) reduce β content; (3) increase Si content. (b) See also Table 6.

of the molybdenum. This has been done in IMI 829 and IMI 834. As maximally α-stabilized alloys with minimal β-stabilizer contents, this family of $\alpha + \beta$ alloys is referred to as "near-α" $\alpha + \beta$ alloys. The picture is completed by noting that the heat resistance increases with increasing silicon content up to the saturation limit of 0.5 wt %.

Table 6 presents an analysis of the α-stabilizer and β-stabilizer contents of numerous U.S. alloys.

6.2. Directions for Advancement

6.2.1. Overall Picture. It seems that the temperature range for conventional IM unprotected (no surface coatings) solid-solution titanium alloys is upper limited to about 600 °C[142] and that future development should include combinations of: (1) inherently oxidation-resistant alloys, (2) coatings, (3) further development of the α_2-phase and γ-phase aluminides and their variants, and (4) the use of rapidly solidified alloys. Figure 56 outlines the manner in which such developments proceed naturally out of the earlier work.

6.2.2. Prognosis for Alpha-Plus-Beta Alloys. The upper temperature limit of the $\alpha + \beta$ alloys is controlled as much by metallurgical stability as by surface or environmental stability (e.g., oxidation). The *service* temperature range designated for the next generation of titanium alloys is actually the temperature range currently being used to *process* the present $\alpha + \beta$ alloys. At temperatures between about 850 and 950 °C (1550 and 1700 °F), the alloy Ti-6242 (Table 17) undergoes volumetric hysteresis during temperature cycling in response to the hysteretic microstructural cycling of β and $(\alpha + \beta)_W$ (where W indicates Widmanstätten precipitation).[143] The use of such an $\alpha + \beta$ alloy in this temperature range would be accompanied by dimensional instability stemming from the volumetric differences between the α and β phases.

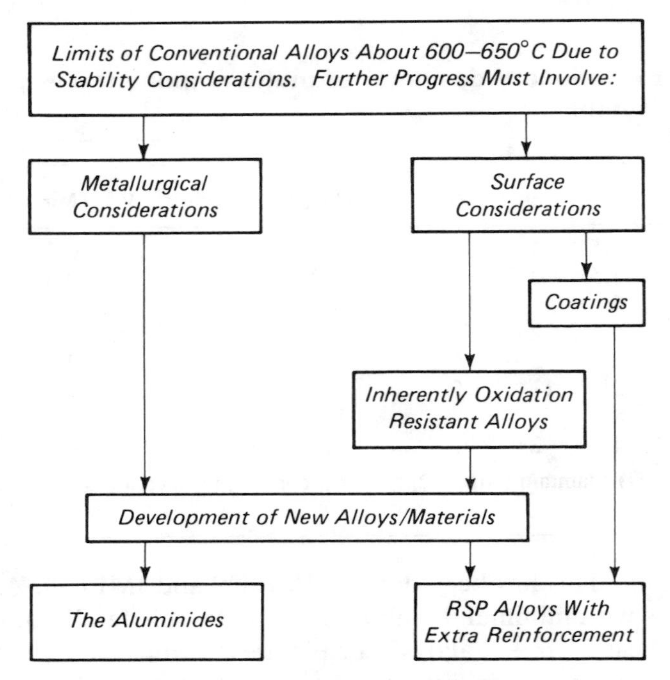

Fig. 56. Directions for further development of titanium-base alloys for very high temperature applications

Accordingly, attempts to develop the $\alpha + \beta$ alloys for very high temperature service are not recommended.

6.2.3. Prognosis for Beta Alloys. The metallurgical properties of the β alloys as they approach equilibrium have been discussed by Collings[141, pp. 72-74,206-210] (see also Section 4.3.3). In alloys such as β-III, so-called solution heat treatment at temperatures of 720 and 790 °C followed by aging at temperatures in the range 370 to 590 °C can be used to control the microstructure and mechanical properties. Likewise, the alloy β-C* can be made to undergo phase separation (at 350 °C) and α-phase precipitation (at 500 °C) by appropriate heat treatment at temperatures through which new-generation high-temperature titanium alloys would be expected to cycle during service. It follows that conventional β-Ti alloys would also be metallurgically unstable in use.

Such difficulties may possibly be averted by choosing very concentrated β alloys. But this tends to erode the density advantages that titanium alloys have over their competitors. It is then natural to ask whether this problem can be solved by selecting vanadium (whose density is only 35% greater than that of titanium) as the β stabilizer†; the answer seems to be no, the $\beta/(\alpha + \beta)$ transus in Ti-V tending to become rather flat at concentrations greater than about 20 wt % V. Indeed, both Ti-34V and Ti-39V begin to precipitate α phase after about 6 h at temperatures above 450 °C.

Again, metallurgical instability considerations militate against the use of β alloys for very high temperature service.

6.2.4. Prognosis for Alpha Alloys. It is well known that the effect of α-stabilizing elements is to preserve (in the case of tin) or substantially increase (in the cases of aluminum and particularly oxygen) the temperature of the $(\alpha + \beta)/\alpha$ transus. With aluminum, at the single-phase solubility limit (at about 800 °C) of 9 wt % (i.e., 15 at. %), the temperature of the $(\alpha + \beta)/\alpha$ transus has increased from 883 °C (α/β for pure titanium) to 1050 °C. The α-stabilizing potency of oxygen is even greater; however, hardness considerations have prevented full advantage from being taken of it, at least for intermediate-temperature operation.

The large negative yield strength temperature dependences of conventional α alloys have prevented them from entering high-temperature service. On the other hand, their intrinsic metallurgical stability at all temperatures up to the $(\alpha + \beta)/\alpha$ transus makes them suitable as bases upon which to build (with the aid of dispersion or fiber strengthening) the next generation of high-temperature alloys.

A logical extension of the process of solid-solution strengthening of titanium by α-stabilizing elements such as interstitial elements or simple metals (aluminum in particular) leads to the formation of the intermetallic compounds. The two most

---

*β-C is Ti-4Mo-8V-6Cr-4Zr-3Al.

†The alloy Ti-15V-6Al and its relatives have been considered as β-Ti candidate alloys and warrant further study as potential RSP (rapid solidification processed) β alloys.

well-known compounds in the Ti-Al system are the so-called "aluminides": α_2 (based on the compound Ti_3Al) and γ (based on the compound TiAl). These compounds are already showing great potential for high-temperature service: Ti_3Al undergoes an order-disorder reaction at 1100 °C and begins to decompose to β phase at about 1150 °C; TiAl melts at about 1450 °C.

It can be concluded that the best bases for the next generation of titanium alloys for elevated-temperature service are α-Ti solid solutions and the α_2-phase and γ-phase aluminides.

6.3. Ingot Metallurgy of Advanced Alpha-Phase Solid-Solution Alloys

Some of the most extensive studies of α-Ti alloys for high-temperature service were carried out in the 1950's at the Baikov Institute of Metallurgy in Moscow.[144] Alloys studied for strength and high-temperature creep resistance included the six-component alloys AT-3, -4, -6, and -8 (whose compositions are listed in Table 18); some developments of them (so-called AT-10 and AT-12) containing seven to eight alloying elements; and, for reference purposes, an alloy based on Ti_3Al. As a result of both tensile and centrifugal (bend-type) creep tests, it was concluded that: (1) the AT-3 and AT-4 alloys were useful up to 400 to 500 °C, the AT-6 up to 550 °C, and the AT-8 up to 500 to 600 °C; and (2) to sustain temperatures in the range 600 to 700 °C it was necessary to increase the number of alloying elements to seven or eight (thereby forming the AT-10 and AT-12 alloy types). Although showing a great improvement over those of the six-component alloys, the properties of the seven- and eight-component α-phase alloys were still inferior to those of an alloy based on Ti_3Al.

The results of work such as this (1) are in accordance with the results of studies conducted in the U.S., which indicate that the aluminides hold great promise as high-temperature structural materials, and (2) indicate that, within the realm of solid-solution alloying, RSP multicomponent materials based on α-phase alloys also have potential for high-temperature applications.

Table 18. Compositions of some experimental Soviet heat-resistant titanium alloys[144]

	Composition, wt %				
Alloy code(a)	Al	Cr	Fe	Si	B
AT-3	3.2	0.84	0.34	0.40	0.01
AT-4	4.8	0.86	0.36	0.29	0.01
AT-6	5.6	0.52	0.30	0.33	0.01
AT-8	6.8	0.98	0.40	0.59	0.01

(a) The six-component alloys specified had considerably lower rates of creep than other investigated industrial alloys (the so-called VT-1, OT-4, VT-5-1, and VT-14) and the experimental alloy OT-4-2 (viz. Ti-4.95Al-1.54Mn).

7. POWDER METALLURGY AND RAPID SOLIDIFICATION PROCESSING

The utility of powder metallurgy stems from its ability to yield relatively low-cost shapes conforming rather closely to those desired in the finished product (so-called "near-net-shape" processing). With the advent of advanced computer-aided design and manufacturing (CAD-CAM) technology, near-net shapes can also be produced in modern forging operations; nevertheless, powder metallurgy still offers certain advantages over ingot metallurgy (IM). These advantages have to do with what might be termed enhanced "microstructural and macrostructural design flexibility." That is to say: (*a*) through the use of blended powders it is possible to adjust and optimize the metallurgical properties of various regions of the part in order to satisfy local mechanical-property requirements; and (*b*) modern methods of producing powder enable hitherto unobtainable alloy compositions and microstructures to be engineered into the finished product. Thus the starting stock for powder metallurgy may be in one of two forms: (1) mixed powder of different compositions—the so-called "blended-elemental," or BE, approach; or (2) powder of uniform composition, identical to that of the intended product—the "prealloyed powder," or PA, approach.

There are numerous methods of producing metallic powders, many of which are in no way related to rapid solidification. However, the generic techniques of powder metallurgy can be applied with advantage to powders produced by rapid solidification processing (RSP). Powder metallurgy (PM) as it applies to titanium alloys has been reviewed by Froes and Eylon.[22, p. 49] Rapid solidification processing of metals and alloys in general has been reviewed by Savage and Froes,[22, p. 60] who have presented a comprehensive listing of RSP techniques along with their estimated cooling rates (ECR).[22, pp. 68,69] Typical examples of free-flight methods are the centrifugal atomization technique with an ECR of 10^5 K·s^{-1} (Fig. 57a), free-flight melt spinning with an ECR of 10^2 to 10^4 K·s^{-1}, and the rotating-electrode method with an ECR of 10^2 K·s^{-1} (Fig. 57b). Typical examples of chill-block processes are chill-block melt spinning (ECR, 10^5 to 10^7 K·s^{-1}) and the crucible and pendant-drop melt-extraction processes (ECR, 10^5 to 10^6 K·s^{-1}) (Fig. 58a and b).

In practice, cooling rate is of course a function of, among several factors, the size of the quenched product—diameter in the case of a gas-cooled particle, and thickness in the case of a chill-block-quenched flake or ribbon. Broderick *et al.*[146] have made a detailed study of the effect of quench rate on the microstructure of Ti-6Al-4V after rapid solidification by the techniques of hammer-and-anvil splat quenching, electron-beam splat quenching, pendant-drop melt extraction, and rotating-electrode processing. Coming out of this work was a relationship between the size of the prior-β grains (L, μm) and the calculated cooling rate ($\dot{T}$, K·s^{-1}) of the form:

$$L = 3.1 \times 10^6/\dot{T}^{0.93\pm0.12} \quad (\mu\text{m}) \qquad \text{(Eq 4)}$$

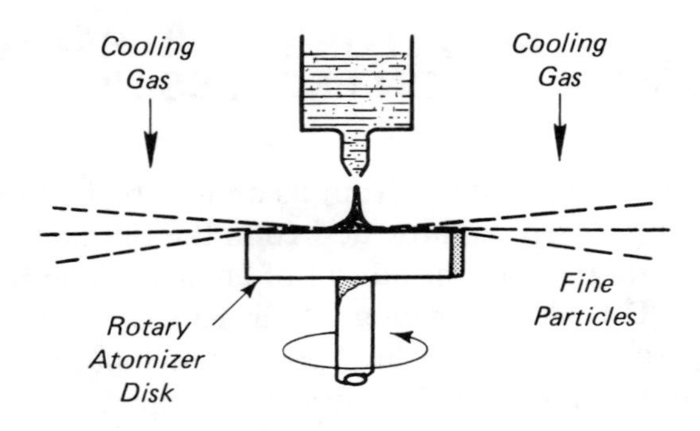

(a) Centrifugal Atomization Technique

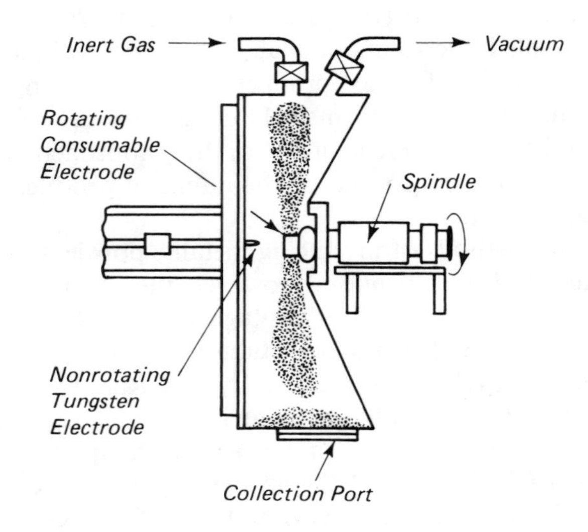

(b) Rotating Electrode Process (REP)

Fig. 57. Schematic representations of two rapid-solidification techniques[145]

For example, measurements of a PDME Ti-6Al-4V fiber (see below) yielded a grain size of 35 μm and hence a cooling rate of 2×10^5 K$\cdot$s^{-1}, in agreement with the general estimates cited above.

After some introductory comments on PM approaches, this section goes on to describe a case study of the microstructural and mechanical properties of pendant-drop melt-extracted Ti-6Al-4V powder and fiber. The purpose of that research was two-fold: (1) to develop a new approach to the production of high-purity titanium-alloy powder for conventional PM processing; and (2) to explore the potential of RSP for the production of high strength/weight titanium-alloy fibers through grain refinement and solution and precipitation strengthening. This section serves as an introduction to Section 8, which goes on to deal with the rapid solidification processing and design of alloys intended for high-temperature applications.

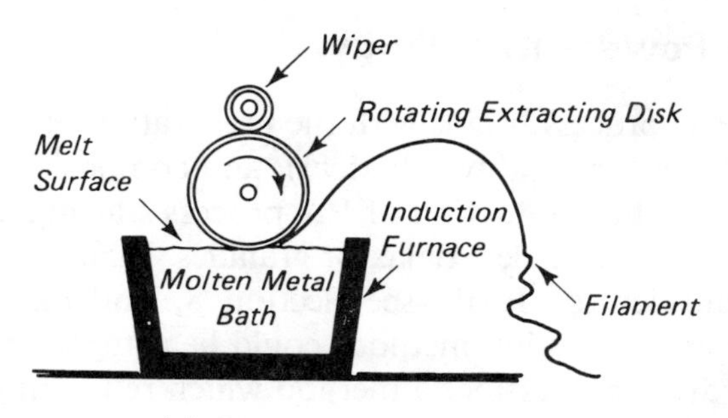

(a) Crucible Melt Extraction

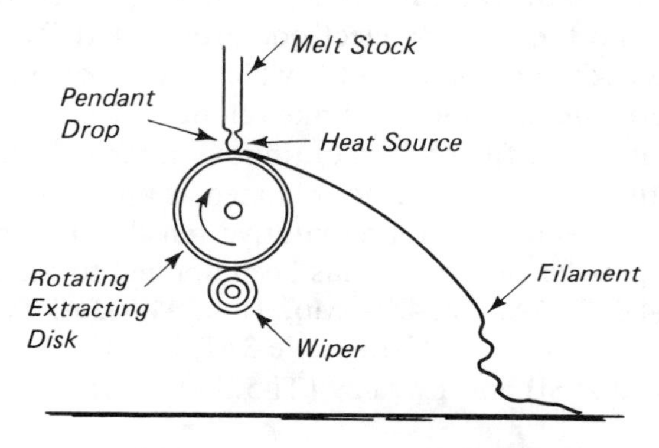

(b) Pendant Drop Melt Extraction

Fig. 58. Schematic representation of two melt-extraction techniques[145]

7.1. Blended-Elemental Powder Metallurgy

In titanium-alloy powder metallurgy (PM), the blended-elemental approach (BE) implies the mixing of fine granular unalloyed titanium with a powdered master alloy. According to Froes *et al.*,[22, p. 50] the method has been applied to alloys Ti-6Al-4V (predominantly), Ti-6Al-6V-2Sn, Ti-6Al-2Sn-4Zr-2Mo, Ti-6Al-2Sn-4Zr-6Mo, and Ti-10V-2Fe-3Al. The BE approach could also imply the use of mixed oxides or salts as prereduced starting materials.[22, p. 52]

The generalized BE approach also finds application in the PM processing of composite materials for high-temperature service—i.e., the term could be applied to the introduction of oxide particles or "unconstitutional precipitates" (e.g., refractory intermetallic compounds) as dispersion strengtheners, or of suitably coated SiC or refractory metallic whiskers for fiber reinforcement.

The BE approach lends itself to the direct production of foil or sheet, in which case the constituent powders are poured into the "pinch" space between a pair of rolls. Other methods of powder consolidation, applicable to both BE and prealloyed (PA) powder metallurgy, are *vacuum hot pressing* (VHP) and *hot isostatic pressing* (HIP), the results of both of which are to be discussed in more detail below.

7.2. Prealloyed Powder Metallurgy

The prealloyed (PA) process begins with the preparation of powder that is uniform in particle size, shape (preferably spherical), and composition. A variety of methods are available for the production of PA-process starting material whether it be in the form of spherical particles, flake, or granules. Many of the available methods fall within the purview of RSP—see Section 8, and also Froes *et al.*[22, p. 60] Briefly, PA powder preparation methods could be categorized under the headings: (1) Hydride-dehydride processing, a method which relies on the severe embrittling effect of dissolved hydrogen to produce small granules of the alloy; (2) liquid-"atomization" methods of various kinds, such as the centrifugal atomization process (Fig. 57a) and the rotating-electrode process (REP) (Fig. 57b); and (3) melt-extraction methods (Fig. 58a and b), which are capable of yielding products in a variety of useful shapes, such as flake, elongated particles (so-called "L/D powder"), and continuous fiber.[147] Continuous fiber is suitable for fiber reinforcement applications—but even if not planned as a final product, it is a convenient form for measurement of representative tensile properties. According to Froes,[22, p. 53] PA powder processing has been applied to the technical titanium-base alloys Ti-6Al-4V, Ti-6Al-2Sn-4Zr-6Mo, Ti-4.5Al-5Mo-1.5Cr ("CORONA 5"), Ti-11.5Mo-6Zr-4.5Sn ("β-III"), Ti-10V-2Fe-3Al, and the British alloys IMI 685 (Ti-6Al-5Zr-0.5Mo-0.25Si) and IMI 829 (Ti-5.5Al-3.5Sn-3Zr-1Nb-0.3Mo-0.3Si).

7.3. Pendant-Drop Melt Extraction of Continuous Ti-6Al-4V Fiber[147,148]

7.3.1. Fiber Production and Microstructures. Long, continuous lengths of Ti-6Al-4V fiber have been produced by pendant-drop melt extraction using a polished-chisel-edged extraction disk. Views of the fiber are given in Fig. 59. Optical micrographs of sections of the fiber, in the as-spun condition and after two heat treatments, are given in Fig. 60(a). For heat treatment the fibers had been wrapped in tantalum foil and enclosed in quartz tubes along with getter-packages of titanium chips; the tubes were sealed off under argon at a pressure adjusted to be atmospheric at the heat treatment temperature.

The as-spun (hence rapidly quenched) structure is martensitic, whereas that of the mill-annealed sample is, in this case, Widmanstätten $(\alpha + \beta)_W$. The recrystallized structure is equilibrium $\alpha + \beta$, with highly elongated α-phase regions. The three classes of microstructures can be more easily distinguished in the replica electron micrographs in Fig. 60(b).

7.3.2. Mechanical Properties of the Ti-6Al-4V Fiber

(a) *Elastic Modulus*: The elastic modulus was determined by combining the results of acoustic-wave velocity and density (aqueous buoyancy) measurements.

(b) *Tensile Strength*: Tensile tests were carried out with the sample held between grips specially designed for accommodating wire specimens. The calculation of modulus and strength from the results of tensile measurements requires a knowl-

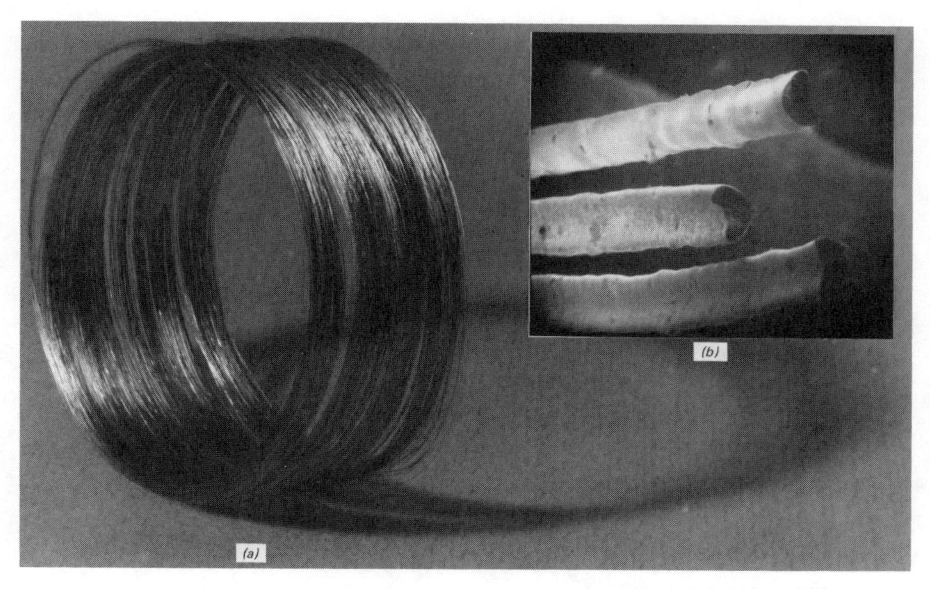

(a) One of several coils (about 14 cm in diameter) produced during the preliminary phase of a so-called "L/D-powder" program. (b) Scanning electron micrograph of the same fiber (width about 30 μm) showing the "wheel surface" and the "free surface."

Fig. 59. Pendant-drop melt-extracted Ti-6Al-4V fiber[147]

edge of the specimen cross-sectional area. This was determined in two different ways: (*a*) from the measured mass-per-unit-length and the density; and (*b*) by comparing the tensile Young's modulus with the acoustically measured value. The results of the mechanical-property measurements are summarized in Table 19.

7.4. Pendant-Drop Melt Extraction and Consolidation of Ti-6Al-4V Powder[147-150]

7.4.1. Powder Production. By introducing notches into the rim of the extraction disk it was possible to interrupt the casting process in such a way as to produce a particulate product (Fig. 61a). By using first widely spaced and then more closely spaced notches, two classes of powder were produced for experimentation. These powders, which were referred to as L/D-1 and L/D-2, are depicted in Fig. 61(b) and (c), respectively. Their properties are defined in the caption of Fig. 61.

7.4.2. Powder Consolidation by Vacuum Hot Pressing (VHP)

(a) *Processing*: Hot pressing of the (elongated) L/D-1 powder was carried out in vacuum under an axial load of 0.56 kg·mm^{-2} (800 psi) for 1 h at 955 °C. The resulting flat disk (after light machining to 5.1 cm diam by 1.1 cm thick), together with a sample of the L/D-1 powder from which it was produced, are depicted in Fig. 62(a).

(b) *Microstructures*: Replica electron micrographs of the vacuum hot pressed

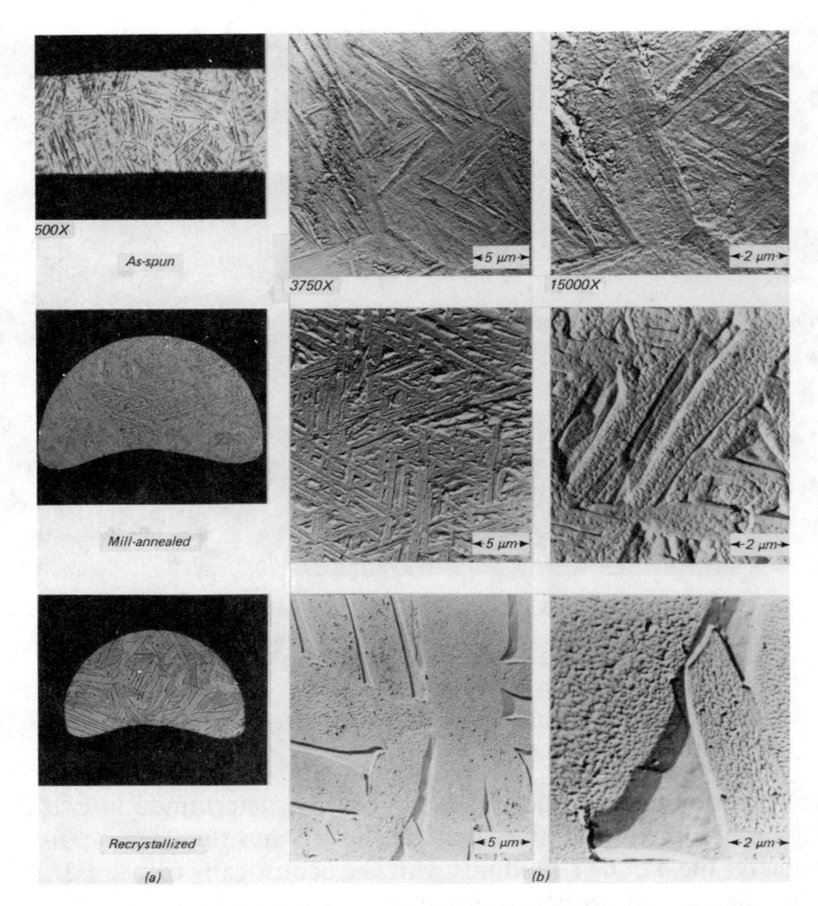

(Top) As-spun (martensitic α'). (Center) Mill-annealed, 2 h at 720 °C, air cooled (Widmanstätten $\alpha + \beta$). (Bottom) Recrystallized, 4 h at 930 °C plus furnace cooled at 55 °C/h to 760 °C plus furnace cooled at 360 °C/h to 480 °C plus air cooled (equilibrium $\alpha + \beta$).

Fig. 60. Optical micrographs (a), and replica electron micrographs (b), of samples of pendant-drop melt-extracted continuous fibers of Ti-6Al-4V in three conditions[147]

(VHP) powder are presented in Fig. 63(a). By comparing this figure with Fig. 60(b) it can be seen that the VHP material is in the "recrystallized" condition. This is to be expected, since the first stage of recrystallization heat treatment is 4 h at 930 °C.

(c) *Mechanical Properties*: Samples of the VHP material were prepared for compression testing and tensile testing. The average tensile strength of the VHP material in the as-VHP condition was 10.4×10^8 N·m^{-2} (151.0 ksi), which may be compared with the 10.7×10^8 N·m^{-2} (155.4 ksi) of the wrought starting material after mill annealing, and the 9.9×10^8 N·m^{-2} (144.3 ksi) of the mill-annealed melt-extracted fiber.

Table 19. Summary of tensile properties of Ti-6Al-4V fibers

Condition	Microstructure	Modulus(a), E, 10^3 ksi(b)	Average tensile strength, ksi(b), based on: A_1(c)	A_2(d)
As-spun	Martensitic α'	13.44	89.0	125.0
Mill-annealed(e)	Reverted α'	15.73	96.9	144.3
Recrystallized(f)	$\alpha + \beta$	15.59	87.8	129.9

(a) From sound-velocity measurement. (b) To convert ksi to MPa, multiply by 6.8948. (c) $A_1 = W/\rho L$. (d) A_2 back calculated from acoustic elastic modulus. (e) Mill anneal (MA): 2 h at 1350 °F, air cool. (f) Recrystallization anneal: 4 h at 1700 °F, furnace cool to 1400 °F (100 °F per hour—no faster), furnace cool to 900 °F (650 °F per hour—no slower), air cool.

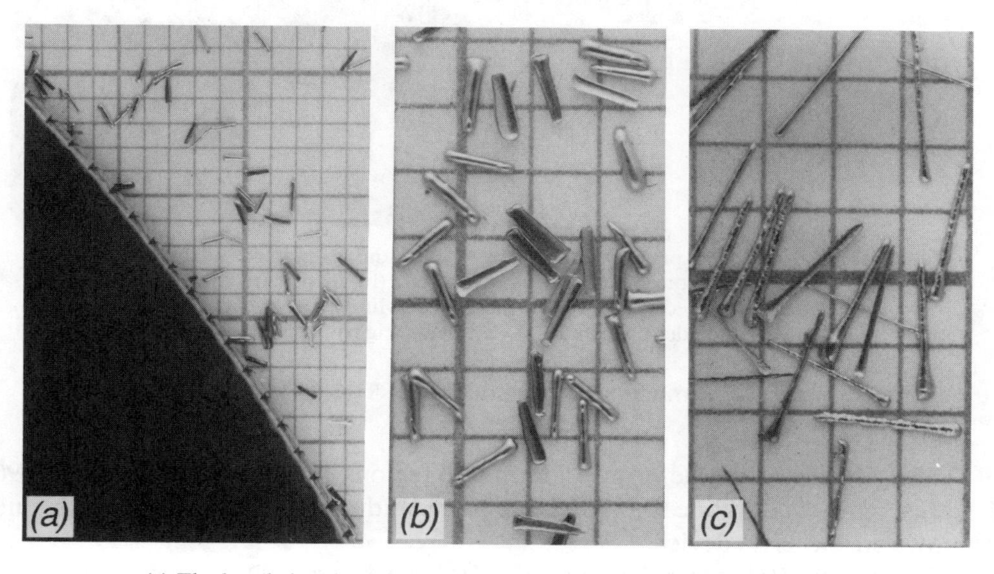

(a) The beveled and notched extraction disk used for producing L/D-2 powder. (b) L/D-2 powder, about 0.08 by 0.2 by 1.3 mm. (c) L/D-1 powder, about 0.05 mm diam by 2.5 mm. All objects photographed against 1-mm-square paper.

Fig. 61. Melt extraction of Ti-6Al-4V in the form of so-called "L/D powder"(147)

7.4.3. Powder Consolidation by Hot Isostatic Pressing (HIP)

(a) *Processing*: Hot isostatic pressing (HIP) of cold pressed pellets of L/D-1 and L/D-2 powders was carried out under conditions of 1 h at 950 °C and 6.9×10^7 $N \cdot m^{-2}$ (1 h at 1750 °F and 10 ksi). The HIP-compacted samples were machined in preparation for compression testing (see, for example, Fig. 62b).

(b) *Microstructures*: Optical micrographs of as-HIP samples prepared from L/D-1 and L/D-2 powders showed the structure to be predominantly α phase—partly equiaxed and partly elongated. The replica electron micrographs in Fig.

(a) A vacuum hot pressed disk (1.1 by 5.1 cm) and an example of the L/D-1 powder from which it was processed. (b) Two pairs of hot isostatically pressed (HIPed) and machined cylinders and examples of the L/D-1 powder (left) and L/D-2 powder (right) from which they were processed.

Fig. 62. Samples of consolidated Ti-6Al-4V L/D powder[147]

63(a) and (b) show that the prior-β interphase regions (β-transus temperature, 996 °C) must have undergone transformation to Widmanstätten $(\alpha + \beta)_W$ during cooldown in the HIP device.

(c) *Mechanical Properties*: Samples of HIP L/D-1 and L/D-2 powders were prepared for compression testing at room temperature. The average compression yield stress (at 0.2% offset) of the as-HIP L/D-1 material was 10.2×10^8 N·m^{-2} (148.2 ksi), and that of the as-HIP L/D-2 material was 10.0×10^8 N·m^{-2} (145.7 ksi). These values may be compared with the 9.9×10^8 N·m^{-2} (144.0 ksi) of the wrought starting material after mill annealing.

7.5. Pendant-Drop Melt Extraction of Precipitation-Strengthened Titanium-Alloy Fiber[147,151,152]

7.5.1. Introduction. Rapid quenching from the melt may lead to extreme grain-size refinement (if not complete amorphousness in suitable alloy formulations). In a pure substance the fine grain size is unstable against growth. But in two-phase material, some degree of elevated-temperature stability is obtainable, especially if

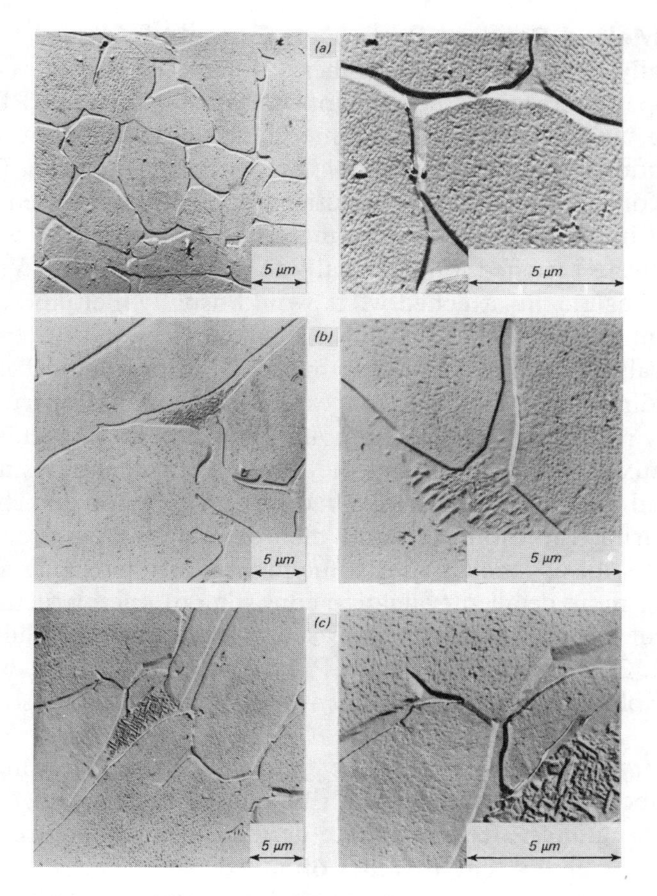

(a) Vacuum hot pressed L/D-1 powder. (b) Hot isostatically pressed L/D-1 powder. (c) Hot isostatically pressed L/D-2 powder.

Fig. 63. Replica electron micrographs of consolidated Ti-6Al-4V L/D powder(147)

the equilibrium phases differ substantially in composition. Reversion to equilibrium is in such cases inhibited by the necessity for a massive flux of atoms over a considerable distance. Although most readily obtainable in a two-phase eutectic or eutectoidal system (in which spinodal decomposition is very effective), a fine (1 to 5 μm), stable, equiaxed grain structure can sometimes be stabilized by grain-boundary-pinning precipitate particles. A stable, refined grain structure is conducive to superplasticity.

An increase in the modulus of elasticity has been observed in some rapidly quenched materials as a result of the presence of a very stable dispersed phase. Increases in strength have also been noted as consequences of: (*a*) fine grain size, (*b*) a fine dispersion of second-phase particles, and (*c*) the generally high concentration of solution strengthener present as a consequence of the extended solubility range offered by the rapid-quench process.

7.5.2. *In-Situ*-Melted Pendant-Drop Melt-Extracted Alloys.[147] For swiftly and economically obtaining large numbers of samples for microstructural and mechanical-property screening, the pendant-drop melt-extraction (PDME) method, operated in the "*in situ alloy-melting*" mode (reminiscent of the BE method of powder preparation), is a useful preparation technique. The *in-situ* PDME method has been used for the preparation of multicomponent alloys based on Ti-6Al-4V. In one series of investigations, alloying materials in the form of wires or rods of various metals were fastened longitudinally to a rod of Ti-6Al-4V. Once melting of the rod had begun, the attached wires would usually melt and alloy smoothly into the pendant drop.

Alloys were also prepared by adding materials in powdered form to the host alloy. Longitudinal slots (about 1 mm wide and 1 mm deep) were cut into a Ti-6Al-4V alloy rod. Intermetallic-compound powders (in particular TiB_2, $TiSi_2$, or TiC) were mixed with alcohol and the resultant thick slurry smeared into the slots and allowed to dry. The dried cake had enough cohesion and strength to bond to the slots during the mounting and melt-extraction operations and alloyed smoothly into the basic stock. After some preliminary testing, several materials were selected for more detailed evaluation consisting of microstructural study, density measurement, acoustic and tensile measurement of elastic modulus, and tensile-strength testing. Some results for *in-situ* PDME Ti-6Al-4V/TiB_2, which of all the compositions explored yielded the highest values of modulus and strength, are presented below.

(a) *Metallography*: Two samples were chosen for metallographic evaluation: a very fine wire about 0.1 mm wide, and a ribbonlike fiber about 0.7 mm wide (Fig. 64a and b). The micron-size precipitate-defined grain structure is presumably responsible for the high strength values obtained.

(b) *Elastic Modulus*: Values of elastic modulus were determined by combining the results of acoustic-wave velocity and density (aqueous buoyancy) measurements (Table 20).

(c) *Tensile Strengths*: Tensile tests were carried out with the sample held between grips specially designed for accommodating wire specimens. The calculation of modulus and strength values from the results of tensile measurements requires a knowledge of the area of the specimen cross section, which was very irregular in some cases. Instead of attempting to use directly measured filament cross sections, it was decided to assign an *effective cross-sectional area*—one which normalized the tensile-test modulus value to that obtained acoustically. The results of the tensile

Table 20. Mechanical properties of Ti-6Al-4V/TiB_2

Average sound velocity, $km \cdot s^{-1}$	5.04
Density, $g \cdot cm^{-3}$	4.50
Acoustic elastic modulus, 10^3 ksi(a)	16.22
Ultimate tensile strength, ksi(a)	255

(a) To convert ksi to MPa, multiply by 6.8948.

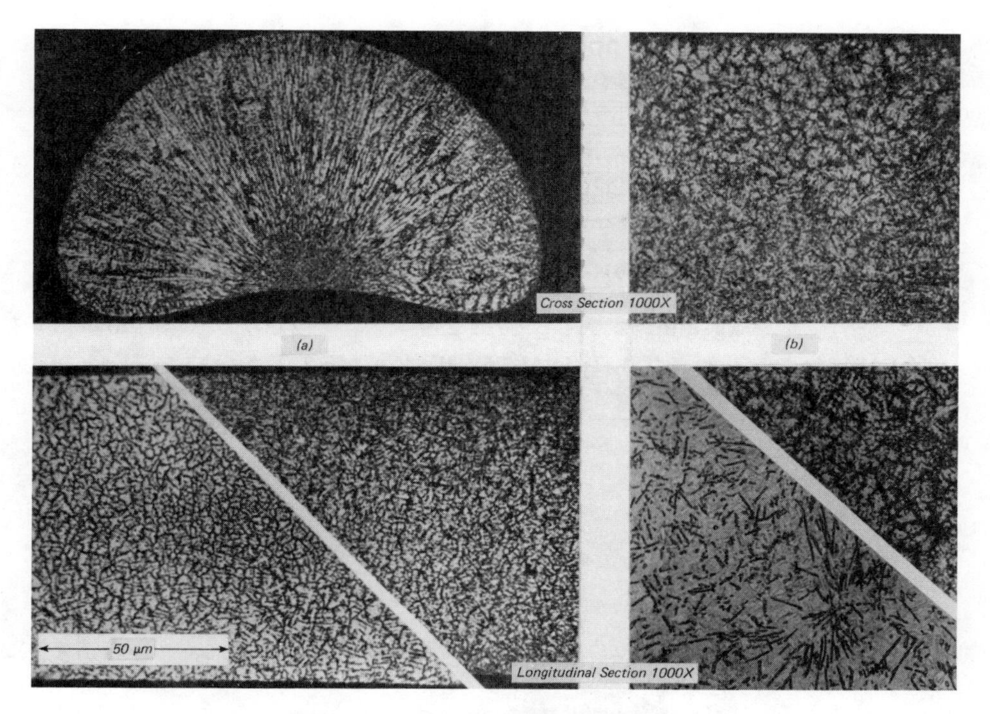

Fig. 64. Optical micrographs of transverse and longitudinal sections of two samples of pendant-drop melt-extracted fibers prepared by *in-situ* alloying of TiB_2 into Ti-6Al-4V[147]

tests are given in Table 20. A comparison of the results presented there with those of Table 19 for the PDME control alloy emphasizes the considerable enhancement in strength that can be obtained by combining the addition of boron to the melt with the use of rapid solidification to achieve solution strengthening, precipitation strengthening, and a fine grain structure.

7.5.3. Advanced Multicomponent Prealloyed Pendant-Drop Melt-Extracted Alloys[151,152]

(a) *Preparation and Basic Properties*: Following the screening study outlined above, several five-component alloys, some based on Ti-6Al-4V, were arc melted and cast into rod form suitable for pendant-drop melt extraction. Three of the resulting materials were selected for detailed evaluation. These alloys, and their process conditions and properties, are listed in Table 21.

(b) *Elastic Modulus*: Values of elastic modulus (Table 21) were determined by combining the results of acoustic-wave velocity and density (aqueous buoyancy) measurements.

(c) *Tensile Strength*: In preliminary tests in which untreated as-spun fibers were mounted in conventional wire-holding grips, fracture occurred at the grips. This difficulty was greatly reduced after the design and construction of a holder spe-

Table 21. Properties of polycrystalline titanium-base alloys prepared by pendant-drop melt extraction

Alloy code(a)	ME-1	ME-2	ME-3
Extraction disk rim speed, $m \cdot s^{-1}$	3 to 4	4 to 13	4 to 9
Vickers hardness number, $kg \cdot mm^{-2}$	588 ± 28	542 ± 30	673 ± 43
Density, $g \cdot cm^{-3}$	4.694	4.686	4.326
Sound velocity, $km \cdot s^{-1}$	4.067	3.768	5.358
Acoustic modulus, 10^3 ksi(b)	11.04	9.46	17.66
Tensile strength: ksi(b)	340	346	242
$kg \cdot mm^{-2}$	239	243	170

(a) Alloy compositions: ME-1, Ti-5.4Al-3.6V-8Fe-3Cu; ME-2, Ti-5.4Al-3.6V-6Fe-5Cu; ME-3, Ti-1.5Al-1V-3Be-2B. (b) To convert ksi to MPa, multiply by 6.8948.

Fig. 65. Small sample of melt-spun ribbon mounted in preparation for tensile testing and photographed against 1-mm-square paper[147]

cially suited for small wire and ribbon samples (Fig. 65). Filamentary wire samples were epoxied into the shanks of "eye-ended" solder lugs, and ribbon samples into the flattened shanks of such lugs. To facilitate handling right up to the time the tensile test actually began, the lugs themselves were attached to a C-shape cardboard yoke. The results of the tensile measurements are summarized in Table 21. Finally, by way of conclusion, Table 22 shows that the specific strengths of the three alloys under discussion compare favorably with data of Tanner and Ray[153] for some commercial metallic glass fibers.

Table 22. Specific strength (strength/density) of polycrystalline fibers compared with those of some commercial metallic glass fibers

Material	Composition	Hardness, kg·mm^{-2}	Strength, kg·mm^{-2}	Density, g·cm^{-3}	Specific strength, 10^5 cm
ME-1	Ti-5.4Al-3.6V-8Fe-3Cu	588	239	4.694	50.9
ME-2	Ti-5.4Al-3.6V-6Fe-5Cu	542	243	4.686	51.9
ME-3	Ti-1.5Al-1V-3Be-2B	673	170	4.326	39.3
Metglas 2615	$Fe_{80}P_{16}C_3B_1$	835	249	7.30	34.1
Metglas 2826A	$Ni_{36}Fe_{32}Cr_{14}P_{12}B_6$	880	278	7.46	36.9
Metglas 2605	$Fe_{80}B_{20}$	1100	370	7.40	50.0
Metglas 2204	$Ti_{50}Be_{40}Zr_{10}$	740	231(a)	4.13	56.0

(a) Calculated from strength = hardness/3.2.

8. RAPID SOLIDIFICATION PROCESSING OF PRECIPITATE- AND DISPERSION-STRENGTHENED TITANIUM ALLOYS

The applicability of conventional titanium alloys in aerospace engineering is limited by their relatively low operating temperature range, which to date is no higher than some 600 °C.[154] Solid-solution strengthening, only partially effective even in the intermediate temperature range, cannot be looked to as a mechanism for high-temperature strengthening. At moderate-to-high temperatures, dislocations are thermally activated around point defects; also as a consequence of thermal activation, the alloy's microstructure or substructure, which is chiefly responsible for strengthening in the intermediate temperature range, is unstable at high temperatures. Dispersion strengthening offers a substitute for solution strengthening at high temperatures, and at the same time tends to stabilize the substructure. Through the use of dispersion strengthening, the operating temperature range of titanium alloys can be substantially increased. Moreover, although the improvement of high-temperature properties is usually the primary goal of dispersion strengthening, the presence of dispersoids also enhances the flow stress throughout the entire temperature range; additional benefits claimed for dispersion strengthening are increases in creep resistance and stress-rupture life.[155,156] In selecting a dispersoid it should be recognized that although its composition, as such, is not important, it should be chemically and physically stable in the matrix at elevated temperatures—i.e., insoluble, nonreactive, and resistant to coarsening. Dispersoids should be incoherent with the matrix crystal and resistant to deformation. Dispersoids should be closely spaced (e.g., number density on a per-unit-area basis, 2.6×10^6 mm^{-2}[157]) and small in size (typical diameters, 0.05 to 0.5 μm[155,158]); as such they are effective barriers to dislocation motion at both ambient and high temperatures. Dispersoids also tend to "pin" grain and subgrain boundaries, thereby stabilizing the alloy's substructure against change and inhibiting recrystallization during high-temperature exposure.[159] The dispersion strengthening discussed in all except the last subsection of this presentation is what might be referred to as "*in situ*," in that the strengthening ingredients are placed in the starting material prior to its first melting.

8.1. Dispersion Strengthening of Titanium Alloys

8.1.1. Dispersion Strengthening in Ingot Metallurgy. The application of dispersion strengthening to conventionally processed (ingot metallurgy, IM) titanium alloys has been successful up to a point. Boron added to titanium alloys segregates to the interdendritic regions during ingot solidification and contributes substantially to grain refinement and stabilization.[157] Small amounts of yttrium and erbium added to IM titanium alloys have been shown to yield a uniform distribution of fine incoherent precipitates of the corresponding sesquioxides, Y_2O_3 and Er_2O_3. But attempts to obtain a suitably high density of fine dispersoids by increasing the levels of yttrium and erbium in the starting ingot have always been unsuccessful. The invariable result is a small number density of large-diameter (>1 μm) particles.[155,158,160] The coarsening which takes place during the relatively slow cooling of the arc-melted ingots renders the precipitate particles ineffective for dispersion strengthening. They are in fact detrimental, degrading both the fracture toughness and the fatigue properties of the alloy. Eutectoid-forming elements such as silicon and the transition elements iron, nickel, and copper might be expected to be candidates for dispersion strengthening through fine intermetallic-compound formation. But these elements tend to segregate to grain boundaries during conventional IM processing.[160,161] The production of dispersoids by solid-state precipitation (precipitation hardening) has resulted in coarse, thermally unstable particles subject to overaging during processing. Thus, because of the difficulties that have been encountered in producing a suitable distribution of sufficiently small stable precipitates, *in-situ* dispersion strengthening did not play a significant role in titanium-alloy metallurgy, particularly in the U.S., until the advent of rapid solidification processing (RSP). It has been discovered that the coarsened precipitates (in the case of the rare-earth oxides, see later) and grain-boundary precipitates (in the case of eutectoid formers) of IM products are completely absent from the products of RSP.

But if the coarsening and grain-boundary segregation of IM is a result of the slow cooling of the ingot ("aging during processing"), how can the products of RSP remain usefully stable at high temperatures, even if produced as fine dispersions in the first place? The answer has to do with the high diffusivities of many elements in β titanium and the need for rapidly solidifying the alloy into the α phase and making sure that its subsequent processing and service conditions never take it above its α transus.

8.1.2. Dispersion Strengthening in Rapid Solidification Processing. The problems encountered in attempts to provide *in-situ* dispersion strengthening or precipitation strengthening in conventional IM titanium alloys can be avoided by RSP.[160] The addition of dispersoid-forming elements to the starting material prior to RSP can lead to microstructural refinement and large number densities of fine dispersoids (e.g., rare-earth oxides) or precipitates. As a result, significant improvements in both room-temperature and elevated-temperature strengths, and in creep properties and stress-rupture lifetimes, have been noted. Table 23 is a comparison of some room-temperature mechanical properties of RSP and IM-processed alloys. Elements which have been added to titanium alloys prior to RSP include: the interstitial elements boron and carbon (which under RSP conditions have contributed significant increases in modulus and yield strength[160]); the eutectoid formers silicon, chromium, manganese, iron, cobalt, nickel, and tungsten; and the rare-earth elements (including yttrium) selected from the "La group" of the periodic table. With regard to the latter, although many rare earths (RE) have low room-temperature solubilities in titanium,[155,163] and have the potential for precipitating in metallic form during solidification, in practice they scavenge dissolved oxygen from the alloy and oxidize to RE_2O_3. The rare earths thus play a useful secondary role in RSP titanium powder metallurgy. Although some dissolved oxygen is desirable for solution strengthening at ordinary temperatures, too much oxygen (easily acquired during PM processing) causes excessive hardening. The inclusion of RE elements in the alloy formulation can control to some extent the final level of oxygen in solid solution. Rapid solidification processing of Ti-RE alloys has resulted in precipitation of fine oxide particles (<0.05 μm in diameter[155]), which are very suitable for dispersion strengthening. In RSP alloys containing boron, TiB is the dispersoid specie; the TiB coarsens rapidly at the grain boundaries, depleting adjacent regions of boron and averting grain-boundary

Table 23. Comparison of mechanical properties of rapid solidification processed (RSP) and ingot-metallurgical (IM) Ti-Al-Er alloys[159]

Alloy	Heat treatment(a)	Yield stress (MPa)		Ultimate tensile stress (MPa)		Total elongation (%)	
		RSP	IM	RSP	IM	RSP	IM
Ti-5Al-2Er	ST	670	469	735	536	27.0	...
Ti-7.5Al-2Er	ST	850	680	920	756	...	7.0
Ti-9Al-2Er	ST	880	750	928	790	11.0	0.1
Ti-5Al-2Er	STA (625 °C)	700	510	763	564	13.8	10.0
Ti-7.5Al-2Er	STA (625 °C)	952	815	973	843	7.7	6.0
Ti-9Al-2Er	STA (625 °C)	931	802	952	824	1.6	0.2
Ti-5Al-2Er	STA (550 °C)	714	515	780	590	54.0	18.0
Ti-7.5Al-2Er	STA (550 °C)	973	830	990	865	12.0	9.0
Ti-9Al-2Er	STA (550 °C)	...	810	...	835	...	0.3

(a) ST = solution treat at 860 °C for 3 h and water quench; STA = ST plus aging at 625 °C for 25 h or 550 °C for 500 h.

embrittlement. Other types of precipitates that have been investigated as potential strengtheners include, for example, Ti_5Si_3, TiC, Al_4La, Al_3La, Ti_2Ni, CeS, and $Ce_2(SO_2)$.[157,161,164]

Although conventional α-Ti alloys such as Ti-5Al-2.5Sn can be improved by RSP-induced dispersion strengthening, it has been claimed that the greatest advantage can be taken of RSP when it is applied to specially formulated alloys.[160] This approach has given rise to the many new alloy compositions discussed in the following sections. Rapid solidification processing is also accompanied by a secondary benefit: When the product is in the form of powder or particles of various kinds, the subsequent consolidation and fabrication steps are accompanied by all of the advantages generally associated with powder metallurgy, with the proviso that any hot consolidation operations should take place at low temperatures and high pressures, rather than conversely, as in conventional PM.

8.2. Systems for Dispersion Strengthening by Rapid Solidification Processing

8.2.1. Review of Recent Advances. Conventional alloys such as Ti-8Al-1Mo-1V, Ti-6Al-4V, and Ti-6Al-2Sn-4Zr-2Mo, after rotating-electrode processing (REP) and plasma rotating-electrode processing (PREP) (both sometimes referred to as "conventional" powder-production methods, but see Section 7), have yielded elongated microstructures after consolidation by hot isostatic pressing (HIP) and vacuum hot pressing (VHP). On the other hand, RSP of the same alloys (into particulate form) has yielded particles of high dislocation density which recrystallize readily into material with a fine equiaxed grain structure, a property conducive to improved room-temperature mechanical properties and to high-temperature superplasticity.[161] But as Sastry has pointed out, advantage should be taken of RSP to produce completely new materials especially designed for *in-situ* dispersion or precipitation strengthening. With this in mind, alloy systems yielding oxide-particle dispersions and intermetallic-compound precipitates, and systems based on the titanium aluminides, have been prepared by RSP and metallurgically examined.[160,161]

The classes of systems which have been examined are listed in Table 24. Alloying elements selected include: (*a*) the interstitial elements boron and carbon (which yield intermetallic-compound precipitates); (*b*) the group IIIB elements scandium and yttrium, and the lanthanides lanthanum, cerium, neodymium, gadolinium, dysprosium, and erbium (which scavenge oxygen from the matrix to form sesquioxide dispersoids); and (*c*) β-eutectoid-forming elements such as silicon, iron, nickel, and copper (which yield intermetallic-compound precipitates or fine lamellar microstructures[165]).

Bases for dispersion strengthening with interstitial elements and RE oxides have been: previously unalloyed titanium,[156,160] Ti-Al alloys,[158,159] and commercial titanium alloys such as Ti-624,[157,166] and Ti-6242 and Ti-633.[166]* Interstitial-

*The compositions of these alloys are: Ti-624, Ti-6Al-2Sn-4Zr; Ti-6242, Ti-6Al-2Sn-4Zr-2Mo; and Ti-633, Ti-6Al-3Sn-3Zr.

Table 24. Microstructural and property improvements of rapidly solidified titanium alloys[161]

Alloy type	Alloy system	Problems with ingot metallurgy	Rapid-solidification microstructural modifications
Dispersion-strengthened alloys	Ti-RE(a)	Coarse particles	Extended solid solutions; fine incoherent dispersoids
Compound formers	Ti-B Ti-C	Limited solid solubility, coarse dispersoids	Grain refinement; titanium boride and carbide dispersoids
Eutectoid formers	Ti-Ni Ti-Si Ti-Fe	Segregation, coarse grains, and precipitates	Controlled eutectoid-decomposition products
Combined precipitates and dispersoids	Ti-Al-RE Ti-Al-Ni Ti-Al-B, C	Coarse dispersoids	Coherent, ordered precipitates and incoherent dispersoids
Conventional titanium alloys	Ti-6Al-4V Ti-8Al-1Mo-1V Ti-6Al-2Sn-4Zr-2Mo	Coarse, elongated grains	Fine martensite structure in as-rapidly-solidified alloys; fine equiaxed $\alpha+\beta$ grains upon annealing in $\alpha+\beta$ field
Amorphous alloys	Ti-M-B(b) or Ti-M-Si	Cannot be made	Amorphous and micro-crystalline structures
Intermetallic compounds	Ti_3Al TiAl	Coarse grains	Grain refinement, incoherent fine dispersions, possible decrease in long-range order

(a) RE = Er, Y, Gd, Nd, Sc, La, Dy. (b) M = Mn, Nb, V, Cr.

element strengthening, particularly with boron, has been applied to unalloyed titanium[160,167] as well as to more complex systems such as Ti-8Al-1.5Er[161] and Ti-6Zr-6Al-1Er.[166] Eutectoid-element strengthening studies have so far been confined principally (but not exclusively) to binary systems such as Ti-Fe,[168] Ti-Co,[169] Ti-Ni,[161,170] and Ti-Cu.[171]

In-situ dispersion and precipitation strengthening of titanium alloys shows great promise for high-temperature applications. Recent work has demonstrated that RSP of RE-containing alloys is capable of yielding ultrafine-grain materials containing dispersoids as fine as 0.02 to 0.05 μm in diameter at number densities (on a per-unit-area basis) as high as 2.6×10^6 mm^{-2}. Some dispersoid species have been found to be stable at temperatures as high as 1000 °C.[157]

8.2.2. Rapid Solidification Processing of Titanium-Base Alloys. As indicated in Table 25, unalloyed titanium with additions of (*a*) interstitial elements, (*b*)

Table 25. Rapid solidification processed dispersion- and precipitation-strengthened previously unalloyed Ti(a)

System	Literature, Ref No.	System	Literature, Ref No.
(*a*) Interstitial-element additions		**(*c*) "Eutectoid-forming" additions**	
Ti-B	164,167,172	Ti-0.6Si	168
Ti-0.5B	173,161,160	Ti-0.9Si	168
Ti-1.0B	159	Ti-2Si	168
Ti-1.0C	160	Ti-Si_6	176
(Ti-2Zr)-1C	160	(Ti-17.9Zr)-3.3Si	177
(*b*) Rare-earth and related-element additions		(Ti-18Zr)-4.4Si	164
		(Ti-Zr_{10})-$Si_{0.8}$	176
Ti-Y	160,167,172	Ti-3Cr	168
Ti-1.0Y	155	Ti-15Cr	168
Ti-1.5Y	155	Ti-30Cr	168
Ti-La	172	Ti-$Mn_{2.5}$	178
Ti-2.0La	155	Ti-Mn_5	178
Ti-3La	167	Ti-Mn_{10}	178
Ti-Ce	167	Ti-$Fe_{2.5}$	178
Ti-1.0Ce	155	Ti-$Fe_{2.8}$	178
Ti-1.5Nd	173,155,156	Ti-$Fe_{3.0}$	178
Ti-2.0Nd	159	Ti-Fe_5	178
Ti-3.0Nd	161,155,156	Ti-3Fe	178
Ti-1.5Gd	155	Ti-16Fe	178
Ti-2.0Dy	155	Ti-22Fe	178
Ti-Er	160	Ti-9Co	169
Ti-$Er_{0.4}$	174	Ti-3Ni	161,179
Ti-0.5Er	155,174	Ti-5.5Ni	169
Ti-$Er_{0.7}$	175	Ti-7Ni	161,179
Ti-1.0Er	155,156	Ti-3W	168
Ti-2.0Er	173,161,155,156,159	Ti-28W	168
		Ti-36W	180
		Ti-40W	168,181
		Ti-7Cu	171

(a) In this and subsequent tables, numerical prefixes indicate composition in weight percent; numerical subscripts indicate atomic percent.

RE elements, and (*c*) eutectoid formers have been used as bases for rapid solidification processing.

Although conventional processing yields coarse boride or carbide precipitates, RSP of Ti-1.0B and Ti-1.0C yielded a large number density of fine dispersoids.[161] Of all the Ti-RE systems investigated,[155] Ti-Er and Ti-Nd showed particularly promising results[156]: Ti-Er yielded closely spaced, thermally stable, incoherent dispersoids less than 0.01 μm in diameter; and Ti-Nd yielded two classes of dispersoid (a "bimodal distribution")—very fine particles less than 0.01 μm in diameter, and coarse particles within the size range 0.1 to 1.0 μm—together with neodymium in solid solution. Sastry regarded Ti-Nd as being strengthened by a combination of dispersion and solution strengthening, and Ti-Er as a purely dispersion-strengthened system. Alloys of titanium with the eutectoid formers are notable for the variety of their microstructures. Depending on solute concentration—i.e., whether hypoeutectoid (solute lean) or eutectoid—it is possible to obtain either a fine-grain material with uniform precipitates (Fig. 66) or a lamellar microstructure.

8.2.3. Rapid Solidification Processing of Ti-Al-Base Alloys

(a) *Binary Ti-Al Alloy Bases*: Table 26 lists many of the binary Ti-Al-base alloys that have been subjected to dispersion or precipitation strengthening by rapid solidification processing. With Ti-Al-B alloys, RSP yielded high-aspect-ratio filamentary dispersions which coarsened during annealing to needle-shape precipitates.[159] This high-aspect-ratio second-phase precipitate, in association with the fine grain size, resulted in significant improvements in modulus and strength.[160] TiB needle formation has also been noted in the heat treatment of Ti-6Zr-6Al-1Er-0.08B.[166] As was the case with the binary alloys, fine incoherent dispersoids associated with fine grain sizes (1 to 5 μm) have been obtained in RSP of RE-containing ternaries.[159] In Ti-Al alloys with eutectoid-forming additions such as silicon and nickel, rapid solidification followed by carefully controlled heat treatment can lead to fine-scale homogeneous microstructures[160]; consolidation temperatures must be kept as low as possible to prevent coarsening. The range of ternary materials in this category which have been studied also includes the so-called "super-α" alloys—ones in which the aluminum content is sufficiently high for some α_2-phase precipitation to take place.[159] Although the presence of α_2 precipitation severely embrittles the IM binary alloy whenever it occurs, the addition of RE solutes in

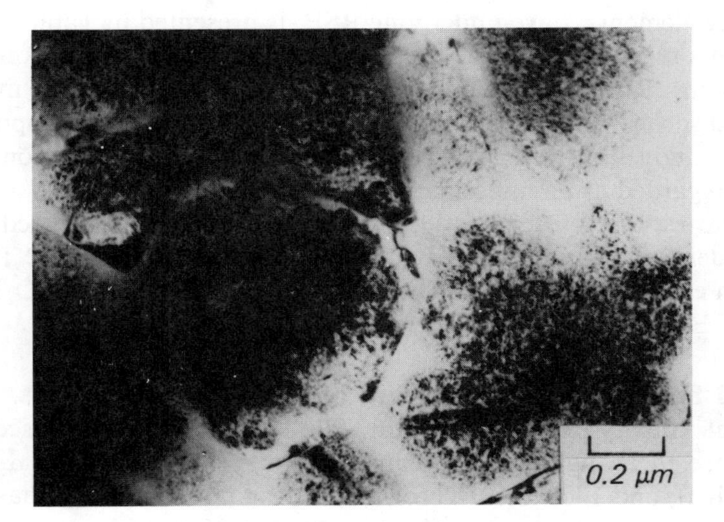

Fig. 66. Transmission electron micrograph of a sample of Ti-Fe(5 at. %) in the as-quenched condition after quenching from the melt by the hammer-and-anvil technique[182]

Table 26. Rapid solidification processed dispersion- and precipitation-strengthened alloys based on Ti-Al

System	Literature, Ref No.	System	Literature, Ref No.
Ti-8Al-2Y	159	Ti-8Al-1B	159
Ti-8Al-4Y	183	Ti-8Al-1.5Er-0.25B	161
Ti-8.5Al-0.5Y	156	Ti-6Zr-6Al-1Er-0.08B	166
Ti-5Al-3La	167	Ti-7.9Zr-3.5Al-1.4B	177
Ti-5Al-4.5La	184	Ti-5Al-2Si	172
Ti-5Sn-3La	167	Ti-8Al-2.0Si	159
Ti-9.5Sn-3La	167	Ti-8.5Al-0.2Si	159
Ti-9.5Sn-5.3La	167	Ti-8.5Al-0.5Si	159
Ti-8Al-2Nd	159	Ti-8.5Al-1.0Si	159
Ti-5Al-2Er	159	Ti-7.7Zr-3.4Al-3.6Si	164
Ti-5Al-5.4Er	158	Ti-6Al-3Ni	161
Ti-7.5Al-2Er	159		
Ti-8Al-2Er	159		
Ti-9Al-2Er	159		
Ti-Al_{10}-$Er_{0.4}$	174		
Ti-Al_{15}-$Er_{0.4}$	174		
Ti-Al_{24}-$Er_{0.4}$	174		

association with RSP is responsible for refining the grain structure and enhancing the post-creep ductility.[163,165] The presence of a finely dispersed α_2-phase precipitate in the RSP material was claimed to improve the high-temperature strengths.

(b) *Multicomponent and Commercial Alloy Bases*: A representative selection of the numerous multicomponent and commercial alloys that, with the addition of strengthening elements, have undergone RSP, is presented in Table 27. Strengthening elements represented in the list are: interstitial elements (boron), eutectoid formers (silicon), RE elements (lanthanum, cerium, and erbium), the metalloid germanium, and sulfur. The presence of oxygen is understood. In an important recent study, cerium and sulfur were added to Ti-6Al-2Sn-4Zr under RSP conditions. Sulfur is often regarded as an undesirable "tramp impurity" which, by segregation to grain boundaries, tends to embrittle some conventional IM-processed alloys. The rationale underlying its introduction, together with cerium, into RSP Ti-624 is that in a titanium environment the heats of formation of CeS and $Ce_2(SO_2)$ are greater than that of Er_2O_3, a favored dispersion hardener.[157]

8.2.4. Rapid Solidification Processing of Titanium Aluminides. Attractive properties of the α_2-phase (based on Ti_3Al) and γ-phase (based on TiAl) aluminides are their high transus temperatures: 1100 °C for the $\alpha_2 \rightarrow \alpha + \alpha_2$ transus of Ti_3Al, and 1400 ± 60 °C for the melting point of the γ phase (although the useful temperature range of the latter tends to be limited by a brittle-to-ductile transition at 700 °C[188]). Thus, at least from a phase-stability standpoint, the aluminides make suitable bases on which to design potentially useful high-temperature alloys. Table 28 refers to some recent studies of RSP and strength-

Table 27. Rapid solidification processed dispersion- and precipitation-strengthened alloys based on multicomponent and commercial titanium alloys

System	Literature, Ref No.
(*a*) Interstitial-element additions	
(Ti-5Al-2.5Sn)-0.2B	177
(Ti-5Al-2.5Sn)-1B	177
(Ti-6Al-4V)-1B	177
Ti-7.5Zr-4Mo-1.3B	177
Ti-8.2Mo-2.3Al-1.4B	164
(Ti-5Al-2.5Sn)-1C	177
(*b*) Rare-earth and interstitial-element additions	
(Ti-5Al-2.5Sn)-2Y	177
(Ti-5Al-2.5Sn)-3La	177
(Ti-5Al-2.5Sn)-3Ce	164
(Ti-6Al-4V)-1Er	166
Ti-6Al-15V-2Er	165
Ti-25V-4Ce-0.6S	165
(Ti-4Zr-5Al-2.5Sn)-3La	164
(Ti-4Zr-6Al-2Sn)-1Er	166
(Ti-4Zr-6Al-2Sn)-1Ce-0.15S	157
(*b*) Rare-earth and interstitial-element additions (contd.)	
(Ti-4Zr-6Al-2Sn-2Mo)-0.08Si	185
(Ti-4Zr-6Al-2Sn-2Mo)-1Er	166
(Ti-4Zr-6Al-2Sn-2Mo)-0.08Si-2Er	185
(Ti-4Zr-6Al-2Sn-2Mo)-0.08Si-3W	185
(Ti-4Zr-6Al-2Sn-2Mo)-0.4Si	185
(Ti-4Zr-6Al-2Sn-2Mo)-0.4Si-2Er	185
(Ti-4Zr-6Al-2Sn-6Mo)-1Er	186
(Ti-4Zr-6Al-2Sn-6Mo)-2Er	185
(Ti-6Zr-6Al)-0.08B-1Er	166
(*c*) "Eutectoid-forming" additions	
(Ti-5Al-2.5Sn)-0.5Ge	187
(Ti-5Al-2.5Sn)-7.5Ge	187
(Ti-5Al-2.5Sn)-0.5Si	187
(Ti-5Al-2.5Sn)-5Si	187
(Ti-6Al-4V)-2.2Si	164
(Ti-7.4Zr)-3.9Mo-3.4Si	177

Table 28. Rapid solidification processed titanium aluminides

System	Literature, Ref No.
$Ti_3Al + Er_{0.4}$	189
$Ti_3Al + Er_{0.6}$	190,191,192
$Ti_3Al + Nb$	193
$Ti_3Al + Nb_5 + Ce_{0.6} + S_{0.2}$	190,191
$Ti_3Al + Nb_{7.5} + Ce_{0.7} + S_{0.2}$	190,191
$Ti_3Al + Nb_5 + Er_{0.6}$	190,191,192
$Ti_3Al + Nb_{7.5} + Er_{0.6}$	190,191,192
$Ti_3Al + Nb_{10} + Er_{0.5}$	192
TiAl	194
TiAl + W	194

ened titanium aluminides. A second important advantage of the α_2- and γ-phase aluminides is their better oxidation resistance compared with conventional titanium alloys, a property which in the case of Ti_3Al has been improved even further by the addition of 5 to 10 wt % Nb.[195] But, as ordered intermetallic compounds, both Ti_3Al and TiAl lack tensile ductility at ordinary temperatures, a property which has severely limited their applicability. Powder metallurgy of Ti_3Al[196] and TiAl[194] has yielded some promising results, and with the advent of RSP metallurgy in association with the introduction of RE and other third-element additions, the aluminides at last seem to be on the threshold of practical application.

Rapid solidification processed Ti_3Al + Nb has been used successfully in experimental studies of Borsic-reinforced metal-matrix composites.[193] The addition of 0.4 at. % Er to Ti_3Al led to a fine Er_2O_3 dispersion which seemed to be quite stable at 900 °C.[189] Erbium added to Ti_3Al + Nb resulted in a refinement of the grain size, and hence to an improvement in ductility. Under extrusion, however, a rapid coarsening of the Er_2O_3 dispersoids was noted.[190] In compounds containing CeS or $Ce_2(SO_2)$, it seems that the coarsening is less pronounced—a reflection of the high stability of these compounds, already considered above in connection with the dispersion strengthening of Ti-624—and hence that Ti_3Al with additions of cerium and sulfur should be considered for inclusion in any list of prospective alloys for high-temperature application.

8.3. Microstructural Stability of Rapid Solidification Processed Titanium Alloys

8.3.1. Precipitate Coarsening. Heat treatment (carefully controlled limited aging) of RSP titanium alloys with interstitial elements (e.g., boron and carbon) generally results in improved properties. In boron-containing alloys, aging results in high-aspect-ratio needle-shape precipitates of TiB. Unless prolonged exposure to elevated temperature allows them to coarsen excessively, these precipitates contribute a large increment of strength.[160] Heat treatment of titanium alloys containing carbon results in an increase in ductility (as carbon becomes removed from solid solution) and an increase in strength as the accompanying reduction in solution strengthening is more than compensated for by an increase in TiC precipitation strengthening.

Precipitates arising from the interstitial elements boron and carbon and the eutectoid-forming elements (including silicon) are all prone to excessive coarsening during prolonged exposure to high temperatures.[160,164,184] Whang has contrasted the behavior of boron, carbon, and silicon with that of the RE element lanthanum under high-temperature aging (at 800 °C). The relative stability of the La_2Sn dispersoids in Ti-5Sn-4.5La as compared with TiSi precipitates in Ti-5Al-2Si was attributed to the lower diffusivity of lanthanum as compared with silicon in α titanium ($\sim 4 \times 10^{-14}$ $cm^2 \cdot s^{-1}$ and $\sim 1.2 \times 10^{-11}$ $cm^2 \cdot s^{-1}$, respectively).[184] Comparisons among the RE dispersoids themselves have been made by Sastry and colleagues.[155] These authors noted that the rare earths could be subdivided into two classes: (1) cerium, neodymium, and gadolinium, which have appreciable solubilities in titanium and which on isothermal aging yield precipitates that become relatively coarse (~0.2 to 2 μm); and (2) yttrium, lanthanum, dysprosium, and erbium, which have negligible solubility in titanium and which under aging yield relatively fine (0.04 to 0.12 μm) dispersoids. The coarse dispersoids of the first group were RE sesquioxides, and the fine dispersoids of the second group were compounds of titanium, rare earths, oxygen, and carbon. Of all the RE elements in titanium, cerium yielded the coarsest dispersoids and erbium yielded the finest.

Sastry's work was followed (or accompanied) by numerous other studies of erbium-containing RSP titanium alloys. Rowe *et al.*[166] selected the system Ti-6Al-6Zr-1Er-0.08B for study. After aging, the usual TiB needles were noted. But of particular interest was the fact that only near the grain boundaries did the Er_2O_3 particles undergo coarsening, presumably as a result of grain-boundary diffusion. Konitzer *et al.*[174] undertook a comprehensive study of Er_2O_3 dispersoids in titanium and Ti-Al(10, 15, 24 at. %) alloys. During the 10-h aging at 900 °C of Ti-Al_{24}-$Er_{0.4}$ (after 10 h at 700 °C to develop the dispersion) it was found that the Er_2O_3 particles were fairly resistant to coarsening. This tended to be true for all the α-Ti alloys below the α-transus temperature. But as indicated in Fig. 67, rapid coarsening could be expected for all Er_2O_3 particles lying within the β regions of a two-phase alloy, due to the much higher diffusivity of oxygen in the β phase. Konitzer and Fraser[189] showed that in RSP Ti_3Al + 0.4 at. % Er a lack of significant oxide-precipitate coarsening was exhibited after a 10-h exposure to temperatures as high as 800 to 900 °C. This performance emphasized the importance of a high transus temperature in dispersion coarsening—oxide precipitates in binary Ti-Er alloys exhibited significant coarsening in response to heat treatment for 10 h at 900 °C. The fineness and stability of Er_2O_3 precipitates in the hexagonal phases of titanium alloys indicate that erbium should always be considered

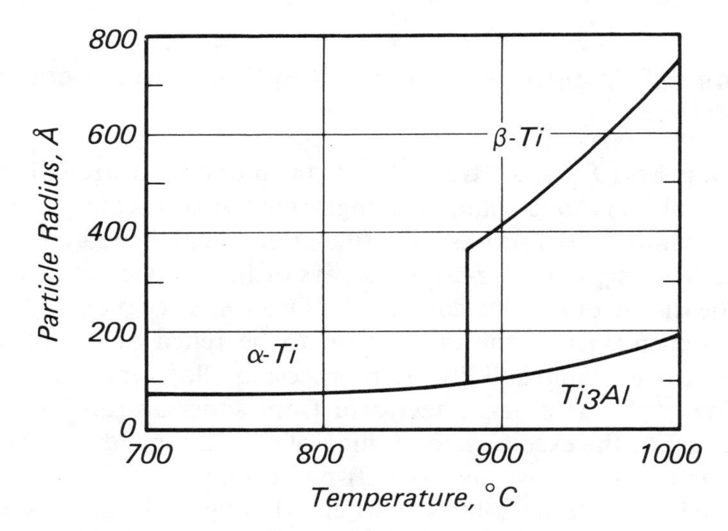

Below about 880 °C (the β transus for titanium) the curves for α-Ti and Ti_3Al are continuous (since the oxygen diffusivity is assumed to be the same in each phase). Above 880 °C, the particles show marked coarsening in the β phase, whereas only a modest change of size in Ti_3Al.

Fig. 67. Simulation of the change in radius of Er_2O_3 particles annealed for 10 h within the temperature range shown[174]

among the possible dispersion-strengthening additives to α-phase and α_2-phase titanium alloys.

Recent work by Rowe and Koch[157] has indicated that other additives besides erbium have important roles to play when, in addition to strengthening, grain refinement must be taken into consideration. As indicated in Section 8.2.3(b), these authors had estimated that sulfides and oxysulfides of cerium were more stable in a titanium environment than was Er_2O_3, the hitherto premier dispersion strengthener. Accordingly, they decided to introduce both cerium and sulfur into a titanium-alloy base; Ti-6Al-2Sn-4Zr was chosen as the test alloy. Sastry's studies had indicated that cerium in titanium yielded the coarsest precipitates of all the RE elements.[155] But CeS and $Ce_2(SO_2)$, according to Rowe *et al.*, resisted coarsening at temperatures as high as 1000 °C except near the grain boundaries, where grain-boundary diffusion was likely to assist in the process. Particles in the grain interiors were about 0.03 to 0.04 μm in size, and those near the grain boundaries, about 0.15 to 0.20 μm.

8.3.2. Grain Growth. Alloys containing erbium were found unable to resist some grain growth during aging, a disadvantage which may outweigh their ability to yield ultrafine dispersions. On the other hand, alloys containing boron or silicon retain their fine as-RSP grain structures to high temperatures. In particular, Rowe and Koch[157] found that their consolidated sulfur-bearing alloy possessed a submicron grain structure which resisted growth at 1000 °C.

8.4. Mechanical Properties of Rapid Solidification Processed Titanium Alloys

8.4.1. Hardness and Tensile Strength. Contributions to strengthening in RSP titanium alloys are: (1) solid-solution strengthening arising from the extended solubilities that accompany the process, (2) fine-grain strengthening, and (3) Orowan strengthening from high-number-density arrays of fine incoherent precipitates. Provided that the dispersoids resist coarsening, Orowan strengthening with its weak temperature dependence is the mechanism to be relied on in alloys for high-temperature service. Rapid solidification processed alloys are generally subject to age hardening[158,167] as precipitates form from supersaturated solid solutions. Overaging refers to the excessive coarsening of the dispersoids, which takes place much more readily in β-phase and $(\alpha + \beta)$-phase alloys than in α alloys due to the higher diffusivities of many solutes (particularly oxygen in this context) in β-Ti. Obviously an alloy's resistance to aging is closely related to its heat resistance—i.e., its ability to withstand high-temperature service conditions.

Many tensile-property studies of RSP alloys have been confined to the room-temperature testing of as-formed and/or age-hardened (moderate-temperature annealed) material. In this regard, Sastry *et al.*[160] have investigated the properties of Ti-C and Ti-B. In the latter study it was found that the strengthening effect of boron also persisted to high temperatures[197] (in spite of the extensive coarsening that has been noted for TiB precipitates during exposure to temperatures in the range 800 to 900 °C[164]). Boron added to Ti-8Al resulted in an alloy with a good combination of low density, high modulus, high room-temperature and elevated-temperature tensile strengths, and a potential for high-temperature applications.

The temperature dependence of the tensile properties of Ti-Nd and Ti-Er have been measured by Sastry *et al.*[156] As pointed out above, neodymium and erbium belong to the "coarse" and "fine," respectively, classes of dispersoid-forming elements, yet at 700 °C the yield strengths of Ti-1.5Nd and Ti-1.0Er (previously aged for 2 h at 700 °C) were the same. Very much greater strengths are exhibited by the ternary alloys based on Ti-Al. The room-temperature properties of various Ti-Al-Nd and Ti-Al-Er alloys have also been measured by Sastry *et al.*[159] It is interesting to note that after Ti_3Al precipitation was caused to form in some of these alloys, the strengthening due to the incoherent dispersoids plus the Ti_3Al was less than that due to the incoherent dispersoids plus aluminum in solid solution. The high-temperature strengths of Ti-Al-Er alloys were anticipated to be greater than those of all conventional titanium alloys.

The relative qualities of erbium and lanthanum dispersoid-forming additions to RSP Ti-Al alloys were investigated by Whang.[158] Both of these RE elements are members of the "fine" class of dispersoid-forming additions. A distinction must be drawn between the room-temperature properties of the age-hardened alloys and their relative performances at elevated temperatures. In the former category, Ti-5Al-4.5La is superior to Ti-5Al-5.4Er after aging for 2 h at all temperatures up to 900 °C. However, in hot hardness tests, due to the rapid softening of Ti-5Al-4.5La at temperatures above about 600 °C, at 900 °C both alloys were equally hard.

For reasons outlined in the previous section on aging, cerium and sulfur in association hold considerable promise as high-temperature strengtheners of titanium alloys. Although tensile testing has not been carried out above 538 °C, metallographic studies of grain and dispersoid growth have been conducted on alloys exposed to temperatures as high as 1000 °C, during which considerable microstructural stability was noted. The tensile work indicated that the sulfide and oxysulfide precipitates provided strengthening over the entire temperature range, yet at the same time permitted adequate room-temperature ductility.[157]

8.4.2. Creep. Relatively little has been written about the creep properties of RSP titanium alloys. They can, however, be qualitatively predicted from those of dispersion-strengthened alloys in general. The usual mechanisms of creep are associated with diffusion, grain-boundary sliding, and dislocation movement. The initial fine grain structure of RSP alloys tends to enhance creep; thus from a creep standpoint a certain amount of deliberately induced grain growth is advantageous. Creep resistance in RSP alloys at high temperatures relies primarily on the ability of the dispersoids to pin dislocations. But severe matrix softening is always to be expected in α-Ti solid-solution alloys at temperatures above about 900 °C.[158] To combat this, the introduction of some form of fibrous reinforcement is recommended.

8.5. Summary

The operating temperature range of conventional IM multicomponent alloys such as Ti-6242 (Ti-6Al-2Sn-4Zr-2Mo) and IMI 834 (Ti-5.5Al-4Sn-4Zr-1Nb-0.3Mo-0.5Si) is limited to 500 to 600 °C, above which microstructural instability becomes a problem. Furthermore, IM alloys are not amenable to *in-situ* dispersion or precipitation strengthening as a consequence of the coarsening which occurs during the alloy's long dwell time in the β-phase field during cooldown. To find a way out of the instability difficulty it is necessary to turn to materials which do not undergo phase transformation within the service-temperature range: stable β-phase alloys (not a practical solution), all-α alloys (especially with high α-transus temperatures), and the aluminides of titanium. The coarsening-during-processing difficulty is eliminated through the use of rapid solidification techniques; coarsening in service is eliminated by turning, again, to the nontransforming class of alloys.

Within the realm of all-α alloys, the requirements of solution strengthening, low density,and high α-transus temperature are simultaneously served if aluminum is selected as a solute. If the aluminum concentration exceeds about 9 wt %, a finely dispersed α_2-phase precipitate will be present (Eq 3). Although this severely embrittles IM alloys whenever it occurs, its presence under RSP conditions has been claimed to improve the high-temperature strength. On the other hand, it has been determined that when incoherent dispersoids are present in it, the single-phase solid solutions are stronger materials than those containing α_2-phase precipitates.

High-temperature creep strength is enhanced through the introduction of a submicroscopic dispersed phase. It has been noted that dispersoids should be insoluble in the alloy matrix, incoherent, nonreactive, fine and closely spaced, resistant to coarsening, and resistant to deformation. Elements that have been considered as ingredients in RSP titanium alloys for dispersion or precipitation strengthening are: (1) the interstitial elements boron and carbon (which yield intermetallic-compound precipitates); (2) the group IIIB elements scandium and yttrium, and the lanthanides lanthanum, cerium, neodymium, gadolinium, and erbium (which scavenge oxygen from the host alloy to form sesquioxide dispersoids); and (3) β-eutectoid-forming elements such as silicon, iron, nickel, and copper (which yield intermetallic-compound precipitates or fine lamellar microstructures). Insufficient information is known about the high-temperature mechanical properties of the alloys with β-eutectoid formers. The chemical reactivity of the rare earths can be turned to advantage—they scavenge excess oxygen from the alloy (which is particularly advantageous in PM) and, in addition, after being converted to RE_2O_3, act as dispersion strengtheners.

Many published studies have focused attention on the interstitial element boron, and several of the RE elements. (1) *Boron additions*: It has been noted that RSP Ti-Al-B alloys contain high-aspect-ratio filamentary dispersoids which coarsen during annealing to needle-shape precipitates ideally suited to matrix reinforcement (see Fig. 64) unless prolonged exposure to very high temperatures causes them to coarsen excessively. The strengthening effect of boron at high temperatures, in spite of coarsening, has been noted: boron-doped Ti-8Al has been identified as an alloy with potential for high-temperature applications. (2) *Rare-earth and other additions*: Comprehensive studies of the stability and effect of RE additions to titanium have indicated that the most promising ones are erbium and neodymium. Both yield very fine dispersions of RE_2O_3 with particle diameters of less than 0.01 μm; but neodymium also yields a crop of larger dispersoids within the size range of 0.1 to 1.0 μm. Cerium, on the other hand, yields the coarsest dispersoids of all the rare earths. As for stability, studies have shown that Er_2O_3 is fairly resistant to coarsening during high-temperature exposure (especially in the grain interiors, as distinct from the grain boundaries within which coarsening seems to be promoted by boundary diffusion). But if Er_2O_3 is fairly stable, the sulfide and oxysulfide of cerium, CeS and $Ce_2(SO_2)$, are even more so. Thus, in spite of the fact that cerium alone in titanium yields the coarsest precipitates of all the RE elements, its inclusion accompanied by sulfur yields a dispersoid system with considerable stability. Recent tensile work has indicated that CeS and $Ce_2(SO_2)$ precipitates are capable of providing strengthening at temperatures approaching 1000 °C.

Both Ti_3Al and TiAl, to which 5 to 10 wt % Nb has been added to improve ambient-temperature ductility and oxide-scale adherence at high temperatures (in the case of Ti_3Al), have assumed considerable importance as potential high-temperature alloys. To these, the addition of dispersoid formers should also be considered—not this time for dispersion strengthening (there is little need for this), but to inhibit grain growth during RSP and in service; the establishment and maintenance of microcrystallinity in this way tends to contribute to ambient-temperature ductility.

ACKNOWLEDGMENTS

The original research content of this review was supported over the years by the U.S. Air Force Office of Scientific Research (Dr. A.H. Rosenstein, Program Monitor) and the Air Force Materials Laboratory, Wright-Patterson Air Force Base (Dr. H.L. Gegel, Program Monitor). Original data presented are the results of research conducted at Battelle with the assistance of Dr. J.C. Ho and Mr. R.D. Smith. The interpretation of some of the results, particularly those related to solid-solution strengthening, was aided by discussions with Dr. H.L. Gegel and Dr. J.E. Enderby.

The technical illustration was performed by Ms. Judith S. Ward.

REFERENCES

1. E.K. Molchanova, *Phase Diagrams of Titanium Alloys* (translation of *Atlas Diagram Sostoyaniya Titanovyk Splavov*), Israel Program for Scientific Translations, Jerusalem, 1965
2. U. Zwicker, *Titan und Titanlegierungen*, Springer-Verlag, 1974

3. M. Hoch, N.C. Birla, S.A. Cole, and H.L. Gegel, "The Development of Heat-Resistant Titanium Alloys," Tech. Report AFML-TR-73-297, Air Force Materials Lab., December 1973
4. H.L. Gegel and M. Hoch, Thermodynamics of α-Stabilized Ti-X-Y Systems, in *Titanium Science and Technology*, Proc. Second Ind. Conf. on Titanium, Boston, edited by R.I. Jaffee and H.M. Burte, Plenum Press, 1973, pp. 923-931
5. R.A. Wood, *Titanium Alloys Handbook*, Metals and Ceramics Information Center, Battelle, Publication No. MCIC-HB-02, December 1972
6. D.R. Salmon, *Low Temperature Data Handbook, Titanium and Titanium Alloys*, National Physical Laboratory, NPL Report QU53 (N 80 23448), May 1979
7. H.L. Gegel, J.C. Ho, and E.W. Collings, An Electronic Approach to Solid Solution Strengthening in Titanium Alloys, *Inst. Met. (London) Monogr. Rep. Ser.*, Vol 1, PAP 116, 1973, pp. 544-548
8. N.V. Ageev and L.A. Petrova, The Theoretical Basis of the Development of the High-Strength Metastable β-Alloys of Titanium, in *The Science, Technology and Application of Titanium*, Proc. First Int. Conf. on Titanium, London, edited by R.I. Jaffee and N.E. Promisel, Pergamon Press, 1970, pp. 809-814
9. T. Nishimura, M. Nishigaki, and H. Kusamichi, Aging Characteristics of Beta Titanium Alloys, in *Titanium and Titanium Alloys, Scientific and Technological Aspects*, Proc. Third Int. Conf. on Titanium, Moscow, edited by J.C. Williams and A.F. Belov, Plenum Press, 1982, pp. 1675-1689
10. F.H. Froes, J.M. Capenos, and M.G.H. Wells, Alloy Partitioning in Beta III and Effects on Aging Characteristics, in *Titanium Science and Technology*, Proc. Second Ind. Conf. on Titanium, Boston, edited by R.I. Jaffee and H.M. Burte, Plenum Press, 1973, pp. 1621-1633
11. V.C. Peterson, F.H. Froes, and R.F. Malone, Metallurgical Characteristics and Mechanical Properties of Beta III, A Heat-Treatable Titanium Alloy, in *Titanium Science and Technology*, Proc. Second Ind. Conf. on Titanium, Boston, edited by R.I. Jaffee and H.M. Burte, Plenum Press, 1973, pp. 1969-1980
12. G. Vigier, J. Merlin, and P.F. Gobin, Decomposition of the Solid Solution in the All-Beta βIII, in *Titanium and Titanium Alloys, Scientific and Technological Aspects*, Proc. Third Int. Conf. on Titanium, Moscow, edited by J.C. Williams and A.F. Belov, Plenum Press, 1982, pp. 1691-1701
13. J.C. Williams, F.H. Froes, and S. Fujishiro, Microstructure and Properties of the Alloy Ti-11.5Mo-6Zr-4.5Sn (Beta III), in *Titanium and Titanium Alloys, Scientific and Technological Aspects*, Proc. Third Int. Conf. on Titanium, Moscow, edited by J.C. Williams and A.F. Belov, Plenum Press, 1982, pp. 1421-1436
14. I.V. Gorynin, B.B. Chechulin, S.S. Ushkov, and O.S. Belova, A Study of the Nature of the Ductile-Brittle Transition in Beta Titanium Alloys, in *Titanium Science and Technology*, Proc. Second Ind. Conf. on Titanium, Boston, edited by R.I. Jaffee and H.M. Burte, Plenum Press, 1973, pp. 1109-1118
15. C. Zener, *Elasticity and Anelasticity of Metals*, University of Chicago Press, 1948
16. E.S. Fisher and D. Dever, Relation of the C′ Elastic Modulus to Stability of b.c.c. Transition Metals, *Acta. Metall.*, Vol 18, 1970, pp. 265-269
17. E.S. Fisher, A Review of Solute Effects on the Elastic Moduli of bcc Transition Metals, in *Physics of Solid Solution Strengthening*, edited by E.W. Collings and H.L. Gegel, Plenum Press, 1975, pp. 199-225
18. E.W. Collings and H.L. Gegel, A Physical Basis for Solid Solution Strengthening and Phase Stability in Alloys of Titanium, *Scripta Met.*, Vol 7, 1973, pp. 437-443
19. H. Margolin and J.P. Nielsen, Titanium Metallurgy, in *Modern Materials, Advances in Development and Application*, edited by H.H. Hausner, Vol 2, Academic Press, 1960, pp. 225-325

20. I.I. Kornilov, Equilibrium Diagrams, Electronic and Crystalline Structures and Physical Properties of Titanium Alloys, in *Titanium and Titanium Alloys, Scientific and Technological Aspects*, Proc. Third Int. Conf. on Titanium, Moscow, edited by J.C. Williams and A.F. Belov, Plenum Press, 1982, pp. 1281-1305
21. *Structural Alloys Handbook*, 1982 Supplement, Produced and Published by Battelle's Columbus Laboratories, Columbus, OH
22. F.H. Froes, D. Eylon, and H.B. Bomberger (eds.), *Titanium Technology: Present Status and Future Trends*, Titanium Development Association, 1985
23. T. Nishimura, T. Mizoguchi, and Y. Itoh, Titanium Materials for Cryogenic Service, *Kobe Steel Engineering Reports*, Vol 34, No. 3, 1984, pp. 63-66
24. H.W. Rosenberg, Titanium Alloying in Theory and Practice, in *The Science, Technology and Application of Titanium*, Proc. First Int. Conf. on Titanium, London, edited by R.I. Jaffee and N.E. Promisel, Pergamon Press, 1970, pp. 851-859
25. A.D. McQuillan and M.K. McQuillan, *Titanium*, Academic Press, 1956
26. A.G. Imgram, D.N. Williams, R.A. Wood, H.R. Ogden, and R.I. Jaffee, "Metallurgical and Mechanical Characteristics of High-Purity Titanium-Base Alloys," WADC Technical Report 59-595, Part II, March 1961
27. R.I. Jaffee, The Physical Metallurgy of Titanium Alloys, *Progr. Met. Phys.*, Vol 7, 1958, pp. 65-163
28. H.R. Ogden, D.J. Maykuth, W.L. Finlay, and R.I. Jaffee, Constitution of Titanium-Aluminum Alloys, *Trans. TMS-AIME*, Vol 191, 1951, pp. 1150-1155
29. E.S. Bumps, H.D. Kessler, and M. Hansen, Titanium-Aluminum System, *Trans. TMS-AIME*, Vol 194, 1952, pp. 609-614
30. I.I. Kornilov, E.N. Pylaeva, and M.A. Volkova, Constitution Diagram of the Binary Titanium-Aluminum System, *Izv. Akad. Nauk SSSR, Otd. Khim. Nauk*, No. 7, 1956, pp. 771-778
31. K. Sagel, E. Schulz, and U. Zwicker, Investigation of the Titanium-Aluminum System, *Z. Metallkund.*, Vol 46, 1956, pp. 529-534
32. T. Sato, Y. Huang, and Y. Kondo, Equilibrium Diagram of the System Ti-Al, *Trans. Jpn. Inst. Metals*, Vol 1, 1959, pp. 456-460
33. E. Ence and H. Margolin, Phase Relations in the Titanium-Aluminum System, *Trans. TMS-AIME*, Vol 221, 1961, pp. 151-157
34. Y.L. Yao, Magnetic Susceptibilities of Titanium-Rich Titanium-Aluminum Alloys, *Trans. ASM*, Vol 54, 1961, pp. 241-246
35. D. Clark, K.S. Jepson, and G.I. Lewis, A Study of the Titanium-Aluminum System up to 40 At.-% Aluminum, *J. Inst. Metals*, Vol 91, 1963, pp. 197-203
36. I.I. Kornilov, E.N. Pylaeva, M.A. Volkova, P.I. Kripyakevich, and V.Ya. Markiv, Phase Structure of Alloys in the Binary Ti-Al System Containing from 0-30 % Al, *Dokl. Akad. Nauk SSSR*, Vol 161, 1965, pp. 843-846
37. F.A. Crossley, Titanium-Rich End of the Titanium-Aluminum Equilibrium Diagram, *Trans. TMS-AIME*, Vol 236, 1966, pp. 1174-1185
38. M.J. Blackburn, The Ordering Transformation in Titanium:Aluminum Alloys Containing up to 25 at. pct. Aluminum, *Trans. TMS-AIME*, Vol 239, 1967, pp. 1200-1208
39. M.J. Blackburn, Some Aspects of Phase Transformations in Titanium Alloys, in *The Science, Technology and Application of Titanium*, Proc. First Int. Conf. on Titanium, London, edited by R.I. Jaffee and N.E. Promisel, Pergamon Press, 1970, pp. 633-643
40. T.K.G. Namboohiri, C.J. McMahon, and H. Herman, Decomposition of the α-Phase in Titanium-Rich Ti-Al Alloys, *Met. Trans.*, Vol 4, 1973, pp. 1323-1331
41. I.I. Kornilov, T.T. Nartova, and S.P. Chernyshova, The Titanium-Aluminum Phase Diagram in the Titanium-Rich Part, *Russ. Met. (Metally)*, No. 6, 1976, pp. 192-198

42. E.W. Collings, Magnetic Studies of Phase Equilibria in Ti-Al(30-57at.%) Alloys, *Met. Trans.*, Vol 10A, 1979, pp. 463-474
43. L.T. Swartzendruber, L.H. Bennett, L.K. Ives, and R.D. Shull, The Ti-Al Phase Diagram: the α-α_2 Phase Boundary, *Mater. Sci. Eng.*, Vol 51, 1981, pp. 1-9
44. R.D. Shull, A.J. McAlister, and R.C. Reno, Phase Equilibria in the Titanium-Aluminum System, in *Titanium Science and Technology*, Proc. Fifth Int. Conf. on Titanium, Oberursel, West Germany, 1984, edited by G. Lutgering, U. Zwicker, and W. Bunk, D.G. für Metallkde., 1985, p. 1495
45. E.W. Collings and H.L. Gegel, Physical Principles of Solid Solution Strengthening in Alloys, in *Physics of Solid Solution Strengthening*, edited by E.W. Collings and H.L. Gegel, Plenum Press, 1975, pp. 147-182
46. P.C. Gehlen, The Crystallographic Structure of Ti-Al, in *The Science, Technology and Application of Titanium*, Proc. First Int. Conf. on Titanium, London, edited by R.I. Jaffee and N.E. Promisel, Pergamon Press, 1970, pp. 349-357
47. E.W. Collings and J.C. Ho, Physical Properties of Titanium Alloys, in *The Science, Technology and Application of Titanium*, Proc. First Int. Conf. on Titanium, London, edited by R.I. Jaffee and N.E. Promisel, Pergamon Press, 1970, pp. 331-347
48. E.W. Collings, Magnetic Investigations of Electronic Bonding and α through γ Phase Equilibria in the Titanium Aluminum System, in *Titanium and Titanium Alloys, Scientific and Technological Aspects*, Proc. Third Int. Conf. on Titanium, Moscow, edited by J.C. Williams and A.F. Belov, Plenum Press, 1982, pp. 1391-1402
49. S. Terauchi, H. Matsumoto, T. Sugimoto, and K. Kamei, Investigation of the Titanium-Molybdenum Binary Phase Diagram, in *Titanium and Titanium Alloys, Scientific and Technological Aspects*, Proc. Third Int. Conf. on Titanium, Moscow, edited by J.C. Williams and A.F. Belov, Plenum Press, 1982, pp. 1335-1349
50. C. Hayman, "Resolution of the Incommensurate Omega Phase Problem by Scattering from Multilayer Films," Thesis Abstract, University of Minnesota, June 1985; see D. de Fontaine, Simple Models for the Omega Phase Transformation, *Met. Trans. A* (in press)
51. M. Hansen, *Constitution of Binary Alloys*, 2nd Ed. (with K. Anderko), McGraw-Hill, 1958
52. T.S. Luhman, "Superconductivity and Constitution of Titanium Base Transition Metal Alloys," Ph.D. Thesis, University of Washington, Seattle, 1970
53. J.C. Williams, Kinetics and Phase Transformations: A Critical Review, in *Titanium Science and Technology*, Proc. Second Ind. Conf. on Titanium, Boston, edited by R.I. Jaffee and H.M. Burte, Plenum Press, 1973, pp. 1433-1494
54. G.R. Love and M.L. Picklesheimer, The Kinetics of Beta-Phase Decomposition in Niobium(Columbium)-Zirconium, *Trans. TMS-AIME*, Vol 236, 1966, pp. 430-435
55. M. Kitada and T. Doi, Precipitation and Superconducting Properties of Nb-40Zr-10Ti Alloy, *Nippon Kinzoku Gakkaishi*, Vol 34, 1970, pp. 369-374
56. B.S. Hickman, Precipitation of the Omega Phase in Titanium-Vanadium Alloys, *J. Inst. Metals*, Vol 96, 1968, pp. 330-337
57. B.S. Hickman, Omega Phase Precipitation in Alloys of Titanium with Transition Metals, *Trans. TMS-AIME*, Vol 245, 1969, pp. 1329-1335
58. A.R.G. Brown, K.S. Jepson, and J. Heavens, High-Speed Quenching in Vacuum. *J. Inst. Metals*, Vol 93, 1965, pp. 542-544
59. K.S. Jepson, A.R.G. Brown, and J.A. Gray, The Effect of Cooling Rate on the Beta Transformation in Titanium-Niobium and Titanium-Aluminum Alloys, in *The Science, Technology and Application of Titanium*, Proc. First Int. Conf. on Titanium, London, edited by R.I. Jaffee and N.E. Promisel, Pergamon Press, 1970, pp. 677-690
60. A.T. Balcerzak and S.L. Sass, The Formation of the ω Phase in Ti-Nb Alloys, *Met. Trans.*, Vol 3, 1972, pp. 1601-1605

61. M. Cohen, The Martensitic Transformation, in *Phase Transformations in Solids*, ed. by R. Smoluchowski, J.E. Mayer, and W.A. Weyl, John Wiley & Sons, 1951, pp. 588-660.
62. B.A. Bilby and J.W. Christian, Martensitic Transformations, in *The Mechanisms of Phase Transformations in Metals*, The Institute of Metals, 1956, pp. 121-172
63. C. Hammond and P.M. Kelly, Martensitic Transformations in Titanium Alloys, in *The Science, Technology and Application of Titanium*, Proc. First Int. Conf. on Titanium, London, edited by R.I. Jaffee and N.E. Promisel, Pergamon Press, 1970, pp. 659-676
64. H.M. Otte, Mechanism of the Martensitic Transformation in Titanium and its Alloys, in *The Science, Technology and Application of Titanium*, Proc. First Int. Conf. on Titanium, London, edited by R.I. Jaffee and N.E. Promisel, Pergamon Press, 1970, pp. 645-657
65. R. Davis, H.M. Flower, and D.R.F. West, Martensitic Transformations in Ti-Mo Alloys, *J. Mater. Sci.*, Vol 14, 1979, pp. 712-722
66. R. Davis, H.M. Flower, and D.R.F. West, The Decomposition of Ti-Mo Alloy Martensites by Nucleation and Growth and Spinoidal Mechanisms, *Acta. Metall.*, Vol 27, 1979, pp. 1041-1052
67. H.M. Flower, R. Davis, and D.R.F. West, Martensite Formation and Decomposition in Alloys of Titanium Containing β-Stabilizing Elements, in *Titanium and Titanium Alloys, Scientific and Technological Aspects*, Proc. Third Int. Conf. on Titanium, Moscow, edited by J.C. Williams and A.F. Belov, Plenum Press, 1982, pp. 1703-1715
68. J.W. Christian, Phase Transformations, in *Physical Metallurgy*, edited by R.W. Cahn, North-Holland, 1965
69. T. Suzuki and M. Wuttig, Analogy Between Spinodal Decomposition and Martensitic Transformation, *Acta. Metall.*, Vol 23, 1975, pp. 1069-1076
70. M. Oka and Y. Taniguchi, Crystallography of Stress-Induced Products in Metastable Beta Ti-Mo Alloys, in *Titanium '80: Science and Technology*, Proc. Fourth Int. Conf. on Titanium, Kyoto, Japan, edited by H. Kimura and O. Izumi, The Metallurgical Society of AIME, 1980, pp. 709-715
71. P. Duwez, The Martensitic Transformation Temperature in Titanium Binary Alloys, *Trans. ASM*, Vol 45, 1953, pp. 934-940
72. M. Hansen, E.L. Kamen, H.D. Kessler, and D.J. McPherson, Systems Titanium-Molybdenum and Titanium-Columbium, *Trans. TMS-AIME*, Vol 191, 1951, pp. 881-888
73. B.A. Hatt and V.G. Rivlin, Phase Transformations in Superconducting Ti-Nb Alloys, *J. Phys. D: Applied Phys.*, Vol 1, 1968, pp. 1145-1149
74. C. Baker and J. Sutton, Correlation of Superconducting and Metallurgical Properties of a Ti-20at.%Nb Alloy, *Phil. Mag.*, Vol 19, 1969, pp. 1223-1255
75. A.T. Balcerzak, "An Electron Microscope Study of the Ti-Nb System," M.S. Thesis, Cornell University, Ithaca, NY, 1971
76. S.L. Sass, The Structure and Decomposition of Zr and Ti b.c.c. Solid Solutions, *J. Less-Common Metals*, Vol 28, 1972, pp. 157-173
77. B.S. Hickman, The Formation of Omega Phase in Titanium and Zirconium Alloys: A Review, *J. Mater. Sci.*, Vol 4, 1969, pp 554-563
78. J.M. Silcock, An X-Ray Examination of the ω Phase in TiV, TiMo, and TiCr Alloys, *Acta. Metall.*, Vol 6, 1958, pp. 481-492
79. Yu.A. Bagariatskii, G.I. Nosova, and T.V. Tagunova, Factors in the Formation of Metastable Phases in Titanium-Base Alloys, *Sov. Phys. Dokl.*, Vol 3, 1959, pp. 1014-1018 (translation of *Dok. Akad. Nauk SSSR*, Vol 122, 1958, pp. 593-596)
80. L.N. Guseva and L.K. Dolinskaya, Metastable Phases in Quenched Titanium Alloys with Transition Elements, in *Titanium and Titanium Alloys, Scientific and Techno-*

logical Aspects, Proc. Third Int. Conf. on Titanium, Moscow, edited by J.C. Williams and A.F. Belov, Plenum Press, 1982, pp. 1559-1565

81. C.A. Luke, R. Taggart, and D.H. Polonis, The Metastable Constitution of Quenched Titanium and Zirconium-Base Binary Alloys, *Trans. ASM*, Vol 57, 1964, pp. 142-149
82. L. Pauling, The Electronic Structures of Metals and Alloys, in *Theory of Alloy Phases*, American Society for Metals, 1956, pp. 220-242
83. E.W. Collings and J.C. Ho, Density of States of Transition Metal Binary Alloys in the Electron-to-Atom Ratio Range 4.0 to 6.0, in *Electronic Density of States*, Proc. 3rd Materials Research Symposium, Nat. Bur. Stand. (U.S.) Spec. Publ. 323, December 1971, pp. 587-596
84. P.C. Clapp, A Localized Soft Mode Theory for Martensitic Transformations, *Phys. Stat. Sol. (b)*, Vol 57, 1973, pp. 561-569
85. A. Jayaraman, W. Klement, and G.C. Kennedy, Solid-Solution Transitions in Titanium and Zirconium at High Pressures, *Phys. Rev.*, Vol 131, 1963, pp. 644-649
86. S.L. Sass, The ω Phase in a Zr-25 at.%Ti Alloy, *Acta. Metall.*, Vol 17, 1969, pp. 813-820
87. C.W. Dawson and S.L. Sass, The As-Quenched Form of the Omega Phase in Zr-Nb Alloys, *Met. Trans.*, Vol 1, 1970, pp. 2225-2233
88. K.K. McCabe and S.L. Sass, The Initial Stages of the Omega Phase Transformation in Ti-V Alloys, *Phil. Mag.*, Vol 23, 1971, pp. 957-970
89. D. de Fontaine, N.E. Paton, and J.C. Williams, The Omega Phase Transformation in Titanium Alloys as an Example of Displacement Controlled Reactions, *Acta. Metall.*, Vol 19, 1971, pp. 1153-1162
90. D. de Fontaine, Mechanical Instabilities in the b.c.c. Lattice and the Beta to Omega Phase Transformation, *Acta. Metall.*, Vol 18, 1970, pp. 275-279
91. J.C. Williams, Precipitation in Titanium-Base Alloys, in *Precipitation Processes in Solids*, edited by K.C. Russell and H.I. Aaronson, The Metallurgical Society of AIME, 1978, pp. 191-224
92. J.B. Vetrano, G.L. Guthrie, H.E. Kissinger, J.L. Brimhall, and B. Mastel, Superconductivity Critical Current Densities in Ti-V Alloys, *J. Appl. Phys.*, Vol 39, 1968, pp. 2524-2528
93. J.C. Williams, B.S. Hickman, and D.H. Leslie, The Effect of Ternary Additions on the Decomposition of Metastable β-Phase Titanium Alloys, *Met. Trans.*, Vol 2, 1971, pp. 477-484
94. P.D. Frost, W.M. Parris, L.L. Hirsch, J.R. Doig, and C.M. Schwartz, Isothermal Transformation of Titanium-Manganese Alloys, *Trans. ASM*, Vol 46, 1954, pp. 1056-1074
95. F.R. Brotzen, E.L. Harmon, Jr., and A.R. Troiano, Decomposition of Beta Titanium, *Trans. TMS-AIME*, Vol 203, 1955, pp. 413-419
96. M.K. McQuillan, Phase Transformations in Titanium and its Alloys, *Metallurgical Reviews*, Vol 8, 1963, pp. 41-104
97. W.M. Parris, L.L. Hirsch, and P.D. Frost, Low Temperature Aging in Titanium Alloys, *Trans. TMS-AIME*, Vol 197, 1953, pp. 178-179
98. V. Chandrasekaran, R. Taggart, and D.H. Polonis, Phase Separation Processes in the Beta Phase of Ti-Mo Binary Alloys, *Metallography*, Vol 5, 1972, pp. 393-398
99. G.H. Narayanan, T.S. Luhman, T.F. Archbold, F. Taggart, and D.H. Polonis, A Phase Separation Reaction in a Binary Titanium-Chromium Alloy, *Metallography*, Vol 4, 1971, pp. 343-358
100. M.K. Koul and J.F. Breedis, Phase Transformations in Beta Isomorphous Titanium Alloys, *Acta. Metall.*, Vol 18, 1970, pp. 579-588
101. M.K. Koul and J.F. Breedis, Reply to Comments on "Phase Transformations in Beta Isomorphous Titanium Alloys," *Scripta Metall.*, Vol 4, 1970, pp. 877-880

102. G.H. Narayanan and T.F. Archbold, Comments on "Phase Transformations in Beta Isomorphous Titanium Alloys," *Scripta Metall.*, Vol 4, 1970, pp. 873-876
103. J.J. Rausch, F.A. Crossley, and H.D. Kessler, Titanium-Rich Corner of the Ti-Al-V System, *Trans. AIME, Journal of Metals*, Vol 8, 1956, pp. 211-214
104. M.T. Donachie, Jr. (ed.), *Titanium and Titanium Alloys Sourcebook*, American Society for Metals, 1982
105. J.A. Feeney and M.J. Blackburn, Effect of Microstructure on the Strength, Toughness, and Stress-Corrosion Cracking Susceptibility of a Metastable Beta Titanium Alloy (Ti-11.5Mo-6Zr-4.5Sn), *Met. Trans.*, Vol 1, 1970, pp. 3309-3323
106. C.C. Chen and J.E. Coyne, Relationships between Microstructure and Mechanical Properties in Ti-6Al-2Sn-4Zr-2Mo~0.1Si Alloy Forgings, in *Titanium '80: Science and Technology*, Proc. Fourth Int. Conf. on Titanium, Kyoto, Japan, edited by H. Kimura and O. Izumi, The Metallurgical Society of AIME, 1980, pp. 1197-1207
107. R.R. Boyer, R. Taggart, and D.H. Polonis, Effect of Thermal and Mechanical Processes on the β-III Titanium Alloy, *Metallography*, Vol 7, 1974, pp. 241-251
108. T.W. Duerig, R.M. Middleton, G.T. Terlinde, and J.C. Williams, Stress Assisted Transformations in Ti-10V-2Fe-3Al, in *Titanium '80: Science and Technology*, Proc. Fourth Int. Conf. on Titanium, Kyoto, Japan, edited by H. Kimura and O. Izumi, The Metallurgical Society of AIME, 1980, pp. 1503-1512
109. H.J. Rack, D. Kalish, and K.D. Fike, Stability of As-Quenched Beta-III Titanium Alloy, *Mater. Sci. Eng.*, Vol 6, 1970, pp. 181-198
110. M.J. Blackburn and J.C. Williams, Phase Transformations in Ti-Mo and Ti-V Alloys, *Trans. TMS-AIME*, Vol 242, 1968, pp. 2461-2469
111. J.R. Toran and R.R. Biederman, Phase Transformation Study of Ti-10V-2Fe-3Al, in *Titanium '80: Science and Technology*, Proc. Fourth Int. Conf. on Titanium, Kyoto, Japan, edited by H. Kimura and O. Izumi, The Metallurgical Society of AIME, 1980, pp. 1491-1501
112. E.W. Collings, Magnetic Studies of Omega-Phase Precipitation and Aging in Titanium-Vanadium Alloys, *J. Less-Common Metals*, Vol 39, 1975, pp. 63-90
113. G.M. Pennock, H.M. Flower, and D.R.F. West, The Control of α Precipitation by Two Step Ageing in β Ti-15%Mo, in *Titanium '80: Science and Technology*, Proc. Fourth Int. Conf. on Titanium, Kyoto, Japan, edited by H. Kimura and O. Izumi, The Metallurgical Society of AIME, 1980, pp. 1343-1351
114. T.S. Luhman, R. Taggart, and D.H. Polonis, The Effect of Omega Phase Reversion on the Superconducting Transition in Titanium-Base Alloys, *Scripta Metall.*, Vol 5, 1971, pp. 81-86
115. V. Chandrasekaran, R. Taggart, and D.H. Polonis, An Electron Microscopic Study of the Aged Omega Phase in Ti-Cr Alloys, *Metallography*, Vol 11, 1978, pp. 183-198
116. V. Chandrasekaran, R. Taggart, and D.H. Polonis, Fracture Modes in a Binary Titanium Alloy, *Metallography*, Vol 5, 1972, pp. 235-250
117. K. Nagai, K. Hiraga, T. Ogata, and K. Ishikawa, Heat Treatments and Low Temperature Fracture Toughness of a Ti-6Al-4V Alloy, *Adv. Cryo. Eng. (Materials)*, Vol 30, 1984, pp. 375-382
118. K. Nagai, K. Hiraga, T. Ogata, and K. Ishikawa, Cryogenic Temperature Mechanical Properties of β-Annealed Ti-6Al-4V, *Trans. Japan Inst. Metals*, Vol 26, 1985, pp. 405-413
119. V.A. Moskalenko and B.H. Pupsova, "Influence of Alloying on the Plastic Deformation of Alpha-Titanium at Low Temperatures" (in Russian), report from the Physico-Technical Institute for Low Temperatures, Ukr. Acad. Sci., 1970
120. V.A. Moskalenko, V.I. Startsev, and V.N. Kovaleva, Low Temperature Peculiarities of Plastic Deformation in Titanium and its Alloys, *Cryogenics*, Vol 20, 1980, pp. 503-508

121. J.T. Ryder and W.E. Witzell, Effect of Low Temperature on Fatigue and Fracture Properties of Ti-5Al-2.5Sn(ELI) for Use in Engine Components, in *Fatigue at Low Temperatures, Proceedings* (Louisville, KY, May 1983), ASTM, 1985, pp. 210-237
122. H. Conrad, B. de Meester, M. Döner, and K. Okasaki, Strengthening of Alpha Titanium by the Interstitial Solutes C, N, and O, in *Physics of Solid Solution Strengthening*, edited by E.W. Collings and H.L. Gegel, Plenum Press, 1975, pp. 1-45
123. H. Conrad and K.K. Wang, Solution Strengthening of Titanium by Aluminum at Low Temperatures, in *Strength of Metals and Alloys*, Fifth International Conference (Aachen, West Germany, August 1979) Proceedings, Vol 2, Pergamon Press, 1980, pp. 1067-1072
124. H. Conrad, Plastic Flow and Fracture of Titanium at Low Temperatures, *Cryogenics*, Vol 24, 1984, pp. 293-304
125. T. Nishimura, T. Mizoguchi, and Y. Itoh, Titanium Materials for Cryogenic Service, *Kobe Steel Engineering Reports*, Vol 34, No. 3, 1984, pp. 63-66
126. C.-A. Yin, M. Döner, and H. Conrad, Deformation Kinetics of Commercial Ti-50A (0.5 At. Pct. O_{eq}) at Low Temperatures ($T < 0.3T_m$), *Met. Trans.*, Vol 14A, 1983, pp. 2545-2555
127. H. Albert and I. Pfeiffer, Temperaturabhängigkeit der Festigkeitseigenschaften des Hochfeldsupraleiters NbTi50, *Z. Metallkde.*, Vol 67, 1976, pp. 356-360
128. H. Hillmann, Werkstoffe für dynamische beanspruchbare supraleitende Magnete, *Forschungsbericht BMFT-FB T 76-13*, Vacuumschmeltze, GmbH, Hanau, June 1976
129. F.F. Lavrentev, Yu.A. Pokhil, and P.P. Dudko, The Evolution of a Defect Structure and its Relation to the Deformation Parameters in Ti-Al-Nb-Zr Alloy at 4.2 K, *Mater. Sci. Eng.*, Vol 56, 1982, pp. 117-124
130. N.M. Madhava and R.W. Armstrong, Discontinuous Twinning of Titanium at 4.2 K, *Met. Trans.*, Vol 5, 1974, pp. 1517-1519
131. C.C. Koch and D.S. Easton, A Review of Mechanical Behavior and Stress Effects in Hard Superconductors, *Cryogenics*, Vol 17, 1977, pp. 391-413
132. V.A. Moskalenko, V.I. Startsev, and V.N. Kovaleva, Low Temperature Peculiarities of Plastic Deformation in Titanium and its Alloys, in *Titanium '80: Science and Technology*, Proc. Fourth Int. Conf. on Titanium, Kyoto, Japan, edited by H. Kimura and O. Izumi, The Metallurgical Society of AIME, 1980, pp. 821-830
133. Z.S. Basinski, The Instability of Plastic Flow of Metals at Very Low Temperatures, *Proc. Roy. Soc.*, Vol A240, 1957, pp. 229-242
134. D.T. Read, Metallurgical Effects in Niobium-Titanium Alloys, *Cryogenics*, Vol 18, 1978, pp. 579-584
135. E.W. Collings, *Applied Superconductivity, Metallurgy, and Physics of Titanium Alloys*, Vol 1, Plenum Press, 1986
136. P. Hochstuhl and B. Obst, Beta-Phase Instability in NbTi-Superconductors, in *Seventeenth International Conference on Low Temperature Physics, LT-17, Proceedings* (conference at Karlsruhe, West Germany, August 1984), North-Holland, 1984, pp. 1369-1370
137. P. Hochstuhl, "Lattice Instability and Metastable Phases in Niobium-Titanium Superconductors," KfK Report No. 3931 (in German), Kernforschungszentrum Karlsruhe, West Germany, July 1985
138. A.R.G. Brown, D. Clark, J. Eastabrook, and K.S. Jepson, The Titanium-Niobium System, *Nature*, Vol 201, 1964, pp. 914-915
139. C. Baker, The Shape-Memory Effect in a Titanium-35 wt.% Niobium Alloy, *Metal Science J.*, Vol 5, 1971, pp. 92-100
140. E.W. Collings, The Metal Physics of Titanium Alloys: A Review, in *Titanium '80: Science and Technology*, Proc. Fourth Int. Conf. on Titanium, Kyoto, Japan, edited by H. Kimura and O. Izumi, The Metallurgical Society of AIME, 1980, pp. 77-132

141. E.W. Collings, *The Physical Metallurgy of Titanium Alloys*, American Society for Metals, 1984
142. P.A. Blenkinsop, Developments in High Temperature Alloys, in *Titanium Science and Technology*, Proc. Fifth Int. Conf. on Titanium, Oberursel, West Germany, 1984, edited by G. Lutgering, U. Zwicker, and W. Bunk, D.G. für Metallkde., 1985, pp. 2323-2338
143. G.D. Lahoti and T. Altan, Research to Develop Process Models for Producing a Dual Property Titanium Alloy Compressor Disc, Tech. Report AFWAL-TR-80-4162, Air Force Wright Aeronautical Laboratory, August 1979 to July 1980, p. 327
144. I.I. Kornilov (ed.), *Titanium and Its Alloys*, Israel Program for Scientific Translation, Jerusalem, 1966, pp. 250-261
145. S.J. Savage and F.H. Froes, Production of Rapidly Solidified Metals and Alloys, *Journal of Metals*, Vol 36, No. 4, 1984, pp. 20-33
146. T.F. Broderick, A.G. Jackson, H. Jones, and F.H. Froes, The Effect of Cooling Conditions on the Microstructure of Rapidly Solidified Ti-6Al-4V, *Met. Trans.*, Vol 16A, 1985, pp. 1951-1959
147. E.W. Collings, R.E. Maringer, and C.E. Mobley, "Amorphous Glassy Metals and Microcrystalline Alloys for Aerospace Applications," Tech. Report AFML-TR-78-70 (for period January 1975 to August 1977), 1978
148. R.E. Maringer, C.E. Mobley, and E.W. Collings, An Experimental Method for the Casting of Rapidly Quenched Filament and Fibers, in *Rapidly Quenched Metals*, Section 1, edited by N.J. Grant and B.C. Giessen, MIT Press, 1976, pp. 29-36
149. R.E. Maringer, C.E. Mobley, and E.W. Collings, Preparation and Properties of Compacts of Cast Staple Fibers, in *A.I.Ch.E. Symposium: Spinning Wire From Molten Metal, Proceedings*, Vol 74, 1978, pp. 111-116
150. R.E. Maringer, C.E. Mobley, and E.W. Collings, Preparation and Properties of Compacts of Cast (Melt Extracted) Staple Fibers of Ti-6Al-4V, Vacuum Metallurgy Conference Proceedings, 1975, p. 336
151. E.W. Collings, C.E. Mobley, R.E. Maringer, and H.L. Gegel, Selected Properties of Melt-Extracted Titanium-Base Polycrystalline Alloys, in *Rapidly Quenched Alloys III*, Vol 1, edited by B. Cantor, The Metals Society (U.K.), 1978, pp. 188-191
152. R.E. Maringer, E.W. Collings, C.E. Mobley, and H.L. Gegel, "High Specific Strength Polycrystalline Titanium-Based Alloys," U.S. Patent No. 4149884, issued April 17, 1979
153. L.E. Tanner and R. Ray, Physical Properties of Ti_{50}-Be_{40}-Zr_{10} Glass, Scripta Met., Vol 11, 1977, pp. 783-789
154. D. Eylon, S. Fujishiro, P.J. Postans, and F.H. Froes, High-Temperature Titanium Alloys—a Review, *J. of Met.*, November 1984, pp. 55-62
155. S.M.L. Sastry, P.J. Meschter, and J.E. O'Neal, Structure and Properties of Rapidly Solidified Dispersion-Strengthened Titanium Alloys: Part I, Characterization of Dispersoid Distribution, Structure, and Chemistry, *Met. Trans.*, Vol 15A, 1984, pp. 1451-1463
156. S.M.L. Sastry, T.C. Peng, and L.P. Beckerman, Structure and Properties of Rapidly Solidified Dispersion-Strengthened Titanium Alloys: Part II, Tensile and Creep Properties, *Met. Trans.*, Vol 15A, 1984, pp. 1465-1474
157. R.G. Rowe and E.F. Koch, Rapidly Solidified Titanium Alloys Containing Cerium Sulfide and Oxysulfide Dispersions, in *Rapidly Solidified Materials*, edited by P.W. Lee and R.S. Carbonara, American Society for Metals, 1985, pp. 115-120
158. C.S. Chi and S.H. Whang, Microstructures and Mechanical Properties of Rapidly Solidified Ti-5Al-4.5La and Ti-5Al-5.4Er Alloys, in *Mechanical Behavior of Rapidly Solidified Materials*, edited by S.M.L. Sastry and B.A. MacDonald, The Metallurgical Society of AIME, 1985, pp. 231-245

159. S.M.L. Sastry, D.M. Bowden, and R.J. Lederich, Dispersion Strengthening of Ti-Al Alloys by Rapid Solidification Technology, in *Titanium Science and Technology*, Proc. Fifth Int. Conf. on Titanium, Oberursel, West Germany, 1984, edited by G. Lutgering, U. Zwicker, and W. Bunk, D.G. für Metallkde., 1985, pp. 435-441
160. S.M.L. Sastry, T.C. Peng, and J.E. O'Neal, Design and Development of Advanced Titanium Alloys by Rapid Solidification Technology, in *Titanium Science and Technology*, Proc. Fifth Int. Conf. on Titanium, Oberursel, West Germany, 1984, edited by G. Lutgering, U. Zwicker, and W. Bunk, D.G. für Metallkde., 1985, pp. 397-404
161. S.M.L. Sastry, T.C. Peng, P.J. Meschter, and J.E. O'Neal, Rapid Solidification Processing of Titanium Alloys, *J. of Met.*, September 1983, pp. 21-28
162. S.R. Seagle, G.S. Hall, and H.B. Bomberger, High Temperature Properties of Ti-6Al-2Sn-4Zr-2Mo-0.09Si, *Met. Eng. Quart.*, February 1975, pp. 48-54
163. H.B. Bomberger and F.H. Froes, Prospects for Developing Novel Titanium Alloys Using Rapid Solidification, in *Titanium Rapid Solidification Technology*, edited by F.H. Froes and D. Eylon, The Metallurgical Society, 1986, pp. 21-43
164. S.H. Whang, Rapidly Solidified Ti Alloys Containing Novel Additives, *J. of Met.*, April 1984, pp. 34-40
165. F.H. Froes and R.G. Rowe, Rapidly Solidified Titanium, in *Rapidly Solidified Alloys and Their Mechanical and Magnetic Properties*, edited by B.C. Giessen, D.E. Polk, and A.I. Taub, Materials Research Society, Vol 58, 1986, pp. 309-334
166. R.G. Rowe, E.F. Koch, T.F. Broderick, and F.H. Froes, Microstructural Study of Rapid Solidified Titanium Alloys Containing Erbium and Boron, in *Rapidly Solidified Materials*, edited by P.W. Lee and R.S. Carbonara, American Society for Metals, 1985, pp. 107-114
167. C.S. Chi and S.H. Whang, Microstructural Characteristics of Rapidly Quenched Alpha-Ti Alloys Containing La (personal communication)
168. S. Krishnamurthy, R.G. Vogt, D. Eylon, and F.H. Froes, Microstructures of Rapidly Solidified Titanium-Eutectoid Former Alloys, in *Rapidly Solidified Metastable Materials*, edited by B.H. Kear and B.C. Giessen, Materials Research Society, Vol 28, Elsevier Science Publishing Co., 1984, pp. 361-366
169. S. Krishnamurthy, I. Weiss, D. Eylon, and F.H. Froes, Aging Response of a Rapidly Solidified Beta-Eutectoid Ti-9 wt.% Co Alloy, in *Strength of Metals and Alloys*, Vol 2, edited by H.J. McQueen, J.-P. Bailon, J.I. Dickson, J.J. Jonas, and M.G. Akben, Pergamon Press, 1985, pp. 1627-1632
170. W.A. Baeslack III, S. Krishnamurthy, and F.H. Froes, Rapid Solidification and Aging of a Near-Eutectoid Titanium Nickel Alloy, in *Strength of Metals and Alloys*, Vol 2, edited by H.J. McQueen, J.-P. Bailon, J.I. Dickson, J.J. Jonas, and M.G. Akben, Pergamon Press, 1985
171. S. Krishnamurthy and F.H. Froes, Secondary Cooling Effects in Rapidly Solidified Titanium Alloys, in *Titanium Rapid Solidification Technology*, edited by F.H. Froes and D. Eylon, The Metallurgical Society, 1986, pp. 111-120
172. S.H. Whang, Rapidly Solidified Titanium Alloys for High-Temperature Applications, *J. Mater. Sci.*, Vol 21, 1986, pp. 2224-2238
173. T.C. Peng, S.M.L. Sastry, and J.E. O'Neal, Rapid Solidification Processing of Titanium Alloys, in *Titanium Science and Technology*, Proc. Fifth Int. Conf. on Titanium, Oberursel, West Germany, 1984, edited by G. Lutgering, U. Zwicker, and W. Bunk, D.G. für Metallkde., 1985, pp. 389-396
174. D.G. Konitzer, B.C. Muddle, H.L. Fraser, and R. Kirchheim, Refined Dispersions of Rare Earth Oxides in Ti-Alloys Produced by Rapid Solidification Processing, in *Titanium Science and Technology*, Proc. Fifth Int. Conf. on Titanium, Oberursel, West Germany, 1984, edited by G. Lutgering, U. Zwicker, and W. Bunk, D.G. für Metallkde., 1985, pp. 405-410
175. D.G. Konitzer, B.C. Muddle, and H.L. Fraser, Formation and Thermal Stability of

an Oxide Dispersion in a Rapidly Solidified Ti-Er Alloy, *Scripta Met.*, Vol 17, 1983, pp. 963-966
176. S.H. Whang, Y.Z. Lu, and Y.W. Kim, Microstructure and Age Hardening of Rapidly Quenched Ti-Zr-Si Alloys, *J. Mater. Sci. Lett.*, Vol 4, 1985, pp. 883-887
177. C.S. Chi and S.H. Whang, Rapidly Solidified Ti Alloys Containing Metalloids and Rare Earth Metals—Their Microstructure and Mechanical Properties, in *Rapidly Solidified Metastable Materials*, MRS 1983 Annual Meeting Symposia Proceedings, Vol 28, edited by B.H. Kear and B.C. Giessen, North-Holland, 1984, pp. 353-366
178. S.H. Whang, Occurrence of Metastable Phases in Binary Ti-Rich Alloys Quenched From the Melt, in *Undercooled Alloy Phases*, edited by E.W. Collings and C.C. Koch, The Metallurgical Society, 1987, pp. 163-183
179. J.E. O'Neal, S.M.L. Sastry, T.C. Peng, and J.F. Tesson, Microstructures of Rapidly Solidified Titanium Alloys, *Microstructural Science*, Vol 11, 1983, pp. 143-151
180. S. Krishnamurthy, A.G. Jackson, I. Weiss, and F.H. Froes, Aging Response of Rapidly Solidified Titanium-Tungsten Alloys with Nickel and Silicon Additions, in *Rapidly Solidified Materials*, edited by P.W. Lee and R.S. Carbonara, American Society for Metals, 1985, pp. 121-127
181. W.A. Baeslack III, S. Krishnamurthy, and F.H. Froes, A Study of Rapidly Quenched Microstructures in a Hypereutectoid Ti-40 wt.% W Alloy, *J. Mater. Sci. Lett.*, Vol 5, 1986, pp. 315-318
182. S.H. Whang, Occurrence of Metastable Phases in Binary Ti-Rich Alloys Quenched from the Melt, in *Undercooled Alloy Phases*, edited by E.W. Collings and C.C. Koch, The Metallurgical Society, 1986, pp. 163-183
183. D.G. Konitzer, B.C. Muddle, and H.L. Fraser, A Comparison of the Microstructure of As-Cast and Laser Surface Melted Ti-4Y, *Met. Trans.*, Vol 14A, 1983, pp. 1979-1988
184. Y.Z. Lu, C.S. Chi, and S.H. Whang, Second Phase Coarsening in Rapidly Solidified Ti-5Sn-4.5La System, in *Rapidly Quenched Metals*, Proc. Fifth Int. Conf. on Rapidly Quenched Metals, Wurzburg, West Germany, Vol 1, edited by S. Steeb and H. Warlimont, Elsevier Science Publishers, 1985, pp. 949-952
185. R.G. Vogt, D. Eylon, and F.H. Froes, Effect of Er, Si, and W Additions on Powder Metallurgy High Temperature Titanium Alloys, in *Titanium Rapid Solidification Technology*, edited by F.H. Froes and D. Eylon, The Metallurgical Society, 1986, pp. 177-194
186. D.B. Snow, Structure and Mechanical Properties of Laser-Consolidated Ti-6Al-4V and Ti-6Al-2Sn-4Zr-6Mo with Rare Earth Element Additions, in *Laser Processing of Materials*, edited by K. Mukherjee and J. Mazumder, The Metallurgical Society, 1984, pp. 83-89
187. A.G. Jackson, T.F. Broderick, and F.H. Froes, Microstructures of Rapidly Solidified Ti-5Al-2.5Sn with Si or Ge Additions, in *Titanium Science and Technology*, Proc. Fifth Int. Conf. on Titanium, Oberursel, West Germany, 1984, edited by G. Lutgering, U. Zwicker, and W. Bunk, D.G. für Metallkde., 1985, pp. 381-387
188. H.A. Lipsitt, D. Shechtman, and R.E. Schafrick, The Deformation and Fracture of TiAl at Elevated Temperatures, *Met. Trans.*, Vol 6A, 1975, pp. 1991-1996
189. D.G. Konitzer and H.L. Fraser, The Production and Thermal Stability of a Refined Dispersion of Er_2O_3 in Ti_3Al Using Rapid Solidification Processing, in *High Temperature Ordered Intermetallic Alloys*, MRS Symposium Proceedings, Vol 39, edited by C.C. Koch, C.T. Liu, and N.S. Stoloff, Materials Research Society, 1985, pp. 437-442
190. R.G. Rowe, J.A. Sutliff, and E.F. Koch, Dispersion Modification of Ti_3Al-Nb Alloys, in *Rapidly Solidified Alloys and Their Mechanical and Magnetic Properties*, edited by G.C. Giessen, D.E. Polk, and A.I. Taub, Materials Research Society, Vol 58, 1986, pp. 359-364; see also J.A. Sutliff and R.G. Rowe, Rare Earth Oxide Dis-

persoid Stability and Microstructural Effects in Rapidly Solidified Ti_3Al and Ti_3Al-Nb, ibid., pp. 371-376

191. R.G. Rowe, J.A. Sutliff, and E.F. Koch, Comparison of Melt Spun and Consolidated Ti_3Al-Nb Alloys With and Without a Dispersoid, in *Titanium Rapid Solidification Technology*, edited by F.H. Froes and D. Eylon, The Metallurgical Society, 1986, pp. 239-248
192. J.A. Sutliff and R.G. Rowe, Rare Earth Oxide Dispersoid Stability and Microstructural Effects in Rapidly Solidified Ti_3Al and Ti_3Al-Nb, in *Rapidly Solidified Alloys and Their Mechanical and Magnetic Properties*, edited by B.C. Giessen, D.E. Polk, and A.I. Taub, Materials Research Society, Vol 58, 1986, pp. 371-376
193. D. Eylon, C.M. Cooke, and F.H. Froes, Production of Metal Matrix Composites from Rapidly Solidified Titanium Alloy Foils, in *Titanium Rapid Solidification Technology*, edited by F.H. Froes and D. Eylon, The Metallurgical Society, 1986, pp. 311-322
194. P.L. Martin, M.G. Mendiratta, and H.A. Lipsitt, Creep Deformation of TiAl and TiAl+W Alloys, *Met. Trans.*, Vol 14A, 1983, pp. 2170-2174
195. M.G. Mendiratta and H.A. Lipsitt, Steady-State Creep Behaviour of Ti_3Al-Base Intermetallics, *J. Mater. Sci.*, Vol 15, 1980, pp. 2985-2990
196. H.A. Lipsitt, D. Shechtman, and R.E. Schafrick, The Deformation and Fracture of Ti_3Al at Elevated Temperatures, *Met. Trans.*, Vol 11A, 1980, pp. 1369-1375
197. S.M. Sastry. T.C. Peng, and J.E. O'Neal, Rapid Solidification and Powder Metallurgical Processing of Titanium Alloys, in *Modern Developments in Powder Metallurgy*, Vol 16, *Ferrous and Nonferrous Materials* (Toronto, Canada, June 1984), edited by E.N. Aqua and C.I. Whitman, Metal Powder Industries Federation, Princeton, NJ, 1985, pp. 577-606

11

Alloying of Nickel

NORMAN S. STOLOFF
Materials Engineering Department
Rensselaer Polytechnic Institute

Nickel alloys in commercial service and under development range from single-phase alloys to precipitation-hardened superalloys, oxide-dispersion-strengthened alloys and composites, and, most recently, intermetallic compounds (see Table 1). Nickel-base superalloys are the most complex, the most widely used for the hottest parts, and, to many metallurgists, the most interesting of all superalloys. Their use extends to the highest homologous temperature of any common alloy system, and they currently comprise over 50% of the weight of advanced aircraft engines.

The principal characteristics of nickel as an alloy base are the high phase sta-

Table 1. Classes of nickel-base alloys

Solid Solutions	DS Eutectics
IN 600	Ni,Cr,Al-TaC (Nitac)
Hastelloy	Ni,Al-Mo (γ/γ'-α)
Hastelloy X	Ni,Co,Al-NbC (Cotac 744)
γ'-Strengthened Superalloys	Fiber-Reinforced Composites
Waspaloy	NiCrAlY-W
Mar-M 200 + Hf	713-W-ThO_2
IN 100	Intermetallics
Udimet 700	Ni_3Al+B
PWA 1480	NiAl
γ''-Strengthened Superalloys	Ni_3Si
IN 718	
IN 901	
Dispersion-Strengthened Alloys	
TD-Ni	
MA 754	
MA 6000	

bility of the fcc nickel matrix and the ability to strengthen by a variety of means, some direct and some indirect. Further, the surface stability of nickel is readily improved by alloying with chromium and/or aluminum. This chapter is concerned primarily with strengthening mechanisms in nickel-base alloys. In order to adequately describe mechanical behavior, however, it is first necessary to consider the compositions and microstructures of the various classes of nickel alloys.

CHEMICAL COMPOSITION

The compositions of several representative nickel-base alloys are listed in Table 2.[1,2] They fall into two categories: Ni-Fe base, where nickel is the *major* solute element; and nickel-base, where at least 50% Ni is present. The most complex are the superalloys, in which as many as a dozen elements are included; in addition, deleterious elements such as silicon, phosphorus, sulfur, oxygen, and nitrogen must be controlled through appropriate melting practice. Other trace elements (e.g., selenium and lead) must be held to very small (ppm) levels in critical parts. Most nickel-base superalloys contain 10 to 20 wt % Cr, up to about 8% Al + Ti, 5 to 10% cobalt, and small amounts of boron, zirconium, and carbon. Other common additions are molybdenum, niobium, tungsten, tantalum, and hafnium, all of which play dual roles as strengthening solutes and carbide formers. Chromium and aluminum, in addition, are necessary to improve surface stability, through the formation of Cr_2O_3 and Al_2O_3, respectively. The functions of the various elements are summarized in Table 3. Other alloys, developed primarily for low-temperature service, often in corrosive environments, are likely to contain chromium and molybdenum, iron, or tungsten in solution, with little or no second phase present. Typical of these alloys are the Hastelloy series and IN 600 (see Table 1).

MICROSTRUCTURE

The major phases that may be present in nickel-base alloys are:

1. *Gamma matrix* (γ). The continuous matrix is an fcc nickel-base magnetic phase which usually contains a high percentage of solid-solution elements such as cobalt, iron, chromium, molybdenum, and tungsten. All nickel-base alloys contain this phase as the matrix.
2. *Gamma prime* (γ'). Aluminum and titanium are added in amounts required to precipitate fcc γ' (Ni_3Al, Ti), which precipitates coherently with the austenitic gamma matrix. Other elements, notably niobium, tantalum, and chromium, also enter γ'. This phase is required for high-temperature strength and creep resistance.
3. *Gamma double prime* (γ''). Nickel and niobium combine, in the presence of iron, to form bct Ni_3Nb, which is coherent with the gamma matrix while

Table 2. Compositions of selected nickel and Fe-Ni-base alloys[1,2]

Alloy	Compositions, wt %									
	Cr	Ni	Co	Mo	W	Nb	Ti	Al	Fe	C	Other
Fe-Ni-Base Alloys											
Incoloy 800	21.0	32.5	...	...	...	...	0.38	0.38	45.7	0.05	0.8 Mn; 0.5 Si
A-286	15.0	26.0	...	1.25	...	...	2.0	0.2	55.2	0.04	0.005 B; 0.3 V
Incoloy 901	12.5	42.5	...	6.0	...	...	2.7	...	36.2	0.10 max	...
Inconel 718	19.0	52.5	...	3.0	...	5.1	0.9	0.5	18.5	0.08 max	0.15 max Cu
Hastelloy X	22.0	49.0	15 max	9.0	0.6	...	...	2.0	15.8	0.15	...
Ni-Base Alloys											
Mar-M 200	9.0	59.0	10.0	...	12.5	1.0	2.0	5.0	1.0	...	0.015 B, 0.05 Zr
Waspaloy	19.5	57.0	13.5	4.3	...	...	3.0	1.4	2.0 max	0.07	0.006 B; 0.09 Zr
Udimet 700	15.0	53.0	18.5	5.0	...	...	3.4	4.3	<1.0	0.07	0.03 B
Astroloy	15.0	56.5	15.0	5.25	...	...	3.5	4.4	<0.3	0.06	0.03 B; 0.06 Zr
René 80	14.0	60.0	9.5	4.0	4.0	...	5.0	3.0	...	0.17	0.015 B; 0.03 Zr
IN 100	10.0	60.0	15.0	3.0	...	...	4.7	5.5	<0.6	0.15	1.0 V; 0.06 Zr; 0.015 B
IN 600	15.5	76.0	...	...	...	...	...	...	8.0	0.08	...
René 95	14.0	61.0	8.0	3.5	3.5	3.5	2.5	3.5	<0.3	0.16	0.01 B; 0.05 Zr
Mar-M 247	8.25	59.0	10.0	0.7	10.0	...	1.0	5.5	<0.5	0.15	0.015 B; 0.05 Zr; 1.5 Hf; 3.0 Ta
Hastelloy C	16.5	56.0	...	17.0	4.5	...	...	...	6.0	0.15 max	...
Hastelloy G	22.0	...	2.5 max	6.5	1.0 max	...	...	...	3.0	0.05 max	2 Cu
Monel 400	...	66.5	...	...	...	...	...	...	1.2	...	31.5 Cu
IN MA 754	20.0	78.5	...	...	...	...	0.5	0.3	...	...	0.6 Y_2O_3
IN MA 6000E	15.0	68.5	...	2.0	4.0	...	2.5	4.5	...	0.05	1.1 Y_2O_3; 2.0 Ta; 0.01 B; 0.15 Zr

Table 3. Effects of several elements in nickel-base superalloys

Element	Effect(a)
Chromium	Oxidation and hot corrosion resistance; solid-solution strengthening
Molybdenum, tungsten	Solid-solution strengthening; form M_6C carbides
Aluminum, titanium	Form γ', $Ni_3(Al,Ti)$, hardening precipitate; Ti forms MC carbides as well; Al enhances oxidation resistance
Cobalt	Raises γ' solvus temperature
Boron, zirconium, hafnium	Improve rupture life through increases in ductility; B also forms borides; Hf forms MC carbides and also promotes eutectic γ-γ' formation in cast alloys
Carbon	Forms MC, M_7C_3, $M_{23}C_6$, and M_6C carbides
Niobium	Forms γ', Ni_3Nb, hardening precipitate; forms δ orthorhombic Ni_3Nb
Tantalum	Solid-solution strengthening; forms MC carbides; enhances oxidation resistance

(a) Not all of these effects necessarily occur in a given alloy.

inducing large mismatch strains. This phase provides very high strength at low to intermediate temperatures, but is unstable at temperatures above about 815 °C.

4. *Carbides*. Carbon, added in amounts of about 0.05 to 0.2 wt %, combines with reactive elements such as titanium, tantalum, and hafnium to form MC carbides. During heat treatment and service these carbides tend to decompose and generate other carbides such as $M_{23}C_6$ and/or M_6C, which tend to form at grain boundaries. Carbides are formed in all superalloys except single crystals.
5. *Grain-boundary γ'*. In the stronger alloys, heat treatments and service exposure produce a film of γ' along the grain boundaries; this is believed to improve rupture properties.
6. *Borides*. Boron segregates to grain boundaries, resulting in the formation of a relatively low density of boride particles.
7. *TCP-type phases*. For some compositions, and under certain conditions, platelike phases such as σ, μ, and Laves may form; these phases cause lowered rupture strength and ductility.

In solid-solution alloys such as IN 600 and Hastelloy C (Tables 1 and 2), only the gamma matrix is present. In nickel-base superalloys, however, most of the above phases (except γ'') are generally present. Several Ni-Fe superalloys such as IN 706 and IN 718, on the other hand, contain γ'' Ni_3Nb as the principal precipitate, as well as γ'. Further, oxide-dispersion-strengthened alloys contain a few vol % of a dispersed phase such as Y_2O_3 in a γ-γ' matrix, while composites (mechanically incorporated) will contain tungsten or tungsten-alloy fibers. Directionally solidified eutectics based on nickel also consist of a γ-γ' matrix, reinforced with refractory metal or carbide fibers.

The Gamma Matrix

Pure nickel does not display an unusually high elastic modulus or low diffusivity (two factors that promote creep-rupture resistance), but the γ matrix is readily strengthened for the most severe temperature and time conditions. Some superalloys can be utilized at 0.85 T_M (melting point), and for times up to 100,000 h at somewhat lower temperatures. These conditions can be tolerated because of[3]:

1. The high tolerance of nickel for solutes without phase instability, due to its nearly filled *d*-shell
2. The tendency, with chromium additions, to form Cr_2O_3 having few cation vacancies, thereby restricting the diffusion rates of metallic elements outward and oxygen, nitrogen, and sulfur inward[4]
3. The additional tendency, at high temperatures, to form Al_2O_3, which displays exceptional resistance to oxidation.

Gamma Prime

Gamma prime (γ') is an intermetallic compound of nominal composition Ni_3Al, and is stable over a relatively narrow range of compositions (Fig. 1). It precipitated as spheroidal particles in early nickel-base alloys, which tended to have low volume fractions of particles (see Fig. 2a). Later, cuboidal precipitates were noted in alloys with higher Al + Ti contents (Fig. 2b). The change in morphology is connected with matrix-precipitate mismatch. It was observed that γ' occurs as spheres at 0 to 0.2% mismatches, becomes cuboidal for mismatches of 0.5 to 1%, and is platelike at mismatches above about 1.25%.

In order to fully understand the vital role of γ' in nickel-base superalloys, it is necessary to consider the structure and properties of this phase in some detail.

Gamma prime is a superlattice, possessing the Cu_3Au ($L1_2$)-type structure, which exhibits long-range order up to its melting point of 1385 °C. It exists over a fairly restricted range of composition, but alloying elements may substitute liberally for either of its constituents to a considerable degree. In particular, most nickel-base alloys are strengthened by a precipitate in which up to 60% of the aluminum can be substituted for by titanium and/or niobium. Further, nickel sites in the superlattice may be occupied by iron or cobalt atoms.

Unalloyed Ni_3Al deforms by $\{111\}\langle 110\rangle$ slip at all temperatures, and at temperatures above 400 °C some slip along $\{100\}$ also is observed; at 700 °C, $\{100\}$ slip predominates.[5]

Three types of stacking faults exist in the $L1_2$ structure[6]:

1. Superlattice (intrinsic or extrinsic) faults
2. Antiphase-boundary faults
3. Complex faults.

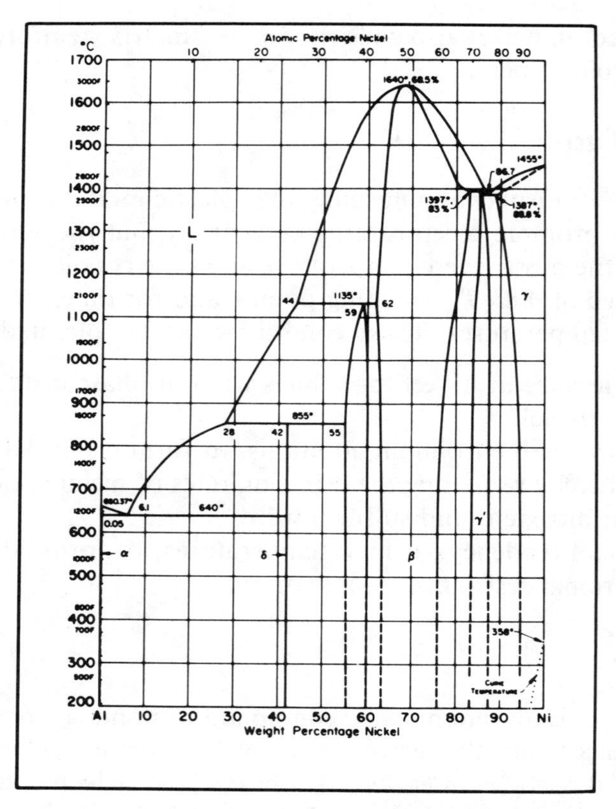

Fig. 1. Nickel-aluminum phase diagram

The large numbers of faults produced in nickel-base alloys may play a significant role in deformation. It will be shown in a later section that several precipitation-hardening models predict a very sensitive dependence of critical resolved shear stress (CRSS) on the fault energy of γ'. Also, creep properties of Mar-M 200 have been related to the nature of the faults in γ' left by the passage of dislocations through the particles.

Both single crystals and polycrystals of unalloyed γ' exhibit a startling, reversible[(7)] increase in flow stress between −196 °C and about 800 °C (which is highly dependent on solute content[(5)]), as shown in Fig. 3. While other superlattices exhibit a modest peak in strength over a rather narrow temperature range near T_c, the critical temperature for ordering, such peaks often are connected with a change in the degree of order with temperature.[(8)] However, several other superlattices—Ni_3Si, Co_3Ti, Ni_3Ge, and Ni_3Ga, all of $L1_2$ structure—display increasing strength over a temperature range comparable to that of Ni_3Al.[(9)] The flow stress in all of these superlattices is fully reversible on changing temperature.

The magnitude and temperature position of the peak in flow stress of γ' may be shifted by alloying elements such as titanium, chromium, and niobium (Fig. 3).[(5)] There is no simple relation between the magnitude of the flow-stress increase and the change in the temperature of the peak. Tantalum, niobium, and titanium are effective solid-solution hardeners of γ' at room temperature. Tungsten and

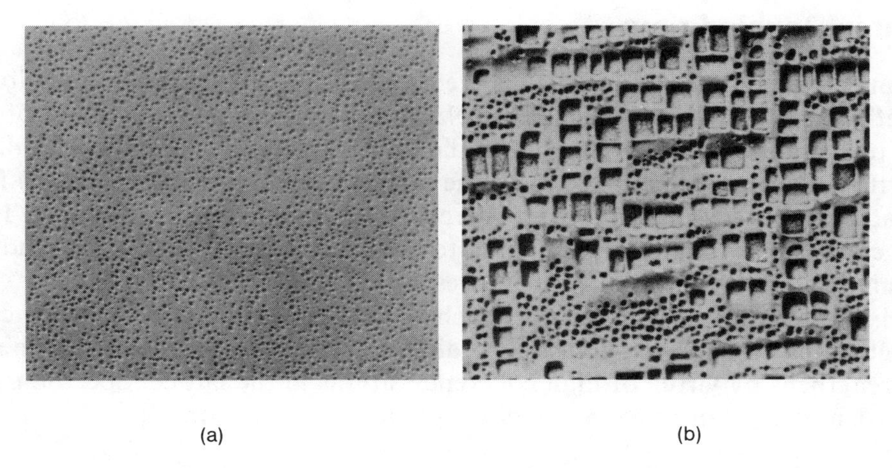

(a) (b)

(a) Spheroids in Waspaloy, typical of early, low-*f* alloys. (b) Cuboids in Udimet 700, typical of later, high-*f* alloys.

Fig. 2. Gamma-prime morphology in wrought nickel-base superalloys

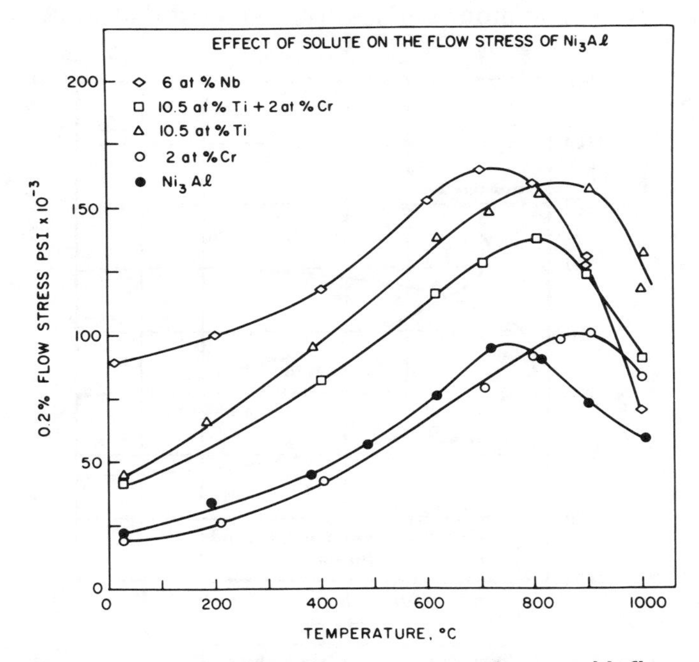

Fig. 3. Flow-stress peak in gamma prime, and influence of several solutes[5]

molybdenum are strengtheners at both room and elevated temperatures, while cobalt does not solid-solution strengthen γ'. All substitutional elements in Ni_3Al single crystals (Mo, Ta, Cb, Ti, W) increase the critical resolved shear stress (CRSS) for (111)[$\bar{1}$01] slip, but decrease the CRSS for (001)[$\bar{1}$10] slip, relative to binary Ni_3Al.[10]

Gamma Double Prime

Gamma double prime (γ''), a bct coherent precipitate of composition Ni_3Nb, precipitates in Ni-Fe-base alloys such as IN 706 and IN 718. In the absence of iron, or at temperatures and times shown in the transformation diagram of Fig. 4,[11,12] an orthorhombic precipitate of the same Ni_3Nb composition (delta phase) forms instead. The latter is invariably incoherent and does not confer strength. Therefore, careful heat treatment is required to ensure precipitation of γ'' instead of δ.

Gamma double prime often precipitates together with γ' in IN 718, but γ'' is the principal strengthening phase under such circumstances. Unlike γ', which causes strengthening through the necessity to disorder the particles as they are sheared, γ'' strengthens by virtue of high coherency strains in the lattice. Specific models of hardening for both phases are discussed in a later section.

Carbides

Carbides serve a number of functions in superalloys. First, carbides often precipitate at grain boundaries in nickel alloys, while in cobalt and iron superalloys, intragranular sites are common. Early work suggested that some grain-boundary

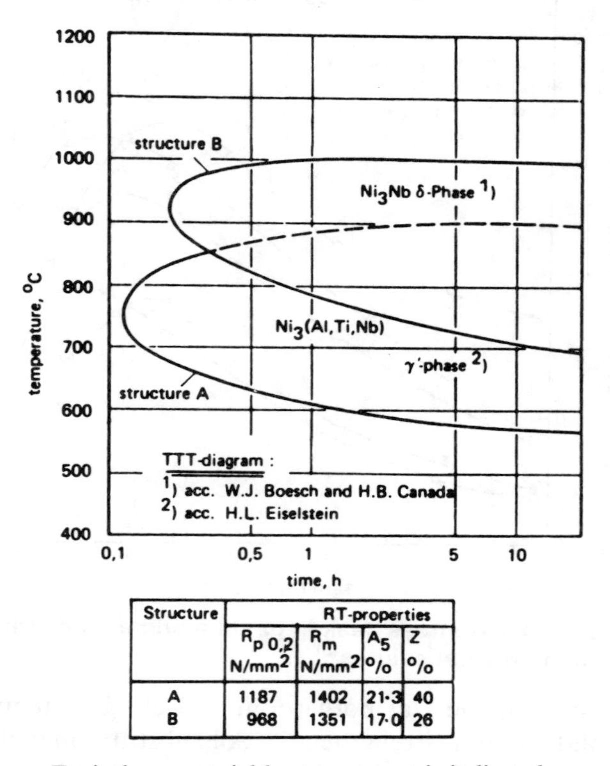

Structure	RT-properties			
	$R_{p\,0,2}$ N/mm²	R_m N/mm²	A_5 %	Z %
A	1187	1402	21·3	40
B	968	1351	17·0	26

Typical commercial heat treatment is indicated.

Fig. 4. Transformation diagram for IN 718 as-hot-rolled bar[11,12]

carbides were detrimental to ductility, but most investigators now believe that carbides exert a beneficial effect on rupture strength at high temperatures.

Carbide Types and Typical Morphologies. The common nickel-base alloy carbides are MC, $M_{23}C_6$, and M_6C.[3] MC usually exhibits a coarse, random cubic or script morphology (Fig. 5a). $M_{23}C_6$ is found primarily at grain boundaries (see Fig. 5b); it usually occurs as irregular, discontinuous, blocky particles, although plates and regular geometric forms have been observed. M_6C also can precipitate in blocky form in grain boundaries, and less often in a Widmanstätten intragranular morphology. Although data are sparse, it appears that continuous grain-

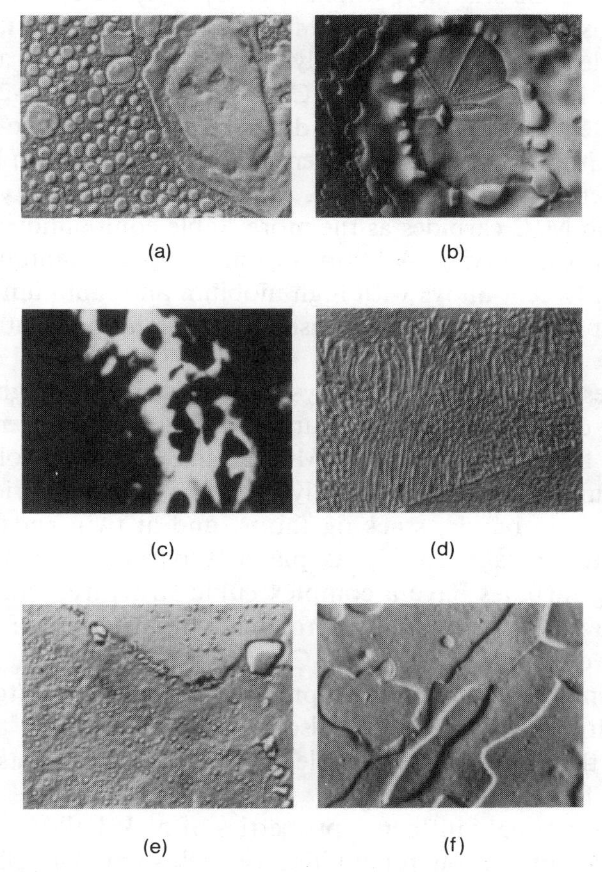

(a) Typical MC particle in γ'-strengthened alloy, degeneration commenced. Magnification, 4900×. (b) Degenerated MC (diamond) surrounded by $M_{23}C_6$ particles, and matrix, in IN 100. Magnification, 2450×. (c) Grain-boundary $M_{23}C_6$ particles in René 80. Transmission electron micrograph. (d) Cellular $M_{23}C_6$ formed in Nimonic 80A at 650 °C (1200 °F). Magnification, 4900×. (e) Fine $M_{23}C_6$ and coarse MC at grain boundary of alloy X 750. Magnification, 4900×. (f) Blocky M_6C_5 surrounded by γ' at grain boundary of AF 1753. Magnification, 4900×.

Fig. 5. Carbides in nickel-base alloys[3]

boundary $M_{23}C_6$ and Widmanstätten M_6C are to be avoided for best ductility and rupture life.[3] Various grain-boundary carbides are shown in Fig. 5(c) to (f).

MC carbides, fcc in structure, usually form in superalloys during freezing; they are distributed heterogeneously throughout the alloy, both in intergranular and transgranular positions, often interdendritically. Little or no orientation relationship with the alloy matrix has been noted. MC carbides are a major source of carbon for subsequent phase reactions during heat treatment and service.[3]

MC carbides—e.g., TiC and HfC—are among the most stable compounds in nature. The preferred order of formation in superalloys for these carbides is: HfC, TaC, NbC, and TiC in order of decreasing stability. This order is not the same as that of thermodynamic stability, which is HfC, TiC, TaC, and NbC.[3] In these carbides, M atoms can readily substitute for each other, as in (Ti,Nb)C. However, the less reactive elements, principally molybdenum and tungsten, also can substitute in these carbides. For example, (Ti,Mo)C is found in U-500, M-252, and René 77. It appears that the change in stability order cited above is due to the molybdenum or tungsten substitution, which so weakens the binding forces in MC carbides that degeneration reactions, discussed below, can occur. This leads typically to formation of $M_{23}C_6$ and M_6C carbides as the more stable compounds in the alloys after heat treatment and/or service. Additions of niobium and tantalum tend to counteract this effect. Recent alloys with high niobium and tantalum contents contain MC carbides that do not break down easily during solution treatment in the range 1200 to 1260 °C.

$M_{23}C_6$ carbides readily form in alloys with moderate-to-high chromium contents. They form during lower-temperature heat treatment and service—that is, 760 to 980 °C—both from degeneration of MC carbides and from soluble residual carbon in the alloy matrix. Although usually seen at grain boundaries, they occasionally occur along twin bands, stacking faults, and at twin ends. In Mar-M 200, $M_{23}C_6$ precipitates intragranularly as platelets parallel to (110) planes of the austenite. $M_{23}C_6$ carbides have a complex cubic structure, which, if the carbon atoms were removed, would closely approximate the structure of the TCP σ phase. In fact, σ plates often nucleate on $M_{23}C_6$ particles.

When tungsten or molybdenum is present, the approximate composition of $M_{23}C_6$ is $Cr_{21}(Mo,W)_2C_6$ although it also has been shown that appreciable nickel can substitute in the carbide; it is possible also that small amounts of cobalt or iron could substitute for chromium.

$M_{23}C_6$ particles strongly influence properties of nickel alloys. Rupture strength is improved, apparently through inhibition of grain-boundary sliding. Eventually, however, failure can initiate either by fracture of particles or by decohesion of the carbide/matrix interface. In some alloys, cellular structures of $M_{23}C_6$ have been noted; these can cause premature failures, but can be avoided by control of chemistry or by proper heat treatment.

M_6C carbides have a complex cubic structure; they form when the molybdenum and/or tungsten content is more than 6 to 8 at. %, typically in the range 815 to 980 °C. M_6C forms with $M_{23}C_6$ in Mar-M 200, B-1900, René 80, René 41, and

AF1753. Typical formulas for M_6C are $(Ni,Co)_3Mo_3C$ and $(Ni,Co)_2W_4C_6$, although a wider range of compositions has been reported for Hastelloy X. Therefore, M_6C carbides are formed when molybdenum or tungsten acts to replace chromium in other carbides; unlike $M_{23}C_6$, the composition can vary widely. Since M_6C carbides are stable at higher temperatures than $M_{23}C_6$ carbides, M_6C is more effective as a grain-boundary precipitate to control grain size during processing of wrought alloys.[4]

Carbide Reactions. MC carbides are a major source of carbon in most nickel-base superalloys below 980 °C. However, MC decomposes slowly during heat treatment and service, releasing carbon for several important reactions.

The principal carbide reaction in many alloys is believed to be the formation of $M_{23}C_6$, as follows[3]:

$$MC + \gamma \rightarrow M_{23}C_6 + \gamma' \qquad \text{(Eq 1)}$$

or

$$(Ti,Mo)C + (Ni,Cr,Al,Ti) \rightarrow Cr_{21}Mo_2C_6 + Ni_3(Al,Ti) \qquad \text{(Eq 2)}$$

This equation cannot be balanced thermodynamically; it was assumed by metallographic observations of phase transformation at grain boundaries by Sims[13] and by Phillips.[14] Reaction 1 begins to occur at about 980 °C and has been observed at temperatures as low as approximately 760 °C. In a few cases, it has been found to be reversible. M_6C can form in a similar manner.

Also, M_6C and $M_{23}C_6$ interact, forming one from the other[3]:

$$M_6C + M' \rightarrow M_{23}C_6 + M'' \qquad \text{(Eq 3)}$$

or

$$Mo_3(Ni,Co)_3C + Cr \rightleftarrows Cr_{21}Mo_2C_6 + (Ni,Co,Mo) \qquad \text{(Eq 4)}$$

depending on the alloy. For example, René 41 and M-252 can be heat treated to generate MC and M_6C initially; long-time exposure then causes conversion of M_6C to $M_{23}C_6$. Conversely, in Mar-M 200,[15] M_6C can be formed from $M_{23}C_6$. The type of refractory-metal atoms present may control the reaction.

These reactions lead to carbide precipitation in various locations, but typically at grain boundaries. Perhaps the most beneficial reaction, and that controlled in many heat treatments, is Reaction 2. Both the blocky carbides and the γ' produced are important, in that they may inhibit grain-boundary sliding; in any case, the γ' generated by this reaction coats the carbides, and the grain boundary will be a relatively ductile, creep-resistant layer.

Carbon also is in solution; at temperatures from 595 to 760 °C its solubility has been exceeded in cooling. Examples of very fine $M_{23}C_6$ precipitating directly on stacking faults have been noted[3]:

$$(\gamma) \rightarrow M_{23}C_6 + \gamma \qquad \text{(Eq 5)}$$

$$(Cr,Mo,C) \rightarrow (Cr_{21}Mo_2)C_6 \qquad \text{(Eq 6)}$$

Mihalisin[16] has suggested that carbon is slowly depleted through the following reaction based on studies of a series of experimental alloys:

$$TiC \rightarrow M_7C_3 \rightarrow Cr_{23}C_6 \rightarrow \sigma \qquad \text{(Eq 7)}$$

Borides

Boron often is present at levels of 50 to 500 ppm in superalloys. It segregates to grain boundaries, prolongs rupture life, and improves rupture ductility. In one alloy, more than 120 ppm boron reacts to form two types of M_3B_2 borides depending on thermal history: one is approximately $(Mo_{0.48}Ti_{0.07}Cr_{0.39}Ni_{0.03}Co_{0.03})_3B_2$ and the other is $(Mo_{0.31}Ti_{0.07}Cr_{0.49}Ni_{0.06}Co_{0.07})_3B_2$.[12] Borides are hard, refractory particles with shapes varying from blocky to half-moon in appearance; they act as a source of boron for the grain boundaries.[3]

TCP Phases

In some alloys, if composition has not been carefully controlled, undesirable phases can form either during heat treatment or, more commonly, during service. These precipitates, known as TCP phases, are composed of close-packed layers of atoms parallel to {111} planes of the gamma matrix. Usually harmful, they may appear as long plates or needles, often nucleating on grain-boundary carbides. Nickel alloys are prone especially to formation of sigma (σ) and mu (μ). The formula for σ is $(Fe,Mo)_x(Ni,Co)_y$, where x and y can vary from 1 to 7. Inhomogeneities in composition, often found in cast alloys, can cause localized TCP phase formation.

The hardness and plate morphology of sigma causes premature cracking, leading to low-temperature brittle failure, although yield strength is unaffected.[3] However, the effect on elevated-temperature rupture strength is particularly serious. Sigma formation must deplete refractory metals in the gamma matrix, causing loss of strength of the latter. Also, high-temperature fracture can occur along σ plates rather than along the normal intergranular path, resulting in sharply reduced rupture life.[17] Platelike μ also can form, but little is known about its detrimental effects.

ALLOYING FOR YIELD OR TENSILE STRENGTH

Solid-Solution Strengthening

Commercial nickel-base alloys always contain substantial alloying additions in solid solution to provide strength, creep resistance, or resistance to surface degradation.

In the case of solid-solution strengthening, it is convenient to discuss several theories of yielding in terms of the effects of solutes on various physical or crystallographic properties—for example, lattice parameter and elastic modulus.

In general, strengthening mechanisms are considered to be independent and additive, although there is considerable controversy as to the means of superposing hardening mechanisms. For the purpose of this chapter, we will treat hardening mechanisms as essentially independent.

Lattice Parameter

Substitutional solutes produce lattice misfit, ϵ, defined as the difference Δa between the lattice parameter of the pure matrix, a_0, and the lattice parameter of the solute atom, a, per unit concentration c:

$$\epsilon = \frac{1}{c}\frac{\Delta a}{a_0} \qquad \text{(Eq 8)}$$

Fleischer[18] has shown, in turn, that lattice-parameter mismatch may be approximated by the relative difference in atomic volume V between solvent S and solute X.

$$\epsilon = \frac{1}{a}\frac{da}{dc} \sim \frac{V_X - V_S}{V_S} \qquad \text{(Eq 9)}$$

While a linear relation between flow stress and solute content is obeyed for several solute elements in nickel (Fig. 6),[19,20] Pelloux and Grant[19] have shown that the change in yield stress for various solutes in nickel is not a single-value function of the lattice parameter, but depends directly on the positions of the solute in the periodic table. N_v is the number of electron vacancies in the third shell of the first long period. For the same lattice strains, the larger the valence difference between solutes and solvent, the greater the hardening. The strengthening influence of alloying elements persists to temperatures at least as high as 815 °C.[19] Fleischer[18] suggests that valency effects may be explained by modulus differences between the various alloys, as discussed in the following section. Alternatively, the effects of valency may be felt through the decrease in stacking-fault energy (SFE) of fcc alloys with increasing electron atom ratio; Beeston *et al.*[21] have correlated

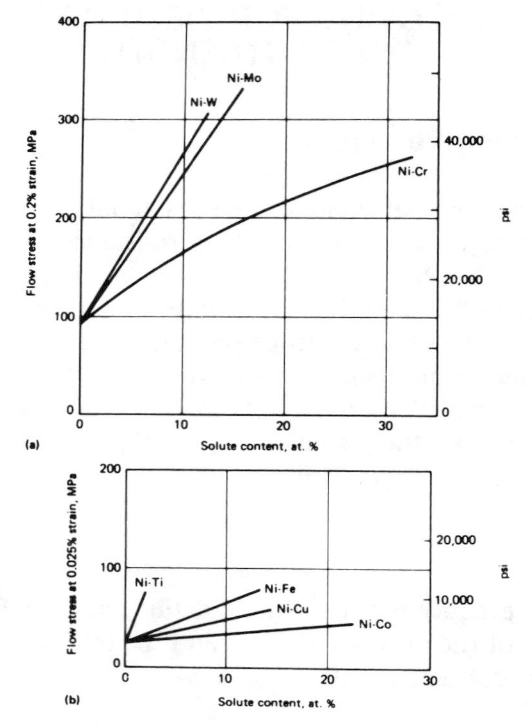

(a) Adapted from Pelloux and Grant.[19] (b) Adapted from Parker and Hazlett.[20]

Fig. 6. Effects of solute atoms on yield strength of single-phase nickel at 25 °C

the electron vacancy number N_v with SFE. It is generally expected that yield stress varies inversely with SFE.

The solid-solution elements typically found in the gamma phase include aluminum, iron, titanium, chromium, tungsten, cobalt, and molybdenum. The difference in atomic diameter from that of nickel varies from less than 1% for cobalt to 17% for titanium.

Modulus

Fleischer's[18] suggestion that modulus differences between solute and solvent may give rise to strengthening is based on the argument that extra work is needed to force a dislocation through hard or soft regions in the matrix.

Fleischer[18,22] concluded that modulus differences and lattice misfit can be incorporated into a single equation, such that where $\epsilon_G' = [(1/G)(dG/dc)]/[1 + (1/2G)(dG/dc)]$ is a modulus misfit parameter and Z is a constant, τ_0 is the critical resolved shear stress of the pure metal:

$$\tau_c = \tau_0 + \frac{G|\epsilon_G - \alpha\epsilon|^{3/2}c^{1/2}}{Z} \qquad \text{(Eq 10)}$$

Labusch[23] extended Fleischer's model to higher concentrations by use of a different type of statistical average of the interaction between solute atoms and dislocations. The resulting expression is:

$$\tau_c = \tau_0 + \frac{G\epsilon^{4/3}c^{2/3}}{550} \qquad \text{(Eq 11)}$$

Although some agreement with Eq 10 was found for copper-base alloys, data for ductile gold-, silver-, and copper-base alloy single crystals showed better agreement with $c^{2/3}$, as predicted by Eq 11.[23] There seem to be no data in the literature specific to nickel-base alloys to verify Eq 10 or 11.

Short-Range Order

Concentrated solid solutions are likely to exhibit appreciable short-range order. The energy required to shear a short-range-ordered crystal causes an increase in the flow stress of the alloy.

The shear stress to move a dislocation as a consequence of short-range order is:

$$\tau = 16\left(\frac{2}{3}\right)\frac{\frac{1}{2}c(1-c)\nu a_s}{a^3} \qquad \text{(Eq 12)}$$

where ν is the interaction energy $[= V_{AB} - \frac{1}{2}(V_{AA} + V_{BB})]$ and a_s is the short-range-order coefficient. Since all terms in Eq 12 are temperature independent, short-range order provides an athermal increment to flow stress. However, a_s increases with decreasing annealing temperature. Consequently, the short-range-order component of flow stress is sensitive to thermal history.

Nordheim and Grant[24] have suggested that short-range order exists in Ni-Cr alloys in the neighborhood of 20 to 25 wt % Cr. This composition range is reached in Hastelloy X and Inconel 625 (22% Cr), and is approached in other γ'-lean alloys such as the early Nimonic series (19.5% Cr). Consequently, short-range-order strengthening may occur in such alloys.[25]

PRECIPITATION HARDENING

Properties of γ-γ Alloys

The major contribution to the strength of precipitation-hardened nickel-base superalloys is provided by the formation of coherent stable intermetallic compounds such as γ' [$Ni_3(Al,Ti)$] and γ'' [$Ni_3(Nb,Al,Ti)$]. Other phases—for example, borides and carbides—provide little additional strengthening at low temperatures due to the small volume fraction present. However, significant effects on creep rate, rupture life, and rupture strain may be provided by these phases.

We are concerned with the properties of γ'-strengthened alloys exclusively in this section, with the experimentally determined strength dependent on such diverse factors as:

1. Volume fraction, f, of γ'
2. Radius, r_0, of γ'
3. Solid-solution strengthening of both γ and γ'
4. Presence of hyperfine γ'.

Particle-Cutting Models

Among the factors that have been suggested to account for observed hardening of nickel-base superalloys by coherent particles are the following:

1. Coherency strains
2. Differences in elastic moduli between particle and matrix
3. Existence of order in the particles
4. Differences in SFE of particle and matrix
5. Energy to create additional particle/matrix interface
6. Increases in lattice resistance of particles with temperature.

While several mechanisms may apply to any single system, theoreticians consider only one mechanism at a time, and then, if necessary, add the increment in shear stress due to each of the various mechanisms. It now appears, however, that the major factors that contribute to strengthening in superalloys are coherency strains and the presence of order in the particles. Consequently, our discussion is limited to these mechanisms and to the Orowan[26] dislocation-bypass model, which limits the strengthening achieved by the other mechanisms.

Rather than attempt a detailed analysis of the various models proposed to account for order and misfit hardening, it is our intention to discuss the principles underlying the most pertinent models. The methods used to treat hardening by solutes and by precipitates are basically similar; they depend on calculating the force of interaction between a moving dislocation and whatever obstacles are in its path.

In order to move through a field of dispersed obstacles, a dislocation must bend to an angle ϕ that is dependent on the obstacle strength (see Fig. 7). For weak obstacles, $\phi \rightarrow \pi$—that is, very little bending is required for the dislocation to escape the obstacle; for strong obstacles, $\phi \rightarrow 0$, as the dislocation is forced to almost double back on itself. The number of obstacles per unit length of dislocation line depends on ϕ; if $\phi \simeq \pi$, the number per unit length is found by computing the number that intersects a random line. As ϕ decreases from π, the dislocation sweeps out more area and therefore meets more obstacles, necessitating expressions for obstacle spacing which depend on the applied stress τ. The most commonly used expression is that of Friedel[28]:

$$L' = \left(\frac{2TL_s^2}{\tau b}\right)^{1/3} \quad \text{(Eq 13)}$$

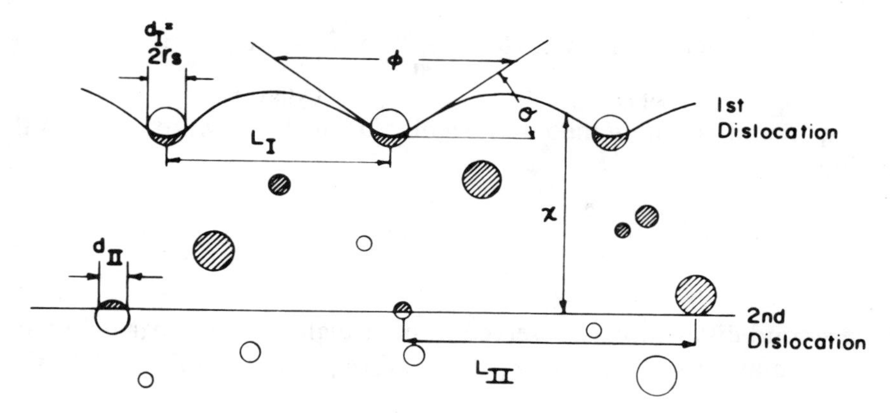

Fig. 7. Dislocation pairs interacting with ordered particles, showing effect of bend angle ϕ on obstacle spacing.[27] Shaded areas represent APB.

where T is the line tension ($\simeq Gb^2/8$ for edge and $Gb^2/2$ for screw dislocations) and L_s is the square lattice spacing ($= n^{1/2}$ where n is the number of particles per unit area of slip plane). To simplify calculations it is generally assumed that dislocations interact with a random array of obstacles of fixed strength. The limits on L' are:

$$L_s \leq L' \leq \frac{4r}{3f} \qquad \text{(Eq 14)}$$

The upper limit, $4r/3f$, represents the spacing of random obstacles along a straight line.

Order Strengthening. Consider the case of superlattice dislocation pairs interacting with particles. The calculation follows the principles first elaborated by Gleiter and Hornbogen[29] but utilizes the specific equations developed by Ham[30] and by Brown and Ham.[27] As the first dislocation is just shearing the particles (see Fig. 7), the second dislocation is pulled forward by the APB remaining in all particles cut by the first dislocation. Provided that the two dislocations assume the same shape and that the separation x between the two dislocations is sufficiently small, but larger than r_s, the second dislocation may lie outside of all the particles. This situation may occur at long aging times.

In general, however, the second dislocation does come into contact with the APB in the sheared particles and is nearly straight. The more APB is cut by the second dislocation, the less effective the particles become as obstacles. Then, referring to Fig. 7, the force balances are as follows, neglecting any new particle/matrix interface formed by shear of the particles:

$$\text{On dislocation 1, } \tau b + \frac{Gb^2}{2\pi kx} - \frac{\gamma_0 d_{\mathrm{I}}}{L_{\mathrm{I}}} = 0 \qquad \text{(Eq 15)}$$

$$\text{On dislocation 2, } \tau b + \frac{\gamma_0 d_{\mathrm{II}}}{L_{\mathrm{II}}} - \frac{Gb^2}{2\pi kx} = 0 \quad \text{(Eq 16)}$$

Solving Eq 15 and 16 simultaneously, we obtain for the forward stress on the first dislocation:

$$2\tau b + \frac{\gamma_0 d_{\mathrm{II}}}{L_{\mathrm{II}}} = \frac{\gamma_0 d_{\mathrm{I}}}{L_{\mathrm{I}}} \quad \text{(Eq 17)}$$

Since the second dislocation is observed to be straight during shear by the first dislocation, we may substitute f for $d_{\mathrm{II}}/L_{\mathrm{II}}$ and $(4\gamma_0 f r_s)^{1/2}/\pi T$ for $d_{\mathrm{I}} L_{\mathrm{I}}$ so that:

$$2\tau b + \gamma_0 f = \left(\frac{4\gamma_0 f r_s}{\pi T}\right)^{1/2} \gamma_0 \quad \text{(Eq 18)}$$

leading to the following relation for the applied stress τ:

$$\tau = \frac{\gamma_0}{2b}\left[\left(\frac{4\gamma_0 f r_s}{\pi T}\right)^{1/2} - f\right] \quad \text{(Eq 19)}$$

When $r_s \leq \pi f/4\gamma_0$, $\tau = 0$, since both dislocations touch the same fraction of APB. Gleiter and Hornbogen(29) also have emphasized the fact that as more and more APB is touched by the second dislocation, the flow stress must drop.

If the line tension T is approximated by $\frac{1}{2}Gb^2$ (screw dislocations), then Eq 19 reduces to:

$$\tau_c = \frac{\gamma_0}{2b}\left[\left(\frac{8\gamma_0 f r_s}{\pi G b^2}\right)^{1/2} - f\right] \quad \text{(Eq 20)}$$

Equation 20 cannot hold for r_s approaching zero, since τ_c cannot be negative. Nevertheless, the negative intercept, $\gamma_0 f/2b$, can be used for an alternate computation of APB energy, as has been done by Martens and Nembach.(31)

The first term of Eq 20, $A\gamma_0^{3/2} f^{1/2} G^{-1/2} b^{-2} r_s^{1/2}$, is similar to an equation for order strengthening that had been proposed earlier by Gleiter and Hornbogen(29):

$$\tau_c = 0.28\gamma_0^{3/2} f^{1/3} G^{-1/2} b^{-2} r_0^{1/2} \quad \text{(Eq 21)}$$

except for a different constant and the dependence of volume fraction. The second term in Eq 20 may be dropped, however, only when the second dislocation can avoid all particles.(27) The flow stress given by Eq 20 would then reduce to one-half of the stress necessary to shear particles by single dislocations. The basic features of this model are summarized schematically in Fig. 8 and have been applied successfully to a variety of austenitic superalloys,(29,31) both nickel- and iron-base. Figure 8 shows two branched curves for either single dislocations or

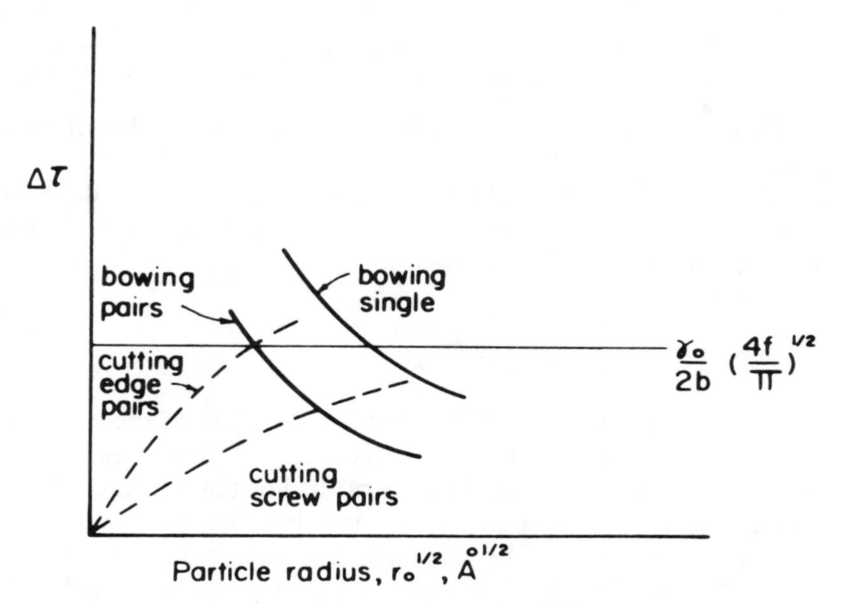

Fig. 8. Schematic age-hardening curves illustrating relation between order hardening and Orowan bowing as particle size increases

superlattice dislocations. In either case, order-induced hardening gives way to Orowan bowing as precipitates grow large and become incoherent. The Orowan mechanism is discussed in a later section.

Copley and Kear[32,33] also modified the Gleiter-Hornbogen theory; the results were applied specifically to the alloy Mar-M 200. Based on extensive electron-microscopic studies, the rate-controlling step for plastic deformation was shown to be moving dislocations from γ into γ'. Instead of setting up a force balance with the first dislocation partly through the particles, as in Fig. 7, they contend that the first dislocation wraps around the particle, assuming its curvature, until forced in by the second dislocation. The conditions of static equilibrium for the leading and trailing dislocations of the superlattice pair about to enter a particle are:

$$\text{Dislocation 1, } (\tau_c - \tau_p)b + \frac{C}{x} + \frac{T}{r_0} - \gamma_0 = 0 \qquad \text{(Eq 22)}$$

$$\text{Dislocation 2, } (\tau_c - \tau_0)b - \frac{C}{x} + \frac{T}{r_0} = 0 \qquad \text{(Eq 23)}$$

where C/x is the force of repulsion between two dislocations, τ_p is the friction stress of particle, and τ_0 is the friction stress of matrix. T/r_0 is the line tension force due to a dislocation assuming the curvature of the particle.

Solving Eq 22 and 23 simultaneously, one obtains for conditions of static equilibrium at 22 °C:

$$\tau_c = \frac{\gamma_0}{2b} - \frac{T}{br_0} + \frac{1}{2}(\tau_0 + \tau_p) \quad \text{(Eq 24)}$$

where $\gamma_0/2b$ is stress to constrict the dislocation pair to the point where particle shear begins.

For dynamic conditions, the CRSS is predicted from the stress dependence of the plastic strain rate, which is related to a derived dislocation velocity-stress function. A very similar equation is obtained for τ_c:

$$\tau_c = \frac{\gamma_0}{2b} - \frac{T}{br_0} + \frac{k}{2}(\tau_0 + \tau_p) \quad \text{(Eq 25)}$$

where k is a constant dependent on the dislocation velocity of the crystal and has a value of 0.823 for Mar-M 200 at room temperature.[33] Penetration is easier for small particles than for large particles because of the line tension force. In any case, the major contribution to τ_c at room temperature is provided by the term $\gamma_0/2b$, which represents about 80% of the total flow stress computed for Mar-M 200. Leverant *et al.*[34] concluded, however, that at high temperatures and high strain rates, where the flow stress of γ' reaches a distinct peak, both the APB energy and the flow stress of γ' are major contributors to τ_c.

In the most general case, γ should be replaced by Γ, the fault energy for shear of the particle, since faults other than APB-type faults may be produced by shear. For example, particle shearing occurs by loosely coupled intrinsic-extrinsic fault pairs in Mar-M 200 at 760 °C (1400 °F).[34] In this model the influence of crystal orientation on flow stress is felt through the variation in the nature of faults produced by different glide mechanisms. Decker[4] has pointed out that, based on the critical temperature for ordering of their respective Ni_3X phases, titanium, niobium, and tantalum in γ' should not appreciably increase APB energy. However, titanium and perhaps tantalum could increase the energy of other fault types. Brown and Ham[27] analyzed several sets of data to calculate APB energy as a function of alloy content, and found that the fault energy may be widely varied (see Table 4). This table is discussed later, in conjunction with alloy-design principles.

Misfit Strengthening. An early attempt[35] to relate the influence of coherency strains to CRSS failed to explain the dependence of CRSS on particle size. A model has been assumed by Gerold and Haberkorn[36] in which the interaction between dislocations and strain fields around the particles plays a dominant role.

The calculation is similar in outline to that of Fleischer[18] for solid-solution hardening and is expected to apply for matrix-particle misfit, ϵ, of approximately 0.01 in the case of spherical, coherent particles ($\epsilon = a_{ppt} - a_{matrix}/a_{matrix}$).

The increase in flow stress due to interaction of single dislocations with strain fields is given by:

$$\Delta\tau = \frac{K}{bL''} \quad \text{(Eq 26)}$$

Table 4. Approximate antiphase-boundary energies of γ' in different alloys

Alloy composition	APB energy, erg/cm^2
Ni-12.7 to 14.0Al (at. %)	153
Ni-18.5Cr-7.5Al (at. %)	104
Ni-8.8Cr-6.2Al (at. %)	90
Fe-Cr-Ni-Al-Ti	
Ti/Al = 1	240
Ti/Al = 8	300
Ni-19Cr-14Co-7Mo-2Ti-2.3Al (at. %)	170 to 220
Ni-33Fe-16.7Cr-3.2Mo-1.6Al-1.1Ti (wt %)	270

where K is the maximum repelling force of the strain field of a single particle on a moving dislocation and L'' is the average distance between the force centers. The problem is to find appropriate expressions for K and L''. K is found to be equal to or less than T, the line tension of an edge dislocation. For L'', the authors use an expression[18] in which the obstacle spacing depends on the bend angle, $\theta = \frac{1}{2}(\pi - \phi)$ (see Fig. 7):

$$L'' = \frac{r_0 \pi^{1/2}}{(\theta f)^{1/2}}, \quad \frac{9\pi f}{16} < \theta < \frac{3}{2} \qquad \text{(Eq 27)}$$

The angle to which a dislocation is bent by the force K before escaping the particle is given by:

$$2 \sin\theta \simeq \frac{K}{2T} \qquad \text{(Eq 28)}$$

The maximum value of K is computed to be:

$$K = 4G|\epsilon|br \qquad \text{(Eq 29)}$$

For bend angles smaller than $9\pi f/16$, the dislocation must be treated as a rigid line; for bend angles near 3/2, the dislocation line is totally flexible and another expression for L'' must be used. Combining Eq 27 and 29, the CRSS is obtained:

$$\Delta\tau = AG\epsilon^{3/2}\left(\frac{r_0 f}{b}\right)^{1/2}, \quad \frac{9\pi f}{16} < \frac{3|\epsilon| r_0}{b} < \frac{1}{2} \qquad \text{(Eq 30)}$$

where $A = 3$ for edge dislocations and $A = 1$ for screw dislocations. This equation predicts that the flow stress should increase slightly more rapidly than ϵ, because increasing misfit bends the dislocation more and makes it interact with more regions of adverse stress.

Gleiter[37] also has discussed the effect of coherency strain field on CRSS in a two-phase alloy, and by following the steps outlined above with different assumptions as to the flexibility of dislocation lines and a different averaging procedure for obstacle distribution, has obtained the following relation of flexible edge dislocations:

$$\Delta\tau = 11.8 G \epsilon^{3/2} f^{5/6} \left(\frac{r_0}{b}\right)^{1/2} \quad \text{(Eq 31)}$$

The main difference between Eq 30 and 31 is the dependence of $\Delta\tau$ on the volume fraction. Data for several Ni-Al alloys have been found to be in agreement with Eq 30.[38]

Nembach and Neite[39] have extensively reviewed the experimental evidence bearing on lattice-misfit effects on the strength of superalloys. It was concluded that there is no convincing experimental proof that misfit affects the flow stress of underaged γ'-hardened alloys, and that lattice misfits of the magnitude used in commercial alloys do not affect τ_p.

Dislocation Bypass Models

All of the dislocation-cutting models previously discussed agree that as particles grow beyond a critical size, bypass may occur by bowing, climb, or other processes. The Orowan bowing model[26] is generally considered to be most applicable for austenitic superalloys. The onset of bowing is accompanied by a loss in strength, as was shown in Fig. 8. The increment in flow stress at low temperature due to bowing is given by consideration of the radius of a curvature ρ to which a flexible dislocation can be bent by an applied stress τ; the line tension is given by[40]:

$$T = \frac{Gb^2}{4\pi} \phi' \ln \frac{L}{2b} \quad \text{(Eq 32)}$$

where $\phi' = \frac{1}{2}[1 + 1/1 - \nu]$ and L is the edge-to-edge spacing of particle, $[(\pi/f)^{1/2} - 2] r_s$. The increment in flow stress $\Delta\tau$ is:

$$\Delta\tau = \frac{Gb}{2\pi L} \phi' \ln \frac{L}{2b} \quad \text{(Eq 33)}$$

The effect of increasing volume fraction f for a given particle size is to decrease L, leading to a prediction of increased strength. Greater hardening should also occur as particle size increases; this effect would be enhanced by coherency strains, producing a larger particle diameter in the path of a dislocation.

Critical Evaluation of Models

The results of aging studies on low-volume-fraction-model γ-γ' alloys are not directly applicable to commercial nickel-base superalloys because of the much higher volume fraction of γ' present in the latter. Also, particle sizes tend to be larger in the superalloys. Another complication is the orientation and strain-rate dependence of stress-strain behavior in high-volume-fraction alloys.[41]

Each of the models outlined suffers from shortcomings, not the least of which is the fact that the microstructures of nickel-base superalloys are too complex to expect a single mechanism to operate over all ranges of stress and service temperatures. We shall distinguish between alloys in which there is little or no mismatch between γ and γ' (that is, Ni-Cr-Al-type alloys) and high-mismatch alloys (that is, Ni-Al-Ti-type alloys).

Alloys With No Lattice Mismatch. The principal elements of the Brown-Ham model of pairs of dislocations interacting with ordered particles have been directly confirmed in high-voltage electron-microscopic experiments on Nimonic PE16.[42] Specifically, the leading dislocation of a pair bows out strongly between γ' precipitates while the trailing dislocation remains nearly straight. The spacing of obstacles along the leading dislocation is in reasonable agreement with Eq 13. When there is little or no mismatch, considerable evidence suggests that the volume fraction of γ' is the most significant variable controlling flow stress and creep resistance. The volume fraction of γ' varies from 0.2% in γ'-lean alloys, such as Nimonic 90A, to 0.6% in Mar-M 200 and 713C. Newer alloys contain up to 70% γ'. The flow stresses of binary Ni-Al aged to peak hardness[7] and of Ni-Cr-Al alloys containing fractions of γ' between 0.4 and 0.6% are remarkably insensitive to temperature. The yield stress of Mar-M 200 is nearly constant from room temperature to 750 °C (1380 °F).[32]

Probably no single model of yielding can be applied over the entire range of volume fraction of γ' and service temperatures of nickel alloys. In polycrystalline alloys containing low volume fractions of coarse particles of γ', the temperature dependence of yielding appears to be controlled by the γ matrix, hardened by nonequilibrium hyperfine γ' precipitates. The rate-controlling step is the stress to move dislocations through the hardened matrix; the large γ' particles are easily sheared. The hyperfine particles redissolve at temperatures over 700 °C (1290 °F), γ is weakened, and the flow stress drops rapidly. Similarly, in single crystals the temperature dependence of yielding depends on the relation between particle size and spacing between cross-slip events.[43]

Alloys containing large volume fractions of γ' behave similarly to pure γ' in that the flow stress increases with increasing temperature. If the alloy contains about 50% primary coarse γ' in γ, the strength characteristics are intermediate. Flow strength is moderately high at low temperatures; a shallow peak in flow stress is reached near 700 °C (1290 °F), and strength falls off at a temperature somewhat higher than for a leaner alloy (see Fig. 9[44]). Note that an intermediate volume

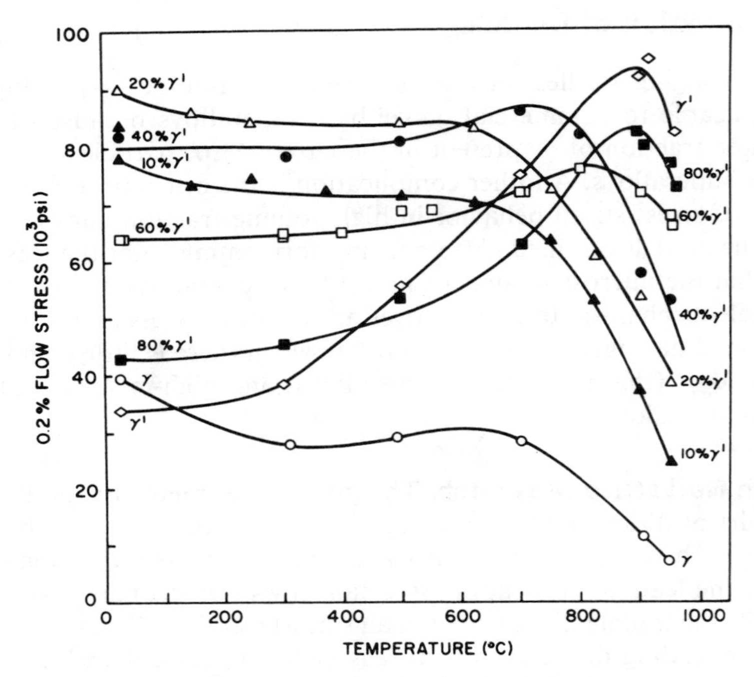

Fig. 9. Flow stress of Ni-Cr-Al alloys containing varying proportions of gamma prime[44]

fraction of 20% produces the highest strength at 21 °C. Most of the lower-temperature yield strength is due to "hyperfine" γ' (50 to 100 Å in diameter).

All of the order-strengthening theories predict an increase in flow stress with increasing particle size, r_0, for constant volume fraction. This has been confirmed in Ni-12.7 at. % Al alloys.[45] However, other evidence on the effect of particle size has been conflicting; it has been shown for an 18Cr-6.5Al-3.3Nb alloy that an increase in size of γ' from 0.05 to 0.5 μm by varying aging time reduced the room-temperature flow stress by about 13%.[46] On the other hand, hardness of Ni-Cr-Ti alloys increases initially with particle size and then decreases (see Fig. 10[47]). Dislocations initially shear the particles; when they grow larger, the particles are avoided by a bypass mechanism. Consequently, we conclude that so long as particles are being cut, the flow stress increases with increasing particle size.

For Ni-Cr-Al-Ti alloys containing 10 to 20% γ', the smallest γ' particle size gave optimum creep resistance at 700 °C(1290 °F); also, size was more important than volume fraction in determining creep life.[50] Small size is achieved in conjunction with small interparticle spacing, which is about 0.05 μm for optimum creep resistance. Consequently, it may be difficult to produce a particle size and spacing that will simultaneously provide good tensile and creep properties.

The Gleiter-Hornbogen-Ham theories are strictly applicable only to low-volume-fraction alloys, but have the advantage of explicitly including f, r_0, and γ_0 in the expressions for flow stress. These models are capable of explaining the observed transition between particle shear and dislocation bowing observed by several inves-

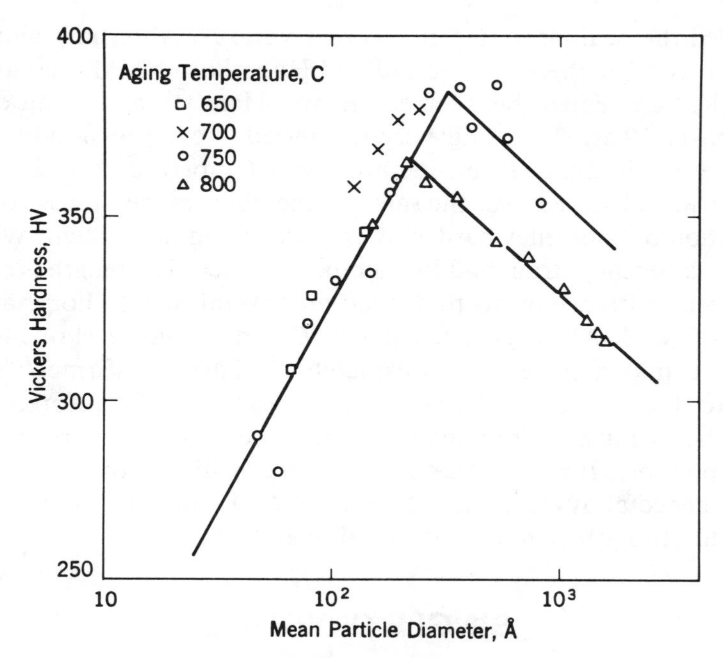

Fig. 10. Effect of particle size on strength of Ni-Cr-Al-Ti alloys[47]

tigators as well as the dependence of flow stress on particle size and APB energy. The Copley-Kear model, on the other hand, is applicable only to high-volume-fraction alloys, and has only been tested with data for Mar-M 200. The major uncertainty involved in the use of all order-hardening theories is that there are no direct means of determining APB energies, and it is difficult to precisely measure the parameters f, r_0, and L', which are so important in applying these models.

Alloys with Lattice Mismatch. It has been suggested that there is a correlation between the titanium-to-aluminum ratios of superalloys and strength or creep resistance. However, there is considerable controversy as to the origins of these effects. Phillips[48] and Raynor and Silcock[49] suggest that increasing the Ti:Al ratio influences strength through an increase in APB energy from approximately 150 ergs/cm^2 (no titanium present) to approximately 240 ergs/cm^2 (Ti:Al = 1) and 300 ergs/cm^2 (Ti:Al = 8), as shown in Table 4.[50] In this view, a difference in lattice parameter between γ and γ' as high as 0.5%, which accompanies high titanium additions, is relatively unimportant as a strengthening mechanism. Rather, mismatch is the driving force in the growth and coalescence of γ' particles. A large mismatch, corresponding to a large interfacial strain energy, may render the γ' precipitate thermally unstable even in the absence of applied stress. Applied stress further lowers the mismatch to stabilize the precipitate, particularly when the stress axis differs from ⟨111⟩.

Conversely, Decker[4] and Decker and Mihalisin[51] argue that a high mismatch can markedly increase peak hardness by aging. Increasing mismatch from 0.2 to

0.8% doubled the peak-aged hardnesses of several Ni-Al ternary alloys, which is in agreement with the theory of Gerold and Haberkorn.[36] Munjal and Ardell[52] found excellent agreement between the Brown-Ham[27] model and experimental results for Ni-12.19 at. % Al single crystals tested in compression between 77 and 373 K. Since misfit changes considerably with temperature, and no significant change in $\Delta\tau$ was observed over the same temperature range, it was concluded that the contribution of coherency hardening is negligible in this system. While the relation between coherency strain and low-temperature tensile strength is still in doubt, optimum creep resistance seems to depend on zero mismatch. For example, creep-rupture life of Ni-Cr-Al alloys tested at 700 °C and a stress level of 146 MPa (21.2 ksi) reaches a maximum at zero mismatch.[53] These confirmed[54] results are attributed to high phase stability at low mismatch. Therefore, the Gerold-Haberkorn theory must be confined to temperatures low enough so that the growth of γ' is not possible. It is clear that misfit strengthening is particularly important in γ''-strengthened alloys (e.g., IN 718 and IN 901), but these alloys are rarely utilized for their strength at temperatures above 815 °C.

SINGLE CRYSTALS

The hardening mechanisms described above do not take into account interactions of dislocations with grain boundaries, and therefore are implicitly applicable only to single crystals.

However, none of them take into account orientation effects, which can be substantial below 760 °C. Schmid's law is not obeyed for octahedral slip, suggesting that other slip systems may be important.[43] Above 760 °C, massive cube slip occurs in γ', so that at lower temperatures limited slip on cube planes may be assumed to occur. Since particles are sheared by dislocations, the yield stress of the γ' (itself dependent on orientation) may be important. Further, a tension-compression anomaly is observed in both Ni_3Al[55] and PWA 1480[43] which may depend on whether the applied stress acts to constrict or extend Shockley partials on {111}.

Shah and Duhl[43] have proposed a model for superalloy single-crystal behavior which accounts for the above experimental observations. The model is shown schematically in Fig. 11. The principal parameters in their model are the particle radius, r, the mean free distance between cross-slip events on the leading dislocation of a superlattice dislocation pair, λ, and the crystal orientation. In general:

$$\sigma \alpha \left(\frac{1}{r} + \frac{1}{\lambda} \right)$$

For ⟨100⟩ oriented crystals, when γ' particles are coarse, $\lambda < r$ at low temperatures, but at high temperatures more cross slip occurs and λ is lowered. This results in a higher stress to shear γ', and the flow stress increases as temperature increases. For fine γ', $\lambda > r$ at low temperatures, and r controls strength to high tempera-

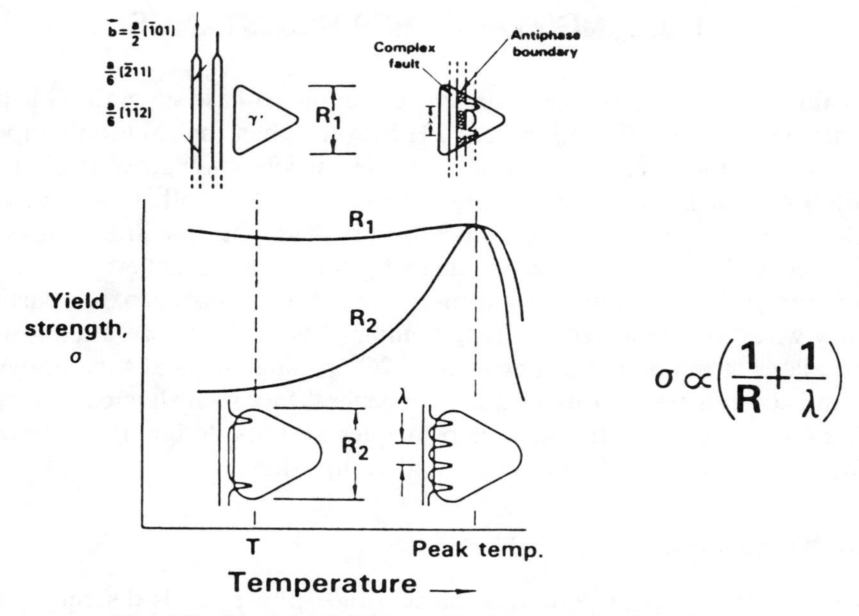

Fig. 11. Model of yield strength as a function of temperature for a single-crystal superalloy with fine and coarse gamma prime, deforming by {111} slip[43]

tures, and therefore a plateau in yield stress with temperature is observed until a high enough temperature is reached for cross slip to occur readily; λ then decreases and controls behavior at high temperatures. For ⟨111⟩ oriented crystals, massive cube slip readily occurs, and thermally activated cross slip of dislocation segments is not relevant. The particle size, r, thus controls at all temperatures, and $\sigma_{ys} = \alpha(1/r)$.

POLYCRYSTALS

Metals and alloys tested at temperatures below about $0.5T_m$, the absolute melting temperature, are further strengthened by the resistance of grain boundaries to dislocation motion. The Hall-Petch relation[56]:

$$\sigma_y = \sigma_0 + k_y d^{-1/2} \quad \text{(Eq 34)}$$

where σ_y is yield stress, σ_0 is a lattice friction stress, d is grain diameter, and k_y is a measure of the grain-boundary resistance, demonstrates that significant strengthening can be obtained for fine-grain alloys when k_y is large. Factors tending to increase k_y are solute hardening and difficult cross slip. Therefore, solutes such as cobalt which lower the stacking fault energy of nickel are expected to increase the contribution of grain boundaries to yield or flow stresses.

ALLOYING FOR CREEP RESISTANCE

The outstanding creep resistance of conventional nickel-base superalloys is primarily a function of the precipitation of γ'. However, when the solution temperature (solvus) of γ' is approached during service, the particles may grow or even go into solid solution. When this occurs, strength is rapidly lost. Another cause of reduced strength is grain-boundary sliding, which occurs readily in fine-grain alloys. However, it is possible to avoid these problems by several approaches: by production of columnar grains or single-crystal materials, or by utilizing oxide particles or refractory wires as supplemental strengthening phases. Therefore, the most creep-resistant alloys at temperatures above 1000 °C are single-crystal superalloys, ODS alloys, and composites. In this section, the general factors influencing creep resistance are considered, and the specific principles of alloy design are reviewed with respect to single crystals, ODS alloys, and composites.

Steady-State Creep

Steady-state creep resistance in crystalline, single-phase solids depends on diffusivity D, stacking-fault energy γ_{SFE}, elastic modulus E, temperature T, and stress σ according to a formula of the form[57,58]:

$$\dot{\epsilon} = A\left(\frac{\sigma}{E}\right)^n f(\gamma_{SFE})e^{-Q/RT} \qquad \text{(Eq 35)}$$

where $f(\gamma_{SFE})$ is a function of SFE and Q is the activation energy for creep. In one model, $\dot{\epsilon}$ is dependent on $(\gamma_{SFE})^{3.5}$, while another formulation incorporates γ_{SFE} into the stress exponent, n, such that as γ_{SFE} increases, n decreases.[59] Typical solid-solution alloys reveal an exponent n with values of 3 to 7 and with Q equal to the activation energy for self-diffusion at temperatures above half the melting point. Consequently, high creep strength is favored by solute additions which raise the modulus or lower the SFE and which lower the diffusivity. Tungsten and molybdenum serve to raise the modulus and lower the diffusivity of nickel-base alloys, while cobalt is effective in lowering SFE.

When second-phase particles are present, the apparent activation energy for creep is much higher than the activation energy for creep (or self-diffusion) of the matrix. Thus the activation energy for steady-state creep of Mar-M 200 and other nickel-base superalloys is as high as twice that of unalloyed nickel and considerably higher than that of solid-solution alloys of nickel. These apparent discrepancies can be eliminated either by considering the temperature dependence of E[60] or by replacing σ in Eq 35 with $(\sigma - \sigma_0)$, where σ_0 is a frictional stress.[61] In either case, the activation energy for creep becomes very close to that for self-diffusion. Similar differences between the activation energy for creep of a multiphase alloy and the activation energy for self-diffusion of the matrix have been noted for dispersion-strengthened alloys such as TD nickel. Grain aspect ratio (GAR) seems to play a role in these alloys, as Q and n both increase with increasing GAR.[62]

Steady-state creep in Mar-M 200 at 760 °C occurs only after appreciable strain hardening due to intersecting {111}⟨112⟩ bands and the development of a substructure during primary creep. Dislocation networks form at γ-γ' interfaces, thereby limiting the mean free path of gliding dislocations to the order of the particle size. These networks reduce the rate of recovery, leading to a low creep rate. The observation that ⟨112⟩ slip is responsible for particle shear suggests that crystal orientations with a low Schmid factor for ⟨112⟩ slip are desirable for good creep resistance. Supporting this conclusion is the observation that single crystals with ⟨111⟩ tensile axes have unusually long creep lives.(63)

Influence of γ' Morphology

The morphology of γ' in nickel-base alloys can be modified by annealing under stress (see Fig. 12(64)). In ⟨100⟩ and ⟨110⟩ orientations, both plates and rods of γ' may be generated, depending on the sense of the applied stress. Tensile annealing produces γ' plates for the ⟨100⟩ orientation, while compressive annealing causes rods to form. In the ⟨110⟩ orientation, the opposite occurs, while ⟨111⟩ oriented crystals show no change in morphology under tension or compression. The sign of the lattice misfit also influences stress-coarsening behavior; the results described above are for alloys with negative misfit. Morphological changes in γ' can affect yielding behavior of U-700 crystals.(65) The yield strength of ⟨100⟩ crystals is increased by rod or plate formation, with plates providing the greater effect to 760 °C (1400 °F) (see Fig. 13). At higher temperatures, morphology has little effect on strength. However, in creep-rupture tests a substantial improvement in properties of ⟨100⟩ crystals has been reported for a Ni-Al-Mo-Ta alloy.(66) Specimens in the solution-treated condition (air cooled) exhibit lower steady-state creep rates and longer rupture lives than material given a standard heat treatment. A prestrain

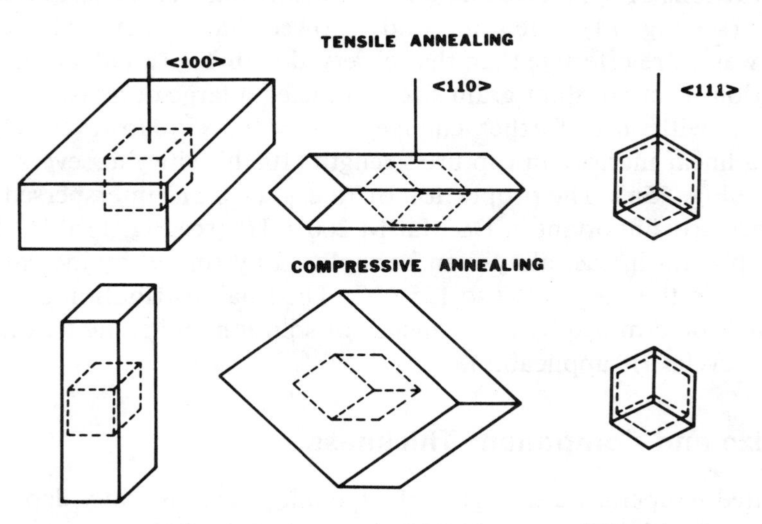

Fig. 12. Morphology of gamma prime annealed under stress(64)

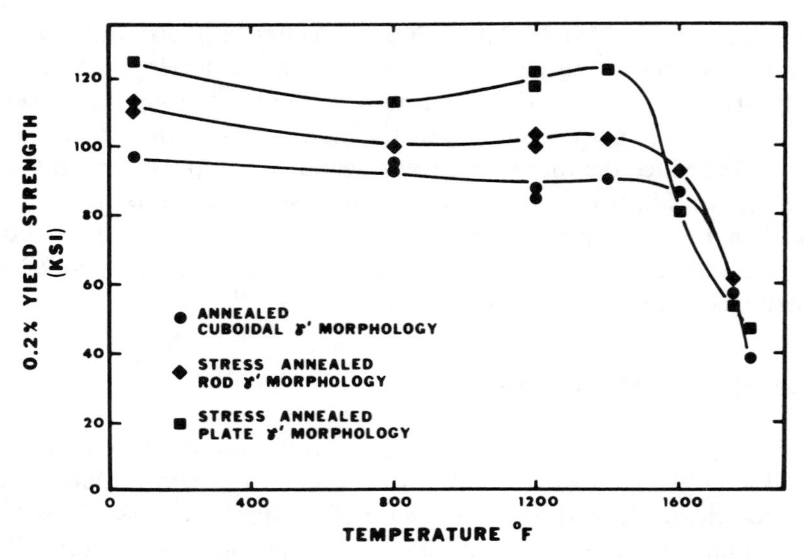

Fig. 13. Influence of gamma prime morphology on yield strength[65]

under creep conditions leads to still further improvement in properties due to the formation of γ' plates or rafts during primary creep. The molybdenum content is critical, with the creep strength maximized at the solubility limit of molybdenum in γ.[67] In summary, optimum strengthening due to γ' rafting in Ni-Al-Mo-X alloys is achieved in homogeneous alloys that are saturated with molybdenum and that exhibit large negative γ/γ' misfit.

The interactions among various microstructural parameters affecting the creep of Ni-20Cr-X alloys have been examined by Gibbons and Hopkins.[68] At high volume fractions of γ', increasing grain size caused a sharp decrease in secondary creep rate (see Fig. 14). The same study showed that hardening by γ' containing Nb + Al was more effective than that observed with Ti + Al alone. Increasing volume fraction, f, at constant grain size, produced a large decrease in creep rate up to $f = 0.2$, with little further change to $f = 0.3$.[68] However, Decker[4] has reported a linear increase in rupture strength (100 h) with f at several test temperatures (see Fig. 15a). The proportion of fine γ' ($<0.5\mu$) interspersed with coarse γ' is particularly important in DS Mar-M 200 + Hf (see Fig. 15b).[69] Much of the coarse γ' present in cast alloys can be replaced by fine γ' by increasing solution temperature in the range 1187 to 1250 °C. The final distribution of γ' size in any superalloy is determined by the sequence of solution and aging treatments as well as coating cycles (if applicable).

Grain Size and Component Thickness

The elevated-temperature strengths of superalloys are also very dependent on the relation of grain size to component thickness. For example, rupture life and creep resistance increase with increasing ratio of component thickness to grain size.[70]

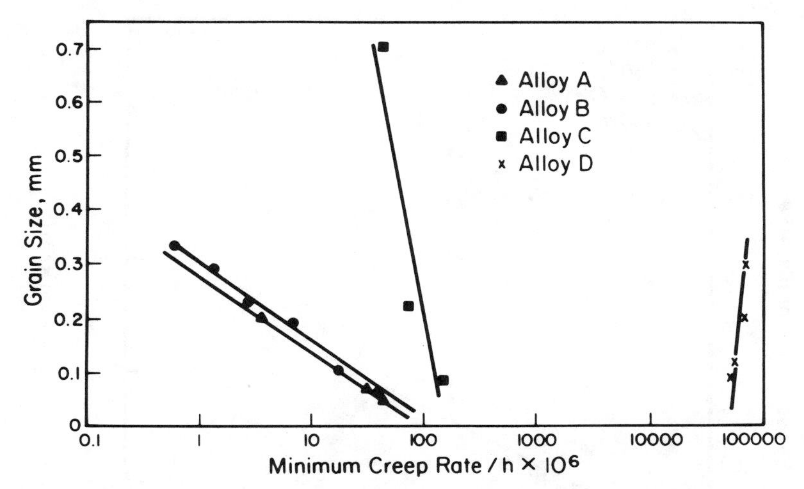

Alloy A: 3Ti-2.2Al. Alloy B: 2.6Ti-1.58Al. Alloy C: 1.63Ti-0.85Al. Alloy D: 1.06Ti-0.65Al.

Fig. 14. Influence of varying grain size on minimum creep rate of Ni-20Cr-Ti-Al alloys(68)

Provided that the ratio is kept constant, life and creep resistance of wrought superalloys increase with grain size.(3) Cast superalloys show the same dependence of life and creep resistance on the ratio of thickness to grain size. As a result, large grains are to be avoided in thin sections to maintain high creep-rupture resistance.

In modern cast superalloys, a balance must be struck to avoid excessively fine grains, which decrease creep and rupture strength, and excessively large grains, which lower tensile strength.(3)

Grain-Boundary Chemistry

Improvement of creep properties by very small additions of boron and zirconium (see Table 2) is a notable feature of nickel-base superalloys. Boron and zirconium can increase the life of Udimet 500 at 870 °C by 13 times, elongation by 7 times, rupture stress by 1.9 times, and n (stress dependence of creep rate) from 2.4 to 9.(4) Magnesium additions from 0.01 to 0.05% also have resulted in improved properties and forgeability in wrought alloys.(4) It is believed that this is due primarily to the magnesium tying up sulfur, a grain-boundary embrittler.

Mechanisms for these property effects are unclear. However, it is believed that boron and zirconium segregate to grain boundaries because of their large size misfit with nickel. Since cracks in superalloys propagate along grain boundaries, the importance of grain-boundary chemistry is apparent. Although early work suggested that boron and zirconium influence rupture properties through their effects on carbide and γ' distribution, recent work on a PM superalloy has revealed no such effects.(71)

Boron and zirconium also improve rupture life of γ'-free alloys, cobalt alloys, and stainless steels, so that microstructural alterations cannot, in any case, apply

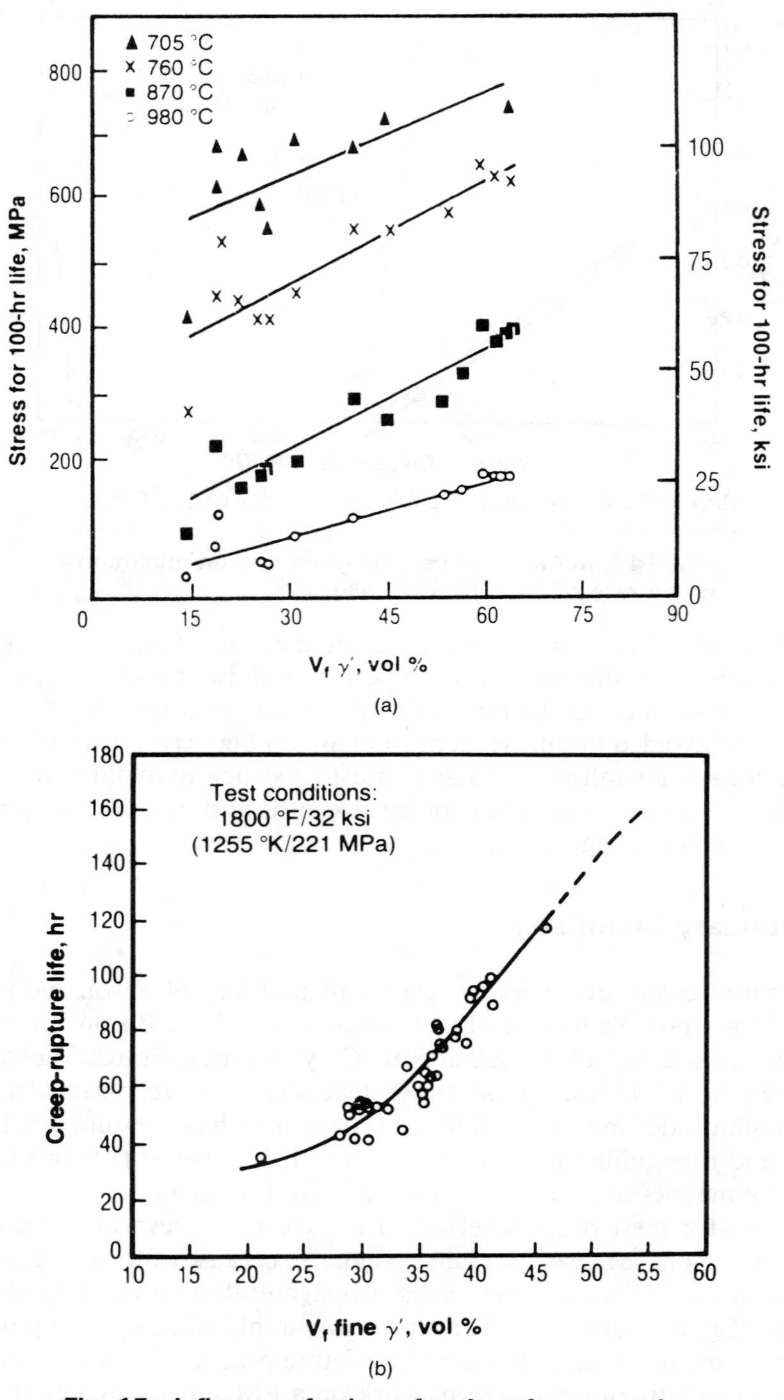

Fig. 15. Influence of volume fraction of gamma prime on (a) stress for 100-h life at several temperatures[4] and (b) creep-rupture life for DS Mar-M 200 + Hf at 982 °C (in this case *f* is for fine γ′)[69]

to all systems. Boron also may reduce carbide precipitation at grain boundaries by releasing carbon into the grains. Magnesium may have a similar effect in a Ni-Cr-Ti-Al alloy in which intragranular MC was noted.[4] Finally, segregation of misfitting atoms to grain boundaries may reduce grain-boundary diffusion rates, consistent with the findings of Tien and Gamble[72] on the formation of denuded zones by Nabarro-Herring-type diffusion. Direct evidence for a lowering of grain-boundary diffusivity by 0.11% Zr in Ni-20Cr alloy over the range 800 to 1200 K has recently been reported by Schneibel *et al.* (see Fig. 16).[73] This effect was accompanied by the precipitation of Ni_3Zr at grain boundaries.

DIRECTIONALLY SOLIDIFIED ALLOYS

Directional solidification (DS) to produce columnar grains served to eliminate a major source of weakness at high temperatures—grain-boundary sliding. Compositions of several columnar-grain alloys are listed in Table 5. Note that hafnium is present in all columnar-grain alloys listed. This is a consequence of two beneficial

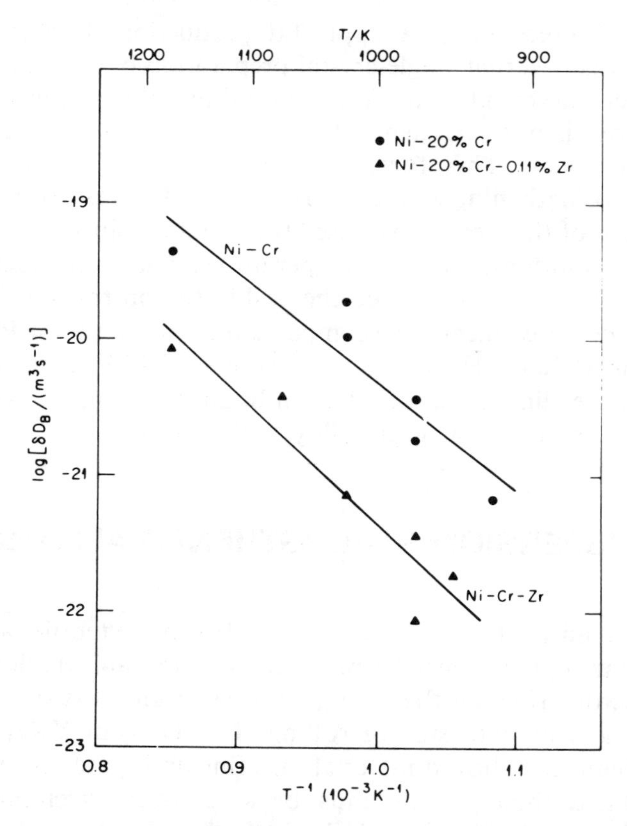

Fig. 16. Effect of 0.11% Zr on grain-boundary diffusivity of Ni-20Cr[73]

Table 5. Compositions of directionally solidified superalloys[82]

Alloy	Composition, %											
	Cr	Co	W	Mo	Ta	Nb	Ti	Al	Hf	B	Zr	C
Columnar-Grain Alloys												
MAR-M 200+Hf	9	10	12.0	...	...	1.0	2.0	5.0	2.0	0.015	0.08	0.14
MAR-M 246+Hf	9	10	10	2.5	1.5	...	1.5	5.5	1.5	0.015	0.05	0.15
MAR-M 247	8.4	10	10	0.6	3.0	...	1.0	5.5	1.4	0.015	0.05	0.15
René 80H	14	9.5	4	4	...	...	4.8	3.0	0.75	0.015	0.02	0.08
Single-Crystal Alloys(a)												
PWA 1480	10	5	4	...	12	...	1.5	5	...	...	...	...
CMSX-2	8	5	8	0.6	6	...	1.0	5.5	...	...	...	...
CMSX-3	8	5	8	0.6	6	...	1.0	5.5	0.15	...	...	...
SRR99	8.5	5	9.5	...	2.8	...	2.2	5.5	...	...	...	...

(a) PWA 1480 is alloy 454.

effects: the suppression of γ/γ' eutectic formation at grain boundaries, and a marked improvement in transverse properties of thin-wall castings (i.e., hollow turbine blades).

At the time the DS technique was applied to production of columnar-grain alloys (1969), it was recognized that single-crystal preparation by the same technique was feasible. However, merely utilizing conventional or columnar-grain chemistry for single-crystal alloys is not sufficiently beneficial to warrant the extra expense of the single-crystal process. Therefore, a new class of alloys containing none of the grain-boundary-strengthening elements (B, C, Zr, Hf) was developed (see Table 5). The elimination of these elements raised the melting points of the alloys, thereby permitting higher solution-treatment temperatures. This in turn led to more complete solutioning of γ', and increased the ability to control the size of the particles produced by heat treatment. The comparison of properties of PWA 1480 (alloy 454) and DS Mar-M 200 + Hf in Fig. 17[74] shows a 25 °C creep advantage of the former. Further, the elimination of all grain boundaries improves thermal-fatigue resistance relative to columnar-grain alloys.

DISPERSION-STRENGTHENED ALLOYS

The Orowan mechanism (Eq 33) is fully applicable to the tensile behavior of equiaxed alloys containing hard particles produced by mechanical alloying. However, an additional factor affecting the strengths of such alloys is the elongated grain structure resulting from extrusion or rolling. The ThO_2 or Y_2O_3 particles in the TD-Ni and mechanically alloyed materials are fine and quite uniformly dispersed. The hardening due to these particles must be added to the strengthening effects of grain boundaries and subgrain boundaries, as well as solid-solution additions. An additional factor is the grain aspect ratio, GAR, the ratio of grain length D to

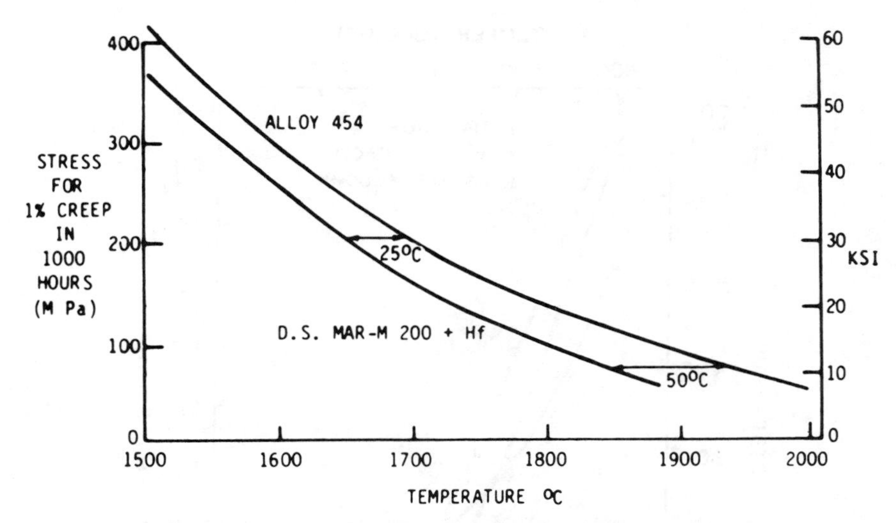

Fig. 17. Comparison of creep strengths of single-crystal alloy 454 and Mar-M 200[74]

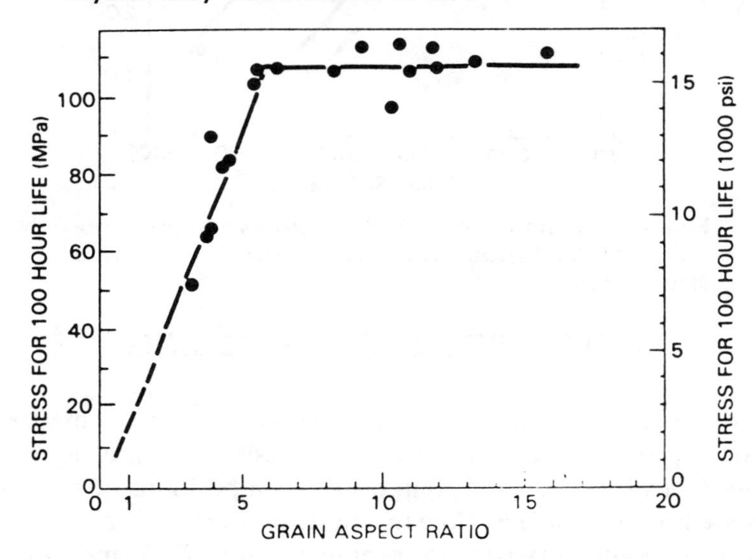

Fig. 18. Influence of grain aspect ratio on stress for 100-h life for MA 753 at 1040 °C[75]

width d. At high temperatures, tensile strength varies approximately linearly with GAR.[62]

Creep and stress-rupture behavior also correlate well with GAR, as shown in Fig. 18 for MA 753 at 1040 °C.[75] Wilcox and Clauer[62] concluded that when grains are elongated the GAR effect swamps any contribution of grain size.

The strength of MA 6000E is greater than that of DS Mar-M 200 + Hf and a eutectic composite, γ/γ'-δ, at temperatures above 1000 °C (Fig. 19).[76] A further advantage of mechanical alloying is a much higher ratio of fatigue resistance to tensile strength relative to precipitation-hardened alloys.

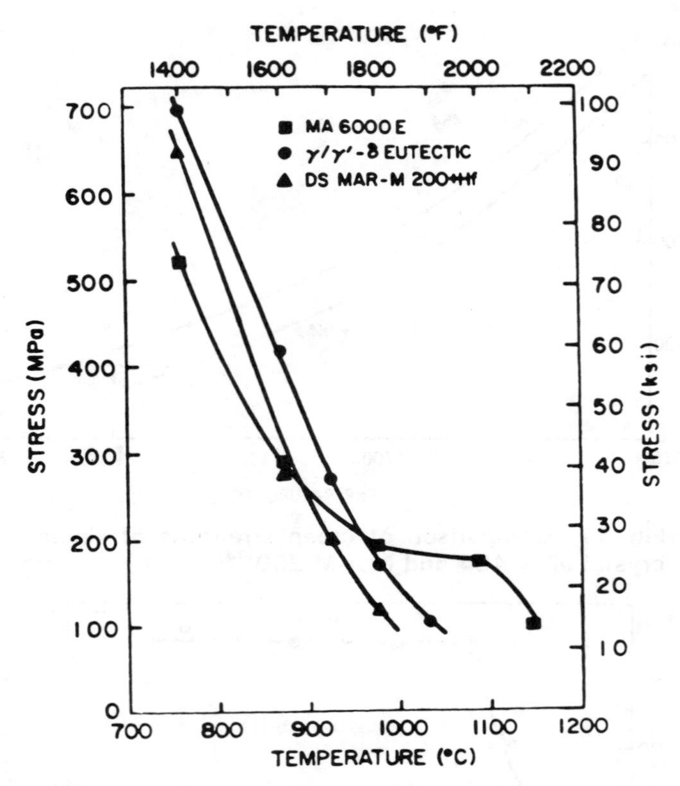

Fig. 19. Comparison of 1000-h stress-rupture properties of ODS MA6000, DS eutectic $\gamma/\gamma' - \delta$, and DS Mar-M 200 + Hf[76]

COMPOSITE STRENGTHENING

Two classes of composite materials with a nickel or nickel-alloy matrix have been developed for improved high-temperature creep resistance: directionally solidified (DS) eutectics and wire-reinforced composites. Compositions of several advanced DS eutectics are listed in Table 6. Note that several different fibers have been incorporated into a superalloy matrix—for example, TaC, NbC, and molybdenum—and that each of these alloys can be heat treated to increase strength by controlled precipitation of γ' in the γ matrix. However, the molybdenum-reinforced alloys are the most responsive to heat treatment.

The high creep-rupture resistance of $\gamma/\gamma' - \delta$, a first-generation lamellar eutectic, was displayed in Fig. 19.[76] Later alloys such as Nitac 13 and Nitac 14B are similar in creep resistance to $\gamma/\gamma' - \delta$ alloys, while exhibiting improved ductility and impact resistance. However, $\gamma/\gamma' - \delta$ alloys remain the most resistant of all eutectic alloys to high-cycle fatigue.[77]

Concurrent with the development of aligned eutectics was a program to develop wire-reinforced nickel-base alloys. Ceramic fibers such as B and SiC proved to be too reactive with superalloys to produce adequately stable microstructures. How-

Table 6. Compositions of some directionally solidified nickel-base eutectic alloys

Alloy	Morphology	f(a)	Solute, wt %								
			Ni	Co	Cr	Al	Nb	Mo	Ta	C	Other
Nitac	F	0.05	69	...	10	5	...	...	14.9	1.1	...
Nitac 13	F	...	63	3.3	4.4	5.4	...	...	8.1	0.54	3.1 W, 6.2 Re, 5.6 V
Cotac-744	F	...	64	10	4	6	3.8	2	...	0.47	10 W
γ/γ'-δ (6% Cr)	L	0.3	71.5	...	6	2.5	20	...	...	...	...
γ/γ'-δ (0% Cr)	L	0.3	76.5	...	...	2.5	21	...	...	...	...
γ'/γ-Mo (AG-34)	F	0.26	62.5	...	...	6.3	...	31.2	...	...	...
γ-δ	L	0.26	66.7	...	...	...	23.3	...	...	...	...
γ'-Ni_3Ta	L	0.35	64.1	...	...	4.9	...	31	...	...	...

(a) Volume fraction.

ever, tungsten and various tungsten-alloy fibers (W-Re-HfC, W-ThO_2) have been effective in improving strength of nickel-base alloys for at least short-term (100-h) exposures (see Fig. 20).[78] A temperature advantage of up to 175 °C is indicated at stress levels up to 350 MPa. Further, the high-cycle fatigue resistance relative to tensile strength of tungsten fiber–reinforced alloys is substantially higher than those of either precipitation-hardened alloys (Nimonic 75, 95, 105) or solid-solution-strengthened Hastelloy X.

INTERMETALLIC COMPOUNDS

A new attempt to achieve superior high-temperature properties in nickel-base alloys comprises the development of alloys based on single-phase intermetallic compounds. Three such nickel-rich systems are of sufficient interest to mention here: Ni_3Al, NiAl, and Ni_3Si. A compilation of physical properties of these alloys appears in Table 7. Note that NiAl has the highest melting point (1640 °C) and the lowest density of the three. Unfortunately, all three alloys are brittle as binary polycrystals. Ni_3Al and Ni_3Si have been made ductile by doping with boron, while repeated attempts to ductilize NiAl at room temperature have been unsuccessful.

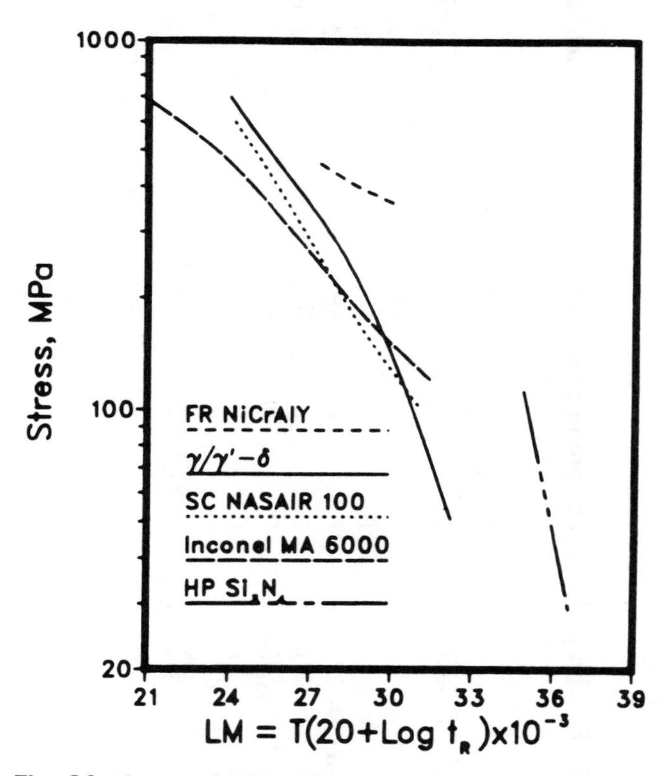

Fig. 20. Larson-Miller plot comparing fiber-reinforced NiCrAlY with other high-temperature materials[78]

Table 7. Physical properties of intermetallic compounds

Compound	T_m, °C	Density, g/cm³	Structure
NiAl	1640	5.86	B2
Ni_3Al	1390	7.5	$L1_2$
Ni_3Si	1284	7.25	$L1_2$

The tensile and creep behavior of NiAl and Ni_3Al have been widely studied; the most remarkable aspect of the strength behavior of the latter is the unusual rise in flow stress with increasing temperature shown in Fig. 3. A series of new alloys containing iron, zirconium, or hafnium has been produced with strengths comparable to those of some early superalloys such as Waspaloy, as shown in Fig. 21.[79] These new alloys are nominally single phase, although small percentages of γ or β phase also may be present. Ni_3Al seems to have a better chance for development as an intermediate-temperature structural material than does NiAl, because of both

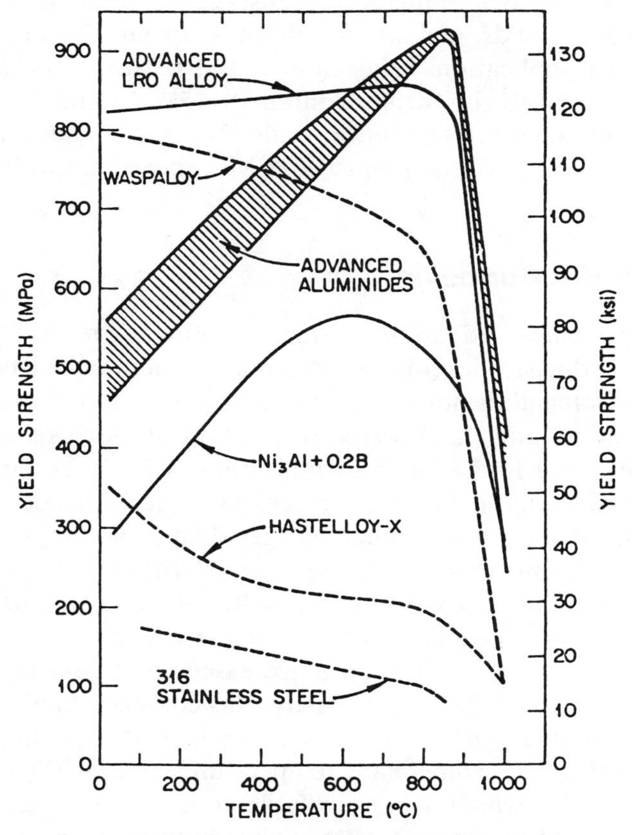

Fig. 21. Comparison of yield strengths of Ni_3Al-base intermetallics and advanced LRO [(Fe,Ni)$_3$V-type] alloys with conventional alloys[79]

higher ductility and the close-packed $L1_2$ structure of the former. The open CsCl structure of NiAl provides very poor creep resistance,[80] and all attempts to provide this alloy with more than 1 or 2% ductility at room temperature have been unsuccessful.

A very new and potentially more promising route to high-temperature structural intermetallics is through reinforcement with filaments. Likely choices for preliminary screening are SiC, Al_2O_3 and tungsten-alloy fibers. While several research groups seem to be working in this area, no reports of specific fiber-matrix combinations have yet been published.

ALLOYING FOR SURFACE STABILITY

Low-Temperature Corrosion

Several elements contribute to the surface stability of nickel-base alloys, depending on the medium (or media) to which any alloy is exposed. High chromium contents are required for low-temperature resistance to corrosive media such as aqueous solutions and acids, and a series of Hastelloy and Inconel alloys has been developed for such applications. Hastelloy C and IN 600 (see Table 2 for exact compositions) are typical; the former contains 16.5% Cr and 17% Mo as principal components for corrosion resistance, while IN 600 contains 15.5% Cr and 8% Fe. Other Hastelloy alloys contain up to 28% Mo, sometimes with small additions of tungsten.

Oxidation and Hot Corrosion

At elevated temperatures, oxidation resistance is provided by Al_2O_3 or Cr_2O_3 protective films. Accordingly, nickel-base alloys must contain one or both of these elements even where strength is not a principal factor. For example, Hastelloy X, one of the most oxidation- and (hot) corrosion-resistant of all nickel-base alloys, contains 22 Cr, 9 Mo, and 15.8 Fe as principal solutes (Table 2). Since Hastelloy X is essentially a solid-solution alloy when placed in service (carbides precipitate after long-term exposure), the alloy is much weaker than superalloys containing γ' or γ'' as strengthening precipitates. Chromium is known to degrade the high-temperature strength and lower the antiphase-boundary energy of γ' (see Fig. 3), so that there has been a strong incentive to lower chromium content in modern superalloys. Thus the level of chromium decreased from 20% in earlier alloys to as little as 9% in Mar-M 200. Unfortunately, this compositional change degraded hot corrosion resistance to the point that superalloys utilized in gas turbines had to be coated. Further, as turbine-blade temperatures exceed 1000 °C, Cr_2O_3 tends to decompose to CrO_3, which is more volatile and, therefore, less protective. To some extent the loss of oxidation resistance has been compensated for by raising aluminum contents, although the latter resides primarily in γ'. (Aluminum in small quantities promotes the formation of Cr_2O_3.) However, Al_2O_3 is less protective than Cr_2O_3 under sulfidizing conditions, so that coatings have become indispensible in both aircraft turbines and more recently in industrial turbines. Other elements that contribute to oxidation and hot corrosion resistance are tantalum, yttrium, and lanthanum. The rare earths appear to improve oxidation resistance by preventing spalling of the oxide, while the mechanism for improvement with tantalum is not known. Yttrium is now widely utilized in overlay coatings of the NiCrAlY type on superalloys.

Finally, it must be pointed out that molybdenum and tungsten are considered to be the most deleterious solutes from the point of view of hot corrosion resistance. Nevertheless, one or both of these elements is required for strength (e.g., most of the γ'-strengthened alloys in Table 2), so that alloying for improved surface stability is often in conflict with alloying for strength. The two most prominent solutes which provide *both* strength and surface stability are aluminum and tantalum.

A comparison of static and cyclic oxidation data for several nickel-, iron-, and cobalt-base alloys is shown in Fig. 22.[81] Note that the best resistance to both cyclic and static oxidation is provided by ODS Ni-Cr alloys and that catastrophic

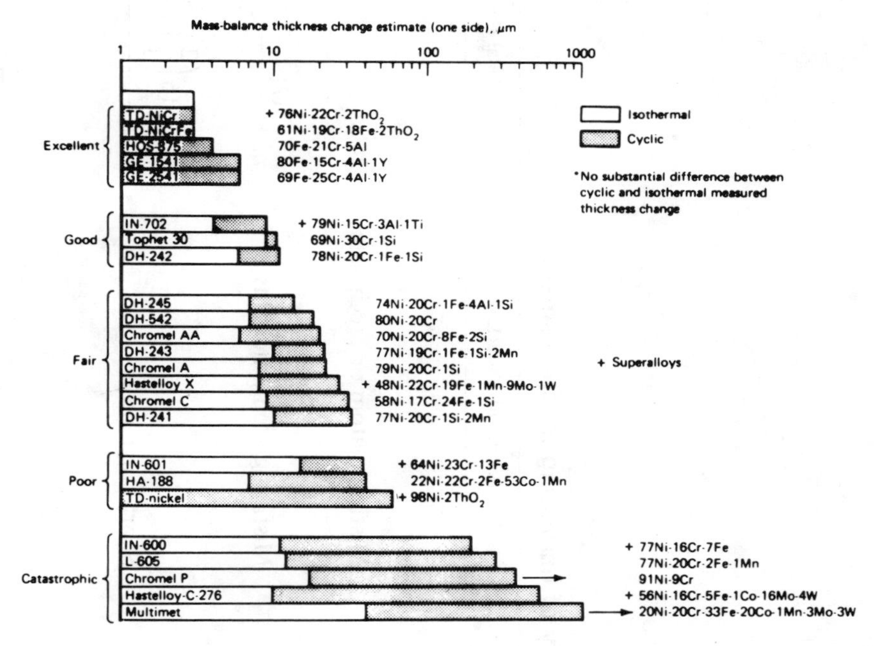

Static tests were conducted in still air for 100 h at 1150 °C. In cyclic tests, samples were heated to 1150 °C and held for 1 h, then cooled to 25 °C. Alloys marked + are considered to be nickel-base superalloys.

Fig. 22. Comparison of static and cyclic oxidation data for several nickel-, iron-, and cobalt-base alloys[81]

Table 8. Summary of hardening mechanisms(a)

Author(s)	Nature of obstacles	Total flow stress	Conditions
Solid Solutions			
Mott-Nabarro[35]	Misfitting atom or precipitate	$2G\epsilon c$	$L \geq \frac{b}{4\lvert\epsilon\rvert f}$
Fleischer[18]	Misfitting atom, modulus	$\tau_0 + \frac{G(\epsilon'_G - \alpha\epsilon)^{3/2} c^{1/2}}{760}$	
Flinn	Short-range order	$\tau_0 + \frac{16\left(\frac{2}{3}\right)^{1/2} c(1-c)\nu\alpha}{a^3}$	
Precipitates			
Copley-Kear[33]	Coherent, ordered, $\epsilon = 0$	$\frac{\gamma_0}{2b} - \frac{T}{br_0} + \frac{(\tau_0 + \tau_p)}{2}$	$f \sim 0.6$
Gleiter-Hornbogen[29]	Coherent, ordered, $\epsilon = 0$	$\tau_0 + \frac{0.28\gamma_0^{3/2} r_0^{1/2} f^{1/3}}{\text{[illegible]}}$	

Brown-Ham[27]	Coherent, ordered, $\epsilon = 0$	$\tau_0 + \frac{[illegible]}{2b}\left[\left(\frac{[illegible]}{\pi T}\right)^{[illegible]} - f\right]$	$\frac{[illegible]}{4\gamma_0} < r_s < \frac{[illegible]}{\gamma_0}$
Brown-Ham[27]	Coherent, ordered, $\epsilon = 0$	$\tau_0 + \frac{\gamma_0}{2b}\left[\left(\frac{4f}{\pi}\right)^{1/2} - f\right]$	$r_s > \frac{T}{\gamma}$
Gerold-Haberkorn[36]	Coherent, ordered, $\epsilon \neq 0$	$\tau_0 + 3G\epsilon^{3/2}\left(\frac{r_0 f}{b}\right)^{1/2}$	Edge dislocation, $\frac{9\pi f}{16} < \frac{3\lvert\epsilon\rvert r_0}{b} < \frac{1}{2}$
Gerold-Haberkorn[36]	Coherent, ordered, $\epsilon \neq 0$	$\tau_0 + G\epsilon^{3/2}\left(\frac{r_0 f}{b}\right)^{1/2}$	Screw dislocation, $\frac{9\pi f}{26} < \lvert\epsilon\rvert r_0 < \frac{1}{2}$
Gleiter[37]	Coherent, ordered, $\epsilon \neq 0$	$\tau_0 + \frac{11.8G\epsilon^{3/2} f^{5/6} r_0^{1/2}}{b^{1/2}}$	
Orowan[26]	Hard particles	$\tau_0 + \frac{Gb}{2\pi L}\phi' \ln \frac{L}{2b}$	$\frac{r_0}{b} > 30$ or incoherent ppt. $\phi' = \frac{1}{2}[1 + (1 - \nu)^{-1}]$

(a) τ_0 is flow stress of matrix without obstacles, r_0 is particle radius, $r_s = (2/3)^{1/2} r_0$, T is line tension, ϵ is misfit. See text for definition of other terms.

behavior is demonstrated by several alloys containing high Mo + W (e.g., Hastelloy C-276) or high iron contents not otherwise compensated for by the presence of strengthening oxide formers (Si or Y).

SUMMARY OF STRENGTHENING MECHANISMS AND ALLOY DESIGN

Table 8 summarizes the various models of low-temperature strengthening which appear to be directly applicable to nickel-base superalloys.(50) For solid solutions, the critical parameters are solute concentration and differences in moduli and atomic radii between solute and matrix. The precipitation of coherent ordered particles offers a potent increment to the strength of austenitic matrices. The critical precipitate parameters are volume fraction, radius, and antiphase-boundary energy; in some cases particle-matrix misfit also is important, particularly at levels of 1% or more. The latter factor controls the strength of IN 718 and IN 901.

The strength of complex nickel superalloys can be analyzed in terms of the basic strengthening mechanisms operative in binary nickel-aluminum alloys, modified by the partitioning of alloying elements to the γ and γ' phases to influence particle-coarsening kinetics, antiphase-boundary energy, and misfit. Analysis of several sets of experiments on austenitic alloys in terms of an order-hardening model, as was summarized in Table 4, showed a marked effect of alloy content on APB energy.* Virtually all superalloys contain both chromium and titanium, yet they produce opposite effects on APB energy. It should be the objective of alloy design to increase γ_0 to the greatest possible extent. Gleiter and Hornbogen(29) have reported evidence for a change in ordering parameters with particle size in Ni-Cr-Al alloys, so that it is possible for strength to change with aging time and temperature due to this factor alone. Alternatively, if the mechanism of particle shear changes with temperature, the fault produced by shear must be taken into account in assessing strength.(34)

Unfortunately, those factors favoring high strength at low temperatures are not necessarily favorable for either good creep-rupture strength or fatigue resistance. While a high volume fraction of γ' undoubtedly is desirable for high yield, tensile, and creep-rupture strengths (e.g., see Fig. 15), the creep strength of pure γ' is very poor, and relative fatigue resistance seems to be lowered with increasing volume fraction of particles. Methods of predicting optimum particle sizes and lattice mismatches for alloy design are in doubt. There is general agreement that the stress to shear particles increases with increasing particle size until the particles grow so large that bypass by dislocation bowing becomes possible. Nevertheless, optimum creep resistance is obtained in Ni-Cr-Al-Ti alloys with small particle sizes. Similarly, large mismatch appears to be favorable for low-temperature strength but is decidedly harmful for good creep resistance. This apparent conflict is easily

*The question of whether APB and misfit strengthening are additive has not yet been resolved.

reconciled, however, if one considers that creep resistance is lowered by any factor that increases the instability of the precipitated phase.

In view of the apparent importance of mismatch, it is necessary to consider the best means of controlling it in nickel-base alloys. Partitioning of solute elements between γ and γ' is perhaps the best means of controlling the lattice parameter. Titanium and niobium partition to γ' and increase its lattice parameter; chromium, molybdenum, and iron partition to γ, resulting in expansion of this phase (the effect will be small for chromium). Tantalum should behave similarly to niobium, and tungsten similarly to molybdenum. Cobalt substitutes primarily in γ and has little effect on lattice parameter. To approach zero mismatch, the elements that partition to γ' should be balanced by those that partition preferentially into γ.

Changes in lattice parameter of γ due to loss of molybdenum and tungsten can occur either through precipitation or transformation of carbides, or by the formation of σ, μ, and other TCP phases. Consequently, alloys that exhibit little or no mismatch prior to service may develop considerable mismatch during exposure to high temperatures, leading to loss of creep resistance. Also, since the coefficient of thermal expansion of γ is greater than that of γ', it is desirable to produce an alloy in which the room-temperature lattice parameter of γ' is somewhat greater than that of γ.

SUMMARY

Nickel alloys are very diverse in type, composition, and methods of processing. Nickel is an extremely stable alloy base, so that many elements can be dissolved in it without changing its crystal structure. For high strength at low and intermediate temperatures (to 815 °C), γ'' or γ' particles are precipitated from solution. For temperatures between 815 and 1000 °C, γ'-strengthened alloys are the strongest, especially in single-crystal form. However, maximum creep and fatigue resistance are achieved in composites and ODS alloys. Intermetallic compounds based on nickel are under development as intermediate- and perhaps high-temperature structural materials, but are not as yet comparable to the stronger superalloys in strength or creep resistance. Alloying for improved mechanical properties at low temperatures is often the cause of reduced creep strength or increased susceptibility to cracking caused by an external environment. Therefore, alloy design is often based on compromises involving the temperature range of service, the applied stress levels, and the ambient environment.

REFERENCES

1. *Metals Handbook*, 9th Ed., Vol 3, ASM, Metals Park, OH, 1980
2. *Metal Progress Material and Processing Databook*, ASM, Metals Park, OH, June, 1982
3. E.W. Ross and C.T. Sims, in *Superalloys II*, Wiley, New York, 1987, p. 97

4. R.F. Decker, "Strengthening Mechanisms in Nickel-Base Superalloys," Climax Molybdenum Company Symposium, Zurich, May 5-6, 1969
5. P.H. Thornton, R.G. Davies, and T.L. Johnston, *Met. Trans.*, Vol 1, 1970, p. 207
6. B.H. Kear, G.R. Leverant, and J.M. Oblak, *Trans. ASM*, Vol 62, 1969, p. 639
7. R.G. Davies and N.S. Stoloff, *Trans. Met. Soc. AIME*, Vol 233, 1965, p. 714
8. N.S. Stoloff and R.G. Davies, *Prog. Mat. Sci.*, Vol 13, No. 1, 1966, p. 3
9. D.M. Wee, O. Moguchi, Y. Oya, and T. Suzuki, *Trans. Japan Inst. Met.*, Vol 21, 1980, p. 237
10. L.R. Curwick, Ph.D. Thesis, University of Minnesota, 1972
11. H.L. Eiselstein, in ASTM STP 369, ASTM, Philadelphia, 1965, p. 62
12. W.J. Boesch and H.B. Canada, *J. Met.*, Vol 21, Oct. 1969, p. 34
13. C.T. Sims, *J. Met.*, Vol 18, Oct. 1966, p. 1119
14. V.A. Phillips (personal communication)
15. B.J. Piearcey and R.W. Smashey, *Trans. AIME*, Vol 239, 1967, p. 451
16. J.R. Mihalisin, *Trans. AIME*, Vol 239, 1967, p. 180
17. E.W. Ross, "Recent Research on IN-100," AIME Annual Meeting, Dallas, Feb. 1963
18. R.L. Fleischer, *Acta Met.*, Vol 11, 1963, p. 203
19. R.M.N. Pelloux and N.J. Grant, *Trans. Met. Soc. AIME*, Vol 218, 1960, p. 232
20. E.R. Parker and T.H. Hazlett, in *Relation of Properties to Microstructure*, ASM, Metals Park, OH, 1954, p. 30
21. B.E.P. Beeston, I.L. Dillamore, and R.E. Smallman, *Met. Sci. J.*, Vol 2, 1968, p. 12
22. R.L. Fleischer, in *The Strengthening of Metals*, Reinhold, New York, 1964, p. 93
23. R. Labusch, *Acta Met.*, Vol 20, 1972, p. 917
24. R. Nordheim and N.J. Grant, *J. Inst. Met.*, Vol 82, 1954, p. 440
25. A. Akhtar and E. Teghtsoonian, *Met. Trans.*, Vol 2, 1971, p. 2757
26. E. Orowan, in Symposium on Internal Stresses in Metals, Institute of Metals, London, 1948, p. 451
27. L.M. Brown and R.K. Ham, in *Strengthening Methods in Crystals*, Elsevier, Amsterdam, 1971, p. 9
28. J. Friedel, *Dislocations*, Pergamon Press, Oxford, 1964
29. H. Gleiter and E. Hornbogen, *Mat. Sci. Eng.*, Vol 2, 1968, p. 285
30. R.K. Ham, *Ordered Alloys: Structural Applications and Physical Metallurgy*, Claitors, Baton Rouge, LA, 1970, p. 365
31. V. Martens and E. Nembach, *Acta Met.*, Vol 23, 1975, p. 149
32. S.M. Copley and B.H. Kear, *Trans. Met. Soc. AIME*, Vol 239, 1967, p. 977
33. S.M. Copley and B.H. Kear, *Trans. Met. Soc. AIME*, Vol 239, 1967, p. 984
34. G.R. Leverant, M. Gell, and S.W. Hopkins, *Proc. Sec. Int. Conf. Strength Met. Alloys*, Vol 3, 1970, p. 1141
35. N.F. Mott and F.R.N. Nabarro, *Rep. Conf. Strength Sol. Phys. Soc.*, 1948, p. 1-9
36. V. Gerold and H. Haberkorn, *Phys. Stat. Sol.*, Vol 16, 1966, p. 675
37. H. Gleiter, *Z. Angew Phys.*, Vol 23, No. 2, 1967, p. 108
38. J.L. Castagne, A. Pineare, and M. Sidzingre, *C.R. Acad. Sci.*, Vol C263, 1966, p. 1465
39. E. Nembach and G. Neite, *Prog. Mat. Sci.* (in press)
40. A. Kelly and R.B. Nicholson, *Prog. Mat. Sci.*, Vol 10, No. 3, 1963, p. 151
41. R.R. Jensen and J.K. Tien, in *Metallurgical Treatises*, edited by J.K. Tien and J.F. Elliott, TMS-AIME, Warrendale, PA, 1981, p. 529
42. E. Nembach, K. Suzuki, M. Ichihara, and S. Takeuchi, *Philos. Mag. A*, Vol 51, 1985, p. 607
43. D. Shah and D. Duhl, in *Superalloys 1984*, TMS-AIME, Warrendale, PA, 1984, p. 105
44. P. Beardmore, R.G. Davies, and T.L. Johnston, *Trans. Met. Soc. AIME*, Vol 245, 1969, p. 1537

45. V.A. Phillips, *Philos. Mag.*, Vol 16, 1967, p. 117
46. R.G. Davies and T.L. Johnston, in *Ordered Alloys: Structural Applications and Physical Metallurgy*, Claitors, Baton Rouge, LA, 1970, p. 447
47. W.J. Mitchell, *Z. Metallkd.*, Vol 57, 1966, p. 586
48. V.A. Phillips, *Scripta Met.*, Vol 2, 1968, p. 147
49. D. Raynor and J.M. Silcock, *Met. Sci. J.*, Vol 4, 1970, p. 121
50. N.S. Stoloff, in *The Superalloys*, Wiley, New York, 1972, p. 79
51. R.F. Decker and J.R. Mihalisin, *Trans. ASM*, Vol 62, 1969, p. 481
52. V. Munjal and A.J. Ardell, *Acta Met.*, Vol 23, 1975, p. 513
53. I.L. Mirkin and O.D. Kancheev, *Met. Sci. Heat Treat.*, Vol 10, 1967, p. 1
54. G.N. Maniar and J.E. Bridge, *Met. Trans.*, Vol 2, 1971, p. 95
55. S.S. Ezz, D.P. Pope, and V. Paidar, *Acta Met.*, Vol 30, 1982, p. 921
56. N.J. Petch, *J. Iron Steel Inst.*, Vol 173, 1953, p. 25
57. O.D. Sherby and P.M. Burke, *Prog. Mat. Sci.*, Vol 13, 1967, p. 325
58. A.K. Mukherjee, J.E. Bird, and J.E. Dorn, *Trans. ASM*, Vol 62, 1969, p. 155
59. A.K. Mukherjee, in *Treatise on Materials Science and Technology*, Vol 6, *Plastic Deformation of Metals*, edited by R.J. Arsenault, Academic Press, New York, 1975, p. 163
60. M. Malu and J.K. Tien, *Scripta Met.*, Vol 9, 1975, p. 1117
61. K.R. Williams and B. Wilshire, *Met. Sci. J.*, Vol 7, 1973, p. 176
62. B.A. Wilcox and A.H. Clauer, *Oxide Dispersion Strengthening*, Gordon & Breach, New York, 1968, p. 323
63. G.R. Leverant and B.H. Kear, *Met. Trans.*, Vol 1, 1970, p. 491
64. J.K. Tien and S.M. Copley, *Met. Trans.*, Vol 2, 1971, p. 543
65. J.K. Tien and R.P. Gamble, *Met. Trans.*, Vol 3, 1972, p. 2157
66. D.D. Pearson, B.H. Kear, and F.D. Lemkey, in *Creep Fracture of Engineering Materials and Structures*, Pineridge Press, Swansea, U.K., 1981, p. 213
67. D.D. Pearson (private communication)
68. T.B. Gibbons and B.E. Hopkins, *Met. Sci. J.*, Vol 5, 1971, p. 233
69. J.J. Jackson, M.J. Donachie, R.J. Herricks, and M. Gell, *Met. Trans. A*, Vol 8A, 1971, p. 1615
70. E.G. Richards, *J. Inst. Met.*, Vol 96, 1968, p. 365
71. T.J. Garosshen, T.D. Tillman, and G.P. McCarthy, *Met. Trans. A*, Vol 18A, 1987, p. 69
72. J.K. Tien and R.P. Gamble, *Met. Trans.*, Vol 2, 1971, p. 1663
73. J.H. Schneibel, C.L. White, and M.H. Yoo, *Met. Trans. A*, Vol 16A, 1985, p. 651
74. M. Gell, D.N. Duhl, and A.F. Gaimei, in *Superalloys 1980*, ASM, Metals Park, OH, 1980, p. 205
75. J.S. Benjamin and M.J. Bomford, *Met. Trans.* Vol 5, 1974, p. 416
76. T.E. Howson, D.A. Mervyn, and J.K. Tien, *Met. Trans. A*, Vol 11A, 1980, p. 1609
77. J.E. Grossman and N.S. Stoloff, *Met. Trans. A*, Vol 9A, 1978, p. 117
78. J.K. Tien and V.C. Nardone, in *Fracture: Interactions of Microstructure, Mechanisms, Mechanics*, TMS-AIME, Warrendale, PA, 1984, p. 321
79. C.T. Liu and C.L. White, in *High Temperature Ordered Intermetallic Alloys*, MRS Symposia, Vol 39, Materials Research Soc., Pittsburgh, PA, 1985, p. 365
80. P.R. Strutt and R.A. Dodd, in *Ordered Alloys: Structural Applications and Physical Metallurgy*, Claitor's, Baton Rouge, LA, 1970, p. 475
81. C.A. Barrett and C.E. Lowell, *Oxid. Metals*, Vol 9, 1975, p. 307
82. D.N. Duhl, in *Superalloys II*, Wiley, New York, 1987, p. 189

12

Alloying of Refractory Metals

R.W. BUCKMAN, Jr.
Westinghouse Electric Corporation
Advanced Energy Systems Division

The high melting temperatures of the refractory metals give them the capability of performing useful engineering functions at temperatures far beyond those possible with conventional alloys. This one important characteristic makes this class of materials technologically important. The metallic elements generally referred to as the "major" refractory metals are niobium (Nb) and tantalum (Ta) of group VA, and molybdenum (Mo) and tungsten (W) of group VIA of the periodic table.

The "pure" refractory metals generally do not exhibit sufficient strength at elevated temperatures for most engineering applications. But through alloying it is possible to improve the high-temperature strength of the refractory metals. Alloy strengthening has been a topic of extensive study, and alloy-strengthening mechanisms have been described in several excellent review papers.[1-5] Alloying effects related to improvement of elevated-temperature strength are generally well understood. While strength is a necessary characteristic, it is not the sole requirement for an alloy to perform useful engineering functions. This chapter discusses alloying of the refractory metals from the perspective of an alloy developer, where requirements in addition to strength are necessary. The factors which have a favorable impact on elevated-temperature strength generally tend to adversely affect low-temperature ductility. Thus, development of a useful alloy is a compromise between the strength level required by the particular application and the need for adequate ductility to allow production of the desired component. Additional mitigating factors may come into play if the end application requires the fabrication of a complex shape such that ductility, fabricability, weldability and resistance to environmental interactions are all important attributes. The various strengthening mechanisms and their limits will be presented along with examples which illustrate the unique features of refractory metals.

REFRACTORY METALS

At the outset, it is appropriate to provide a definition of "refractory metals." Originally, melting temperature was used as the single determinant; in earlier works, refractory metals were metals melting above 1925 °C (3500 °F).[6,7] However, the use of melting temperature alone would include hafnium from group IVA, generally identified as one of the reactive metals, and rhenium, ruthenium, iridium, osmium, and technetium from groups VIIA and VIIIA. If this latter list were included, a rather heterogeneous class of materials, from crystallographic, chemical, and deformation aspects, would result. By defining refractory metals to include not only crystallography (bcc) and a melting point minimum (1925 °C or 3500 °F), but to also specify that the ratio of oxide melting temperature be less than one, then Nb, Ta, Mo, and W emerge as the only elements that share this unique combination of features (see Table 1). The refractory metals under this definition all exhibit poor oxidation resistance and cannot be utilized in air at even moderately elevated temperatures. The subject of oxidation resistance of refractory metals and alloys is a separate topic and will not be covered in this chapter.

The key physical characteristics of the refractory metals are listed in Table 2 for comparison. Mo and W have higher elastic moduli and thermal conductivities, and lower linear coefficients of thermal expansion, than do Nb and Ta. One key significant difference between the group VA and group VIA refractory metals is the solubilities of interstitial elements (C, O, N, H). Nb and Ta exhibit high solubilities for C, O, N, and H, while Mo and W exhibit very low solubilities for these elements. Robins[8] and Hahn *et al.*[9] have advanced the hypothesis that the electronically stable configurations exhibited by the group VIA elements give very low lattice solubilities for these interstitial elements. In addition to its high-temperature capabilities, Nb also exhibits superconductivity behavior. This property makes Nb useful from cryogenic near absolute zero to very high temperatures.

When compared at equivalent fractions of their melting temperatures (T/T_m), Mo and W are significantly stronger in creep-rupture than Nb and Ta (see Fig. 1). The creep-rupture-strength advantage of the group VIA refractory metals can be qualitatively explained on the basis of their higher moduli and lower diffusivities.[2,10] The temperature dependence of the flow stress of the refractory metals is shown schematically in Fig. 2. Below about 0.2 T_m, the flow stress increases rapidly, and as the temperature is further decreased, the character of fracture changes from ductile to brittle. Twinning is observed at low temperatures in the refractory metals.[11-16] Bechtold *et al.*[3] conclude that twinning may not be an absolute prerequisite for brittle fracture; however, microcracks have been observed in conjunction with twinning. The ductile-to-brittle transition (DBTT) behavior of recrystallized polycrystalline refractory metals is illustrated in Fig. 3. The group VA refractory metals exhibit lower temperatures for DBTT behavior than the group VIA refractory metals. The temperature at which this change in fracture behavior occurs is a function of metallurgical variables such as grain size, impu-

Table 1. Comparison of melting points of metal elements and their oxides

Element (stable crystal structure at room temperature)	Melting point, K — Metal	Melting point, K — Oxide	$R_{0/m}$
Ti (hcp)	1943	2098	1.1
Zr (hcp)	2125	2688	1.3
Hf (hcp)	2500	3173	1.3
V (bcc)	2175	963	0.4
Nb (bcc)	2740	1763	0.6
Ta (bcc)	3287	2045	0.6
Cr (bcc)	2130	2540	1.2
Mo (bcc)	2890	1068	0.4
W (bcc)	3680	1773	0.5
Re (hcp)	3453	570	0.2

(a) $R_{0/m} = \dfrac{T_m \text{ (oxide)}}{T_m \text{ (metal)}}$

Table 2. Physical properties of the refractory metals

Property	Nb	Ta	Mo	W
Melting point, °C	2470	3000	2610	3410
Density, gm/cm^3	8.57	16.6	10.2	19.3
Elastic modulus, 10^{-6} psi (at RT)	16	27	42	52
Linear thermal expansion, 10^{-6}/°C	7.1	5.9	5.4	4.5
Thermal conductivity, cal/cm/°C/s (near RT)	0.125	0.130	0.34	0.397
Thermal neutron cross section, barns/atom	1.1	21.3	2.4	19.2
Superconducting transition temperature, K	9.46	...	...	...
Interstitial solubility, ppm by weight(a):				
Oxygen	1000	300	1.0	1.0
Nitrogen	300	1000	1.0	<0.1
Carbon	100	70	<1.0	<0.1
Hydrogen	9000	4000	0.1	N.D.

(a) Based on estimates of the equilibrium solubility at the temperature where diffusion $D = 10^{-11}$ cm^2/s.

rity content, degree of cold working, as well as extrinsic conditions such as strain rate, specimen surface condition, and loading direction vs. drawing or worked direction. In polycrystalline Ta and Nb, the DBTT is well below room temperature, with Ta exhibiting ductile fracture at temperatures down to ~4 K. For Mo (and W), the DBTT is near or above room temperature. The room-temperature ductility controls alloy design since this is a practical consideration that cannot be ignored in developing alloys for applications in real systems.

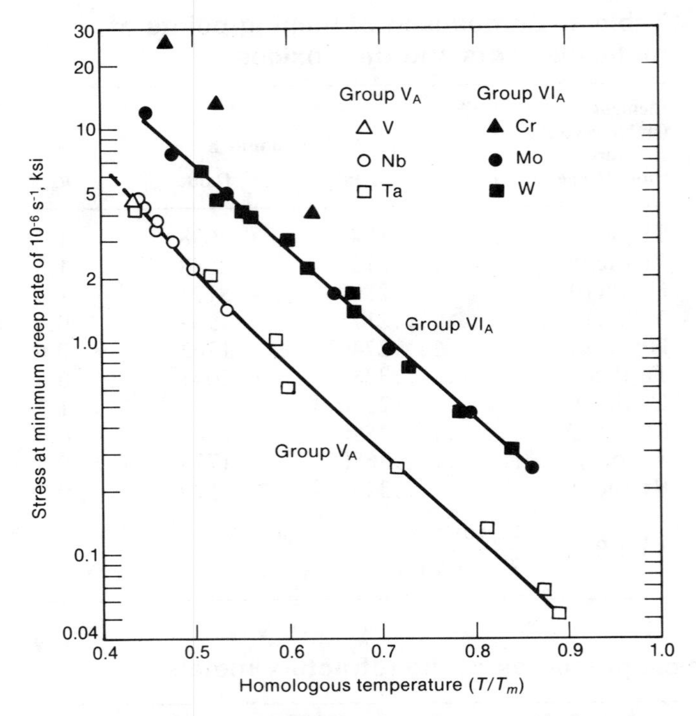

Fig. 1. Creep strength of refractory metal elements as a function of homologous temperature[2]

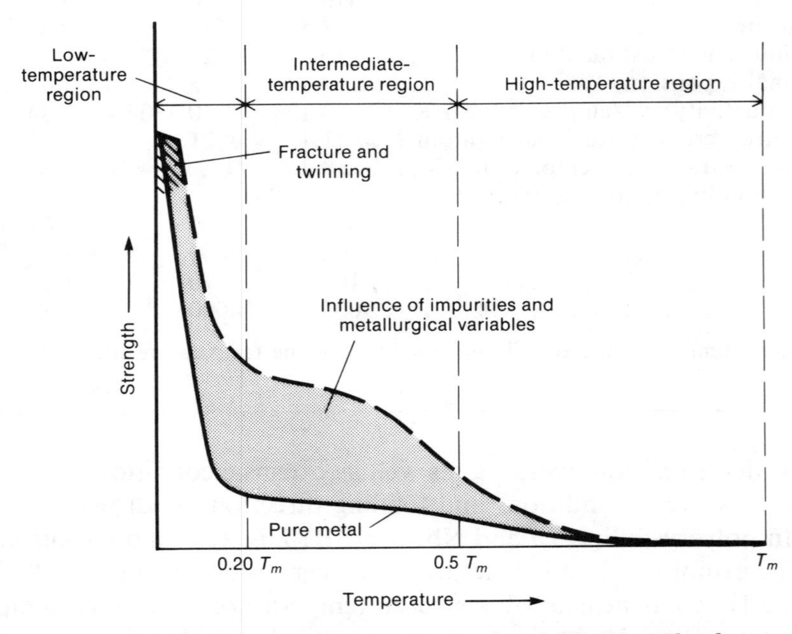

Fig. 2. Effect of temperature on the yield strength of bcc refractory metals[3]

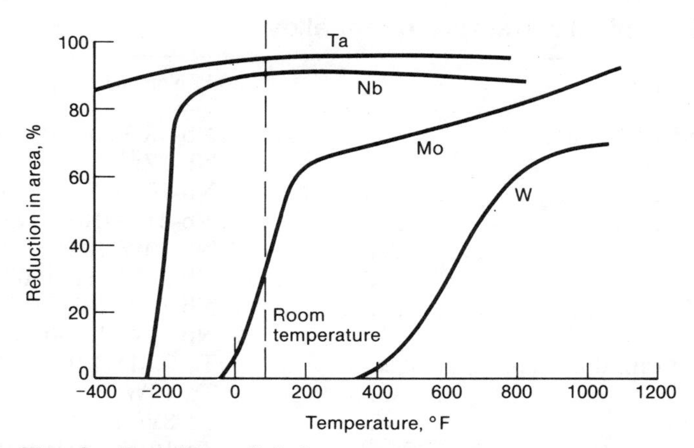

Fig. 3. Ductile-to-brittle transition temperatures for recrystallized polycrystalline refractory metals[11]

STRENGTHENING MECHANISMS

Wilcox[1] has provided an excellent review of the various strengthening mechanisms applicable to the refractory metals. Nb, Ta, Mo, and W are isomorphic and as such can be strengthened by (*a*) cold working, (*b*) solid-solution alloying additions, (*c*) second-phase particles, and (*d*) additions of strong fibers in conveniently high volume fractions. Solid-solution strengthening and second-phase particles (dispersion hardening) are the predominant mechanisms used for increasing the elevated-temperature strengths of Nb and Ta. Mo and W are strengthened primarily by second-phase particles in combination with cold working. The use of fiber reinforcement is a special case and will be treated separately.

A complete review of the literature on refractory-metal strengthening is beyond the scope of this chapter. The intent here, as indicated earlier, is to discuss strengthening from the viewpoint of the alloy developer. The prime interest is in developing potential alloy compositions for useful purposes.

The design of a refractory-metal alloy composition should be tailored to the end application requirements—i.e., to whether the design criteria are based on time-independent or time-dependent deformation criteria. The alloying behavior of Nb and Ta is distinctly different from that of Mo and W. The currently available refractory-metal alloy compositions are listed in Table 3. With the exception of a few of these alloy compositions, the majority have compositional designs of the general type proposed by Begley *et al.*[2]:

$$M + M_{ss} + M_{R.E.} + M_I = \text{Alloy}$$

where M stands for base element (Nb, Ta, Mo, or W), M_{ss} for substitutional solutes, $M_{R.E.}$ for reactive elements (Hf, Ti, Zr), and M_I for interstitial solutes (C, N, O).

Listed in Table 4 are the M_{ss}, $M_{R.E.}$, and M_I, for the base refractory metals.

Table 3. Typical refractory-metal alloys

Alloy designation	Nominal composition, wt %
Unalloyed niobium	Nb-0.030O-0.01C-0.03N(a)
Nb-1Zr	Nb-1Zr
WC-103	Nb-10Hf-1Ti
FS-85	Nb-27Ta-10W-1Zr
SCb-291	Nb-10W-10Ta
B-88	Nb-28W-2Hf-0.07C
WC-129Y	Nb-10W-10Hf-0.24Y
D-43	Nb-10W-1Zr-0.1C
Unalloyed tantalum	Ta-0.015O-0.01C-0.01N(b)
Ta-10W	Ta-10W
T-111	Ta-8W-2Hf
T-222	Ta-10W-2.5Hf-0.01C
ASTAR 811C	Ta-8W-1Re-1Hf-0.025C
Unalloyed molybdenum	Mo-0.04C-0.003O-0.001N(c)
Mo-TZM	Mo-0.5Ti-0.1Zr-0.03C
Mo-42Re	Mo-42Re
Mo-50Re	Mo-50Re
Unalloyed tungsten	W-0.01C-0.006O-0.005N(d)
W-3Re	W-3Re
W-5Re	W-5Re
W-25Re	W-25Re
W-0.3Hf-0.025C	W-0.3Hf-0.025C
W-4Re-0.3Hf-0.025C	W-4Re-0.3Hf-0.025C
W-24Re-0.3Hf-0.025C	W-24Re-0.3Hf-0.025C

(a) ASTM B-391. (b) ASTM B-364. (c) ASTM B-387. (d) ASTM B-410.

Table 4. Additions in refractory-metal alloys

Alloy base	Substitutional solutes	Reactive elements	Interstitial elements
Nb	W, Mo, Ta, Hf, Zr, Ti	Hf, Zr, Ti	C, N
Ta	W, Re, Hf	Hf	C, N
Mo	W, Re, Ti, Zr	Ti, Zr, Hf	C, N
W	Re, Hf	Hf, Th	C, N, O

Alloy design evolved in part by necessity, since the base-element starting materials available in the 1950's and 1960's contained significant levels of interstitial C, O, and N. Thus, elements from the reactive group IVA (Ti, Zr, Hf) interacted to form stable oxides, carbides, and/or nitrides which removed the interstitials from solid solution as second-phase particulates that either were acting as innocuous tramp elements or contributed to high-temperature strength.

Mechanistic strengthening theories have been proposed using room-temperature data. However, this becomes very complicated, since at room temperature, two components of the flow stress—an athermal component and a temperature-dependent component—are acting, influenced by the DBTT phenomena. The most recent solid-solution alloying theory includes not only the traditional atom-size mismatch between the solute and solvent elements but also the difference in elastic moduli between the solvent and solute atoms.[17,18] Applying these theoretical treatments to the refractory metals at room temperature is complicated because identifying which mechanism is having a dominant effect on yield strength is difficult. For example, from room temperature (298 K) to 200 K, the yield strength of Nb increases at a rate of 140 psi/°C (78 psi/°F) and Ta at a rate of 245 psi/°C (136 psi/°F).[19] Also, alloying additions which change the melting temperature significantly result in room temperature being a different homologous temperature, and, as shown in Fig. 2, the flow stress is strongly temperature-dependent below 0.2 T_m. For example, Ti and W form a continuous series of solid solutions with Nb. Thus, for Ti additions, the melting point of Nb is decreasing at a rate of 8 °C/at. % (14 °F/at. %) addition, while W is increasing the melting temperature of Nb at a rate of 9.4 °C/at. % (17 °F/at. %) addition. The rate of change in strength is also a function of concentration, and so, to better correlate theory with data, it would be more appropriate to test at equivalent homologous temperatures and above the temperature where the DBTT phenomenon contributes to the temperature dependent component of the flow stress. Thus, room temperature strength data should be applied to refractory metal alloys with caution when applying theories of strengthening.

Qualitatively, the alloying additions which tend to increase elastic modulus and decrease diffusivity should improve creep strength, as proposed by Sherby[10]:

$$\dot{\epsilon} = S\left(\frac{\sigma}{E}\right)^{\eta} D$$

where $\dot{\epsilon}$ is steady-state creep rate, S is a structure term (grain morphology, dislocation density, distribution, etc.), σ is stress, E is elastic modulus, D is self-diffusivity, and η is a stress exponent.

The data of McAdam[20] on Nb (Fig. 4) and Buckman and Goodspeed[21] on Ta (Fig. 5) are comparable in that for creep strength, those elements that increase the elastic modulus and decrease diffusivity increase the elevated-temperature strength. For Mo, Semchyshen[22] showed that a Mo-1 at. % Ti alloy exhibited a 100-h rupture stress of 138 MPa (20 ksi). Increasing the Ti content to 4 at. % reduced the stress for 100-h rupture by a factor of about two. Again, adding an element that decreased the modulus and increased diffusivity decreased the time-dependent deformation resistance. However, the creep behavior of W at 1925 °C (3500 °F) is affected considerably more by Hf than by Re (see Fig. 6). This behavior is contrary to what one would expect for solid-solution alloying effects based on the Sherby hypothesis. Klopp *et al.*,[24] Raffo and Klopp,[23] and Raffo *et al.*[25] did demonstrate that these effects could be attributable to substitutional solid-solution alloying effects since the interstitial contents of the base material were quite low and did not give rise to any significant solute-interstitial interaction to form a dispersed second phase. Also, there was a one-to-one correlation between elevated-temperature tensile strength (time independent) and creep strength (time

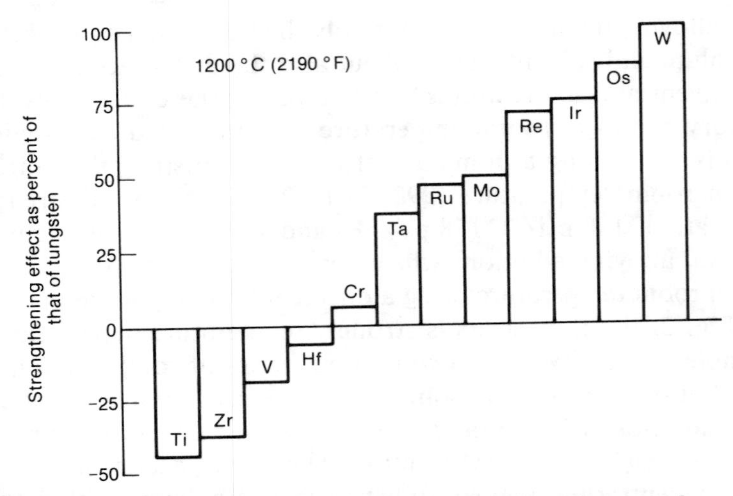

Fig. 4. Strengthening effect of alloying elements in Nb[20]

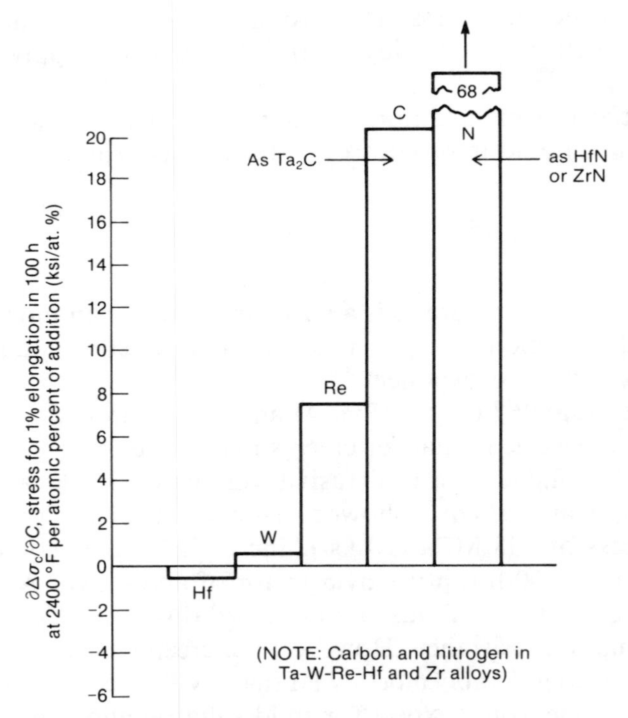

Fig. 5. Creep-strengthening indices for various solutes in Ta-base alloys[21]

dependent). This is not generally the case with the group VA elements, where there is not a good correlation between elevated-temperature tensile strength and creep strength.[2] The effect of binary alloy additions on tensile properties is shown in Fig. 7 for Nb and Fig. 8 for Ta. The alloy additions having the greatest effect on

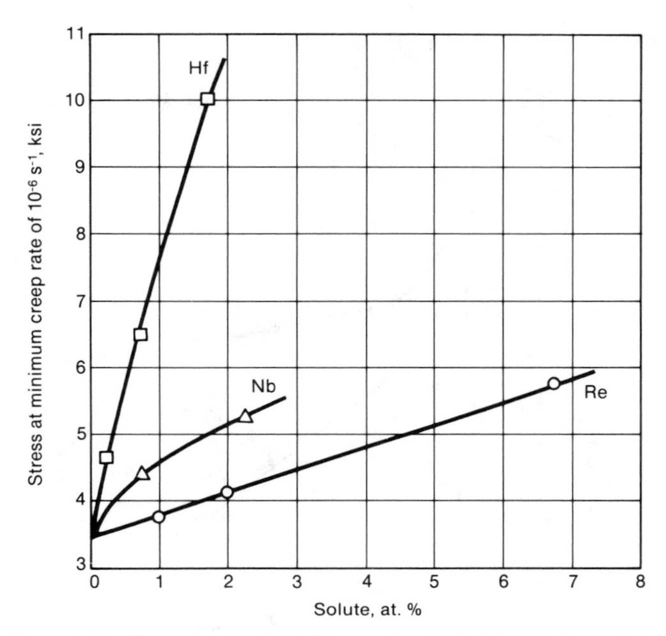

Fig. 6. Influence of alloying on the 1925 °C (3500 °F) creep strength of binary W alloys[23]

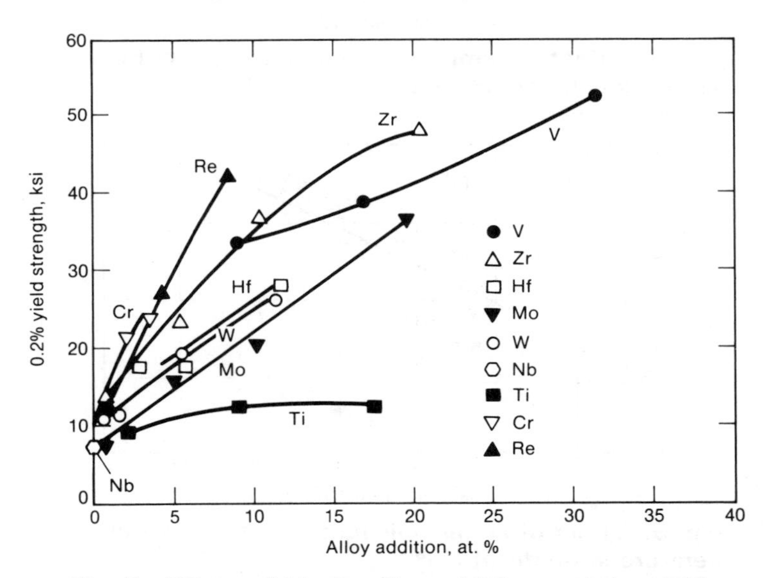

Fig. 7. Effect of binary alloy additions on the yield strength of Nb at 1095 °C (2000 °F)[26,27]

short-time tensile strength are generally those elements having the greatest atom-size mismatch with the solvent. This is shown in Fig. 9 for Nb alloys.

In evaluating elevated-temperature strength, it is important that the test environment does not influence the property being measured. This is particularly true when measuring time-dependent properties. Because of their reactive nature, environmen-

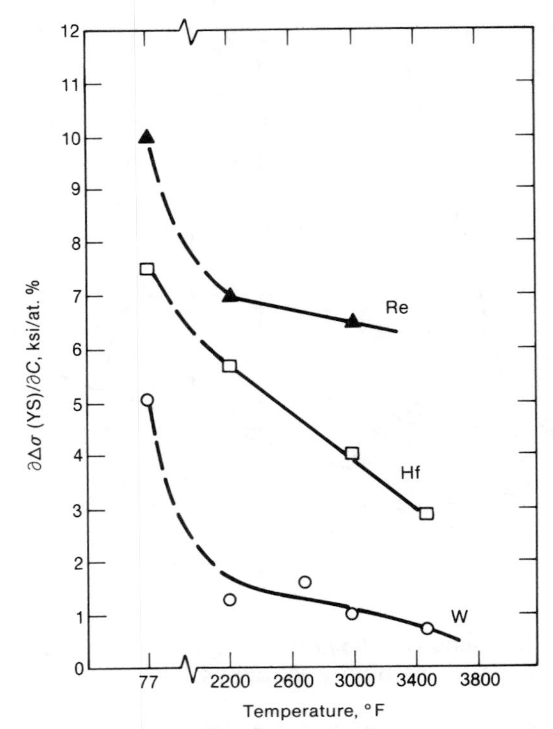

Fig. 8. Effect of temperature on solute strengthening of substitutional elements in Ta[28]

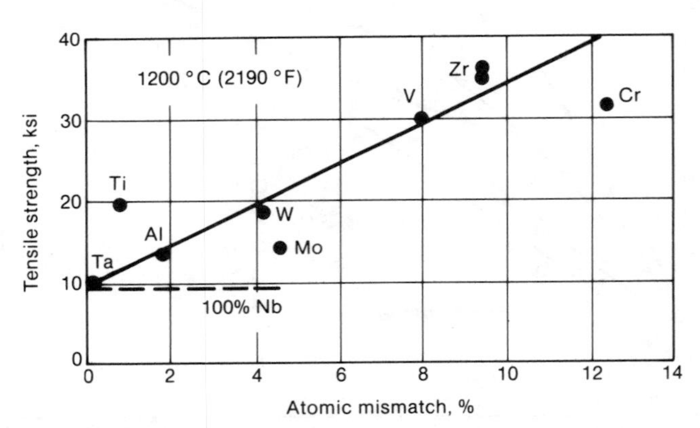

Fig. 9. Effect of atomic mismatch on elevated-temperature strength of Nb[5]

tal interactions can strongly influence creep behavior. When testing the Ta-base alloy T-111 (Ta-8W-2Hf) in an unbaked polymer sealed vacuum chamber at a test pressure on the order of 1×10^{-6} torr, Buckman[27] showed that the creep strength was significantly better than that of material tested under ultrahigh-vacuum conditions ($<1 \times 10^{-8}$ torr), as shown in Fig. 10. The increased resistance to creep was attributed to the Hf oxides and carbides formed (see Fig. 11) during exposure to the relatively high partial pressures of oxygen and hydrocarbons pres-

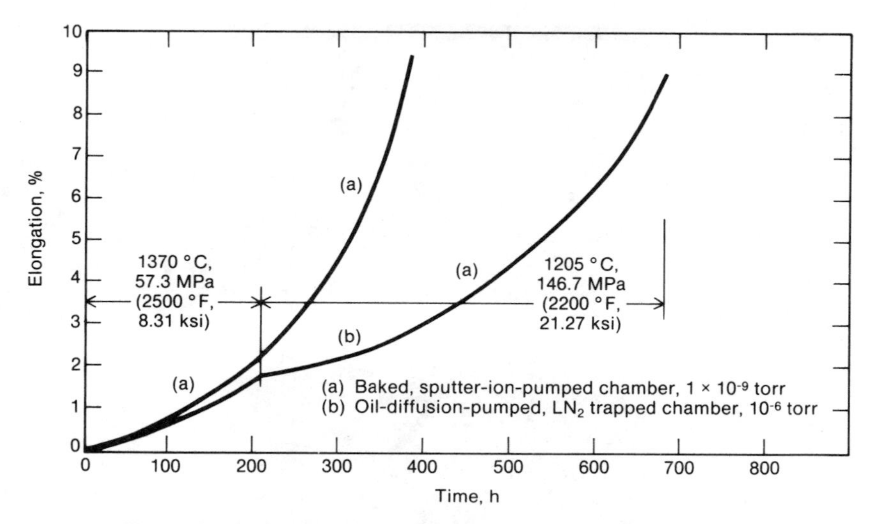

Fig. 10. Elongation-time curves for Ta alloy T-111, showing effect of vacuum environment[27]

Table 5. Effect of vacuum environment on contamination of Ta alloy (T-111) test specimens[27]

	Temperature		Duration,	Test pressure, Torr		Chemical analysis, wt %					
						Pretest			Posttest		
Specimen(a)	°C	°F	h	Initial	Final	C	O	N	C	O	N
No. 1	1370	2500	213(b)	2.3×10^{-6}	4×10^{-7}	0.0026	0.0015	0.001	...	...	...
	1200	2200	459	1.5×10^{-8}	1.4×10^{-9}	...	...	...	0.029	0.0178	0.001
No. 2	1370	2500	213	6×10^{-9}	9.6×10^{-10}	0.0026	0.0015	0.001	...	...	...
	1200	2200	170	2×10^{-8}	7.2×10^{-10}	...	...	...	...	...	...
	1150	2100	125	7.2×10^{-10}	8.6×10^{-11}	...	...	...	0.0019	0.0034	0.001

(a) Specimen thickness, 1.0 mm (0.040 in.). (b) Tested initially in commercial stress-rupture unit, oil-diffusion pumped with LN_2 trap.

ent in the unbaked vacuum chamber. The chemical-analysis results in Table 5 indicate the change in oxygen and carbon contents for tests in the oil-diffusion-pumped chamber and essentially no change in oxygen or carbon when exposed to the ultrahigh-vacuum conditions. The effect of residual gas species in the test environment has not been restricted to alloys containing reactive elements (Hf, Zr, Ti). Inouye[29] has observed significant effects of oxygen partial pressure on the creep behavior of W at 1800 and 2000 °C (3270 and 1090 °F) (see Fig. 12). The influence of oxygen partial pressure on the creep rate was related to a combination of the formation of internal voids and sublimation. The author found these processes apparently dependent on stress and oxygen partial pressure.

Bonesteel *et al.*[30] showed the effects of internal oxidation on strength of Nb-1Zr. From Table 6 it can be seen that relatively small amounts of oxygen pickup have significant effects on the room-temperature flow stress of Nb-1Zr. The

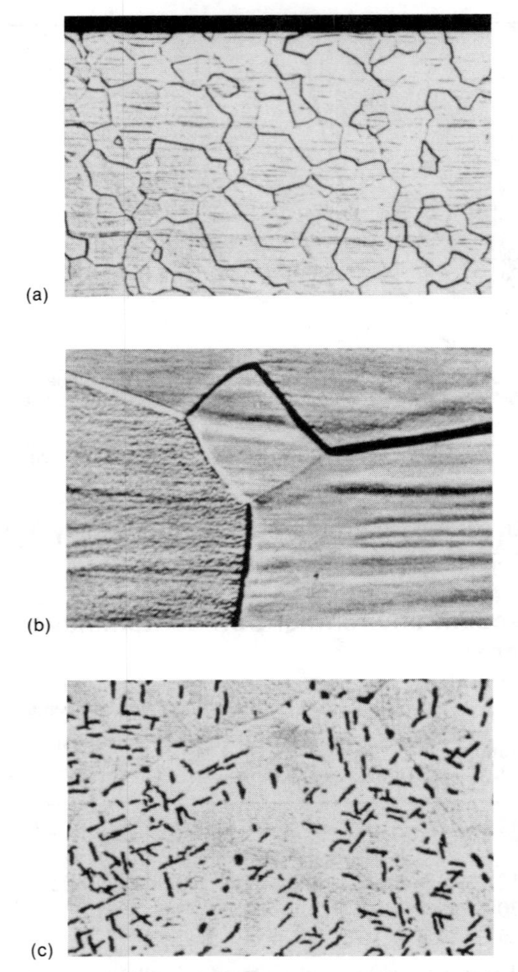

(a) Pretest microstructure. Magnification, 200×. (b) Microstructure after testing at $<10^{-8}$ torr. Magnification, 1500×. (c) Microstructure after testing at 10^{-5} torr. Magnification, 1500×.

Fig. 11. Microstructures of Ta-8W-2Hf (T-111) alloy tested 200 h at 1370 °C (2500 °F)[29]

effect is strongly temperature dependent, with the room-temperature flow stress unaffected when the internal oxidation temperature was ≥850 °C (1560 °F). Transmission electron microscopy showed that when maximum strengthening occurred, coherent ZrO clusters were formed. At the highest internal oxidation temperature, noncoherent ZrO_2 particles precipitated and resulted in very little change in the original room-temperature flow stress. However, the high-temperature creep behavior of an internally oxidized Nb-1W-1Zr alloy was shown by Bonesteel *et al.*[30] to be significantly enhanced when tested at 1200 °C (2190 °F)—well above the internal-oxidation temperature of 800 °C (1470 °F) (see Fig. 13). At the test temperature of 1200 °C, strengthening by coherent ZrO_2 particles would not be expected. Thus, the observed strength increase must be attributable to the dispersed

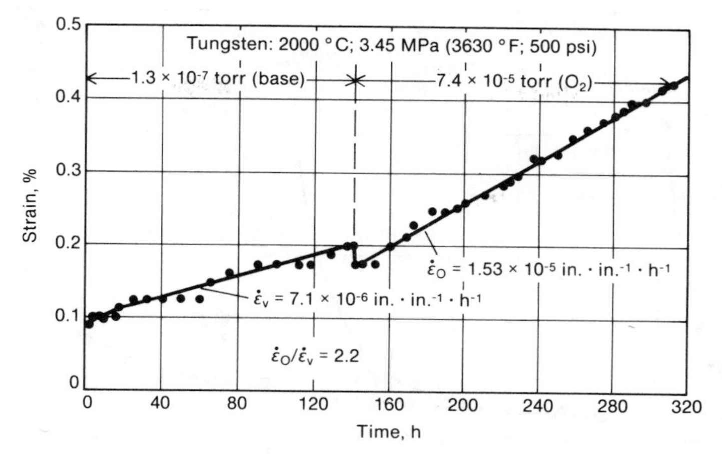

Fig. 12. Effect of 7.4 × 10^{-5} torr O_2 on creep rate of W at 2000 °C and 3.45 MPa (3630 °F and 500 psi)[29]

Table 6. Room-temperature flow stresses of a Nb-1%Zr alloy internally oxidized for 20 h. After Bonesteel *et al.*[30]

Internal-oxidation treatment	Oxygen content, wppm	Flow stress at $\epsilon = 0.002$, ksi(a)	Flow stress at $\epsilon = 0.08$, ksi(a)
Untreated (as-recrystallized)	80	24.5(b)	35.4
750 °C at 2 to 6 × 10^{-6} torr	180	28.3(b)	41.1
800 °C at 2 to 6 × 10^{-6} torr	230	27.1	55.1
800 °C at 4 to 8 × 10^{-6} torr	380	45.6	63.3
800 °C at 1 to 3 × 10^{-5} torr	290	42.1	61.1
850 °C at 4 to 8 × 10^{-6} torr	210	25.6	42.0

(a) To convert ksi to MPa, multiply by 6.8948. (b) Value represents upper yield strength, as specimen exhibited in homogeneous deformation.

oxide phase. Buckman and Goodspeed[21] observed similar behavior in an internally oxidized Ta-8W-2Hf alloy (see Fig. 14). At relatively low levels of internal oxidation, the strengthening increment of the HfO_2 precipitate was relatively short lived. Significant strengthening persisted at the higher oxygen content for longer times, but when the oxide particles grow beyond a critical size, it is expected that their effectiveness will diminish, but will not be eliminated.

As strengthening mechanisms for Nb and Ta, oxide dispersions are not generally considered to be useful alloy ingredients. But the effectiveness of oxygen when added in a controlled (temperature, time, pressure) fashion can significantly alter both the time-independent as well as the time-dependent strength characteristics of these two metals. It is for this reason that creep testing under ultrahigh-vacuum conditions ($<10^{-8}$ torr) is preferred for refractory-metal alloys, to eliminate (or minimize) contamination from the residual atmosphere.

The use of oxides in strengthening of W has been demonstrated with thoria. Tho-

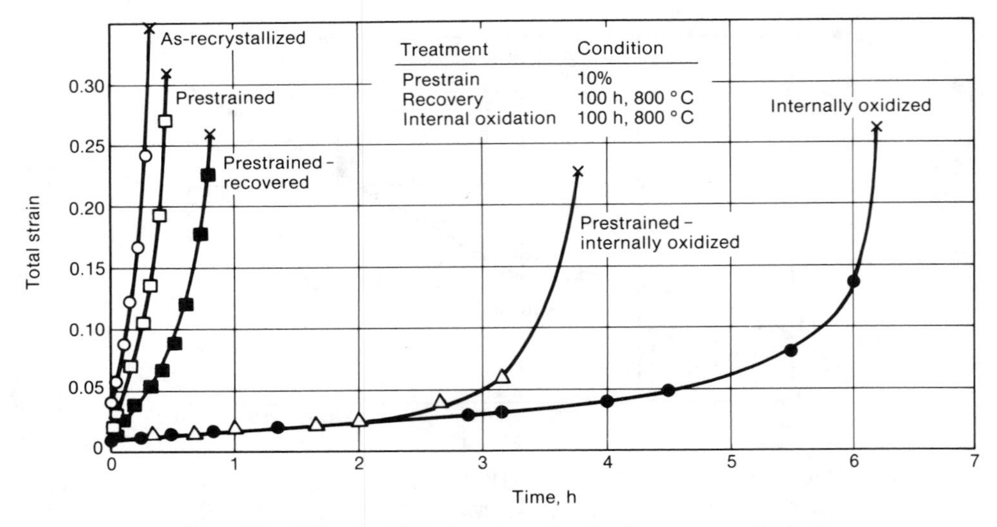

Fig. 13. Effects of thermomechanical treatments and dispersed phases on creep of Nb-1W-1Zr at 1200 °C (2190 °F)[30]

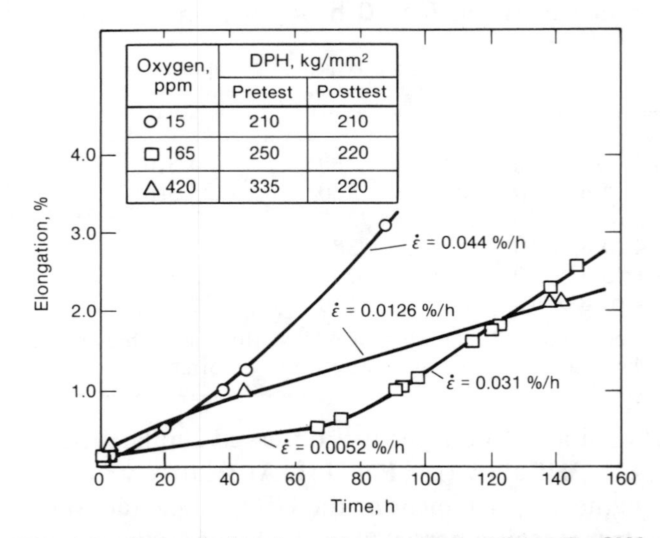

Fig. 14. Creep behavior of internally oxidized Ta-8W-2Hf (T-111) alloy at 1205 °C and 145 MPa (2200 °F and 21 ksi)[21]

rium oxide particles, when present in the proper particle-size distribution, are very effective in stabilizing a worked microstructure in W and hence impart high-temperature strength.[1,31] Through the coprecipitation of the thoria during W oxide preparation, thoria particles on the order of 40 nm are produced,[31] and are effective pins for retarding dislocation motion during recovery and restrain grain growth during recrystallization. Bubbles formed during processing of "doped" W also stabilize a heavily cold worked microstructure.[1,31-34] The stringerlike distribution of 10 to 100 nm dia. bubbles of K inhibit the lateral movement of grain boundaries, giving rise to the elongated interlocking grain structure(35) typically found in "nonsag" or "doped" W lamp filaments.

Nitrogen and C are by far the most potent additions for achieving high strength in the refractory metals.(1,2,21,23,36-43) The strength achievable by C additions is presented in Fig. 15, where the stress-rupture properties of selected refractory alloys are plotted versus the time-temperature compensating Larson-Miller parameter. In creep rupture, the W alloys containing a dispersed HfC precipitate are by far the strongest metals at high temperatures that are currently available. An analog of this composition has been produced in Mo, but its strength decreases rapidly as the test temperature is increased above 1400 °C (2550 °F).(43) The large increment of strength obtained in the carbide-strengthened W and Mo alloys is from cold working. The cold worked structure is stabilized by a very fine stable HfC precipitate. The 1-h recrystallization temperature for a heavily worked W-Hf-C alloy is reported to be above 2650 °C (4800 °F). At higher temperatures, the carbide precipitate does go into solution.(39) Pinning of dislocation substructure by second-phase particles or precipitates in Mo and W is well documented in the literature.(1,31,48,49) However, in Nb and Ta alloys, recovery kinetics are much more

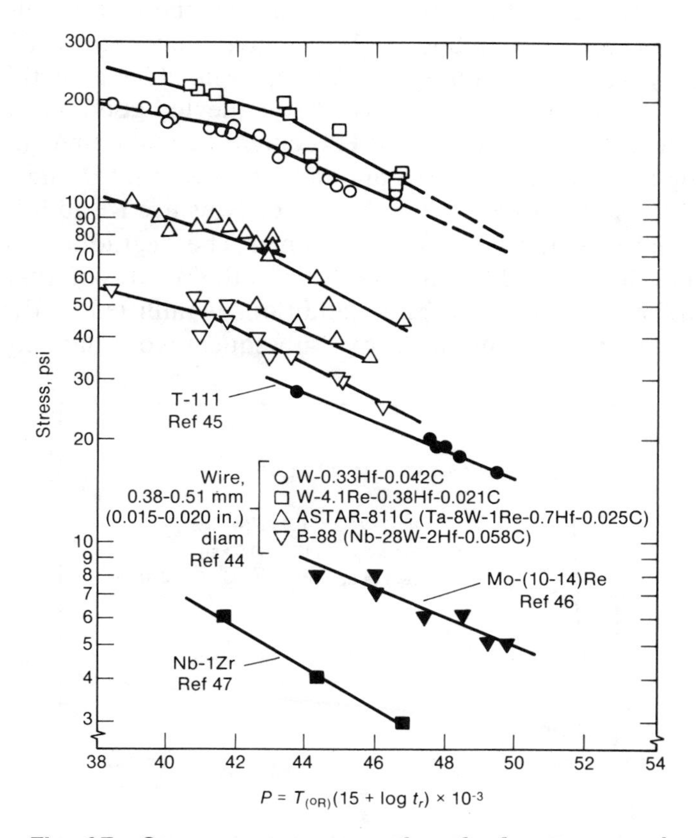

Fig. 15. Stress-rupture properties of refractory-metal wire, sheet, and bar

rapid, and stabilization of cold worked structures is not effective for time-dependent deformation.[48] A cold worked Nb alloy containing a carbide precipitate displays creep-rupture properties inferior to those of recrystallized material (see Fig. 16).

The improvement in high-temperature strength normally results in a concomitant drop in low-temperature ductility. Although dramatic increases in the strength of Ta can be achieved by substitutional and interstitial solute additions, the level that can be tolerated is therefore limited by the effect on low-temperature ductility. This latter property is the most attractive characteristic of Ta. Since welding increases the DBTT, tests on welded joints are an excellent measure of fabricability—that is, welded joints which show good low-temperature ductility are indicative of a highly fabricable base metal. However, base metal which exhibits good ductility does not always exhibit a low DBTT after welding.

The DBTT data plotted in Fig. 17 are for sheet 0.75 to 1.0 mm (0.03 to 0.04 in.) thick tested in bending in the as-recrystallized or inert gas tungsten-arc (GTA) as-welded condition. Schmidt *et al.*[50] have shown that Ta can tolerate up to 19 at. % W before the base-metal bend DBTT is raised above room temperature. They also reported that GTA welding increased the DBTT approximately 360 °C (650 °F), which would limit a room temperature ductile weld to alloys containing less than 13 at. % W. Ammon and Begley[51] have shown that a Ta-11.2W-2.8Hf (14 at. % solute) composition retained a GTA as-welded bend ductility to −155 °C (−250 °F). However, they also reported that achieving good as-welded ductility required that the ratio of W content to Hf content be maintained at about 4 to 1. It is interesting to note here that the anomalous effect that Re has on the ductility of Mo and W is not observed in either Ta or Nb since Re additions drastically increase the DBTT of the latter two elements. The degrading effect of Re on low-temperature ductility led Schmidt *et al.*[50] to the conclusion that any elevated-temperature advantage imparted by Re additions is nullified by the concomitant degradation of low-temperature ductility. Subsequent work has shown that at the

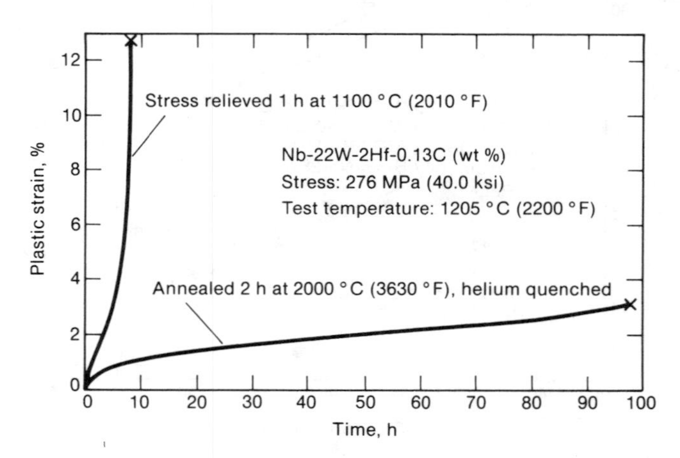

Fig. 16. Effects of thermomechanical treatments on the creep behavior of a Nb-W-Hf-C alloy

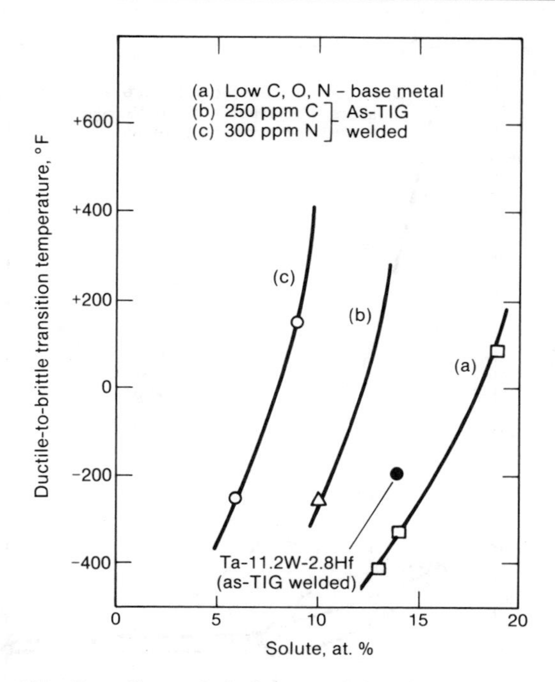

Fig. 17. Ductile-to-brittle transition behavior of Ta-base alloys[21,50]

10 at. % total solute level, Re additions should be maintained at less than 3 at. % to retain low-temperature ductility and weldability.[21,52,53]

Additions of C and N to an alloy matrix further restrict the substitutional solute level before GTA as-welded ductility is seriously limited, reflecting the solubility relationships of the additions with the matrix. Although C additions of 1 at. % to Ta-10 (W+Mo+Zr) and Ta–10 to 12(W+Hf) (both in at. %) do not seriously degrade the low-temperature ductility of recrystallized sheet material, during a welding cycle, C is dissolved and the cooling rate of the solidified metal is sufficiently high to retain a large amount of C in solid solution.[50,53] The potent low-temperature strengthening of the Ta matrix by C and N in solid solution is shown in Fig. 18. By proper postweld annealing treatments, low-temperature ductility can be restored in C-containing alloys. This may prove impractical, however, in large, complex fabricated structures.

Fortunately, the trade-offs between elevated-temperature strength and low-temperature ductility may not represent an insurmountable problem. The data of Schmidt *et al.*[50] plotted in Fig. 19 show that increasing the solute level in Ta above 13 at. % is not warranted since the increase in the elevated-temperature strength is not significant enough to offset the drastic decrease in low-temperature ductility. Thus, for adequate low-temperature weld ductility, substitutional-solute-strengthened Ta alloys should contain less than 12 to 14 at. % total alloy addition. If C is added at the 200 to 300 wppm level, the substitutional solute addition should be restricted to 10 at. % or less. An equivalent amount of N would require a substitutional solute level of less than 8 at. %. Generally, the adverse effects that N

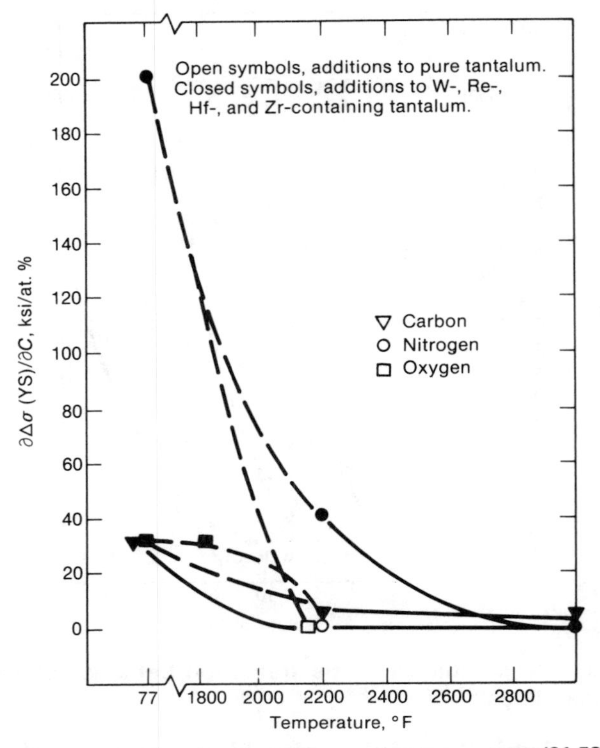

Fig. 18. Interstitial solute strengthening inTa[21,53]

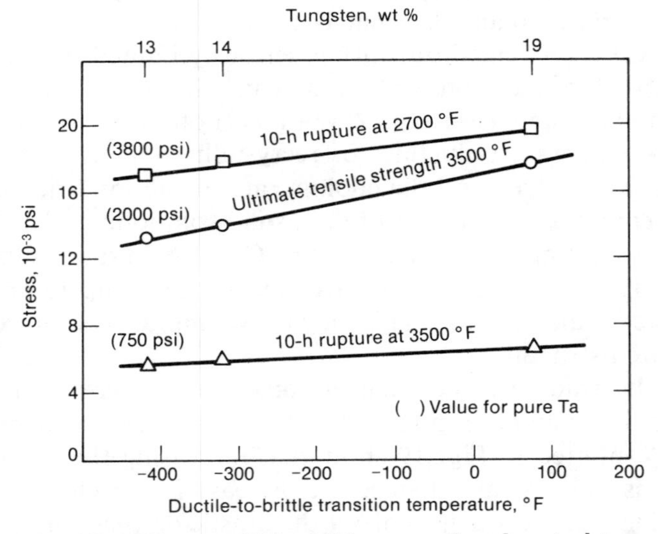

Fig. 19. Effects of W additions on the elevated-temperature strength and DBTT of Ta[50]

has on low-temperature weld ductility will limit its usefulness as a strengthening addition in weldable sheet alloys.

The observations on Ta regarding low-temperature ductility are generally applicable to Nb. However, the absolute magnitudes of solute additions required to produce equivalent changes in Nb will be less since pure Nb exhibits a higher DBTT than Ta.

The lack of widespread structural application of Mo- and W-base alloys is generally related to their poor room-temperature ductility. The DBTT of unalloyed polycrystalline Mo is close to or below room temperature, while that of unalloyed polycrystalline W is significantly above room temperature. As discussed earlier, the DBTT is affected by factors such as alloying, strain rate, grain size, purity, and surface condition. The most remarkable alloying effects on the low-temperature ductility of Mo and W are imparted by Re additions, and have been studied extensively.(46,54,55) Two mechanisms are thought to be responsible. At low Re contents (5 to 15 at. %), improvements in low-temperature ductility are attributed to solution softening, while at concentrations near the solubility limit for Re in Mo or W, improved ductility is attributed to the so-called "rhenium ductilizing effects."(24,55) The DBTT for cold worked and recrystallized Mo-Re and W-Re are shown in Fig. 20. Table 7 gives bend ductility values for W and W alloy sheet. Strain rate also has a significant effect on low-temperature ductility of Mo and W, much greater than, say, for Nb or Ta. For Mo, the DBTT is ≤200 K at strain rates of 10^{-4} to 10^{-5} s^{-1} (strain rates typical of slow bend or tensile testing). However, at impact strain rates of 10^2 to 10^3 s^{-1}, the DBTT is >500 K.(43) Similar behavior is also obtained for W. DBTT's are generally quoted for bend and/or tensile tests conducted at low strain rates. However, critical times during service life occur during handling, assembly, and fabrication when loading under high-strain rate conditions will probably be present. Fabrication of a structure generally requires fusion welding, particularly if leak tightness is a requirement. Thus, the low-temperature ductility of weldments is particularly critical.

Tungsten-rhenium alloys show significant increases in DBTT after welding, as do the dilute Mo-14Re alloys (see Table 8 and Fig. 21). However, the Mo-50Re alloys display ductility well below room temperature in the welded condition.

Recently Wadsworth *et al.*(58) reviewed the so-called work-softening of dilute Mo-Re alloys and concluded that the work-softening theory has not been verified conclusively. These investigators have presented data which show that the DBTT of unalloyed Mo is strongly influenced by the C/O ratio. For C/O ratios greater than 2:1, unalloyed recrystallized Mo exhibits 30 to 60% elongation at room tem-

Table 7. W and W alloy base-metal bend ductilities(56)

	4*t* bend DBTT				
	Longitudinal		Transferse		
Metal or alloy(a)	°C	°F	°C	°F	As-received condition(b)
AC tungsten	220	425	135	275	SR 1 h at 925 °C (1700 °F)
AC W-25Re (wt %)	−130	−200	−60	−75	SR 1 h at 1400 °C (2550 °F)
PM W-25Re-30Mo (at. %)	−100	−150	−45	−50	SR 1/2 h at 1150 °C (2100 °F)
AC W-25Re-30Mo (at. %)	<−195	<−320	−155	−250	SR 1/2 h at 1050 °C (1920 °F)

(a) AC = arc cast. PM = powder metallurgy. (b) All as-received material was in the wrought condition. SR = stress relieved.

Table 8. Ductile-to-brittle transition temperatures of welded W and W alloys(56)

Alloy	Structure	Weld preheat, °C (°F)	1-h anneal temperature, °C (°F)	Change in 4t bend transition temperature(a), °C (°F)	Lowest DBTT(b), °C (°F)
W	GTA weld	None	1405 (2560)	+55 (+100) (L)	370 (700)
	GTA weld	290 (550)	1405 (2560)	−55 (−100) (L)	480 (900)
	GTA weld	None	1405 (2560)	Increased (L)	370 (700)
W-25Re	GTA weld	290 (550)	1405 (2560)	+110 (+200) (L)	425 (800)
	4 GTA welds	3–290 (550); 1–none	1405 (2560)	Increased ductility implied	425 (800)
	4 GTA welds	1–290 (550); 3–none	1405 (2560)	Decreased ductility implied	425 (800)
	3 GTA welds	760 (1400)	1800 (3270)	−220 (−400) max (L and T)	315 (600)
	GTA weld	760 (1400)	1800 (3270)	Questionable	540 (1000)
	11 EB welds	None	1405 (2560)	−280 (−500) max (T)	260 (500)
W-Re-Mo (PM)	GTA welds	None	1315 (2400) 1540 (2800)	−28 to −55 (−50 to −100) (L)	175 (350)
	GTA weld	None	1760 (3200)	+14 (+25) (L)	220 (425)
	EB welds	None	1315 (2400)	−14 (−25) (L) } (T), no change	(L) 79 (175)
			1540 (2800)	>−14 (>−25) (L) } (T), no change	(T) 205 (400)
			1760 (3200)	+14 (+25) (L) } (T), no change	
	Base metal	. . .	1540 (2800)	+69 (+125) (L) +97 (+175) (T)	−100 (−150) −59 (−75)
W-Re-Mo (AC)	EB welds	760 (1400)	1315 (2400)	+28 (+50) (L); +110 (+200) (T)	(L) 65 (150)
			1540 (2800)	+28 (+50) (L); +140 (+250) (T)	(T) 93 (200)
			1760 (3200)	+55 (+100) (L); +140 (+250) (T)	

(a) Bend type: (L) = longitudinal; (T) = transverse. (b) DBTT for annealed or unannealed, whichever is lower.

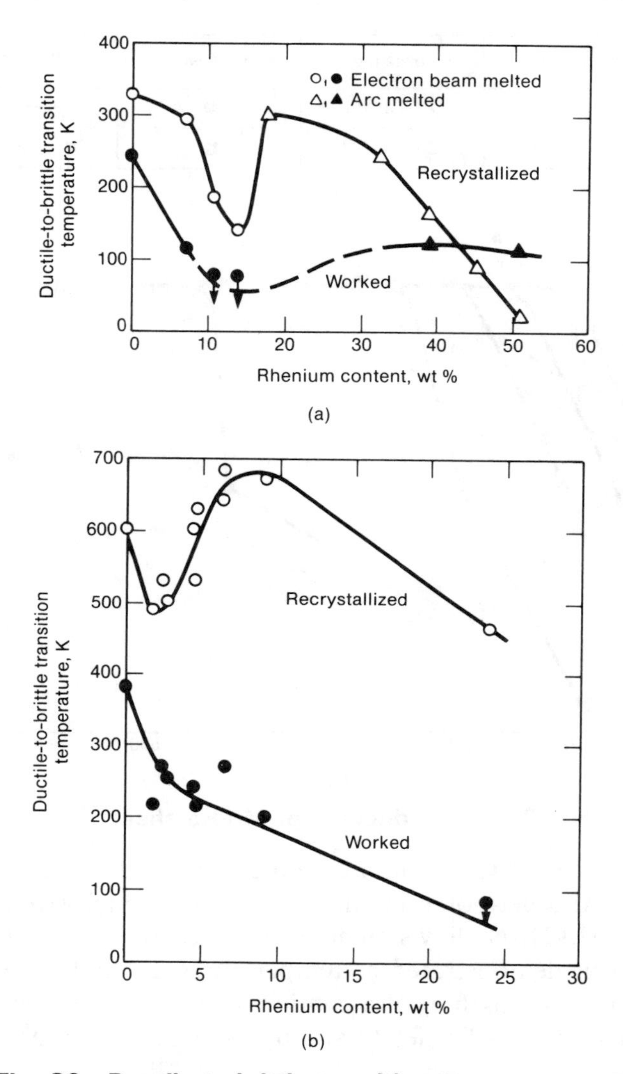

Fig. 20. Ductile-to-brittle transition temperatures of (a) Mo-Re and (b) W-Re alloys[24,43,46]

perature, while at C/O ratios less than 2:1, room-temperature ductility is near zero. For the data presented by Lundberg *et al.*[59] to support the solution-softening concept, the unalloyed Mo and dilute Mo-Re alloys exhibiting low ductility all had C/O ratios less than 2:1. The as-welded bend DBTT of a dilute Mo-Re alloy was shifted to above room temperature (see Fig. 21). This material had a C/O ratio greater than 2:1, and thus the work-softening theory for dilute Mo-Re alloys is an area where further study would appear to be necessary before firm conclusions can be drawn.

One additional strengthening mechanism of interest in refractory metals is the use of continuous, strong refractory-metal fibers in a weak refractory-metal alloy matrix. The strongest high-strength fiber currently known is the W-Re-Hf-C alloy.

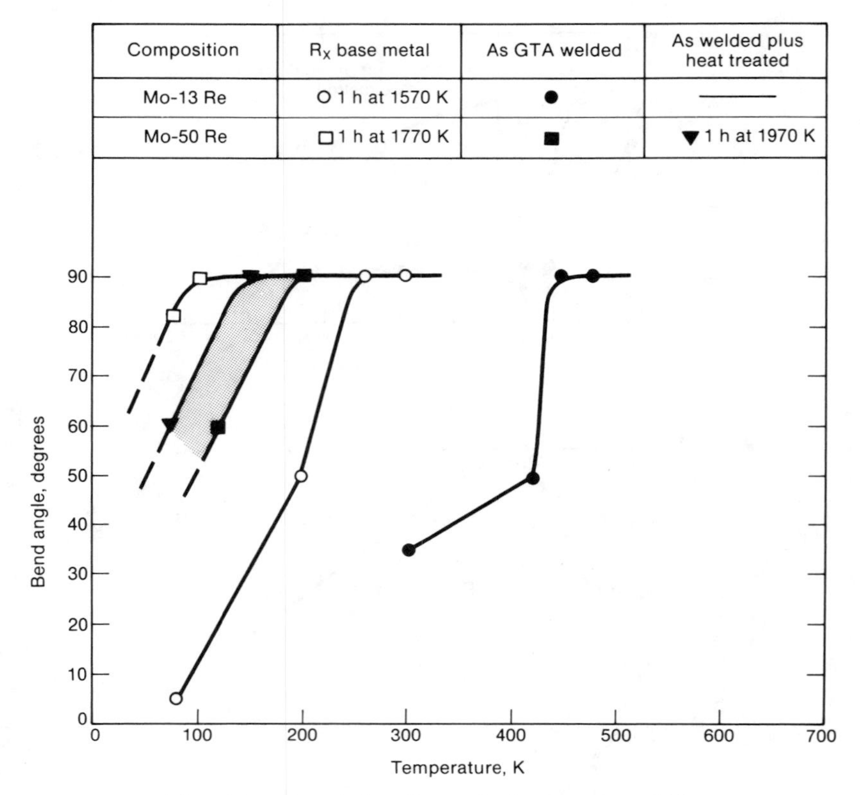

Fig. 21. Bend ductility of Mo-Re sheet[57]

Tensile properties are shown in Fig. 22, and creep-rupture properties of wire are shown in Fig. 15. Another potential high-temperature, high-strength fiber candidate would be the Mo-Hf-N alloy system, where strengths on the order of 793 MPa (115 ksi) have been demonstrated at temperatures up to 1500 K.[62] The basic equations relating the strength σ_c of a continuous-fiber-reinforced composite have been reviewed by Kelly and Davies[63] and by Kelly and Tyson,[64] and σ_c is given by:

$$\sigma_c = \sigma_f \cdot V_f + \sigma'_m(1 - V_f)$$

where σ_f is the strength of the continuous fiber, V_f is the volume fraction of fiber, and σ'_m is the stress in the matrix at the composite failure strain.

Using this relation and the data for the W-Hf-C alloy wire, the creep-rupture properties of Nb reinforced with a 50 vol % W alloy fiber results in a significantly stronger material than can be achieved by traditional alloying methods (see Fig. 23). Westfall *et al.*[65] have recently produced a thoriated W fiber reinforced Nb composite and indeed achieved properties in the composite predicted by the rule-of-mixtures equation. This is a new composite technology which will find applications in space power systems.

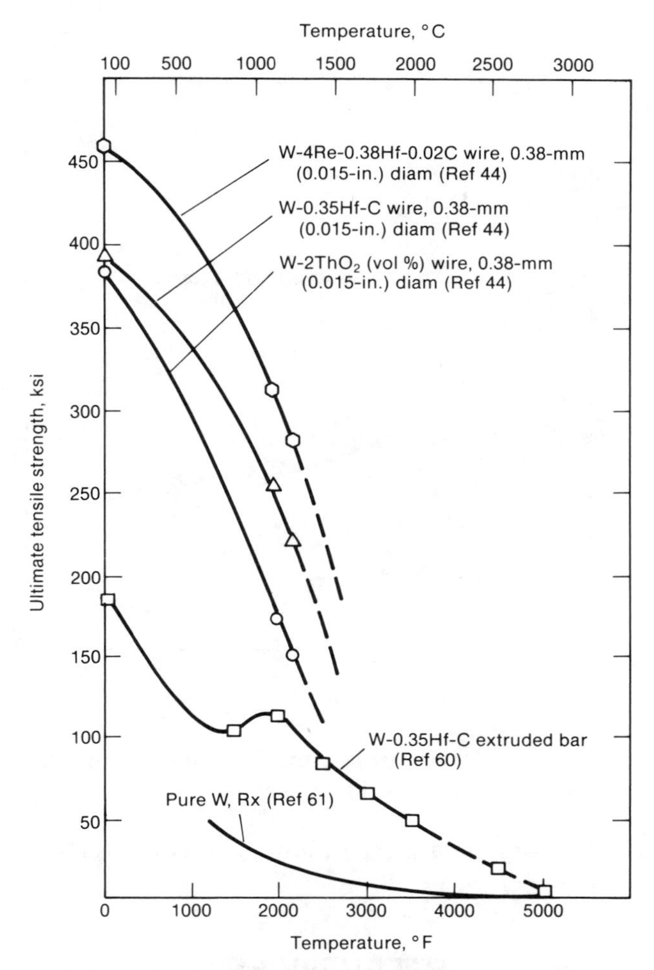

Fig. 22. Tensile properties of W and W alloys as functions of temperature[44,60,61]

SUMMARY

This chapter has examined the alloying behavior of refractory metals from the perspective of having to design an alloy composition to fulfill a number of specific design requirements. This generally results in a compromise between strength and room-temperature ductility. The metallurgical and mechanical effects of alloying on the refractory metals have been extensively studied, but behavioral mechanisms which were once accepted are now open to question in the light of an increasing data base. Thus, there are opportunities for study of fundamental issues relative to purity, and effects on recovery and recrystallization behavior, low-temperature ductility, the DBTT phenomena, and high-temperature strength. As exploration of space progresses, the need for high-temperature materials will increase, and the

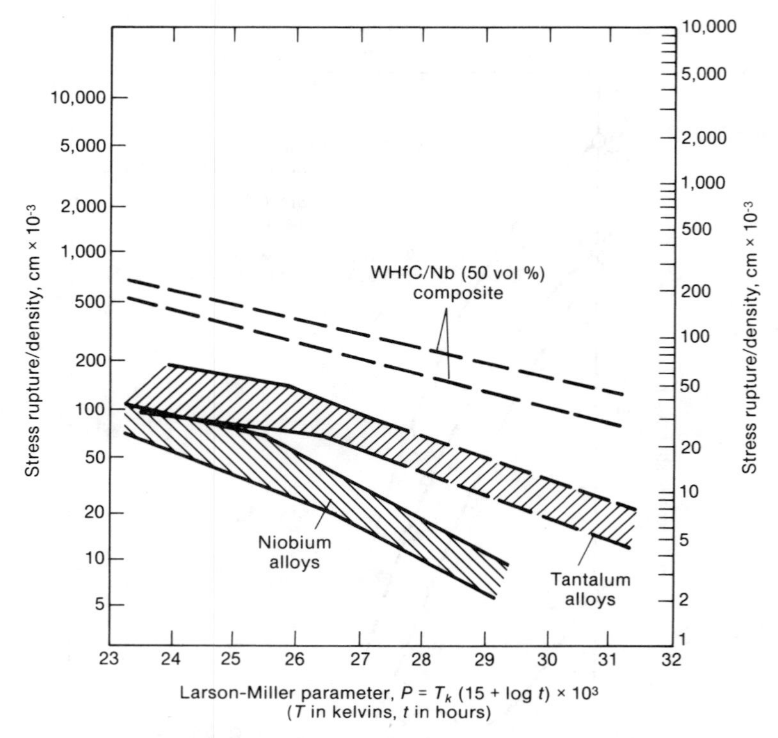

Fig. 23. Calculated stress-rupture properties of Nb-matrix composite materials

specialized role of refractory metals, their alloys, and composites will be an expanding one.

REFERENCES

1. B.A. Wilcox, "Basic Strengthening Mechanisms in Refractory Metals," in *Refractory Metal Alloys*, edited by I. Machlin, R.T. Begley, and E.D. Weisert, Plenum Press, New York, 1968, pp. 1-34
2. R.T. Begley, D.L. Harrod, and R.E. Gold, "High Temperature Creep and Fracture Behavior of the Refractory Metals," in *Refractory Metal Alloys*, edited by I. Machlin, R.T. Begley, and E.D. Weisert, Plenum Press, New York, 1968, pp. 41-84
3. J.H. Bechtold, E.T. Wessel, and L.L. France, "Mechanical Behavior of the Refractory Metals," in *Refractory Metals and Alloys*, edited by M. Semchyshen and J.J. Harwood, Vol 11, Interscience Publishers, New York, 1961, pp. 25-81
4. R.W. Armstrong, J.H. Bechtold, and R.T. Begley, "Mechanisms of Alloy Strengthening in Refractory Metals," in *Refractory Metals and Alloys*, AIME Metall. Soc. Conferences, Vol 17, 1963, pp. 159-190
5. W.H. Chang, "Strengthening of Refractory Metals," Refractory Metals and Alloys, *AIME Metall. Soc. Conferences*, Vol 11, 1961, pp. 83-117
6. I. Machlin, R.T. Begley, and E.D. Weisert, *Refractory Metal Alloys Metallurgy and Technology*, Plenum Press, New York, 1968, p. V

7. M. Semchyshen and I. Perlmutter, eds., *Refractory Metals and Alloys II*, Interscience Publishers, New York, 1963, p. V
8. D.A. Robins, An Interpretation of Some of the Properties of the Transition Metals and Their Alloys, *J. Less-Common Metals*, Vol I, 1959, pp. 396-410
9. G.T. Hahn, A. Gilbert, and R.I. Jaffee, "The Effects of Solutes on the Ductile-to-Brittle Transition in Refractory Metals," in *Refractory Metals and Alloys II*, edited by M. Semchyshen and I. Perlmutter, Interscience Publishers, New York, 1963, pp. 23-63
10. O.D. Sherby, Factors Affecting the High Temperature Strength of Polycrystalline Solids, *Acta Met.*, Vol 10, 1962, pp. 135-147
11. W.R. Clough and A.S. Pavolic, *Trans. Am. Soc. Metals*, Vol 52, 1960
12. E.T. Wessel, L.L. France, and R.T. Begley, in *Columbium Metallurgy*, Interscience, New York-London, 1951, pp. 459-502
13. M.J. Marcinkowski, WADC Technical Report 59-294, ASTIA Document No. AD216360, June 1959
14. J.H. Bechtold, *Trans. AIME*, Vol 197, 1953, p. 1469
15. J.H. Bechtold and P.G. Shewmon, *Trans. Am. Soc. Metals*, Vol 46, 1954, p. 397
16. C.S. Barrett and R. Bakish, *Trans. AIME*, Vol 212, 1958, p. 122
17. R.L. Fleischer, *Acta Met.*, Vol 9, 1961, p. 996
18. B. Harris, *Phys. Stat. Sol.*, Vol 18, 1966, p. 715
19. W.A. Spitzig, C.V. Owen, and T.E. Scott, *Met. Trans. A*, Vol 17A, May 1986, p. 853
20. G.D. McAdam, Substitutional Niobium Alloys of High Creep Strength, *J. Inst. Metals*, Vol 93, 1964-1965, pp. 559-564
21. R.W. Buckman and R.C. Goodspeed, "Considerations in the Development of Tantalum-Base Alloys," in *Refractory Metal Alloys*, edited by I. Machlin, R.T. Begley, and E.D. Weisert, Plenum Press, New York, 1968, pp. 373-394
22. M. Semchyshen, "Development and Properties of Arc-Cast Molybdenum Base Alloys," in *The Metal Molybdenum*, edited by J.J. Harwood, Chapter 14, ASM, Cleveland, 1958
23. P.L. Raffo and W.D. Klopp, "Mechanical Properties of Solid Solution and Carbide Strengthened Arc-Melted Tungsten Alloys," NASA-TND-3248, February 1966
24. W.D. Klopp, W.R. Witzke, and P.L. Raffo, "Mechanical Properties of Dilute Tungsten-Rhenium Alloys," NASA-TND-3483, September 1966
25. P.L. Raffo, W.D. Klopp, and W.R. Witzke, "Mechanical Properties of Arc-Melted and Electron Beam Melted Tungsten Base Alloys," NASA-TND-2561, January 1965
26. G.D. Gemmel, "Some Effects of Alloying on the Strength Properties of Niobium," *Trans. AIME*, Vol 215, No. 6, December 1959, p. 898
27. R.W. Buckman, Jr., "Operation of Ultra-High Vacuum Creep Testing Laboratory," Transactions of Vacuum Metallurgy Conference, 1966, American Vacuum Society, 1967, pp. 25-37
28. R.T. Begley, W.N. Platte, A.I. Lewis, and R.L. Ammon, "Development of Niobium Base Alloys," WADC Technical Report 57-344, Part V, January 1961
29. H. Inouye, "The Effect of Low Pressure Oxygen on the Creep Properties of Tungsten," Transactions of Vacuum Metallurgy Conference, 1969, American Vacuum Society, 1968, p. 99
30. R.M. Bonesteel, J.L. Lytton, D.J. Rowcliffe, and T.E. Tietz, "Recovery and Internal Oxidation of Columbium and Columbium Alloys," AFML-TR-66-253, August 1966
31. H.G. Sell, "Advanced Processing Technology and High Temperature Mechanical Properties of Tungsten Base Alloys," in Refractory Metal Alloys, edited by I. Machlin, R.T. Begley, and E.D. Weisert, Plenum Press, New York, 1968, pp. 395-450
32. R.C. Koo, *Trans. AIME*, Vol 239, 1967, p. 1996

33. G. Das and S.V. Radcliffe, "Internal Void Formation in Tungsten," 97th AIME Annual Meeting, New York, Feb 26-29, 1968
34. R.C. Koo and D.M. Moon, "Nucleation and Growth of Bubbles in Doped Tungsten," 97th AIME Annual Meeting, New York, Feb 26-29, 1968
35. J.L. Walter, *Trans. AIME*. Vol 239, 1967, p. 272
36. W.H. Chang, "Influence of Heat Treatment on Microstructure and Properties of Columbium-Base and Chromium-Base Alloys," ASD-TDR-62-211, Part IV, March 1966
37. R.T. Begley, J.A. Cornie, and R.C. Goodspeed, "Development of Columbium Base Alloys," AFML-TR-67-116, November 1967
38. G.D. McAdam, The Influence of Carbide and Boron Additions on the Creep Strength of Niobium Alloys, *J. Inst. Metals*, Vol 96, 1968, pp. 13-16
39. L.S. Rubenstein, "Effects of Composition and Heat Treatment on High-Temperature Strength of Arc-Melted Tungsten-Hafnium-Carbon Alloys," NASA-TN-D-4379, February 1968
40. R.T. Begley, J.L. Godshall, and R. Stickler, "Precipitation Hardening Columbium-Hafnium-Nitrogen Alloys," Fifth Plansee Seminar, June 22-26, 1964, Reutte/Tyrol, pp. 401-420
41. A.K. Mukherjee and J.W. Martin, The Effect of Nitriding Upon the Creep Properties of Some Molybdenum Alloys, *J. Less-Common Metals*, Vol 5, 1963, pp. 403-410
42. J.S. Kane, "Creep of Internally Nitrided Molybdenum-Based Alloy in a Nitrogen Environment," UCRL-71429, Rev. II, Lawrence Radiation Laboratory, Univ. of California, October 2, 1969
43. W.D. Klopp, "Technology Status of Molybdenum and Tungsten Alloys," Transactions of Fourth Symposium on Space Nuclear Power Systems, Albuquerque, NM, January 15, 1987
44. D.W. Petrasek, "High-Temperature Strength of Refractory-Metal Wires and Consideration for Composite Applications," NASA-TND-881, August 1972
45. D.R. Stoner and R.W. Buckman, Jr., "Development of Large Diameter T-111 (Ta-8W-2Hf) Tubing," NASA-CR-72869, Westinghouse Electric Corp., December 1970
46. W.D. Klopp and W.R. Witzke, "Mechanical Properties of Electron Beam Melted Molybdenum and Dilute Molybdenum-Rhenium Alloys," NASA-TM-X-2576
47. Teledyne-Wah Chang Technical Information Bulletin, January 1968
48. R.A. Perkins, "The Effect of Thermal-Mechanical Treatments on the Structure and Properties of Refractory Metals," in *Refractory Metal Alloys*, edited by I. Machlin, R.T. Begley, and E.D. Weisert, Plenum Press, New York, 1968, pp. 85-120
49. L.L. Siegle, "Structural Considerations in Developing Refractory Metal Alloys," in *The Science and Technology of Selected Refractory Metals*, edited by N.E. Promisel, AGARD Conference on Refractory Metals, Oslo, Norway, June 23-26, 1963, pp. 63-93
50. F.F. Schmidt, E.S. Bartlett, and H.R. Ogden, "Investigation of Tantalum and Its Alloys," Technical Documentary Report No. ASD-TDR-62-594, Part II, May 1963
51. R.L. Ammon and R.T. Begley, "Pilot Production and Evaluation of Tantalum Alloy Sheet," Summary Phase Report Part II, WANL-PR-M-009, Westinghouse Advanced Energy Systems Div., Pittsburgh, July 1, 1964
52. L.H. Amra and G.D. Oxx, Jr., "Tantalum Base Alloys with High Strength above 3000°F," Metallurgical Society Conferences, Vol 34, *High Temperature Refractory Metals*, Part II, edited by R.W. Fountain, J. Maltz, and L.S. Richardson, 1964
53. R.W. Buckman, Jr., and R.C. Goodspeed, "Precipitation Strengthened Tantalum Base Alloys," NASA Contractors Report NASA-CR-1642, May 1971
54. G.A. Geach and J.E. Hughes, "The Alloys of Rhenium with Molybdenum or with

Tungsten and Having Good High Temperature Properties," in Plansee Seminar, *De Re Metallic*, 2nd Ed., Pergamon Press, London, 1956, pp. 245-253

55. R.I. Jaffee, D.J. Maykuth, and R.W. Douglas, "Rhenium and the Refractory Platinum-Group Metals," in *Refractory Metals and Alloys II*, Interscience Publishers, New York, 1961, pp. 383-463
56. G.G. Lessmann and R.E. Gold, "The Weldability of Tungsten Base Alloys," *Welding Journal Research Supplement*, December 1969
57. R.L. Ammon and R.W. Buckman, Jr., "Observations on the As-Welded (GTAW) Bend Ductility of Mo-13%Re Alloy," in *Space Nuclear Power Systems, 1986*, edited by M.S. El-Genk and M.D. Hoover, Orbit Book Co., Malabar, FL, 1987, pp. 283-289
58. J. Wadsworth, T.G. Nieh, and J.J. Stephens, "Dilute Mo-Re Alloys—A Critical Evaluation of Their Comparative Mechanical Properties," *Scripta Met.*, Vol 20, Pergamon Press, Ltd., 1986, p. 637-642
59. L.B. Lundberg, E.K. Ohrimer, S.M. Tuominen, E.P. Whalen, and J.A. Shields, Jr., in *Physical Metallurgy and Technology of Molybdenum and its Alloys*, p. 71, edited by K.H. Miska *et al.*, AMAX, Ann Arbor, MI, 1985
60. A. Ossin and L. Aronin, "Refractory Metal Air Vane Leading Edge for Ballistic Missiles Defense Interceptor," *AIAA Journal*, Vol 14, June, 1976, pp. 781-788
61. Anon., "The Engineering Properties of Tungsten and Tungsten Alloys," DMIC Report 191, September 1963
62. J.L. Mitchell, "Development of High-Creep-Strength Molybdenum and Tungsten Alloys by the Internal Nitriding Process," Transactions of Fourth Symposium on Space Nuclear Power Systems, Albuquerque, NM, Jan 15, 1987, pp. 39-43
63. A. Kelly and G.J. Davies, *Met. Rev.*, Vol 10, 1965, p. 1
64. A. Kelly and W. Tyson, in *High Strength Materials*, edited by V.F. Zackay, Wiley, New York, 1965, p. 578
65. L.J. Westfall, D.W. Petrasek, L. McDanels, and T.L. Grobstein, "Preliminary Feasibility Studies of Tungsten/Niobium Composite for Advanced Space Power Systems Applications," NASA Technical Memorandum 87248, March 1986

13

Sialons and Related Ceramic Alloys

K.H. JACK
Emeritus Professor of Applied Crystal Chemistry
University of Newcastle upon Tyne

Ceramic alloys are not really new. The traditional ceramics are silicates, and in all silicates the structural unit is the $(SiO_4)^{4-}$ tetrahedron (see Fig. 1) carrying four negative valency charges. These tetrahedra occasionally occur separately as in magnesium silicate, Mg_2SiO_4, where the four negative valencies are balanced by the four positive valencies on the two Mg^{2+} cations. But the tetrahedra are usually joined together by sharing oxygen corners to form rings, chains, double chains, sheets, or three-dimensional networks to give the tremendous variety of mineral silicates that make up the earth's lithosphere.

Aluminum plays a special role in these minerals because the $(AlO_4)^{5-}$ tetrahedron—this time with five negative charges on it—has almost the same

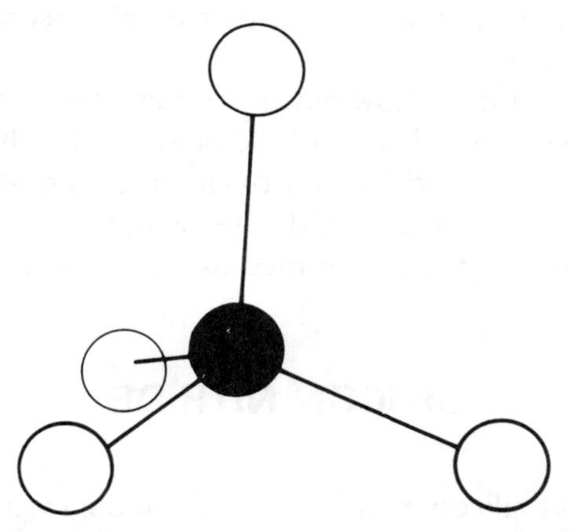

Fig. 1. The $(SiO_4)^{4-}$ tetrahedron, the structural unit of the silicates

dimensions as $(SiO_4)^{4-}$, and so aluminum can always replace silicon in the rings, chains, sheets, and networks, provided that compensation is made somewhere else in the structure for the change of valency from Si^{IV} to Al^{III}.

Tremolite is a typical amphibole double-chain silicate where a unit length of the chain is

$$[Si_4O_{11}]^{6-}$$

There are $(OH)^{1-}$ anions outside the chains, i.e.:

$$[Si_8O_{22}]^{12-}(OH)_2^{2-}$$

and the chains and the hydroxyl groups are held together in the structure, and the 14 negative valencies are balanced by the 14 positive valencies on $2Ca^{2+}$ and $5Mg^{2+}$ cations:

$$(Ca_2Mg_5)^{14+}[Si_8O_{22}]^{12-}(OH)_2^{2-}$$

However, up to one-fourth of the Si^{IV} atoms can be replaced by Al^{III} and, to compensate for the valency change, $2Mg^{2+}$ outside the chains can be replaced by $2Al^{3+}$ or by $2Fe^{3+}$; or additional alkali or alkali-earth cations can be introduced. Hydroxyl anions can be replaced by fluoride or by oxide anions, and so the composition of a tremolite mineral silicate must be represented by the chemical formula:

$$(O,OH,F)_2^{2-:4-}(Ca,Na)_2^{2+:4+}(Na,K)_{0:2}^{0:2+}(Mg,Fe)_1^{2+}(Mg,Fe;Al,Fe)_4^{8+:12+}$$

$$[(Al,Si)_2Si_6O_{22}]^{12-:14-}$$

There is no change in the structure, despite the changes in chemical composition, and so it seems not unreasonable to describe tremolite as an alloy with an extensive range of homogeneity!

In this chapter I hope to show how new materials have been developed by applying the following two very simple and obvious ideas: (1) the principles of silicate chemistry can be applied to nitrides and oxynitrides; and (2) chemical reactions between strongly bonded covalent solids are usually reconstructive and require either a vapor-phase or a liquid-phase medium in order to occur.

SILICON NITRIDE

The properties that make silicon nitride such a good engineering ceramic—strength, hardness, wear resistance, stability up to 1800 °C, oxidation resistance, and excellent thermal-shock properties—are achieved only in fully dense material but, being covalently bonded, the nitride cannot be densified merely by firing it. Self-diffusivity becomes appreciable only at temperatures where the silicon nitride begins to decompose. This difficulty has been partly overcome by sintering, hot pressing, or hot isostatic pressing with additives, and these additives are usually oxides. As shown schematically in Fig. 2, the metal-oxide additive such as MgO or Y_2O_3 reacts with a small amount of nitride and the surface silica that is always present on each nitride particle to give, at high temperatures, a metal-silicon-oxygen-nitrogen (M-Si-O-N) liquid that allows densification by liquid-phase sintering. α-Si_3N_4 dissolves in the liquid and β-Si_3N_4 is precipitated from it. However, the liquid cools to give glassy or secondary crystalline phases in the grain boundaries, and these degrade the properties of the densified product.

A sample of silicon nitride hot pressed with magnesium oxide is fully dense, and its room-temperature properties are excellent. The intergranular Mg-Si-O-N glass, formed from the liquid on cooling, softens above 1000 °C, and so the high-temperature creep resistance is very poor. With yttria additive (Y_2O_3), the Y-Si-O-N liquid cools to give one or more of the four crystalline quaternary oxynitrides shown in the phase diagram in Fig. 3. The intergranular phases that are formed depend on the amount of added yttria and the amount of surface silica on the silicon nitride, but all of them oxidize with an increase in volume. In particular, the oxynitride $Y_2Si_3O_3N_4$ is isostructural with the melilite silicate, Åkermanite, $Ca_2MgSi_2O_7$, and oxidizes with a 30% increase in specific volume (see Fig. 4). In an oxidizing environment at about 1000 °C, this expansion opens up grain boundaries and exposes fresh surfaces for further attack, and so the oxidation becomes catastrophic (see Fig. 5).

There are similar limitations with pressureless-sintered silicon nitride. Hot isostatic pressing (HIP) uses the least amount of additive and so the degradation of properties is minimized, but its cost is likely to be uneconomical for many appli-

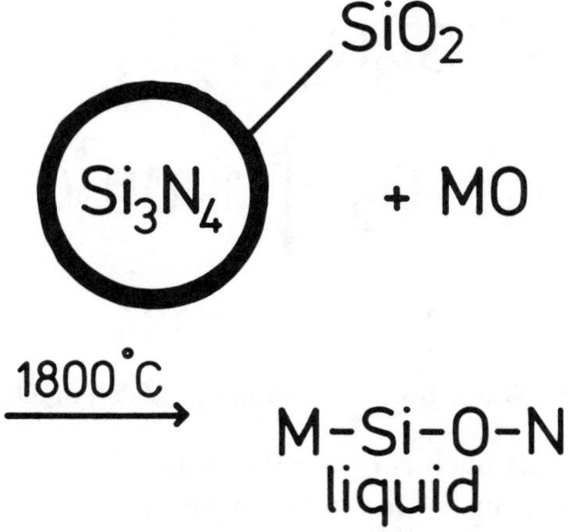

Fig. 2. Reaction of metal-oxide additive with silicon nitride and its surface layer of silica

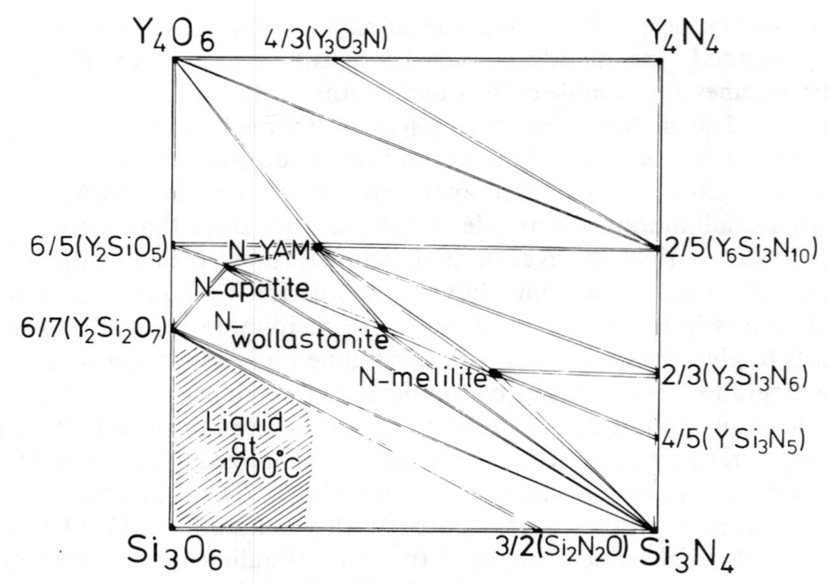

Fig. 3. Behavior diagram of the Y-Si-O-N system

Åkermanite

$$Ca_2Mg[Si_2O_7]^{6-}$$

$$Y_2Si[Si_2O_3N_4]^{10-}$$

$$\equiv \; Si_3N_4 . Y_2O_3$$

$$O_2 \downarrow 1000\,^\circ C$$

$$\Delta V = 30\%$$

Fig. 4. Oxidation of N-melilite, $Y_2Si[Si_2O_3N_4]$

cations. An alternative approach is to use the principles of ceramic alloying inherent in the production of "sialons"—the acronym given[1] to phases in the Si-Al-O-N and related systems that were discovered independently in Japan[2] and at the University of Newcastle upon Tyne.[3]

Fig. 5. Silicon nitride bar hot pressed with 15 wt% Y_2O_3 and oxidized for 120 h at 1000 °C

THE Si-Al-O-N SYSTEM

β′-Sialon

As shown in Fig. 6, silicon nitride is built up of SiN_4 tetrahedra joined in a three-dimensional network by sharing corners in the same way that SiO_4 units are joined in a silicate. Indeed, the atomic arrangement in β-Si_3N_4 is the same as in Be_2SiO_4 or in Zn_2SiO_4, and the same principles of crystal chemistry apply. Thus, up to two-thirds of the silicon in β-Si_3N_4 can be replaced by aluminum without a change in structure provided that an equivalent concentration of nitrogen is replaced by oxygen:

$$Si^{4+}N^{3-} \rightleftharpoons Al^{3+}O^{2-} \quad \text{(Eq 1)}$$

In the Si-Al-O-N phase diagram at 1800 °C presented in Fig. 7, aluminum concentration is plotted in equivalents on the x-axis (the balance being silicon) and oxygen concentration on the y-axis (balance, nitrogen). The β'-sialon phase, in which the β-Si_3N_4 crystal structure is retained, extends over a range of hexagonal unit-cell contents:

$$Si_{6-z}^{(24-4z)+} Al_z^{3z+} O_z^{2z-} N_{8-z}^{(24-3z)-} \quad \text{(Eq 2)}$$

with z varying between 0 and 4.

The relationship between the sialon and silicon nitride is analogous to that between brass and copper. Pure copper is soft and weak, but up to 40% of the copper atoms can be replaced by zinc without changing the structure, to give a harder, stronger alloy that melts at a lower temperature and so can be fabricated more easily than copper.

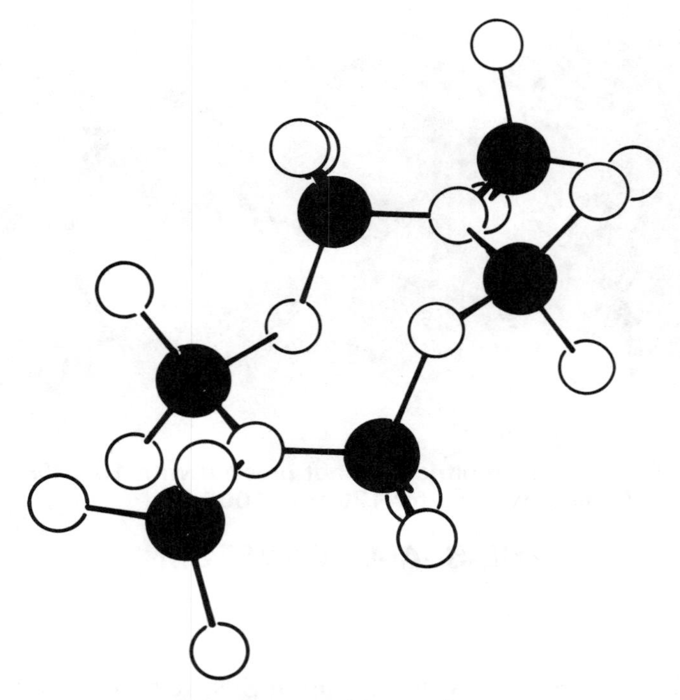

• Metal atom. ○ Nonmetal atom.

Fig. 6. The crystal structure of β-Si_3N_4 and β'-$(Si,Al)_3(O,N)_4$

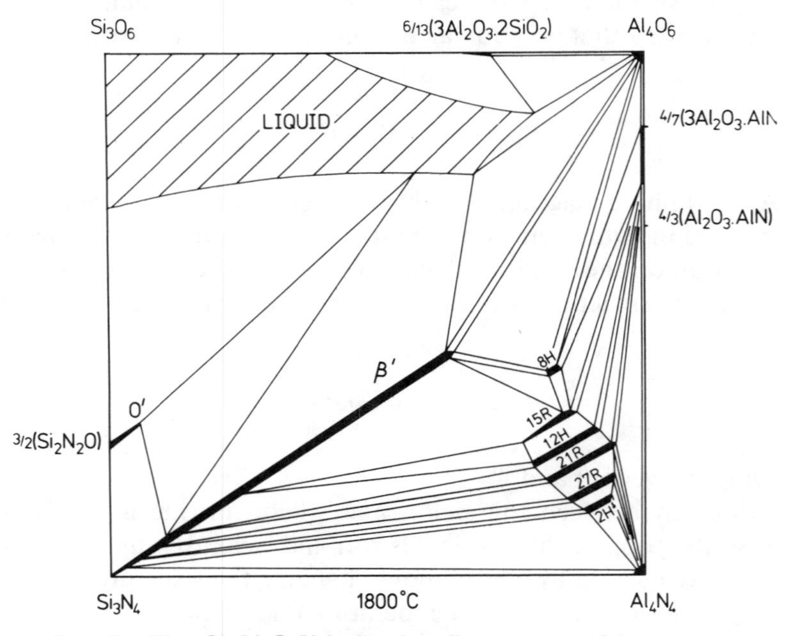

Fig. 7. The Si-Al-O-N behavior diagram at 1800 °C

As might be expected, phase relationships in the Si-Al-O-N system change with temperature; the behavior diagrams at 1700 and 1400 °C are shown in Fig. 8 and 9, respectively. The extensive chemical replacement is of Si-N by Al-O; these bonds are about the same length (1.74 and 1.75 Å, respectively), and so, as in other alloys, one of the determiners for formation is a size factor. The extent of homogeneity in other directions of the phase diagram is small because the bond lengths of Si-O (1.62 Å) and Al-N (1.87 Å) are appreciably different from one another. It should be noted that Si-C (1.89 Å) is similar to Al-N and so solid solutions between silicon carbide and aluminum nitride are not unexpected. It is also relevant that the Al-O bond strength ($\sim$500 kJ·mol^{-1}), even for six-fold coordinated aluminum, is higher than that of Si-N (440 kJ·mol^{-1}).

Because of its atomic arrangement, β'-sialon has mechanical and physical properties similar to those of silicon nitride—e.g., high strength and a small coefficient of thermal expansion; for β' with $z = 3$, $\alpha = 2.7 \times 10^{-6}$/°C compared with 3.5×10^{-6}/°C for β-Si_3N_4. Chemically, however, β'-sialon has some of the characteristics of aluminum oxide, but with important modifications. No matter how much aluminum and oxygen are substituted for silicon and nitrogen, the aluminum is 4-coordinated by O (AlO_4) and not 6-coordinated (AlO_6) as in alumina. The Al-O interatomic bond strength in β' is therefore about 50% stronger than in Al_2O_3.

The β'-sialon is a solid solution and, like all solutions, its vapor pressure is lower than that of the pure solvent—i.e., silicon nitride (see Fig. 10). Thus, compared with silicon nitride, β'-sialon forms more liquid at a lower temperature with oxide additives such as magnesia and yttria. Control of the volume of liquid allows the

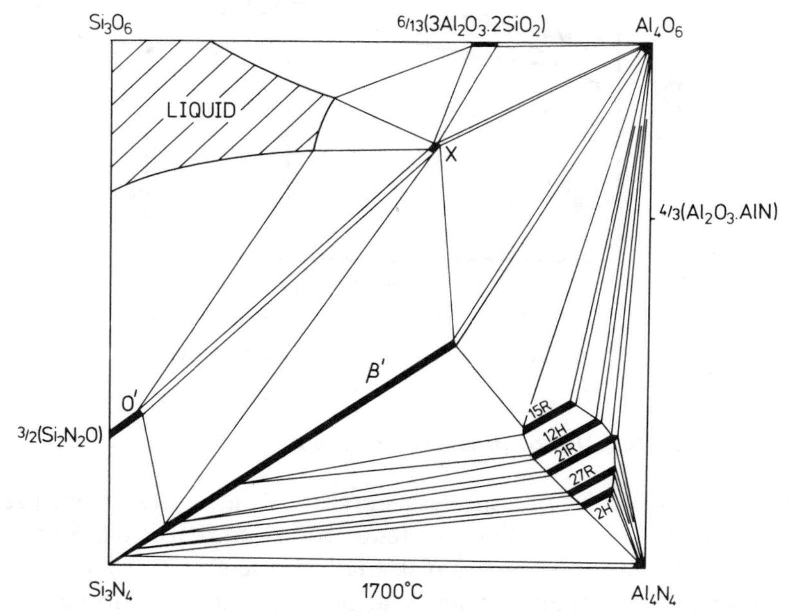

Fig. 8. The Si-Al-O-N behavior diagram at 1700 °C

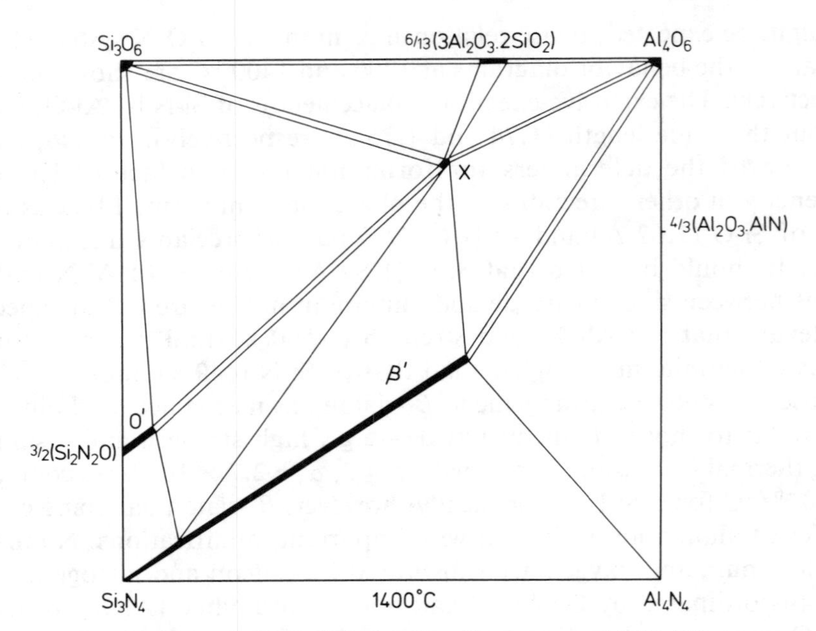

Fig. 9. The Si-Al-O-N behavior diagram at 1400 °C

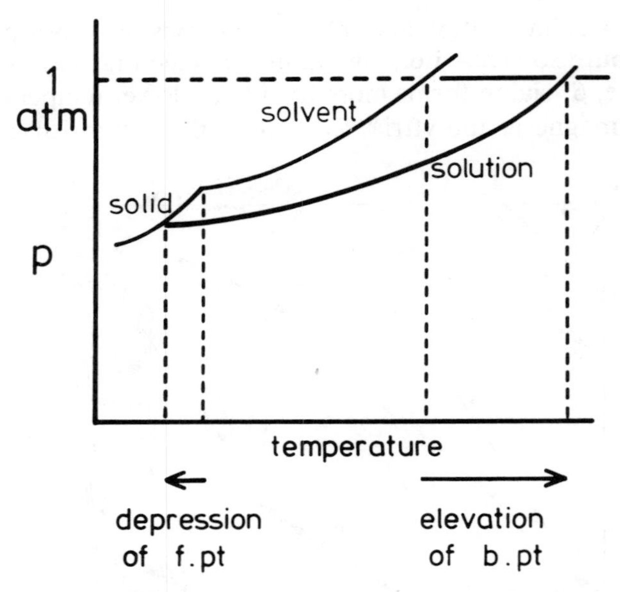

Fig. 10. Vapor pressures of solution and solvent

material to be densified by pressureless sintering—i.e., without hot pressing and like a conventional ceramic. The lower temperature of densification avoids excessive grain growth and so the fine-grain strength of the material is retained. Finally, the lower vapor pressure reduces volatilization and decomposition at high temperatures; under normal preparative conditions, β'-sialon is thermodynamically more stable than β-silicon nitride.

Densification with Yttria

After reaction sintering with yttria or yttria plus alumina, the mixed oxide and nitride powders corresponding to the required sialon composition give β' and the yttrium-sialon liquid necessary for densification. The latter cools to produce an intergranular glass. As shown later, this oxynitride glass is more refractory and stronger than an oxide glass but, by postpreparative heat treatment at about 1400 °C, or by controlled cooling, it can be reacted to give crystalline yttrium-aluminum-garnet ("YAG", $Y_3Al_5O_{12}$) together with a slightly changed β'-sialon composition:

$$\underset{\beta'\text{-sialon}}{Si_5AlON_7} + \underset{\text{glass}}{\text{Y-Si-Al-O-N}} \rightarrow \underset{\beta'\text{-sialon}}{Si_{5+x}Al_{1-x}O_{1-x}N_{7+x}} + \underset{\text{"YAG"}}{Y_3Al_5O_{12}} \quad \text{(Eq 3)}$$

The silicon and nitrogen from the glass go back into the sialon, but the range of sialon homogeneity is so large that the small change in its composition does not affect its properties. The remaining glass components—yttrium, aluminum, and oxygen—crystallize out as a grain-boundary oxide. These relationships are shown schematically in Fig. 11, where the line of the β'-sialon phase and the position of the intergranular phase "P" define a plane in the Jänecke prism (see Fig. 12) representing the Y-Si-Al-O-N system. In order to effect the reaction given by Eq 3, the overall composition "X" and the composition of the liquid and the glass formed from it on cooling must also lie on this same plane. Phase relationships and the products of glass devitrification change markedly with changes in heat treatment time and temperature, and so it is not always easy to meet the objective of producing a two-phase ceramic, β' + YAG.

Commercial Sialon Ceramics

Based on the principles described, extensive process development[4,5] resulted in the commercial production of two groups of β'-sialons with different main micro-

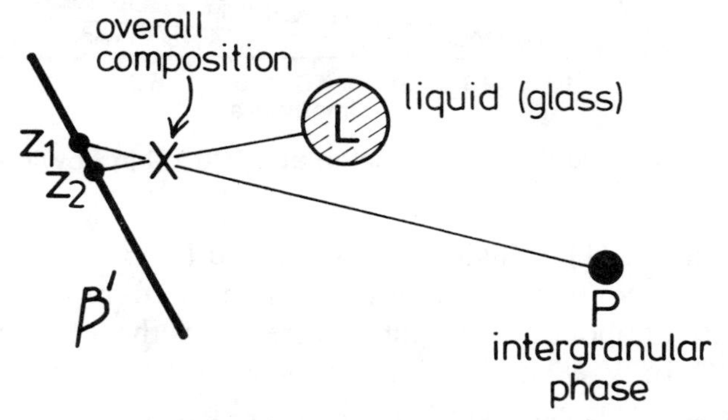

Fig. 11. Postpreparative heat treatment of β' + glass to give β' + YAG

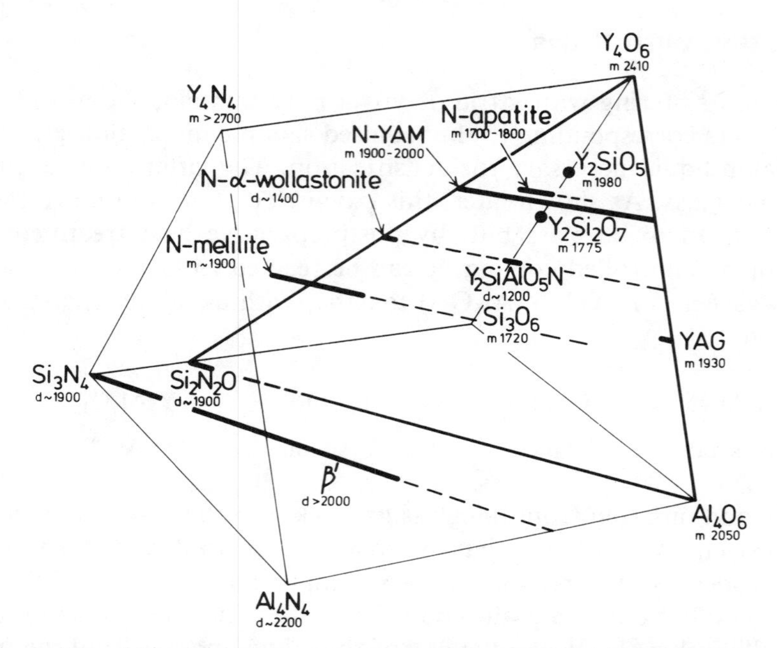

Fig. 12. Jänecke prism behavior diagram showing some crystalline phases in the Y-Si-Al-O-N system

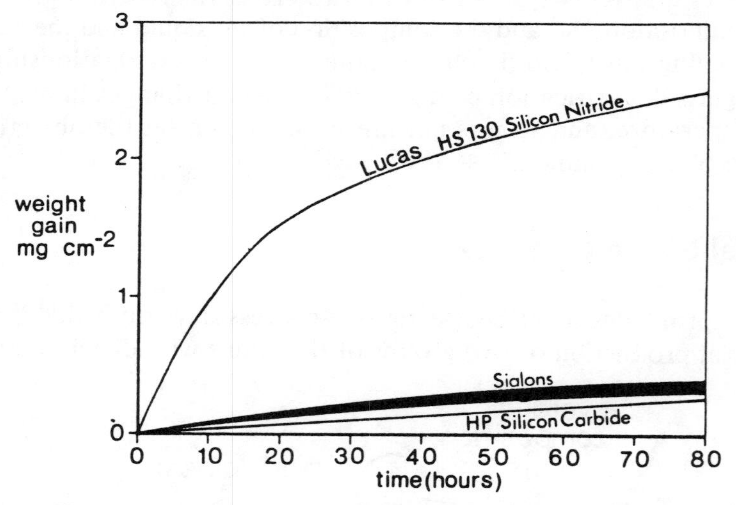

Fig. 13. Oxidation of β′-sialon at 1400 °C in flowing dry air. After Arrol.[(6)]

structural constituents: (1) β′-sialon plus glass and (2) β′-sialon plus YAG. The strength of the glass-containing materials is high at room temperature, with a modulus of rupture of about 1 GPa, but it decreases as the intergranular glass softens above 1000 °C. In type 2 material, since the grain-boundary phase is an oxide, the oxidation resistance is excellent (Fig. 13), and because there is a minimum of intergranular glass it has good creep properties (Fig. 14). Its strength is similar to

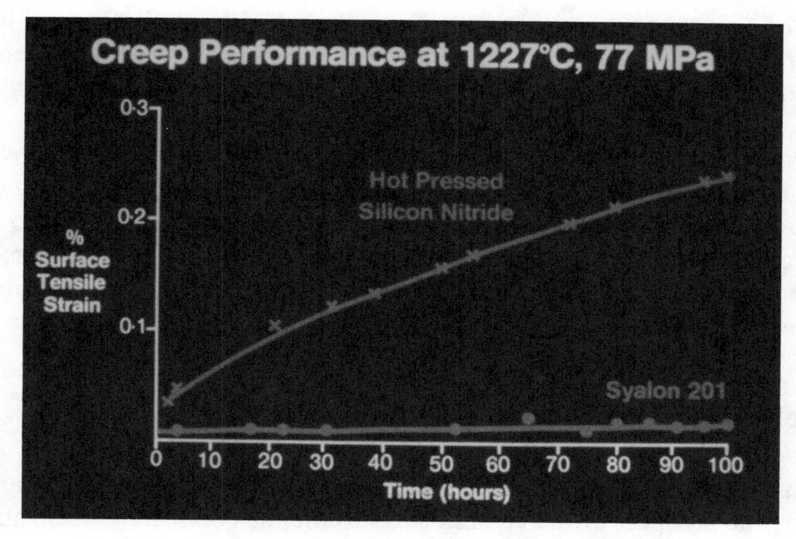

Fig. 14. Creep of β'-sialon at 1227 °C and 77 MPa. After Lumby.[7]

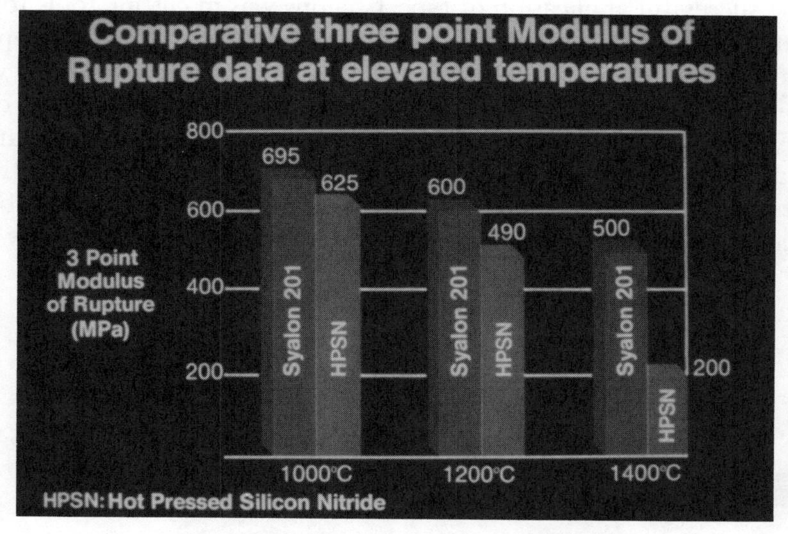

Fig. 15. Strength of β'-sialon at 1000 to 1400 °C. After Lumby.[7]

that of type 1 material at room temperature and, as shown in Fig. 15, it retains about one-half this strength at 1400 °C (modulus of rupture, about 500 MPa).

Forming, Sintering, and Machining

The appropriate mixes of nitride, oxide, and oxynitride powders can be cold or warm formed prior to sintering by using all the shaping methods normally employed for oxide ceramics. These include: (*a*) isostatic pressing and, for large numbers of components, uniaxial pressing with and without binders and lubricants; (*b*) warm or cold extrusion with addition of plasticizers to produce continuous sections of preformed material; (*c*) injection molding for intricate shapes; and (*d*) slip casting in aqueous media.

Preforms may be machined in the green state and, after debonding where necessary in air at up to 500 °C, are then sprayed with a protective coating of refractory oxides before being sintered in a nitrogen atmosphere at 1750 to 1850 °C. Reproducible linear shrinkage occurs, and, because this is allowed for, the final diamond machining of the hard, fully dense component to precise dimensional tolerances is reduced to a minimum.

Engineering Applications

Although the main motivation for the development of engineering ceramics was, and still is, the prospect of the ceramic gas turbine, its commercial realization might well take another decade and, meanwhile, the unique properties of β'-sialons and their ease of fabrication are being applied in other directions.

The first successful application of type 1 sialons was in cutting tools for machining of metals. Figure 16 shows a selection of tool tips, and Table 1 compares their performances in cutting cast iron, hardened steel, and a nickel-base alloy with those of cobalt-bonded tungsten carbide and of alumina. The hot hardness of the sialon (see Fig. 17) is higher than that of the other materials, and so it can be most effective at very high cutting speeds where the working temperature at the tool tip may exceed 1000 °C. The "lead time" for machining some of the Rolls Royce RB211 aeroengine turbine disks was reduced to less than one-quarter by using sia-

Fig. 16. Selection of sialon cutting-tool tips

Table 1. Performances of cobalt-bonded WC, Al_2O_3, and a β′-sialon in cutting cast iron, hardened steel, and Incoloy 901

Performance	Cast iron	Hardened steel EN31	Incoloy 901
WC			
Cutting speed, m/min	250	5	20
Depth of cut, mm	6.5	...	...
Feed rate, mm/rev	0.50	...	...
Al_2O_3			
Cutting speed, m/min	600	[Impossible	300
Depth of cut, mm	6.5	to	No second
Feed rate, mm/rev	0.25	cut]	entry(a)
β′-sialon			
Cutting speed, m/min	1100	120	300
Depth of cut, mm	10.0	0.5	2.0
Feed rate, mm/rev	0.50	0.25	0.25

(a) If cut is interrupted by removing tool from the workpiece, it cannot be restarted without tool failure.

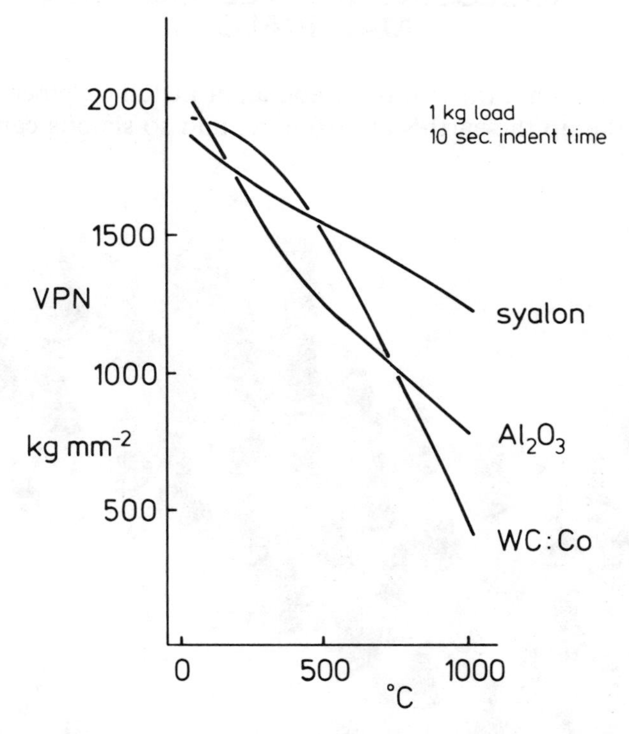

Fig. 17. Hot hardness of sialon, Al_2O_3, and WC:Co cutting-tool tips

lon inserts. They are being manufactured by two major tool companies, one in Sweden and one in the United States.

Excellent thermal-shock resistance, high-temperature mechanical strength, and electrical insulation combine to make sialon unexcelled in welding operations. Figure 18 shows location pins for resistance welding of captive nuts on vehicle chassis. The usual hardened steel pins in alumina insulating sleeves last for 7000 operations—i.e., a working shift of 8 h; sialon pins have completed 5 million operations—i.e., one year—without showing signs of wear.

Ten years ago, hot pressed silicon nitride was predicted to be an ideal material for ball and roller bearings, but the cost of machining spheres from the simple hot pressed shapes was prohibitive. With a β'-sialon sintered to almost the final required dimensions, and with even better wear resistance, hardness, and tribological properties, these applications are again possible.

Sialon die inserts used in both hot and cold extrusion of brass, copper, bronze, aluminum, titanium, and steel have provided remarkable improvements in surface finish and dimensional accuracy, and higher extrusion speeds (see Fig. 19). This material also copes with a wide range of wear environments in contact with metals, with or without lubrication.

PROCESSING ROUTES AND RAW MATERIALS

Silicon and aluminum are the two most abundant metallic elements, while oxygen and nitrogen make up the earth's atmosphere, and so sialons can be made from

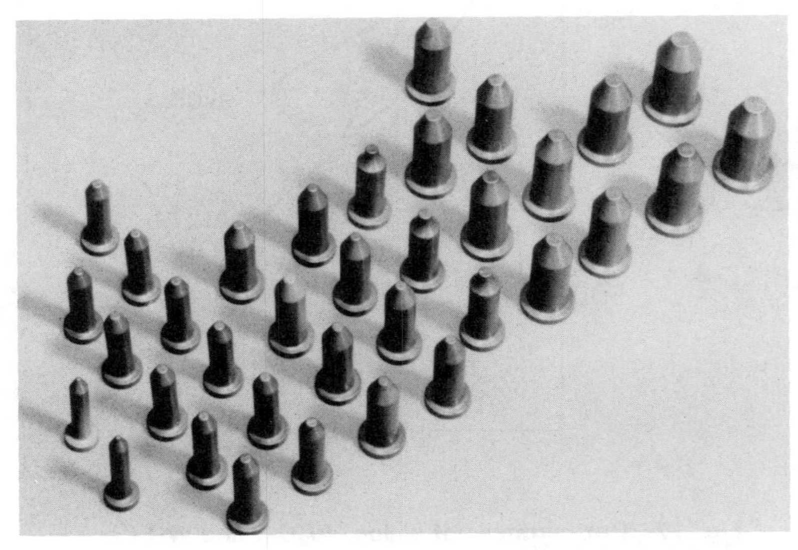

Fig. 18. Sialon location pins for welding of captive nuts

Fig. 19. Sialon die and mandrel plug for metal tube drawing

cheap, readily available, nonstrategic raw materials. Apart from the β'-phase, other ceramic alloys with equal promise are derived from α-silicon nitride and silicon oxynitride, Si_2N_2O.

Lee and Cutler[8] produced β'-sialon powder by heating pelletized mixtures of clay and coal in nitrogen (see Fig. 20) according to:

$$\underset{\text{clay}}{3Al_2[Si_2O_5](OH)_4} + \underset{\text{coal}}{15C} + \underset{\text{nitrogen}}{5N_2} \rightarrow \underset{\beta'\text{-sialon } (z=3)}{2Si_3Al_3O_3N_5} + 6H_2O + 15CO \quad \text{(Eq 4)}$$

and the similar production of other nitrogen ceramics by the carbothermal reduction of mixed oxides in nitrogen is now being extensively explored. In Japan, Umebayashi[9] obtained β' by reaction in nitrogen of volcanic ash (impure silica) and aluminum powder:

$$\underset{\text{volcanic ash}}{2SiO_2} + \underset{\text{aluminum powder}}{4Al} + 2N_2 \rightarrow \underset{\beta'\text{-sialon } (z=4)}{Si_2Al_4O_4N_4} \quad \text{(Eq 5)}$$

Rice husks, of which 13 million tons are produced annually in India alone, consist mainly of cellulose and silica. Their pyrolysis gives "black ash," an intimate

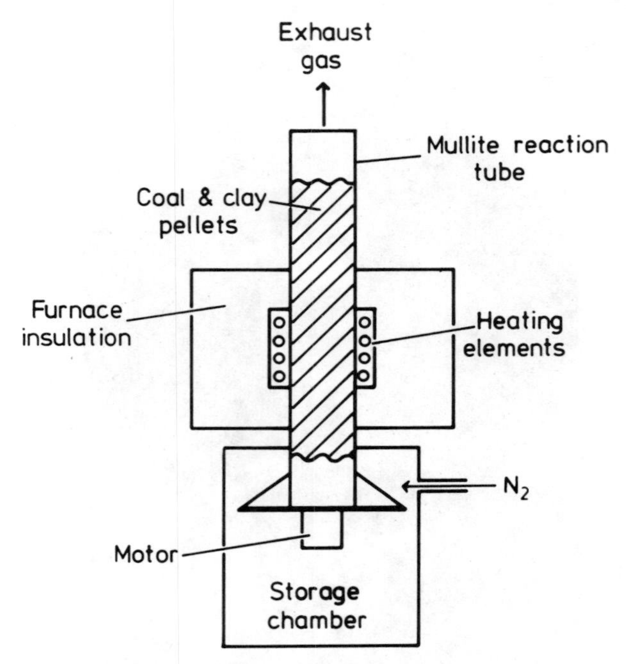

Fig. 20. Shaft kiln for production of sialon from clay and coal. After Lee and Cutler.[8]

mixture of carbon and silica, which is a useful starting material for silicon nitride and sialon production. Clay, coal, volcanic ash, and rice husks are not going to produce pure sialons for sophisticated applications such as the ceramic turbine, but they will provide useful refractory bricks, furnace linings, and materials resistant to molten metals. Just as there are many grades of alumina ranging from single-crystal sapphire, through transparent polycrystalline lucalox and 99.7% recrystallized alumina, to the debased alumina refractories, so there will be a variety of sialon grades, each having its specific applications and its appropriate methods of manufacture.

α'-SIALONS

The α and β Structures

The "idealized" silicon nitride structures can be described as a stacking of Si-N layers in either an ABAB.... β sequence or an ABCD.... α sequence, as shown in Fig. 21. This gives, in the hexagonal β unit cell containing Si_6N_8, long, continuous channels running parallel with the c-direction and centered at the x and y coordinates 2/3, 1/3. In α the layers CD are identical with layers AB but are inverted and translated with respect to them by a c-glide plane. The continuous channels of β are replaced in α by large, closed interstices at 2/3, 1/3, 3/8 and 1/3, 2/3, 7/8. Thus, in the α hexagonal unit cell containing $Si_{12}N_{16}$ there are two sites

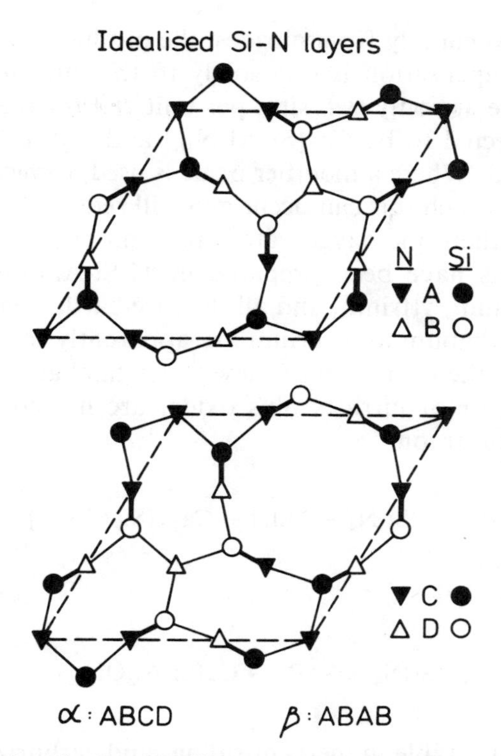

Fig. 21. Idealized Si-N layers in α and β silicon nitrides

large enough to accommodate other atoms or ions. Although the Si-N layers in the actual β structure are almost identical with the "ideal" configuration, those in α are distorted and nitrogen atoms at heights of approximately 3/8 and 7/8 are pulled in toward the centers of the two respective interstices.[10]

α′ Structures

The existence of two interstitial sites in the α-silicon nitride structure invokes another simple principle of silicate chemistry that is illustrated by the formation of "stuffed quartz" structures. These occur when Al^{3+} replaces Si^{4+} and positive valency deficiencies are compensated by "stuffing" Li^{1+} or Mg^{2+} into interstitial sites:

$$\underset{\text{quartz}}{Si_2O_4} \rightarrow \underset{\beta\text{-eucryptite}}{Li[SiAlO_4]} \qquad \text{(Eq 6)}$$

$$Si^{4+} \rightarrow Li^{1+}Al^{3+} \qquad \text{(Eq 7)}$$

In exactly the same way, α′-sialons are derived from α-$Si_{12}N_{16}$ by partial replacement of Si^{4+} with Al^{3+}, with valency compensation by "modifying" cations such as Li^{1+}, Ca^{2+}, and Y^{3+} occupying the interstices in the (Si,Al)-N network.

When α' is synthesized entirely from nitrides, the products should contain no oxygen, and valency compensation is due solely to the introduction of interstitial cations. Because there are only two sites per unit cell for these, limiting compositions might be expected to be $Ca_2[Si_8Al_4N_{16}]$ and $Y_2[Si_6Al_6N_{16}]$; these limits have not been achieved. Where a modifier oxide is used, oxygen also replaces nitrogen, but the extent to which this can occur and still retain the α'-structure is probably not much more than one oxygen atom per unit cell.

Alpha-prime sialons have been prepared in M-Si-Al-O-N systems where M includes lithium, calcium, yttrium, and all the rare-earth elements from neodymium to lutetium. Lanthanum and cerium are apparently too large, and so again there is a size factor in the formation of these "interstitial alloys." Appropriate mixtures of the nitrides, or of nitrides plus oxide, are heated without pressure at 1750 °C in nitrogen or argon:

$$0.5Ca_3N_2 + 3Si_3N_4 + 3AlN \rightarrow Ca_{1.5}[Si_9Al_3N_{16}] \quad \text{(Eq 8)}$$

or

$$CaO + 3Si_3N_4 + 3AlN \rightarrow Ca[Si_9Al_3ON_{15}] \quad \text{(Eq 9)}$$

The α' products are stable in inert, nitriding, and carburizing atmospheres up to 1750 °C and, like β'-sialons, can be prepared by carbothermal reduction of mixed oxides in nitrogen. They have good oxidation resistance up to 1350 °C and coefficients of thermal expansion similar to that of silicon nitride ($\sim 3 \times 10^{-6}$/°C at 0 to 1250 °C), and prospects for their technological application are as good as those for β'-sialons.

One of the major advantages of an α'-sialon is that the additive, such as calcia or yttria, necessary for liquid-phase densification can subsequently be incorporated into the structure and so reduce the amounts of intergranular glass or secondary crystalline phases in the product. Moreover, α-phases are stronger than β and about 50% harder.

The phase relationships between α' and β' are represented in Fig. 22. The transformations between them,

$$\alpha' \rightleftharpoons \beta'$$

are chemical as well as structural and occur only with a liquid intermediary. For example, α' reacts with Al_2O_3 at 1750 °C to give β':

$$\underset{\alpha'}{Y_{0.4}[Si_{9.4}Al_{2.6}O_{1.3}N_{14.7}]} + 2.7Al_2O_3 \rightarrow \underset{\beta'\,(z = 2.8)}{2.85Si_{3.2}Al_{2.8}O_{2.8}N_{5.2}} + 0.2Y_2Si_2O_7 \quad \text{(Eq 10)}$$

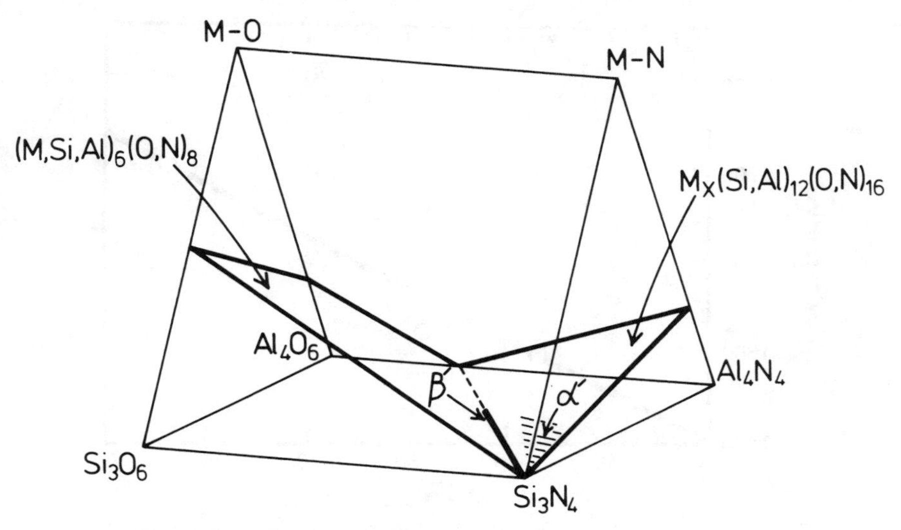

Fig. 22. Phase relationships between α′- and β′-sialons

and β′ reacts with AlN plus an appropriate nitride moderator to give α′:

$$\underset{\beta'\,(z\,=\,1.2)}{1.45\mathrm{Si_{4.8}Al_{1.2}O_{1.2}N_{6.8}}} + 3.34\mathrm{AlN} + 0.56\mathrm{Ca_3N_2} \rightarrow \underset{\alpha'}{\mathrm{Ca_{1.2}[Si_{7.0}Al_{5.0}O_{1.7}N_{14.3}]}} \quad \text{(Eq 11)}$$

The two phases are ideally compatible, and composites of them can be prepared with different α′:β′ ratios from the appropriate oxide-nitride powder mixes by pressureless sintering in a single-stage process. The rates of α′ and β′ formation differ, and so by varying the composition and the reaction sintering conditions it is possible to vary and control the microstructure in a way that is impossible for a single-phase sialon. These composites are already being used in preference to β′ in cutting-tool applications because, as shown in Fig. 23, the hardness increases linearly with increasing α′ content while strength remains unchanged. The hardness increment remains almost constant with increasing temperature even though the overall values decrease, and so at 1000 °C the 50α′:50β′ composite is about twice as hard as the pure β′-sialon (see Fig. 24). Tool lives are compared in Fig. 25 for a 50α′:50β′ sialon (Sandvik CC680), an alumina ceramic, and a WC:Co cemented carbide in cutting a nickel-base superalloy; the advantage of the sialon composite is obvious.

O′-SIALONS

There is a limited solubility[11] of about 10 mol % Al_2O_3 in silicon oxynitride, Si_2N_2O, to give an O′-sialon solid solution without a change in structure (see

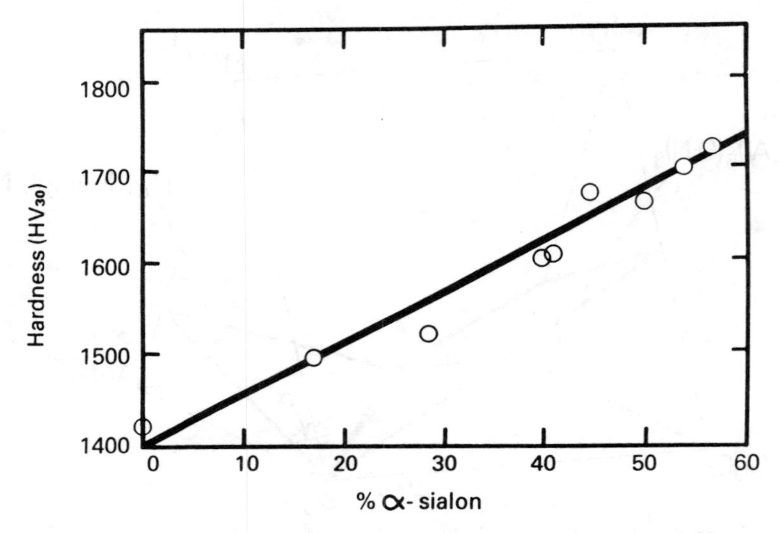

Fig. 23. Variation of hardness with α′ content in α′:β′ sialon composites. After Sandvik Hard Materials Ltd.

Fig. 7). A suitable additive such as Y_2O_3 lowers the solidus temperature and, by increasing the volume of liquid, allows densification by pressureless sintering in the same way as in the processing of β′-sialons.

If the overall compositions lie on the Si_2N_2O-Al_2O_3-$Y_2Si_2O_7$ plane of the Y-Si-Al-O-N system shown in Fig. 26, mixed powders of Si_3N_4, SiO_2, Al_2O_3, and Y_2O_3 react and densify at 1600 to 1800 °C to produce O′-sialon ceramics with an intergranular glass phase that can be devitrified by postpreparative heat treatment to give $Y_2Si_2O_7$. Once again, the products have low coefficients of thermal expansion ($\alpha \sim 2.9 \times 10^{-6}$/°C) and good thermal-shock properties. They are oxidation resistant to 1350 °C and should be as useful as α′ and β′ in engineering applications.[12]

O′:β′ Composites

Dense O′:β′ ceramics are obtained[13] by pressureless sintering of the appropriate powder mixes and devitrifying the intergranular glass to yield $Y_2Si_2O_7$ and YAG; the strength of β′ is combined with the oxidation resistance of O′. Microstructures are changed by varying the proportions of the two compatible sialon phases, and with approximately 50% O′–50% β′ the material is oxidation resistant to about 1350 °C. Much more research and development are required, however, before the potentials of these and other composites can be assessed.

Electrical Properties

Silicon oxynitride is built up of SiN_3O tetrahedra joined by sharing corners, and the orthorhombic crystal structure consists of irregular but parallel sheets of covalently bonded silicon and nitrogen atoms linked by Si-O-Si bonds. It can be

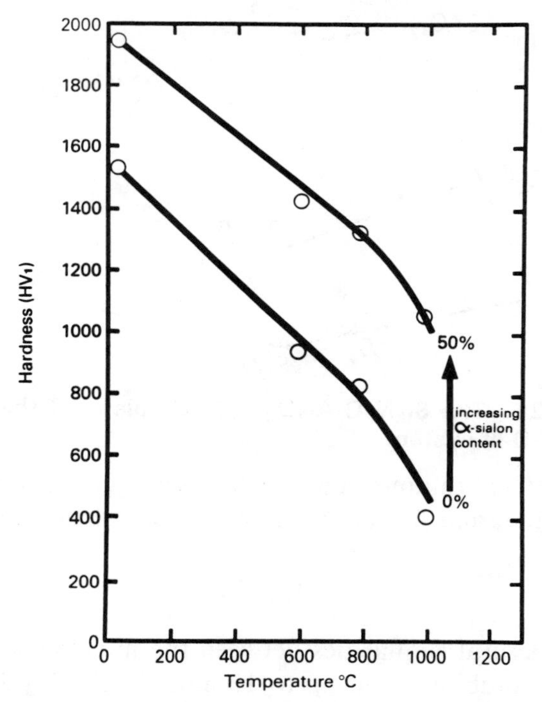

Fig. 24. Variation of hardness with temperature for $50\alpha':50\beta'$ composite compared with 100% β'. After Sandvik Hard Materials Ltd.

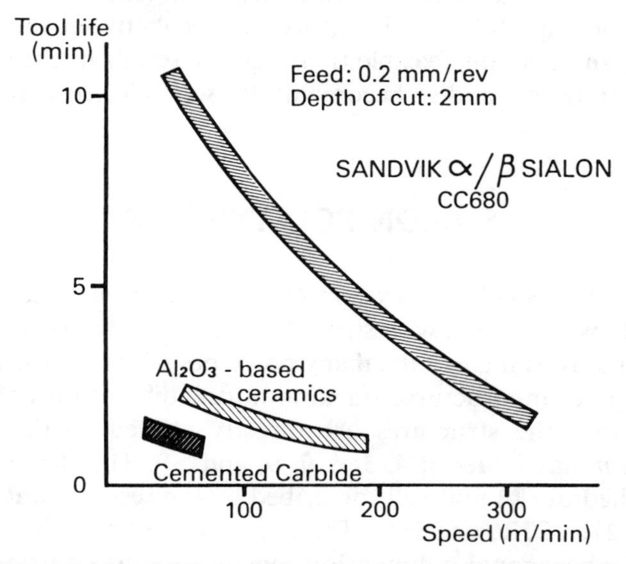

Fig. 25. Tool life as a function of cutting speed for $\alpha':\beta'$ sialon composite, alumina, and WC:Co cemented carbide. After Sandvik Hard Materials Ltd.

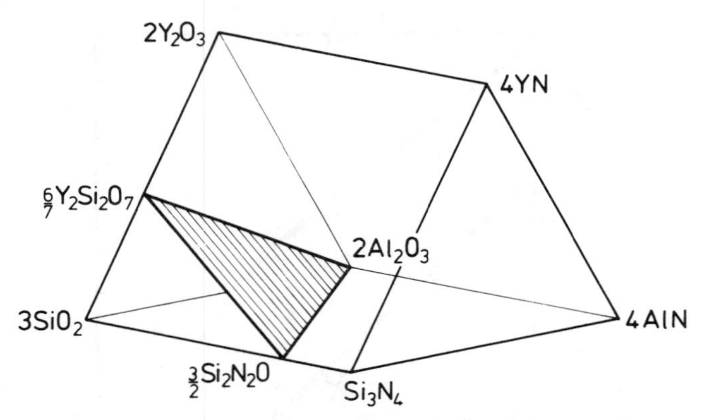

Fig. 26. The Si_2N_2O-Al_2O_3-$Y_2Si_2O_7$ plane of the Y-Si-Al-O-N system

regarded as a defective lithium-silicon nitride (see Fig. 27) in which one nitrogen is replaced by oxygen, and lithium is removed for valency compensation:

$$LiSi_2N_3 \rightarrow \square Si_2N_2O \tag{Eq 12}$$

Because of the structural similarities between the nitride and oxynitride there is some mutual solid solubility, and up to 12 mol % $LiSi_2N_3$ dissolves in Si_2N_2O. Thus, in the O′-sialons the replacement

$$O^{2-} \rightleftharpoons Li^{1+}N^{3-} \tag{Eq 13}$$

occurs separately or simultaneously with those represented by Eq 1 and 7. The Li^{1+} ions, and perhaps other cations, are located between the parallel silicon-nitrogen sheets, and it seems feasible that these materials might provide ceramics with fast cation transport and so become useful solid electrolytes similar to the β-aluminas.

SIALON POLYTYPOIDS

Compositions in the Si-Al-O-N system (Fig. 7) between β' and the aluminum nitride corner show six phases with structures based on that of wurtzite-type AlN. They are polytypoids, similar to the many polytypes of silicon carbide, SiC, in that they exhibit changes in structural stacking in one dimension only, but differing from them in that the structures are directly related to their compositions, M_mX_{m+1}, when m has values of 4, 5, 6, 7, 9, and >9. They have either hexagonal (H) or rhombohedral (R) unit cells described by the respective Ramsdell symbols 8H, 15R, 12H, 21R, 27R, and $2H^{\delta}$. The end member, AlN, is 2H, and throughout the series the hexagonal a-dimension remains almost constant. Simple examples of the Ramsdell notation for SiC polytypes are presented in Fig. 28, and cell projections are shown in Fig. 29. In the sialon polytypoids, as shown in Table 2,

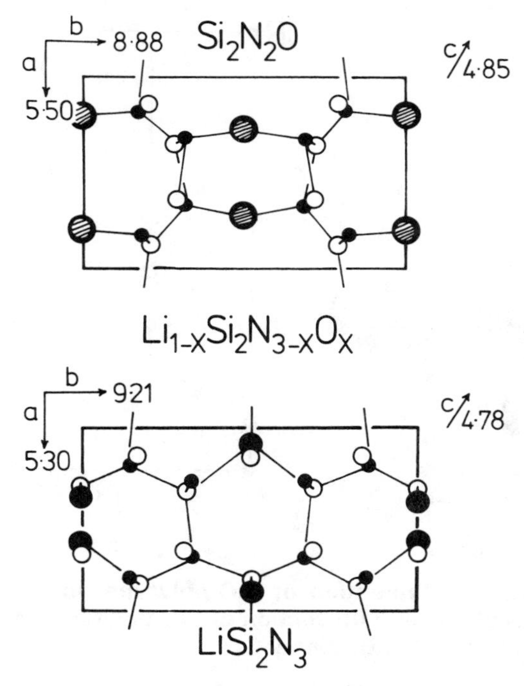

Fig. 27. Crystal structures of Si_2N_2O and $LiSi_2N_3$

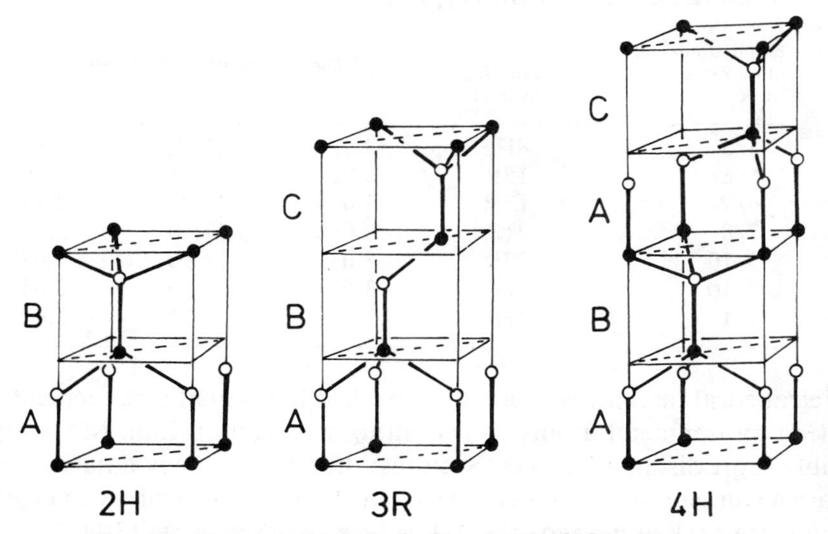

Fig. 28. Ramsdell symbols for simple SiC polytypes

there are n double layers of metal and nonmetal atoms (MX) stacked along the c-dimension of the cell, where n is the Ramsdell numeral and where $n_H = 2m$ and $n_R = 3m$. One in each block of m double layers contains an extra nonmetal atom X and so becomes an MX_2 layer; this is shown schematically for the 12H polytypoid in Fig. 30. This sialon structure consists of two blocks, symmetry related, each comprised of six double layers — five MX layers and one MX_2 layer. Occupation

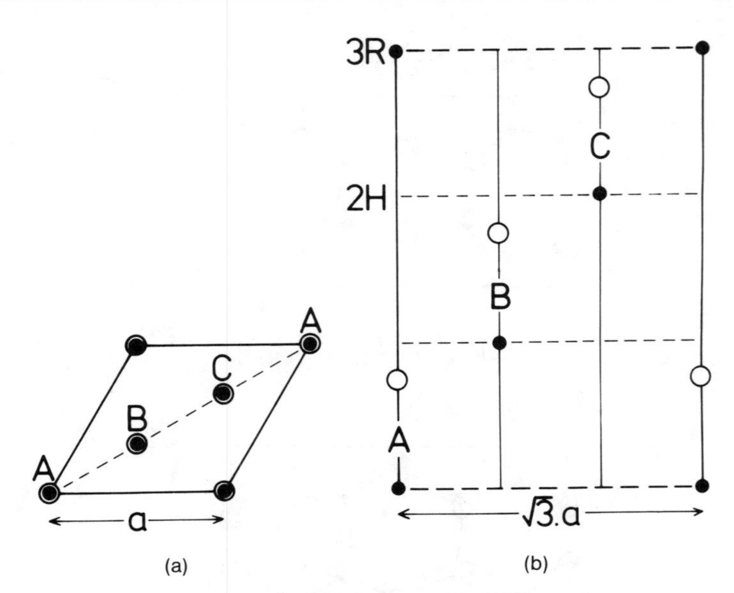

Fig. 29. (a) Projection of SiC-polytype unit cell on (0001) plane, and (b) section of SiC-polytype unit cell through the (11$\bar{2}$0) plane

Table 2. Si-Al-O-N polytypoids

Metal:nonmetal atom ratio, M/X	Ramsdell symbol	Hexagonal unit-cell dimensions, Å		
		a	*c*	*c/n*
4/5	8H	2.988	23.02	2.88
5/6	15R	3.010	41.81	2.79
6/7	12H	3.029	32.91	2.74
7/8	21R	3.048	57.19	2.72
9/10	27R	3.059	71.98	2.67
>9/10	$2H^{\delta}$	3.079	5.30	2.65
1/1	2H	3.114	4.986	2.49

of the tetrahedral interstices of a closed-packed hexagonal arrangement of metal atoms (M) by nonmetal atoms (X) resulting in a composition MX_2 requires an impossibly short distance between X atoms; they must occupy adjacent tetrahedra that share a common base. This can be avoided only by a localized change in stacking from close-packed hexagonal (2H) to face-centered cubic (3R). It was at first proposed that each polytypoid block of composition M_mX_{m+1} contained such a stacking discontinuity as well as an MX_2 layer. The structures proposed for 15R and 12H are shown in Fig. 31. The one-dimensional electron-microscope lattice image of 12H in Fig. 32 offers supporting evidence since it is obvious that one layer in six is different from the other five.

However, subsequent x-ray powder diffraction refinement[15] showed the struc-

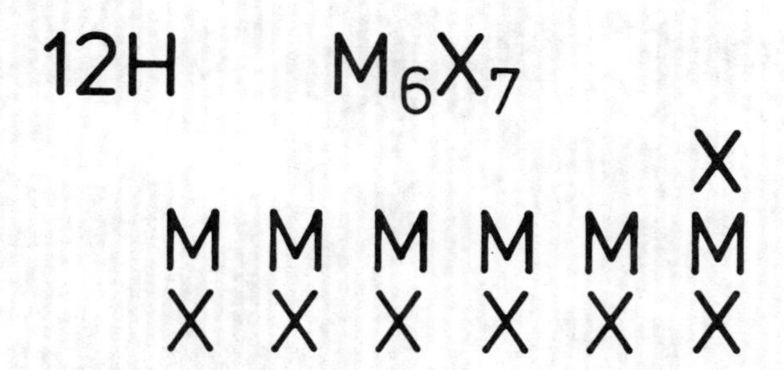

Fig. 30. Representation of six double layers in the 12H polytypoid

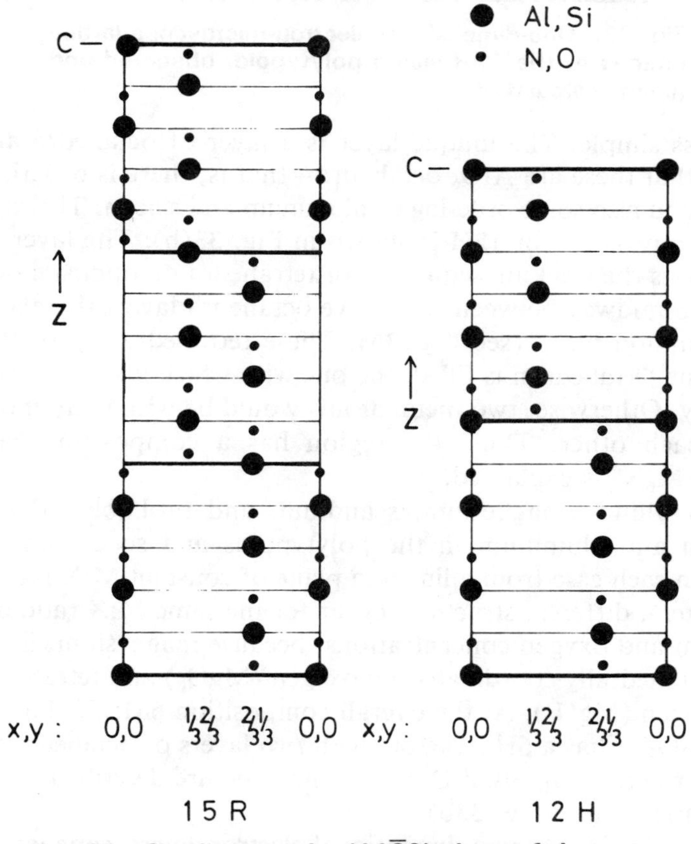

Fig. 31. Sections on the (11$\bar{2}$0) plane of the structures first proposed[14] for the 15R and 12H sialon polytypoids

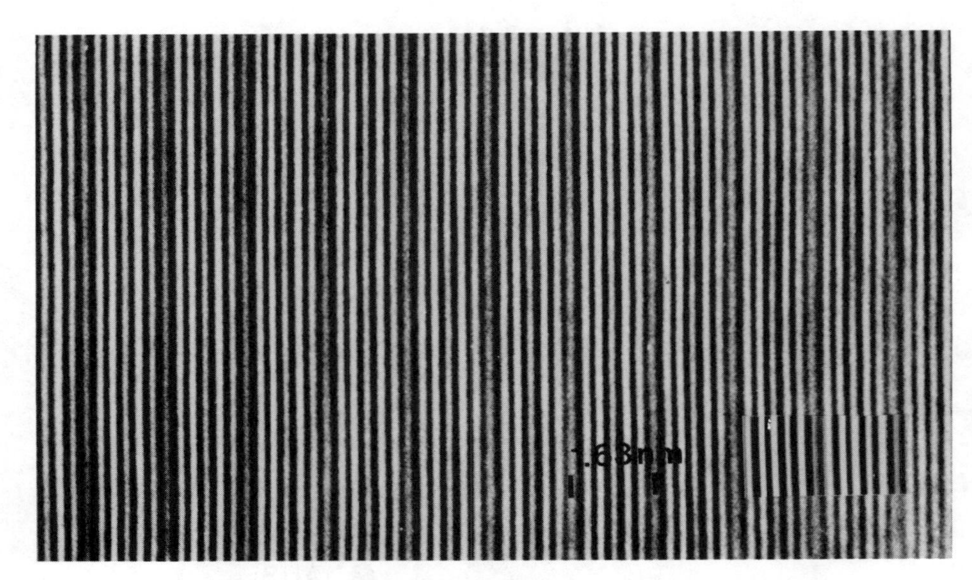

Fig. 32. One-dimensional electron-microscope lattice images of the 12H sialon polytypoid, observed and (inset) calculated

tures to be less simple. The unique layer is a layer of octahedra and the bond lengths show that these are AlO_6 octahedra—that is, there is ordering of oxygen and nitrogen, and also some ordering of aluminum and silicon. The structure determined by Thompson[15] for 12H is shown in Fig. 33(b). The layer of nonmetal octahedra inverts the stacking sequences of tetrahedra on opposite sides of it (see Fig. 34) and so midway between successive octahedra layers there are tetrahedra that share common bases (see Fig. 35). These tetrahedra layers are only half-occupied; if one tetrahedron is filled, the one with which it shares a common base remains empty. Otherwise, two metal atoms would be within an impossibly close distance of each other. Thus, the region has a composition MX_2 and the stoichiometry M_6X_7 is explained.

Lithium, beryllium, magnesium, scandium, and probably other metals can replace silicon and aluminum in the polytypoids and so extend the range of homogeneity in each case from a line to a plane of constant M:X ratio. In the Mg-Si-Al-O-N system, different structures occur for the same M:X ratio depending on the magnesium and oxygen concentrations, because magnesium, like aluminum, tends to be octahedrally coordinated by oxygen (MgO_6) and tetrahedrally coordinated by nitrogen (MgN_4). At the overall composition 6M:7X, Fig. 33(a) shows that $Mg_3Si_1Al_2O_5N_2$ has a 6H structure with *two* layers of octahedra to five layers of tetrahedra, whereas $Mg_2Si_2Al_2O_3N_4$ has the structure described above with only one octahedra layer (see Fig. 33b).

As shown in Fig. 36, the two-dimensional electron-microscope lattice image for the 12H structure of $SiAl_5O_2N_5$ is in complete agreement with the image calculated from the x-ray structure refinement.

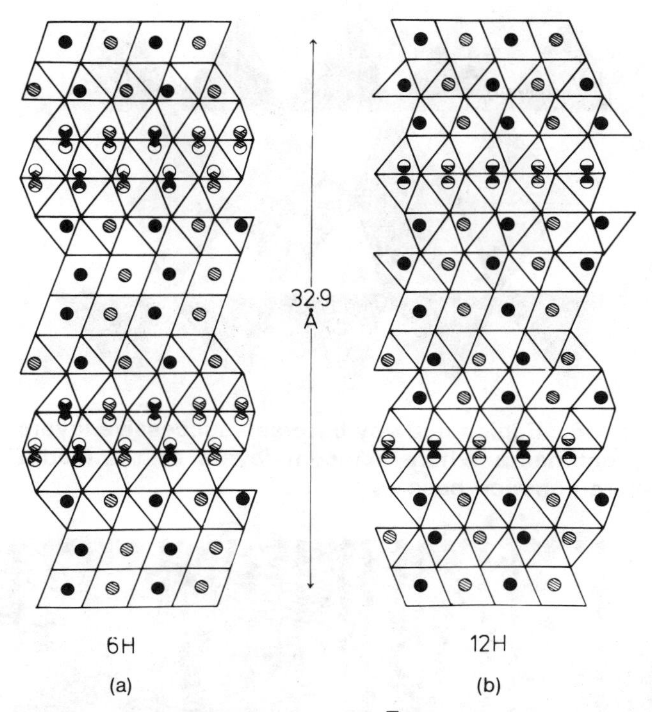

The structures are projected on the $(11\bar{2}0)$ plane with metal atoms shown as circles. Filled and shaded circles represent metal atoms at heights of 0 and 1/2, respectively, relative to the distance between two successive $(11\bar{2}0)$ planes. The partly filled circles represent the extent of occupation of a site.

Fig. 33. Atomic arrangements in (a) 6H and (b) 12H magnesium sialons

Fig. 34. Inversion of tetrahedra stacking on opposite sides of a layer of octahedra

Fig. 35. Region midway between successive layers of octahedra where adjacent layers of tetrahedra share common bases

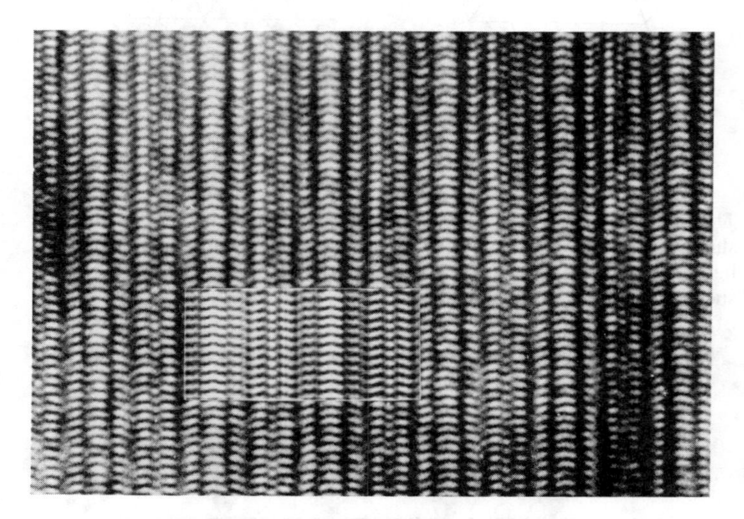

Fig. 36. Two-dimensional lattice images of the 12H sialon polytypoid, observed and (inset) calculated, taken with the electron beam normal to [0001]

MASNMR Data

Although the two-dimensional lattice images and the x-ray evidence are in agreement, the structures proposed for the polytypoids are sufficiently unusual to require confirmation by a third technique—magic angle spinning nuclear magnetic resonance (MASNMR).

The compositions of the polytypoids lie approximately along part of the join in the Si-Al-O-N behavior diagram (Fig. 7) between SiO_2 and AlN. Thus, the general composition M_mX_{m+1} is effectively $SiO_2 \cdot (m-1)AlN$. The respective compositions for 12H and 15R are then $SiO_2(AlN)_5$ and $SiO_2(AlN)_4$, which can be rearranged as $AlO_2 \cdot SiAl_4N_5$ and $AlO_2 \cdot SiAl_3N_4$. Assuming that one aluminum is octahedrally coordinated by oxygen—the AlO_2 layer—the likely distributions of silicon and aluminum in the remaining tetrahedral layers are as shown in Fig. 37. This model gives two principal environments for silicon, $SiON_3$ and SiN_4, and two for aluminum, octahedral AlO_6 and tetrahedral AlN_4. Figures 38 and 39 show that the observed MASNMR spectral peaks ^{29}Si and ^{27}Al correspond in both their chemical shifts and their intensities[16,17] with the model and so add further confirmation to the polytypoid structures deduced initially by Thompson[15] from x-ray powder diffraction data.

Other Polytypoids and Their Uses

Sialon polytypoid phases are refractory and oxidation resistant, but no exploration has yet been made of their potential as ceramics. Some of them (15R, 21R) are used instead of aluminum nitride as components of the oxide-nitride mixes for the synthesis of α'- and β'-sialons since they are less readily hydrolyzed and so are more easily handled as fine powders. Because they can dissolve appreciable amounts of magnesia, scandia, and perhaps other oxides, they may be useful in accommodating densifying additives that would otherwise form grain-boundary glasses.

With decreasing temperature, the polytypoids with smaller M:X ratios become increasingly unstable, and, as shown by the behavior diagram presented in Fig. 8,

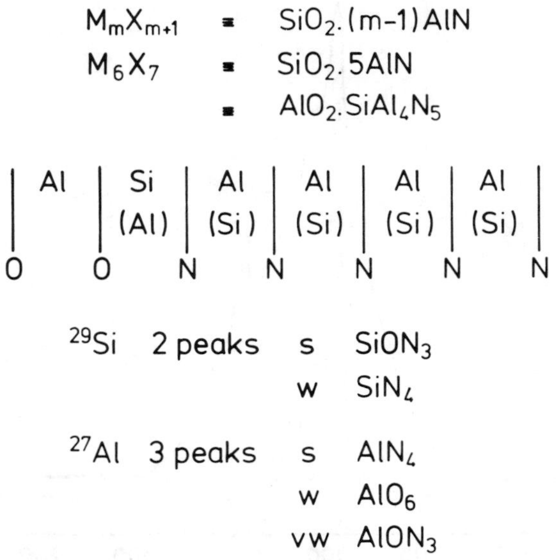

$$M_mX_{m+1} \equiv SiO_2.(m-1)AlN$$
$$M_6X_7 \equiv SiO_2.5AlN$$
$$\equiv AlO_2.SiAl_4N_5$$

Al	Si (Al)	Al (Si)	Al (Si)	Al (Si)	Al (Si)

O O N N N N N

^{29}Si	2 peaks	s	$SiON_3$
		w	SiN_4
^{27}Al	3 peaks	s	AlN_4
		w	AlO_6
		vw	$AlON_3$

Atomic symbols in parentheses represent minor concentrations. The 15R sequence has one less tetrahedral layer.

Fig. 37. The suggested ordering arrangement in the 12H sialon polytypoid

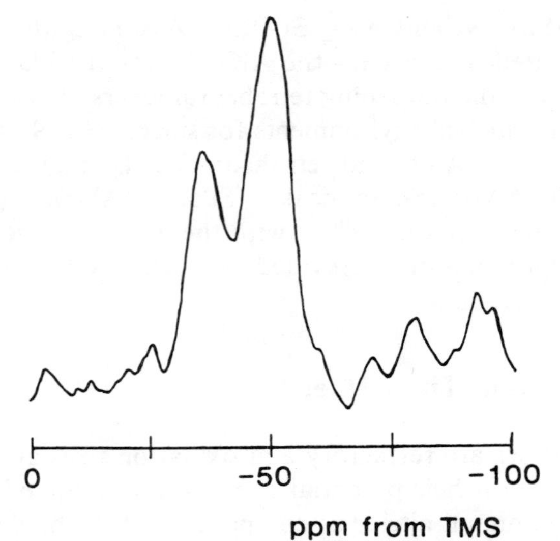

Chemical shifts correspond with SiN_4 (−32 ppm) and $SiON_3$ (−45 ppm).

Fig. 38. ^{29}**Si MASNMR spectrum of 15R. After Klinowski *et al.*[16]**

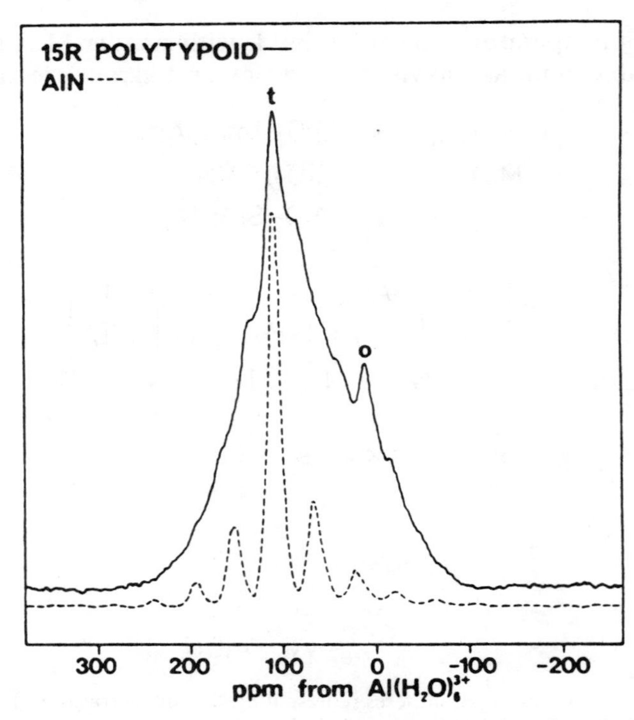

Chemical shifts for 15R correspond with AlO_6 (zero) and AlN_4 (+110 ppm).

Fig. 39. ^{27}**Al MASNMR spectra of 15R and AlN. After Butler *et al.*[17]**

8H does not occur at 1700 °C. Conversely, with increasing temperature the phases extend to reach the AlN-Al_2O_3 edge of the Si-Al-O-N system. The sialon polytypoids therefore exist in the Al-O-N system under conditions defined by the phase diagram in Fig. 40, determined by McCauley and Corbin.[18]

CERAMIC ALLOYS OF SILICON CARBIDE WITH ALUMINUM NITRIDE AND NITROGEN

About ten years ago it was found at Newcastle that if Ca-α′-sialon was heated with carbon at not less than 1800 °C a solid solution of SiC and AlN was obtained with the 2H wurtzite structure:

$$\underset{\alpha'\text{-sialon}}{Ca[Si_9Al_3ON_{15}]} + \underset{\text{carbon}}{10C} \xrightarrow{1800^\circ C} \underset{\text{2H alloy}}{9SiC \cdot 3AlN} + \uparrow Ca + \uparrow CO + \uparrow 6N_2 \qquad \text{(Eq 14)}$$

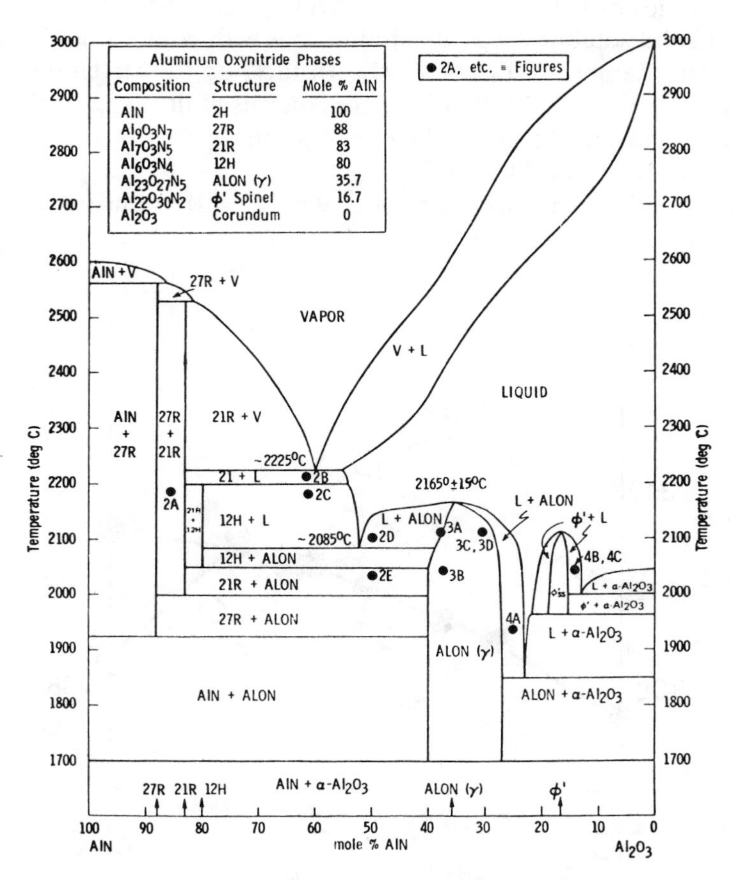

Fig. 40. Phase diagram for the Al_2O_3-AlN system. After McCauley and Corbin.[18]

It was then found that it was unnecessary to prepare the initial α'-phase; a mixture of the appropriate oxide and nitrides was equally effective:

$$CaO + 3Si_3N_4 + 3AlN + 10C \xrightarrow{1800\,°C} 9SiC \cdot 3AlN + \uparrow Ca + \uparrow CO + \uparrow 6N_2 \qquad \text{(Eq 15)}$$

Moreover, the composition of the mixture could be varied to give a wide range of solid solutions. At about the same time, the late Professor Cutler and his colleagues at Utah found similar SiC:AlN solid solutions by carbothermal reduction in nitrogen of very finely divided mixtures of silica and alumina, the latter being produced *in situ* by hydrolysis of aluminum chloride:

$$\underset{\text{cabosil}}{6SiO_2} + Al_2O_3 + \underset{\text{starch}}{21C} + N_2 \xrightarrow{1650\,°C} 6SiC \cdot 2AlN + 15CO \qquad \text{(Eq 16)}$$

The two groups published their results jointly and claimed initially that there was a continuous series of solid solutions from SiC to AlN.[19] This apparent continuous range of solid solubility was also found, but with compositional inhomogeneities, by Ruh and Zangvil[20] (see Fig. 41). In fact, up to the temperature limit of experiments at Newcastle (2100 °C), there is a miscibility gap in the range 40 to 80 mol % AlN, and Fig. 42 shows the observed phase relationships; complete miscibility occurs only above 2400 °C.

By preparing SiC from Si_3N_4 at about 1900 °C with addition of 1 wt % CaO

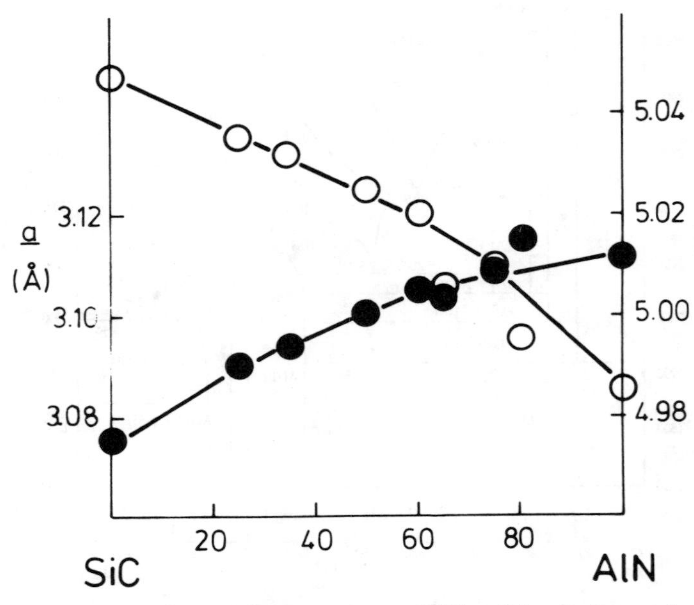

Fig. 41. Unit-cell dimensions of 2H solid solutions of SiC-AlN. After Ruh and Zangvil.[20]

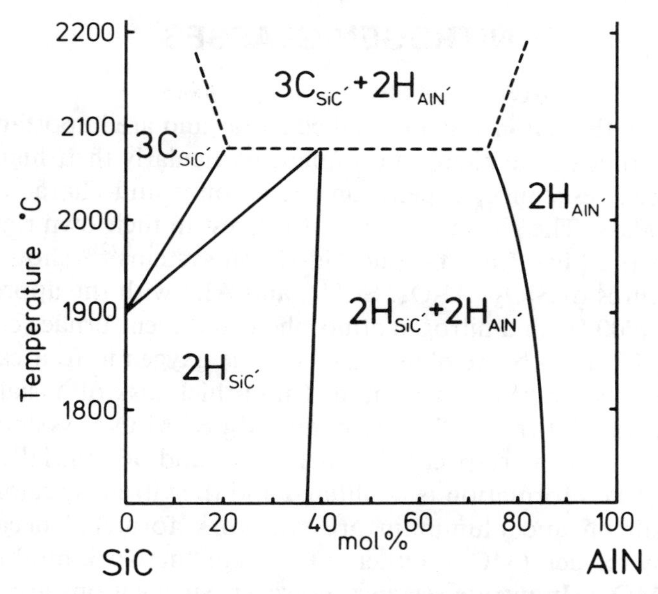

Fig. 42. Phase relationships in the SiC-AlN system. After Patience *et al.*[21]

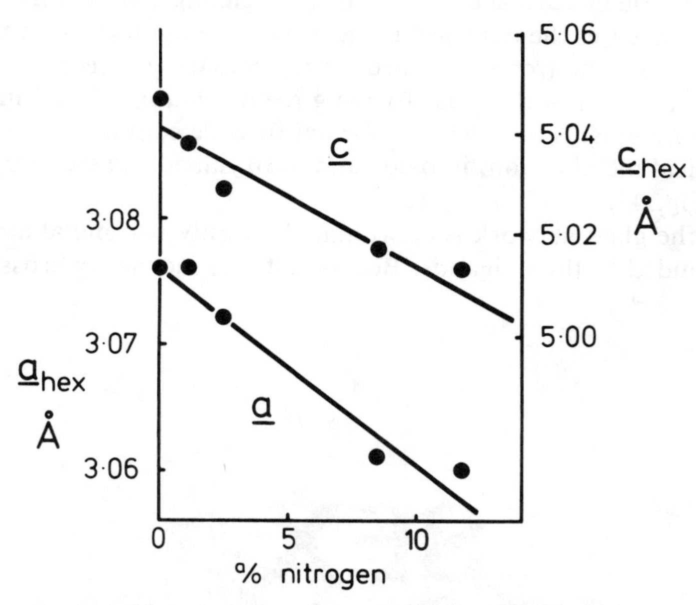

Fig. 43. Unit-cell dimensions of 2H Si(C,N) alloys. After Patience *et al.*[21]

to provide a Ca-Si-O-N liquid, the product is a 2H silicon carbide containing up to about 10 at. % N, i.e., a solid solution of SiC and Si_3N_4 with a wurtzite structure. The unit-cell dimensions decrease with increasing nitrogen concentration, as shown in Fig. 43. These "alloys" must contain silicon vacancies and should therefore show interesting electrical properties.

NITROGEN GLASSES

Glasses occur in all the sialon systems studied so far and are important because the mechanical properties of the nitrogen ceramics, particularly their high-temperature strength and creep resistance, depend on the amount and characteristics of the grain-boundary glass. The glasses are also of interest in their own right, and a systematic study of the Mg-, Ca-, Y-, and Nd-sialon systems[22] shows that by fusing powder mixtures of SiO_2, Al_2O_3, Si_3N_4, and AlN with the appropriate metal oxide at 1600 to 1700 °C in a nitrogen atmosphere and then furnace cooling, glasses containing up to 15 at. % N are obtained—i.e., one oxygen in four can be replaced by nitrogen. Figure 44 shows a transparent 1-mm-thick disk of Y-sialon glass. The glass-forming regions of the Y-Si-Al-O-N and Mg-Si-Al-O-N systems on cooling from 1700 °C are shown, respectively, in Fig. 45 and 46. Initially, as nitrogen replaces oxygen, glass formation is facilitated and the vitreous region expands. In oxide glasses, silicon and aluminum are "network formers" because they are 4-coordinated by oxygen (MO_4). Other cations are "network modifiers" and are 6-coordinated (MO_6). In nitride systems, however, such cations are 4-coordinated (e.g., MgN_4) and become "network formers." With increasing replacement of oxygen by nitrogen, the increasing covalency of the bonding gives it a more directional character and increases the tendency toward crystallinity in the structure. The net effect of these opposing trends, as nitrogen replaces oxygen, is first an expansion and then a contraction of the glass-forming region. Ultimately a limit is reached beyond which no glass is obtained by normal furnace cooling. The roles of magnesium, a typical metal cation, in oxide and in oxynitride glasses are shown schematically in Fig. 47.

Oxygen in the glass network is coordinated by only two metal atoms whereas nitrogen is bonded to three ligands. Because of this increasing cross linking, the

Fig. 44. Transparent disk of Y-sialon glass, 1 mm thick

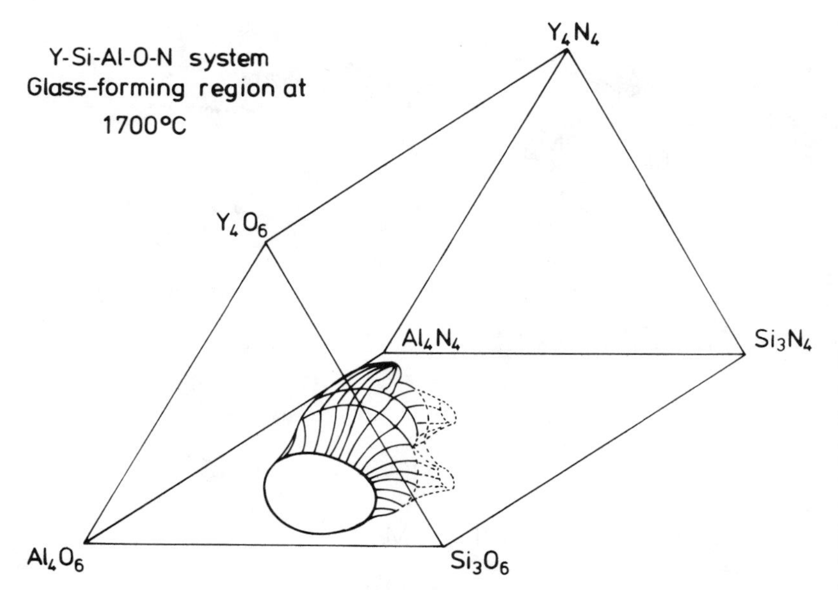

Fig. 45. Glass-forming region in the Y-Si-Al-O-N system on cooling from 1700 °C

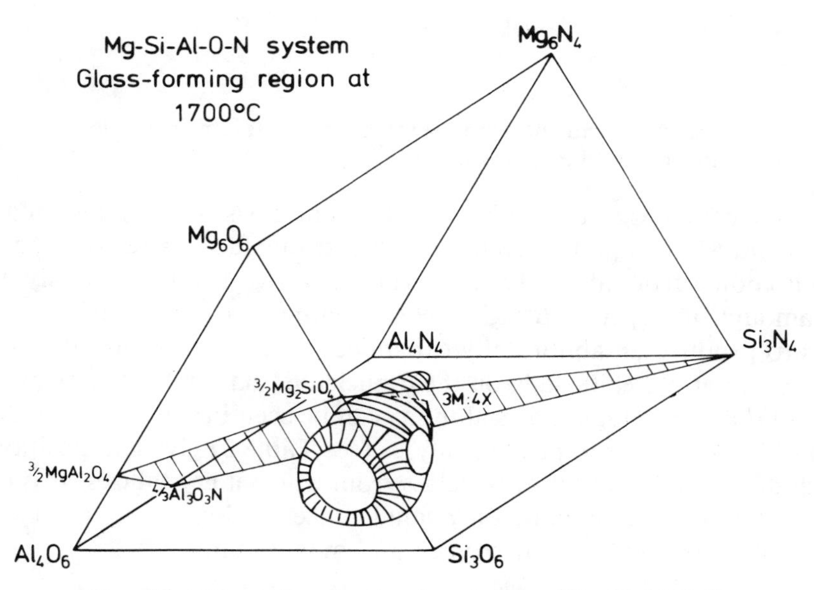

Fig. 46. Glass-forming region in the Mg-Si-Al-O-N system on cooling from 1700 °C

viscosity (Fig. 48), hardness (Fig. 49), and density (Fig. 50) of the glass all increase with increasing nitrogen concentration. Thus, the nitrogen-containing glasses are more refractory and more erosion resistant than their oxide counterparts.

Devitrification of glasses often gives rise to phases that are unobtainable by other means, and this is as true for nitrogen-containing systems as it is for oxides. The

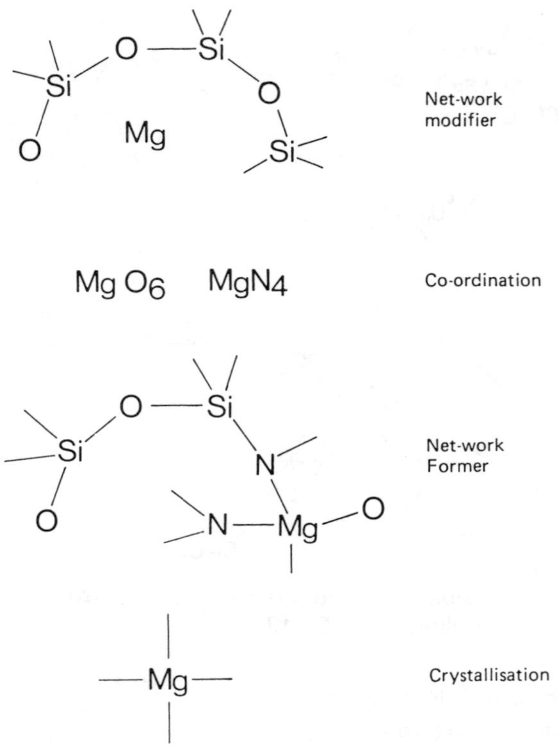

Net-work modifier

MgO_6 MgN_4

Co-ordination

Net-work Former

Crystallisation

Fig. 47. The different roles of magnesium in oxide and oxynitride glasses

3M:4X plane of the Mg-Si-Al-O-N system and its glass-forming region are shown in Fig. 46 and 51. If liquid of composition just outside this region (marked + in Fig. 51) is cooled from about 1700 °C, the resulting product is mainly glass with a small amount of dispersed β-Si_3N_4. Heat treatment at 900 to 1000 °C devitrifies the glass to produce an abnormally expanded β'-phase, designated as β'', which crystallizes and grows epitaxially on the β nuclei. β'' has unit-cell dimensions much larger than those of a high-z β'-sialon, as can be seen in the x-ray powder photographs in Fig. 52. Electron probe analyses (see Table 3) give compositions for β'' not much different from that of the glass from which it crystallizes. It is essentially Forsterite (Mg_2SiO_4) containing only small concentrations of nitrogen and aluminum by mutual replacement of oxygen and magnesium:

$$Mg^{2+}O^{2-} \rightleftharpoons Al^{3+}N^{3-} \qquad \text{(Eq 17)}$$

This replacement is sufficient to transform the Forsterite to the phenacite (Be_2SiO_4) structure, suggesting that the lattice energies of the two types must not be too different. Taking accepted values of bond length, Table 3 shows that the observed unit-cell dimensions of a typical β''-phase are in reasonable agreement with those calculated from its composition.

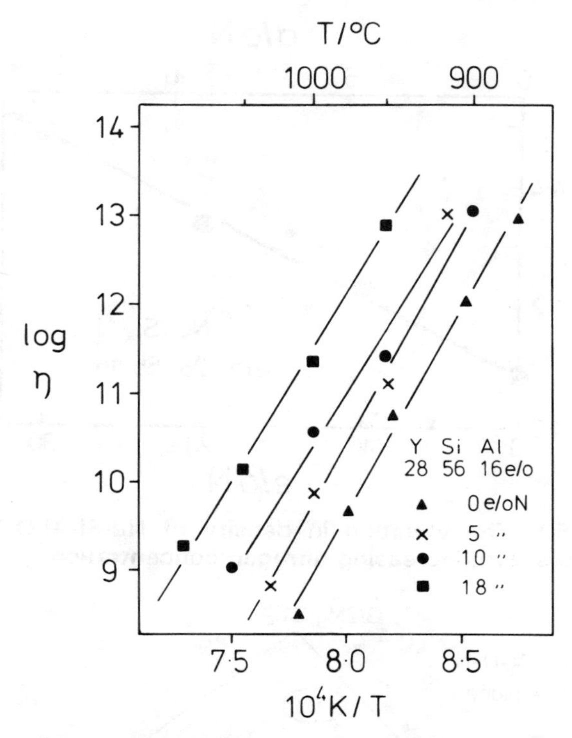

Fig. 48. Viscosities of Y-Si-Al-O-N glasses with increasing nitrogen concentration

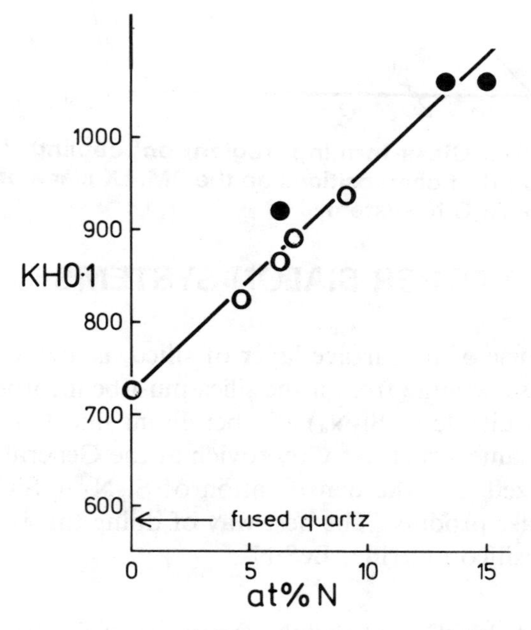

○ Shillito *et al.*[23] ● Messier and Broz.[24]

Fig. 49. The hardness of Y-Si-Al-O-N glasses

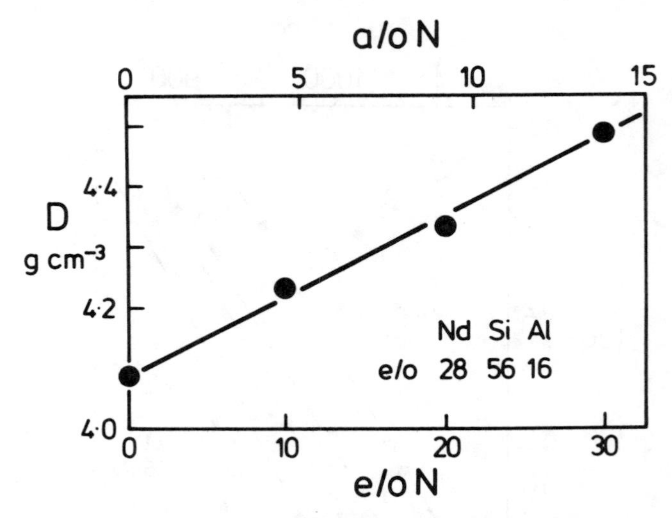

Fig. 50. The variation in density of Nd-Si-Al-O-N glasses with increasing nitrogen concentration

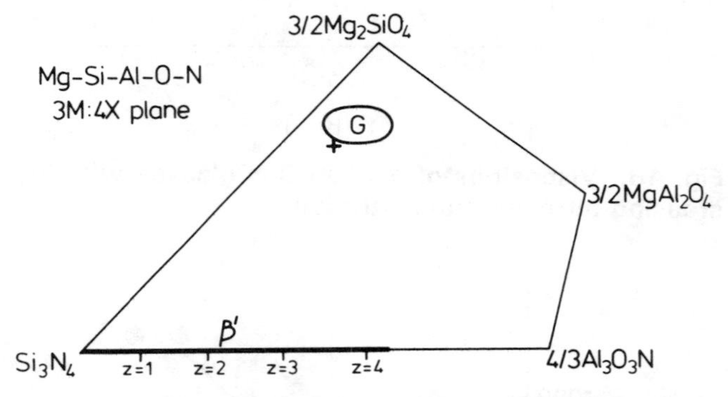

Fig. 51. Glass-forming region on cooling from 1700 °C of compositions on the 3M:4X plane of the Mg-Si-Al-O-N system

OTHER SIALON SYSTEMS

Silicon nitride, because of its surface layer of silica, is always two-phase, and to produce a single-phase ceramic from it the silica must be incorporated into its structure. Because silicon nitride (β-Si_3N_4) and beryllium silicate (Be_2SiO_4, phenacite) have essentially the same structure, Greskovich at the General Electric Company, among others, realized that the densification of $Si_3N_4 + SiO_2$ with BeO would result in a single-phase product. The best way of doing this is by a balanced reaction with beryllium-silicon nitride, $BeSiN_2$:

$$0.9Si_3N_4 + 0.1SiO_2 + 0.1BeSiN_2 \rightarrow (Be_{0.1}Si_{2.9})(O_{0.2}N_{3.8}) \quad \text{(Eq 18)}$$

"silicon nitride"

Table 3. Calculated and observed unit-cell dimensions for β''-Mg-sialon(a)

	Unit-cell dimensions, Å		
	a	*c*	V
β-Si_3N_4	7.61	2.91	145.9
β'', observed	7.92	3.10	168.4
β'', calculated	7.98	3.05	168.4

(a) Composition, at. %: 20.7Mg-18.1Si-3.6Al-47.9O-9.7N. Bond lengths, Å: Si-N, 1.74; Si-O, 1.62; Al-N, 1.87; Al-O, 1.75; Mg-N, 2.13; Mg-O, 1.97.

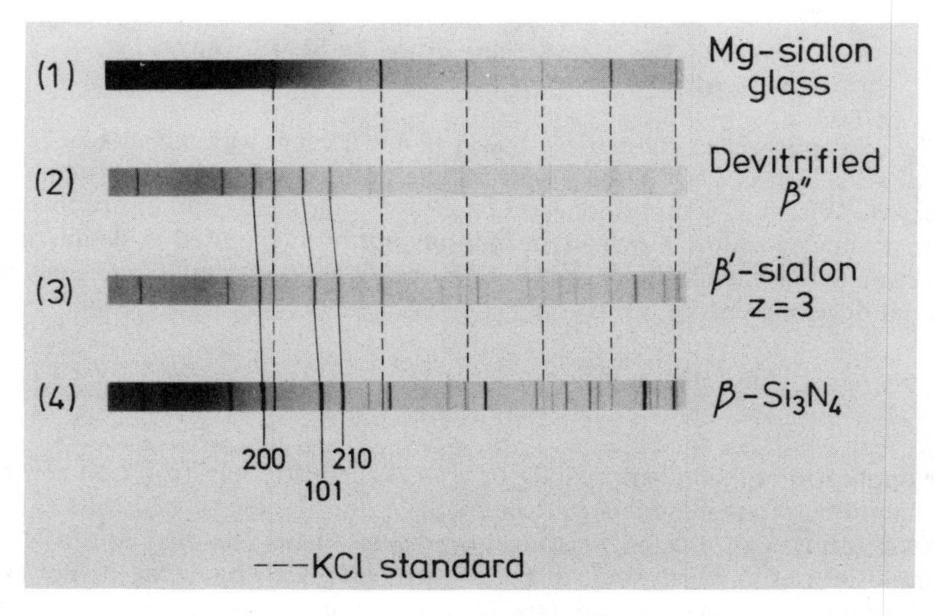

Fig. 52. X-ray powder photographs showing the isostructural phases β-Si_3N_4, β'-sialon, and β''-Mg-sialon

The resulting single-phase beryllium-silicon oxynitride has no intergranular glass and so its creep resistance is excellent, but the temperature to effect the reaction given by Eq 18 is so high that nitrogen gas overpressures are necessary to prevent decomposition. Even so, the large grain growth at the elevated preparative temperature impairs the strength of the product.[25]

The alternative Newcastle approach is to incorporate the silica as a sialon:

$$0.9Si_3N_4 + 0.1SiO_2 + 0.2AlN \rightarrow (Si_{2.8}Al_{0.2})(O_{0.2}N_{3.8}) \qquad \text{(Eq 19)}$$

"silicon nitride"

but to facilitate this reaction (Eq 19) a liquid such as Y-Si-Al-O-N is necessary and so other oxides (Y_2O_3, MgO) are also added.

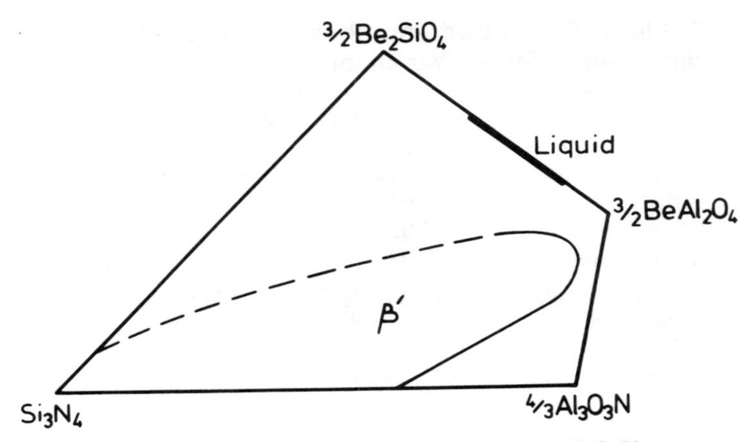

Fig. 53. The 3M:4X plane of the Be-Si-Al-O-N system. After Gauckler *et al.*[26]

The two approaches summarized by Eq 18 and 19 can, with advantage, be combined. As shown in Fig. 53, the β'-phase region of the Be-Si-Al-O-N system is very extensive. Within it, compositions can be melted congruently and it is probably the most promising sialon alloy system that has *not* been explored in detail.[27] It is unfortunate that the health hazard posed by beryllium compounds deters any commercial development.

CONCLUSION

The application of simple principles of crystal chemistry has produced new crystalline and vitreous materials by "ceramic alloying," some of which have useful chemical, physical, and engineering properties. These "sialons" are essentially aluminosilicates in which some of the oxygen is replaced by nitrogen. For example, in the yttrium-sialon system (see Fig. 12) the phases include nitrogen analogs of melilite ($Ca_2MgSi_2O_7$), wollastonite ($CaSiO_3$), yttrium aluminate ($Y_4Al_2O_9$), and apatite [$Ca_5(PO_4)_3OH$]. There is no reason why many more of the wide variety of mineral silicates should not also have sialon analogs. Whether they should be called "ceramic alloys" is unimportant; their further exploration offers the possibility of tailoring crystalline and vitreous materials to meet some of the needs of modern technology.

ACKNOWLEDGMENTS

This chapter is not a comprehensive or authoritative account of the development of nitrogen ceramics. Principles of this development are illustrated mainly by the work of the Wolfson Research Group for High-Strength Materials at the University of Newcastle upon Tyne, and only a limited number of references to the research conducted there and in other laboratories are cited. Full references and acknowledgments are given in more formal publications on which this chapter is based. Its presentation has been assisted by the award of a Leverhulme Emeritus Fellowship. The following illustrations are reproduced with permission: Fig. 14, 15, 16, 18, and 19 by Lucas-Cookson Syalon Ltd., and Fig. 23, 24, and 25 by Sandvik Hard Materials Ltd.

REFERENCES

1. K.H. Jack, *Trans. and J. Brit. Ceram. Soc.*, Vol 72, 1973, p. 376
2. Y. Oyama and O. Kamigaito, *Japan. J. Appl. Phys.*, Vol 10, 1971, p. 1637
3. K.H. Jack and W.I. Wilson, *Nature*, Vol 238, 1972, p. 28
4. M.H. Lewis, A.R. Bhatti, R.J. Lumby, and B. North, *J. Mater. Sci.*, Vol 15, 1980, p. 103
5. R.J. Lumby, E. Butler, and M.H. Lewis, in *Progress in Nitrogen Ceramics*, edited by F.L. Riley, Martinus Nijhoff, The Hague, 1983, p. 683
6. W.J. Arrol, in *Ceramics for High Performance Applications*, Proceedings of the Second Army Materials Technology Conference, Hyannis, 1973, edited by J.J. Burke, A.E. Gorum, and R.N. Katz, Brook Hill Publishing Co., Chestnut Hill, U.S.A., 1974, p. 729
7. R.J. Lumby, Lucas-Cookson Syalon Limited (private communication, 1986)
8. J.G. Lee and I.B. Cutler, *Ceram. Bull.*, Vol 58, 1979, p. 869
9. S. Umebayashi, in *Nitrogen Ceramics*, edited by F.L. Riley, Noordhoff, Leyden, 1977, p. 323
10. K.H. Jack, in *Progress in Nitrogen Ceramics*, edited by F.L. Riley, Martinus Nijhoff, The Hague, 1983, p. 45
11. M.B. Trigg and K.H. Jack, *J. Mater. Sci. Lett.*, Vol 6, 1987, p. 407
12. M.B. Trigg and K.H. Jack, in *Proceedings of the First International Symposium on Ceramic Components for Engine*, Hakone, 1983, edited by S. Somiya, E. Kanai, and K. Ando, KTK Scientific Publishers, Tokyo, 1984, p. 199; *J. Mater. Sci.*, Vol 23, 1988, p. 481
13. W.Y. Sun, D.P. Thompson, and K.H. Jack, in *Tailoring Multiphase and Composite Ceramics*, Proceedings of the Twenty-First University Conference on Ceramic Science, Pennsylvania State University, 1985, edited by R. Tressler, G.L. Messing, C.G. Pantano and R.E. Newnham, Plenum Press, New York, 1986, p. 93
14. K.H. Jack, *J. Mater. Sci.*, Vol 11, 1976, p. 1135
15. D.P. Thompson, in *Nitrogen Ceramics*, edited by F.L. Riley, Noordhoff, Leyden, 1977, p. 129
16. J. Klinowski, J.M. Thomas, D.P. Thompson, P. Korgul, K.H. Jack, C.A. Fyfe, and G.C. Gobbi, *Polyhedron*, Vol 3, 1984, p. 1267
17. N.D. Butler, R. Dupree, and M.H. Lewis, *J. Mater. Sci. Lett.*, Vol 3, 1984, p. 469
18. J.W. McCauley and N.D. Corbin, in *Progress in Nitrogen Ceramics*, edited by F.L. Riley, Martinus Nijhoff, The Hague, 1983, p. 111
19. I.B. Cutler, P.D. Miller, W. Rafaniello, H.K. Park, D.P. Thompson, and K.H. Jack, *Nature*, Vol 275, 1978, p. 434
20. R. Ruh and A. Zangvil, *J. Amer. Ceram. Soc.*, Vol 65, 1982, p. 260
21. M.M. Patience, P.J. England, D.P. Thompson, and K.H. Jack, in *Proceedings of the First International Symposium on Ceramic Components for Engine*, Hakone, 1983, edited by S. Somiya, E. Kanai, and K. Ando, KTK Scientific Publishers, Tokyo, 1984, p. 473

22. R.A.L. Drew, S. Hampshire, and K.H. Jack, in Special Ceramics 7, edited by D. Taylor and P. Popper, *Proc. Brit. Ceram. Soc.*, Vol 31, 1981, p. 119
23. K.R. Shillito, R.R. Wills, and R.B. Bennett, *J. Amer. Ceram. Soc.*, Vol 61, 1978, p. 537
24. D.R. Messier and A. Broz, *J. Amer. Ceram. Soc.*, Vol 65, 1982, p. C-123
25. J.A. Palm and C.D. Greskovich, *Amer. Ceram. Soc. Bull.*, Vol 59, 1980, p. 447
26. L.J. Gauckler, H.L. Lukas, and T.Y. Tien, *Mater. Res. Bull.*, Vol 11, 1976, p. 503
27. K.H. Jack, in "Role of Additives in the Densification of Nitrogen Ceramics," Final Technical Report to the European Research Office, U.S. Army, under Grant No. DAERO-76-G-067, November 1977

14

Thin-Film Compound Formation: Kinetics and Effects of Volume Changes

U. GÖSELE
Department of Mechanical Engineering and Materials Science
School of Engineering
Duke University

In solid-state electronics, a variety of thin films are used for metallization or passivation purposes, as dielectrics or as the semiconductor material itself such as in quantum-well structures. The importance of the thin-film area is evidenced by the fast-growing number of publications, books, and conferences on thin films and the closely related subject of interfaces.[1-6] In the present paper we will deal with phenomena associated with the growth kinetics of planar thin films of *binary* compounds which form as a result of the reactions between the constituent elements, one of which should be present in solid form. Examples are the formation of silicides by the reaction of deposited metals with the silicon substrate, or the formation of a silicon dioxide film on silicon by the reaction of solid silicon with oxygen gas. Although many of the specific examples will deal with silicides and SiO_2, the general concepts are also applicable to thin-film metal-metal compounds and other growing thin compound films such as oxide or sulfide scales.

We will not deal with compound formation which occurs *during* deposition as a result of the reaction between the deposited elements. Examples of growth methods based on this formation mechanism are molecular beam epitaxy and metal-organic-chemical vapor deposition, which are adequately covered elsewhere.[7]

In Section 1 we will review the mathematical description of the growth kinetics of a single compound layer before treating the more complex case of various compound layers in Section 2. Experimentally it is well known that even if various compound phases exist in the equilibrium phase diagram, in thin-film diffusion couples these phases form and appear *sequentially* rather than simultaneously as in bulk diffusion couples. We will discuss the various models which have been suggested to explain this thin-film phenomenon and to predict which compound forms first. The concept of a "critical thickness" which the first-forming compound has to exceed before a second compound can start to grow is then applied to the formation of *amorphous* solids from multilayer binary diffusion couples.

Section 3 is devoted to various phenomena associated with volume changes during compound formation. Examples are the coupling of diffusion fluxes in solid-phase epitaxy through compound transport layers and the injection of intrinsic point defects into the substrate during surface oxidation of silicon or of some specific metals. The effects of oxidation-induced point defects on dopant diffusion allowed a long-standing controversy to be settled on intrinsic point defects in silicon. Finally we will suggest similar experiments for metals and speculate on the nature of the point defects which are generated during corrosion and cause corrosion-enhanced fatigue.

1. GROWTH OF A SINGLE COMPOUND LAYER

In this section we review the growth kinetics of a single compound layer, say $A_\beta B$, between the two neighboring phases in the equilibrium phase diagram—e.g., $A_\alpha B$ and $A_\beta B$—as indicated in Fig. 1 and 2. The growth of the compound layer in-

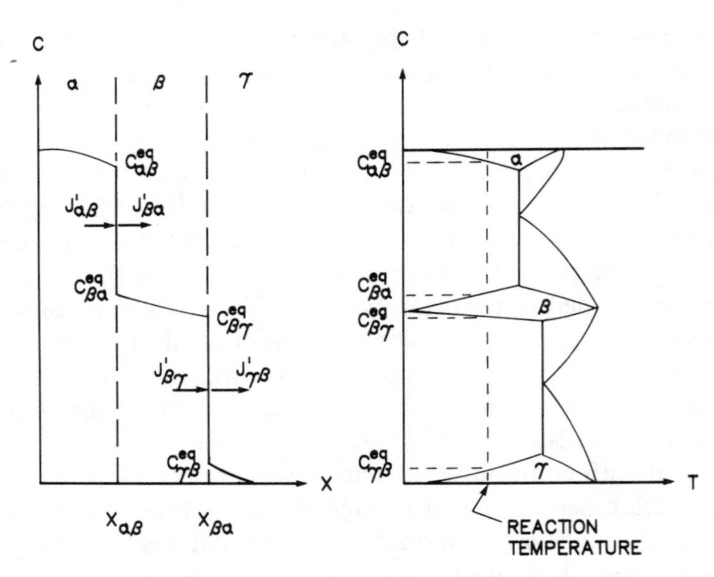

Fig. 1. Schematic binary phase diagram and concentration profile of a diffusing element during growth of β phase between α and γ phases

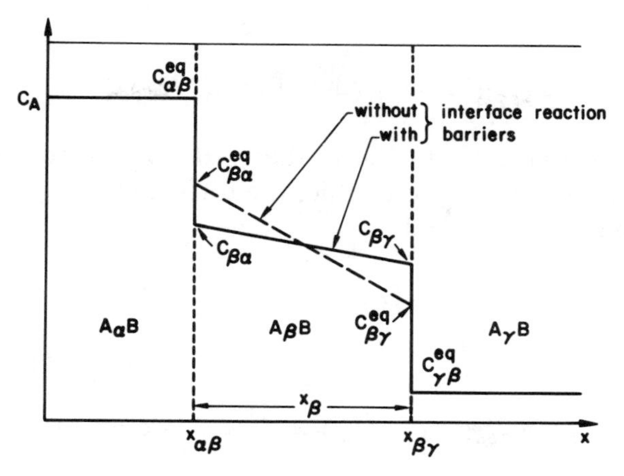

The phases $A_\alpha B$ and $A_\gamma B$ are assumed to be saturated.

Fig. 2. Schematic concentration profile of A atoms across an $A_\alpha B/A_\beta B/A_\gamma B$ diffusion couple with and without interface reaction barriers, where $\alpha > \beta > \gamma$ (according to Gösele and Tu[8])

volves the *diffusion* of A and/or B across the layer and subsequent *reaction* at one of the interfaces, which involves rearrangement of atoms and therefore in general is associated with a reaction barrier. If the diffusion is the slower process and limits the growth rate, the process is called *diffusion-controlled* (or diffusion-limited). If the interfacial reaction determines the growth rate, the growth kinetics is termed *interface-controlled* or (reaction-controlled). In the presence of interfacial reaction barriers the concentrations of the components A and B deviate from their equilibrium concentrations C^{eq} at the interfaces as indicated in Fig. 2.

1.1. Saturated Terminal Phases

Let us first calculate the growth kinetics of a compound layer $A_\beta B$ between two saturated phases $A_\alpha B$ and $A_\gamma B$, with $\alpha > \beta > \gamma$ for which the A concentration profile C_A is schematically shown in Fig. 2, taking into account the presence of reaction barriers at both interfaces. For simplicity we assume that the diffusional transport of A or B across the $A_\beta B$ layer may be described in terms of an essentially constant chemical interdiffusion coefficient $\tilde{D}_\beta$, which in the most simple case is related to the component diffusion coefficients D^A_β and D^B_β via

$$\tilde{D}_\beta = D^A_\beta/(\beta + 1) = \beta D^B_\beta/(\beta + 1) \qquad \text{(Eq 1)}$$

Following Kidson,[9] the changes in the position of the $\alpha\beta$-interface and the $\beta\gamma$-interface are given by

$$(C^{eq}_{\alpha\beta} - C_{\beta\alpha})dx_{\alpha\beta}/dt = \tilde{D}_\beta(\partial C^A_\beta/\partial x)_{\beta\alpha} \qquad \text{(Eq 2)}$$

and

$$(C_{\beta\gamma} - C_{\gamma\beta}^{eq})dx_{\beta\gamma}/dt = \tilde{D}_\beta(\partial C_\beta^A/\partial x)_{\beta\gamma} \qquad \text{(Eq 3)}$$

where C_β^A is the concentration of A in the $A_\beta B$ layer. Under the assumption of steady state of the concentration profile, the diffusion flux j_β^A is equal at the two interfaces and may be expressed as

$$j_\beta^A = -\tilde{D}_\beta(dC_\beta^A/dx)_{\beta\alpha} = -\tilde{D}_\beta(dC_\beta^A/dx)_{\beta\gamma} = -\tilde{D}_\beta(C_{\beta\alpha} - C_{\beta\gamma})/x_\beta \qquad \text{(Eq 4)}$$

and also as

$$j_\beta^A = \kappa_{\beta\alpha}(C_{\beta\alpha}^{eq} - C_{\beta\alpha}) = \kappa_{\beta\gamma}(C_{\beta\gamma} - C_{\beta\gamma}^{eq}) \qquad \text{(Eq 5)}$$

In terms of the time-independent equilibrium value

$$\Delta C_\beta^{eq} = C_{\beta\alpha}^{eq} - C_{\beta\gamma}^{eq} \qquad \text{(Eq 6)}$$

the diffusion-flux j_β^A may be expressed as

$$j_\beta^A = \Delta C_\beta^{eq}\kappa_\beta^{eff}/(1 + x_\beta\kappa_\beta^{eff}/\tilde{D}_\beta) \qquad \text{(Eq 7)}$$

where κ_β^{eff} is an effective interfacial reaction barrier composed of the two interfacial reaction barriers $\kappa_{\beta\alpha}$ and $\kappa_{\beta\gamma}$ according to

$$1/\kappa_\beta^{eff} = 1/\kappa_{\beta\alpha} + 1/\kappa_{\beta\gamma} \qquad \text{(Eq 8)}$$

The change of the layer thickness $x_\beta = x_{\beta\gamma} - x_{\alpha\beta}$ with time is given by

$$dx_\beta/dt = [1/(C_{\alpha\beta}^{eq} - C_{\beta\alpha}) + 1/(C_{\beta\gamma} - C_{\gamma\beta}^{eq})]j_\beta^A \qquad \text{(Eq 9)}$$

The quantities $C_{\beta\alpha}$ and $C_{\beta\gamma}$ are time-dependent and are approaching $C_{\beta\alpha}^{eq}$ and $C_{\beta\gamma}^{eq}$ with increasing time and layer thickness. Provided that the compound $A_\beta B$ has a narrow range of homogeneity, we may neglect this time-dependence and even approximate $C_{\beta\alpha}$ and $C_{\beta\gamma}$ in the square brackets of Eq 9 by the same value $\beta/[\Omega_0(1 + \beta)]$. Ω_0 is the volume per A or B atom, which for simplicity we assume to be constant throughout the sample. The quantity β is the number of A atoms per B atom in the $A_\beta B$ compound. We thus arrive at

$$dx_\beta/dt = H_\beta\Delta C_\beta^{eq}\kappa_\beta^{eff}/(1 + x_\beta\kappa_\beta^{eff}/\tilde{D}_\beta) \qquad \text{(Eq 10)}$$

where H_β is a constant determined by the compositions of the three phases $A_\alpha B$, $A_\beta B$, and $A_\gamma B$ via

$$H_\beta = \Omega_0(1 + \beta)^2\left(\frac{1}{\alpha - \beta} + \frac{1}{\beta - \gamma}\right) \qquad \text{(Eq 11)}$$

Again, α and γ represent the numbers of A atoms per B atom in the $A_\alpha B$ and $A_\gamma B$ compounds. Equation 10 is basically the same as those given by Deal and Grove[10] for SiO_2 layer growth on silicon and by Farrell *et al.*[11] and Dybkov[12] for one diffusing species. If there is a volume difference between A and B and if a volume change occurs upon reaction (see also Section 2), $\Omega_0(1+\beta)$ in Eq 11 has to be changed to Ω_β, which is the volume per $A_\beta B$. Equation 10 has to be modified slightly if the diffusion coefficient $\tilde{D}_\beta$ depends on composition, but the basic structure remains the same.

Integration of Eq 10 yields

$$(x_\beta - x_\beta^0) + (x_\beta - x_\beta^0)^2 \kappa_\beta^{\text{eff}}/(2\tilde{D}_\beta) = H_\beta \Delta C_\beta^{\text{eq}} \kappa_\beta^{\text{eff}} t \qquad \text{(Eq 12)}$$

where x_β^0 is the starting thickness of the $A_\beta B$ layer at $t = 0$. The growth-kinetics changes from interface- to diffusion-controlled at a *changeover thickness*

$$x_\beta^* = \tilde{D}_\beta/\kappa_\beta^{\text{eff}} \qquad \text{(Eq 13)}$$

leading to

$$dx_\beta/dt = H_\beta \Delta C_\beta^{\text{eq}} \kappa_\beta^{\text{eff}} \quad \text{for } x_\beta \ll x_\beta^* \qquad \text{(Eq 14)}$$

and

$$dx_\beta/dt = H_\beta \Delta C_\beta^{\text{eq}} \tilde{D}_\beta/x_\beta \quad \text{for } x_\beta \gg x_\beta^* \qquad \text{(Eq 15)}$$

In integrated form and for $x_\beta^0 = 0$, this leads to

$$x_\beta = H_\beta \Delta C_\beta^{\text{eq}} \kappa_\beta^{\text{eff}} t \quad \text{for } x \ll x_\beta^* \qquad \text{(Eq 16)}$$

and

$$x_\beta = (2H_\beta \Delta C_\beta^{\text{eq}} \tilde{D}_\beta t)^{1/2} \quad \text{for } x \gg x_\beta^* \qquad \text{(Eq 17)}$$

Experimental examples for these growth laws for compound layers have been compiled in many books and review articles. We mention here only the especially well investigated growth kinetics of many silicides as discussed, for example, by Tu and Mayer[13] and by Ottaviani.[14]

The growth kinetics of insulator films may be strongly affected by electrical-charge effects. Under these circumstances, growth laws of the form

$$x_\beta \propto t^{1/3} \qquad \text{(Eq 18)}$$

and

$$x_\beta \propto \ln t \qquad \text{(Eq 19)}$$

have been observed. Electrical effects are beyond the scope of this chapter, and for more details the reader is referred to the excellent book on solid-state reactions by Schmalzried.[15] We will also not deal with nonlinear effects in the diffusion behavior when the diffusion distances get extremely small, since these effects have been discussed extensively in recent papers by Tu[16] and Geer.[17]

1.2. Unsaturated Terminal Phases

In Fig. 2 the terminal phases $A_\alpha B$ and $A_\gamma B$ have been assumed to be saturated. If this is not the case, as indicated in Fig. 1, then the diffusion fluxes in these phases also have to be taken into account. We will give here only the treatment for the *diffusion-controlled* case based on the treatment by Kidson.[9] Instead of Eq 9 we get

$$\frac{dx_\beta}{dt} = (j_\beta^A - j_\alpha^A)/(C_{\alpha\beta}^{eq} - C_{\beta\alpha}^{eq}) + (j_\beta^A - j_\gamma^A)/(C_{\beta\gamma}^{eq} - C_{\gamma\beta}^{eq}) \qquad \text{(Eq 20)}$$

where j_α^A and j_γ^A are the diffusion fluxes of A in the $A_\alpha B$ and $A_\gamma B$ compounds, respectively. For an infinitely extended diffusion couple it may be shown that the concentration profiles and diffusion fluxes depend only on the variable

$$\lambda = x/t^{1/2} \qquad \text{(Eq 21)}$$

which leads, for quasi steady-state conditions and negligible starting thickness of x_β, to

$$x_\beta = 2\left[\frac{\tilde{D}_\beta K_{\beta\alpha} - \tilde{D}_\alpha K_{\alpha\beta}}{C_{\alpha\beta}^{eq} - C_{\beta\alpha}^{eq}} + \frac{\tilde{D}_\beta K_{\beta\gamma} - \tilde{D}_\gamma K_{\gamma\beta}}{C_{\beta\gamma}^{eq} - C_{\gamma\beta}^{eq}}\right] t^{1/2} \qquad \text{(Eq 22)}$$

$$x_\beta = Bt^{1/2} \qquad \text{(Eq 23)}$$

where values of

$$K_{ij} = dC_{ij}/d\lambda \qquad \text{(Eq 24)}$$

are constants to be taken at the appropriate interfaces, and $\tilde{D}_\alpha$ and $\tilde{D}_\gamma$ are the chemical interdiffusion coefficients in the $A_\alpha B$ and $A_\gamma B$ layers, respectively.

If the $A_\beta B$ compound layer consists of two sublattices in which A and B may move independently, then the diffusion fluxes of A and B are independent of each other and may be written in the form

$$j_\beta^A = -D_\beta^A \frac{dC_\beta^A}{dx} \qquad \text{(Eq 25)}$$

$$j_\beta^B = -D_\beta^B \frac{dC_\beta^B}{dx} \qquad \text{(Eq 26)}$$

where C_β^A and C_β^B are the concentrations of A and B in the $A_\beta B$ compound layer. Equations 25 and 26 lead to the same basic results described by Eq 16 and 17.

1.3. Relation to Changes in Chemical Potential

It is often convenient to express the usually unknown quantity ΔC_β^{eq} in Eq 16 or 17 approximately in terms of the chemical potential changes $\Delta\mu_\beta^A$ over the composition range of the $A_\beta B$ compound. Instead of Eq 4 we may use the thermodynamically more fundamental expression

$$j_\beta^A = -\frac{\tilde{D}_B}{k_B T} C_\beta^A \frac{d\mu_\beta^A}{dx} \qquad \text{(Eq 27)}$$

and approximate it by

$$j_\beta^A = -\frac{\tilde{D}_B}{k_B T} C_\beta^A \frac{\Delta\mu_\beta^A}{x_\beta} \qquad \text{(Eq 28)}$$

where k_B denotes Boltzmann's constant and T is the absolute temperature. For a narrow composition range of the compounds involved, $\Delta\mu_\beta^A$ may be expressed in terms of G_α^0, G_β^0, and G_γ^0, which are the Gibbs' free energies of formation of the respective compounds from the elements A and B. If G_α^0, G_β^0, and G_γ^0 are given per atom of the compound (and not per molecule, as usually found in tables), then one obtains

$$\Delta\mu_\beta^A = \frac{\gamma + 1}{\beta - \gamma}(G_\beta^0 - G_\gamma^0) + \frac{\alpha + 1}{\alpha - \beta}(G_\beta^0 - G_\alpha^0) \qquad \text{(Eq 29)}$$

An analogous expression has been given by Schmalzried.[15] For the growth of an $A_\beta B$ compound between two phases A and B with negligible solubility for the other element, Eq 29 reduces to

$$\Delta\mu_\beta^A = \frac{1 + \beta}{\beta} G_\beta^0 \qquad \text{(Eq 30)}$$

Using Eq 28 and $\Delta\mu_\beta^A$, we may rewrite Eq 16 and 17 in the forms

$$x_\beta = H_\beta' \kappa_\beta^{eff} \Delta\mu_\beta^A t/k_B T \quad \text{for } x_\beta \ll x_\beta^* \qquad \text{(Eq 31)}$$

and

$$x_\beta = [2H_\beta' \Delta\mu_\beta^A t/k_B T]^{1/2} \quad \text{for } x_\beta \gg x_\beta^* \qquad \text{(Eq 32)}$$

with

$$H_\beta' = H_\beta \beta/[\Omega_0(1 + \beta)] \qquad \text{(Eq 33)}$$

Equations 31 and 32 hold independently of whether the diffusion processes of A and B are independent of each other or related via a chemical interdiffusion coefficient. They constitute good approximations only as long as $\Delta\mu_\beta^A \ll k_BT$. These equations will be used in Section 2 in the context of predicting the first-growing compound phase in the case of multiple-layer growth.

2. GROWTH OF MULTIPLE COMPOUND LAYERS

2.1. Diffusion-Controlled Layer Growth

For a binary system with more than one compound occurring in the equilibrium phase diagram, such as for the Pt-Si system shown in Fig. 3(a), one expects all of

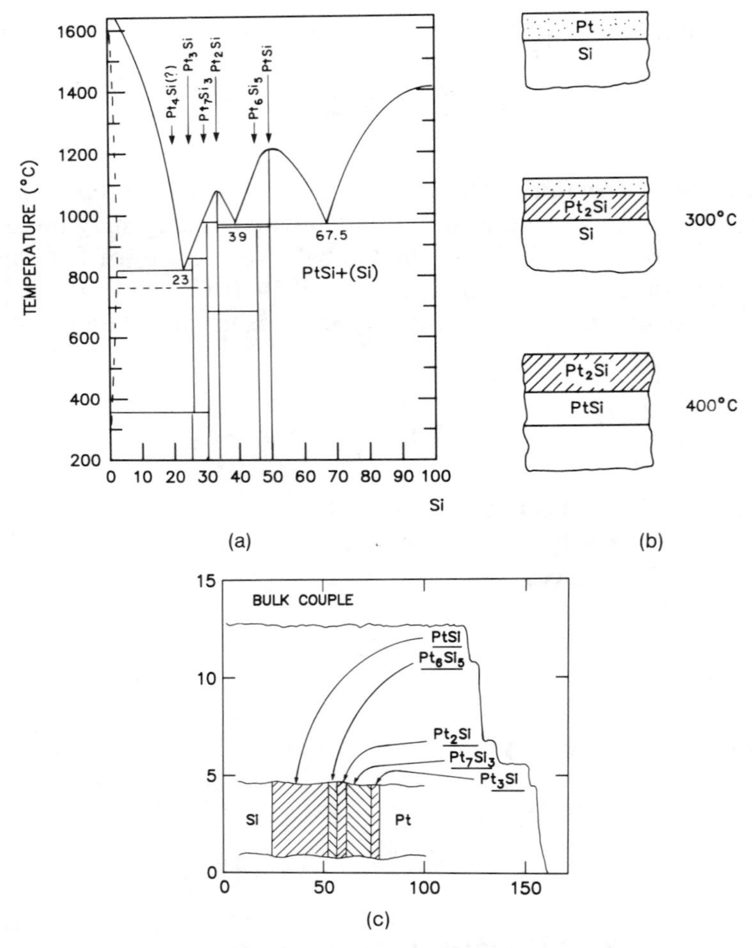

Fig. 3. (a) Platinum-silicon phase diagram, (b) phase sequence for thin-film Pt/Si diffusion couple (schematic)[14]; and (c) electron microprobe scan across a Pt/Si bulk diffusion couple annealed 96 h at 700 °C (according to Ottaviani[14] and Claassen[23])

these phases to show up in a diffusion couple starting from the elemental phases—e.g., from platinum and silicon. A straightforward treatment of diffusion-controlled multiple compound layer growth by Kidson[9] starts from equations analogous to Eq 20 for each compound phase. The same concept of a variable λ may be applied, leading finally to

$$x_i = B_i t^{1/2} \quad \text{(Eq 34)}$$

for all phases predicted by the phase diagram (except for the terminal phases, which of course shrink). Similar treatments for diffusion-controlled multiple compound layer growth have been presented by other authors.[18-22] All of them predict the growth of all intermediate compound phases in accordance with the equilibrium phase diagram. No compound can shrink away within these treatments basically because a vanishing compound thickness would lead to an infinitely large diffusion flux across the layer, which prevents the disappearance of phases.

2.2. Differences Between Thin-Film and Bulk Diffusion Couples: Experimental Observations

In bulk diffusion couples, most of the compound phases predicted by the equilibrium phase diagram usually may be found. As an example, the results for a Pt-Si bulk diffusion couple[23] are shown in Fig. 3(c). In contrast, it is well established experimentally that in planar binary thin-film diffusion couples not all of the compound phases predicted by the equilibrium phase diagram are present simultaneously. For silicides it is even the common case that only *one* compound phase grows between the components silicon and metal, with no indications of the presence of other phases.[13,14] Only if either the metal or the silicon is consumed by the growing compound will a second compound start to grow as indicated in Fig. 3(b) for the case of platinum. In this case Pt_2Si forms as the first phase and PtSi as the second phase when all the platinum has been consumed. Basically all equilibrium phases may be finally reached provided that the proper metal-to-silicon ratio is used for the starting diffusion couple, as indicated for the Pt-Si system in Fig. 4. Analogous behavior may be observed for metal-metal thin-film diffusion couples, as shown in Fig. 5 for an Al-Au diffusion couple.[25]

The observed absence of phases was first attributed to experimental limitations to detect thin compound phases, but this has been ruled out with the advent of high-resolution electron microscopy. The absence of certain phases was then attributed to a nucleation barrier—e.g., to the creation of two interfaces instead of just one by the nucleation of a new phase. Ottaviani and coworkers[26-28] have shown that this is generally not the main reason for the absence of phases. They used as an example a Pt/Si diffusion couple in which the first-growing phase Pt_2Si had consumed all the platinum and the second phase PtSi had in turn consumed all the Pt_2Si. On such a PtSi/Si sample, platinum was deposited. In spite of the well-developed PtSi layer, the normal *first phase* Pt_2Si started to nucleate and grow between the platinum and the PtSi (Fig. 6). The PtSi layer shrank away. This

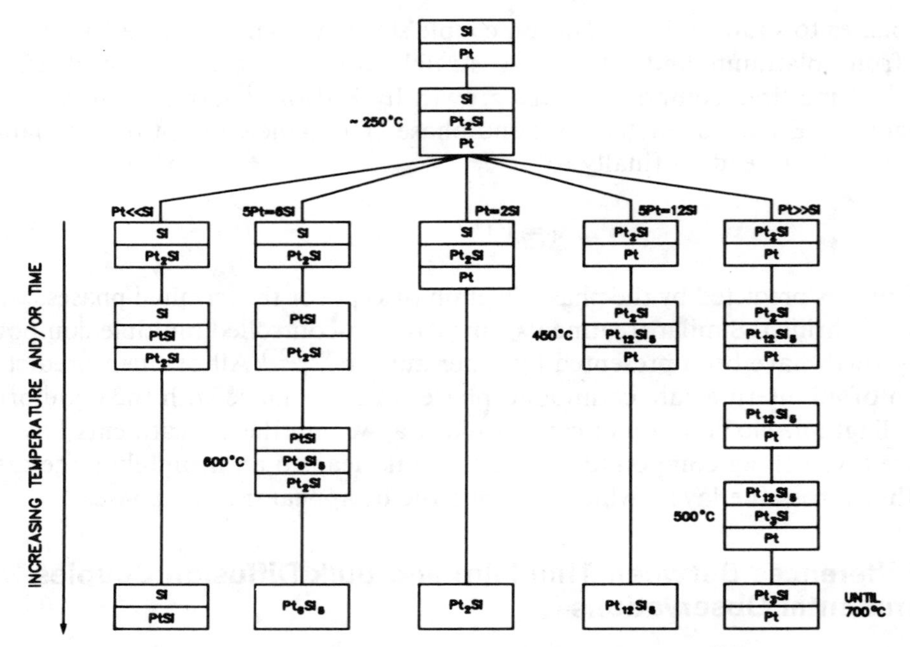

Fig. 4. Phase sequence in the Pt-Si system[24]

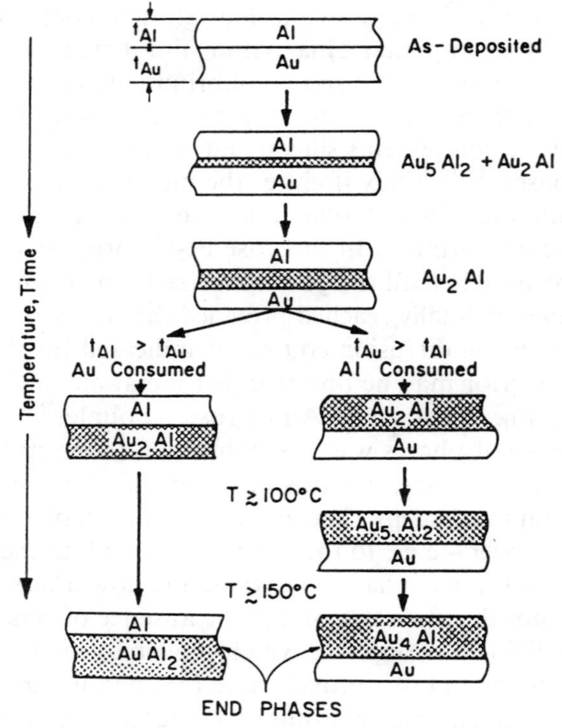

Fig. 5. Phase sequence in the Al-Au system[25]

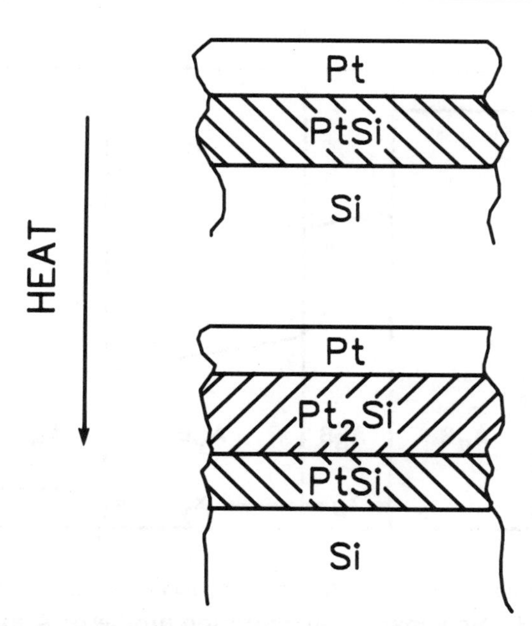

Fig. 6. Schematic illustration of experiments by Ottaviani and coworkers[26-29] showing that second-growing phase (PtSi here) starts to shrink away and the usual first-growing phase (Pt_2Si) nucleates and grows when supply phase (Pt) is redeposited

experiment clearly showed that the instability of the PtSi layer (the second-growing phase) is not due to a typical nucleation barrier (the interfaces had already successfully been created earlier) but rather is *kinetic* in nature. We therefore term this behavior a *kinetic growth instability*. In the following subsection we will describe two models for this kinetic growth instability and the resulting sequential growth of phases.

2.3. Models for Kinetic Phase Suppression

2.3.1. Concept of Critical Thickness (Serial Model)

(a) *Mathematical Treatment*: Since, within the model of diffusion-controlled growth, no compound phases predicted by the equilibrium phase diagram can disappear because otherwise the diffusion flux would grow to infinity, it was noticed that a limitation of the diffusion flux even across extremely thin layers due to *interface reaction barriers* could lead to a kinetic growth instability of these compound phases. The concept of a kinetic phase instability induced by interface reaction barriers has been developed independently by several groups[29-34] in different mathematical forms and in different contexts. In the following, we will give the essentials of a treatment by Gösele and Tu,[8] who concentrated on a simple model system consisting of two layers $A_\beta B$ and $A_\gamma B$ between two saturated phases $A_\alpha B$ and $A_\delta B$ with $\alpha > \beta > \gamma > \delta$, as indicated in Fig. 7. The starting thicknesses of the two

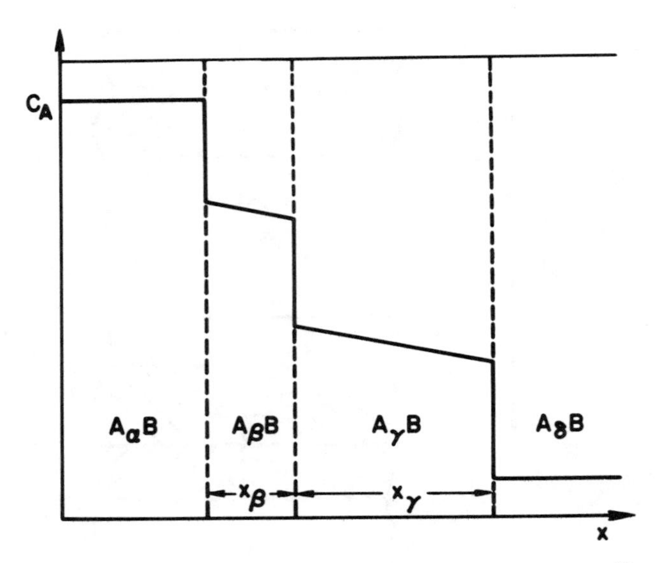

The phases $A_\alpha B$ and $A_\delta B$ are assumed to be saturated.[8]

Fig. 7. Schematic concentration profile of A atoms in an $A_\alpha B/A_\beta B/A_\gamma B/A_\delta B$ diffusion couple, where $\alpha > \beta > \gamma > \delta$

layers are parameters, and it is shown that under certain circumstances one of the layers will shrink away, which is the predicted kinetic growth instability.

The model system is described by chemical interdiffusion coefficients $\tilde{D}_\beta$ and $\tilde{D}_\gamma$ in the $A_\beta B$ and $A_\gamma B$ layers, respectively. Similarly as in Section 1, an effective reaction barrier κ_γ^{eff} is introduced for the $A_\gamma B$ layer, in addition to the effective reaction barrier κ_β^{eff} for the $A_\beta B$ layer. In both cases the effective reaction barriers are composed of the individual reaction barriers at the two interfaces, as in Eq 8. The changes in the layer thicknesses x_β and x_γ may be calculated using the same principles as discussed in Section 1. The resulting growth equations are

$$dx_\beta/dt = H_\beta j_\beta^A - H_{\beta\gamma} j_\gamma^A \quad \text{(Eq 35)}$$

and

$$dx_\gamma/dt = H_\gamma j_\gamma^A - H_{\gamma\beta} j_\beta^A \quad \text{(Eq 36)}$$

where the (positive) diffusion fluxes of A atoms in the two layers are given by

$$j_\beta^A = \Delta C_\beta^{eq} \kappa_\beta^{eff}/(1 + x_\beta \kappa_\beta^{eff}/\tilde{D}_\beta) \quad \text{(Eq 37)}$$

and

$$j_\gamma^A = \Delta C_\gamma^{eq} \kappa_\gamma^{eff}/(1 + x_\gamma \kappa_\gamma^{eff}/\tilde{D}_\gamma) \quad \text{(Eq 38)}$$

The quantities H_β, $H_{\beta\gamma}$, H_γ, and $H_{\gamma\beta}$ account for changes in composition and have been determined as

$$H_\beta = \Omega_0(1+\beta)^2\left[\frac{1}{\alpha-\beta}+\frac{1}{\beta-\gamma}\right] \quad \text{(Eq 39)}$$

$$H_\gamma = \Omega_0(1+\gamma)^2\left[\frac{1}{\beta-\gamma}+\frac{1}{\gamma-\delta}\right] \quad \text{(Eq 40)}$$

$$H_{\beta\gamma} = H_{\gamma\beta} = \Omega_0(1+\beta)(1+\gamma)/(\beta-\gamma) \quad \text{(Eq 41)}$$

As in Section 1, for simplicity the volume Ω_0 per A or B atom is assumed to be constant in all phases. As mentioned in Section 1.3, the quantities ΔC_β^{eq} and ΔC_γ^{eq} may be approximately expressed in terms of chemical potential differences across the layers. *The presence of interface reaction barriers limits the maximally attainable diffusion flux across the layers and enables one of the layers to shrink away completely.*

The conditions for the growth or shrinkage of the two layers may be investigated most conveniently in terms of the positive ratio r of the diffusion fluxes j_β^A and j_γ^A:

$$r = j_\beta^A / j_\gamma^A \quad \text{(Eq 42)}$$

The $A_\beta B$ layer grows if

$$r > r_1 = (1+\gamma)(\alpha-\beta)/(1+\beta)(\alpha-\gamma) \quad \text{(Eq 43)}$$

The $A_\gamma B$ layer grows correspondingly if

$$r < r_2 = (1+\gamma)(\beta-\delta)/(1+\beta)(\gamma-\delta) \quad \text{(Eq 44)}$$

where $r_2 > r_1$. Both r_1 and r_2 depend not on time or temperature but only on the composition factors α, β, γ, and δ. Since $r_2 > r_1$, in principle a combination of parameters x_β, x_γ, $\tilde{D}_\beta$, $\tilde{D}_\gamma$, κ_β^{eff}, and κ_γ^{eff} can exist, which allows both compound layers to grow simultaneously. If the flux ratio is *not* in the range between r_1 and r_2, one of the layers will grow whereas the other one will shrink. The various growth and shrinkage regimes are illustrated schematically in Fig. 8.

If both layers are above their change-over thicknesses x_β^* and x_γ^* ($= \tilde{D}_\gamma/\kappa_\gamma^{eff}$) their layer thicknesses tend toward a steady-state ratio as expected from the treatment of diffusion-controlled multiple layer growth. None of them will shrink away under these circumstances.

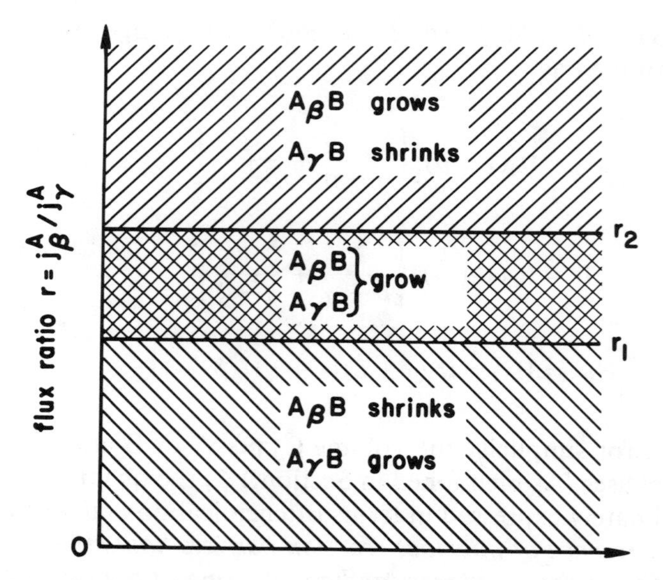

Fig. 8. Schematic illustration of various growth and shrinkage regimes for diffusion couple in Fig. 7 in dependence of ratio r of diffusion fluxes of A atoms in $A_\beta B$ layer (j^A_β) and $A_\gamma B$ layer (j^A_γ) (according to Gösele and Tu[8])

If the thicknesses of $A_\beta B$ and $A_\gamma B$ are below x^*_β and x^*_γ, respectively, then the flux ratio r is given by

$$r = \Delta C^{eq}_\beta \kappa^{eff}_\beta / \Delta C^{eq}_\gamma \kappa^{eff} \quad \text{(Eq 45)}$$

The ratio r does not depend on time. If by chance the parameters are such that r lies between r_1 and r_2, then *both* layers will grow simultaneously. In the more likely case that r lies outside this range, one of the phases will shrink away completely and thus will be *kinetically unstable*. The kinetic instability of this phase ends if the flux ratio r is shifted into the region between r_1 and r_2. This will finally happen when the growing layer, say $A_\gamma B$, exceeds its changeover thickness x^*_γ and starts to grow in a diffusion-controlled manner, with a correspondingly *decreasing* diffusion flux. This situation where $A_\gamma B$ grows diffusion-controlled is illustrated schematically as case (c) in Fig. 9. As long as x_γ is below a *critical thickness* x^{crit}_γ:

$$x^{crit}_\gamma = (r_1 \Delta C^{eq}_\gamma \tilde{D}_\gamma / \Delta C^{eq}_\beta \kappa^{eff}_\beta) \quad \text{(Eq 46)}$$

the growth of the $A_\beta B$ layer, which is controlled by an interface reaction barrier, is still suppressed, but as soon as x_γ exceeds this critical thickness the $A_\beta B$ layer can start to grow within this model. In the more general case where the growth of the $A_\gamma B$ layer is not yet completely diffusion-controlled, x^{crit}_γ is given by

$$x^{crit}_\gamma = (r_1 \Delta C^{eq}_\gamma \tilde{D}_\gamma / C^{eq}_\beta \kappa^{eff}_\beta) - \tilde{D}_\gamma / \kappa^{eff}_\gamma \quad \text{(Eq 47)}$$

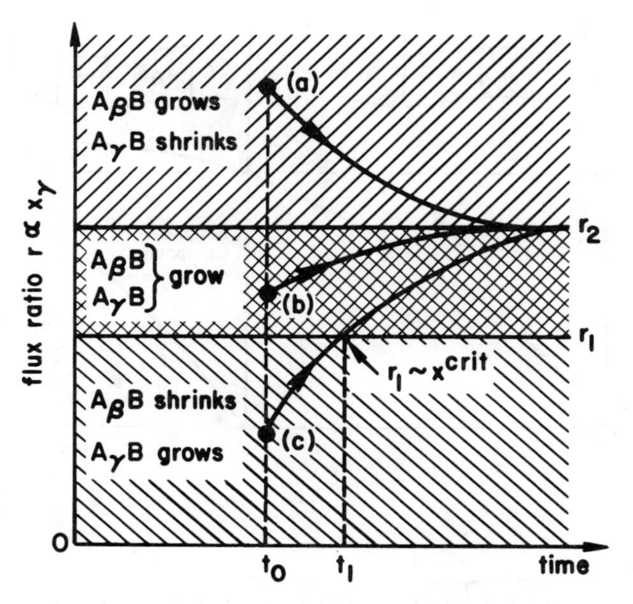

It is assumed that the transport across $A_\gamma B$ is diffusion-controlled ($x_\gamma \gg x_\gamma^*$), whereas the transport across the $A_\beta B$ layer is interface-controlled ($x_\beta \ll x_\beta^*$). At time t, in case (c) the growing $A_\gamma B$ layer has reached its critical thickness x_γ^{crit}, above which the $A_\beta B$ layer can grow.[8]

Fig. 9. Schematic illustration of growth/shrinkage behavior of diffusion couple in Fig. 7 in dependence of time

Alternatively, the second phase may start to grow if one of the supply phases has been used up, so that the flux across the first-growing layer—e.g., $A_\gamma B$—vanishes and Eq 35 and 36 reduce to

$$dx_\beta/dt = H_\beta j_\beta^A \qquad \text{(Eq 48)}$$

and

$$dx_\gamma/dt = -H_{\gamma\beta} j_\beta^A \qquad \text{(Eq 49)}$$

In thin-film growth, nucleation of the second phase is usually induced by *supply limitation* and not because the first-growing phase has exceeded its critical thickness. These two different cases are illustrated schematically in Fig. 10.

(b) *Comparison With Experimental Results*: The basic features of the serial model just discussed are in agreement with experimental results. Normally the phases grow sequentially as indicated in Fig. 4 and 5, which is the case of supply-limitation-induced second- (third-, ...) phase growth (Fig. 10b). The supply limitation may also be triggered by the presence of impurities, which may slow down the diffusion flux through the first-growing phase.[35] The observation of a criti-

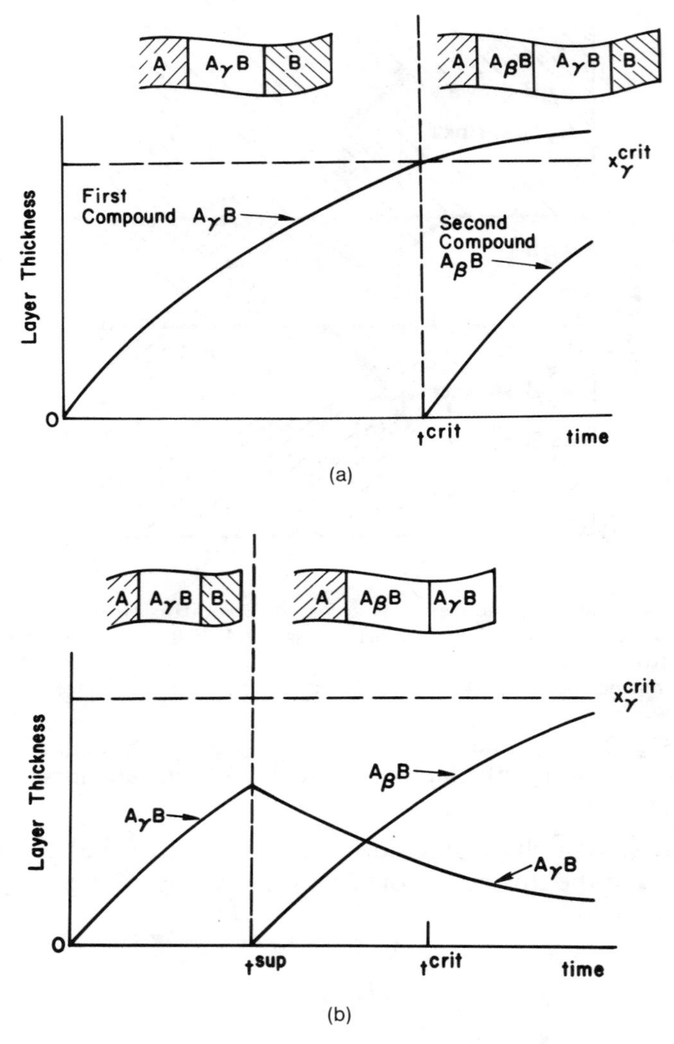

In (a), the $A_\gamma B$ layer reaches its critical thickness x_γ^{crit} at time t^{crit}, where unreacted A and B are still available. In (b), B is transformed completely into $A_\gamma B$ at time $t_0 < t^{crit}$, a case of second-phase formation induced by supply limitation.[8]

Fig. 10. Schematic illustration of growth/shrinkage behavior of first-forming compound ($A_\gamma B$) and growth of second-forming compound ($A_\beta B$) in an A/B diffusion couple, starting from the components A and B

cal thickness for the first-growing phase before a second phase starts to grow is not as straightforward, since bulk diffusion couples tend to form very rugged interfaces and finally to crack. Co/Si[36,37] and Rh/Si[38] diffusion couples show the growth of a second phase with the first-forming phase still growing. These cases have not been analyzed in terms of a critical thickness. Tu *et al.*[39] have shown that in a Ni/Si bulk diffusion couple Ni_2Si actually grows first, and NiSi grows later on.

Comparison with theoretical predictions of x^{crit} requires a knowledge of the parameters in Eq 46 or 47. Since it is not obvious whether in an A/B diffusion couple the second-growing phase is more A-rich or more B-rich than the first-growing phase, in principle one has to calculate x^{crit} for both cases. These parameters may be obtained from the growth kinetics of the first-growing phase and from the initial linear growth kinetics of the second phases under supply-limitation conditions. Suppose that $A_\gamma B$ is growing first and the diffusion-controlled kinetics may be described by

$$x_\gamma = Bt^{1/2} \qquad \text{(Eq 50)}$$

The second-growing phase (e.g., $A_\beta B$) under supply-limitation conditions should first grow linearly in time:

$$x_\beta = At \qquad \text{(Eq 51)}$$

as shown in Fig. 11 for the case of PtSi, which is the second-growing phase in a Pt/Si thin-film diffusion couple when all the metal has been consumed. Application of Eq 36 then yields

$$x_\gamma^{crit} = \frac{H_{\beta\gamma}}{H\gamma}\frac{B^2}{2A} \qquad \text{(Eq 52)}$$

The composition factor $H_{\beta\gamma}/H_\gamma$ arises because the first-growing phase grows with one different neighboring phase than in the case when both phases $A_\beta B$ and $A_\gamma B$ are present. Neglecting the composition factor $H_{\beta\gamma}/H_\gamma$, Panini *et al.*[34] deter-

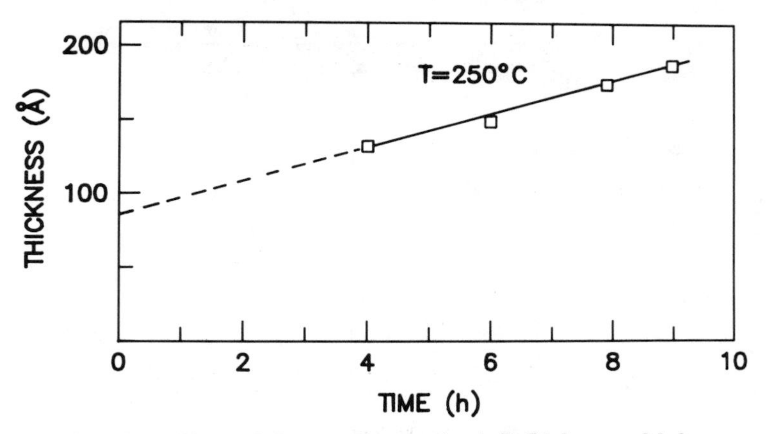

Fig. 11. Plot of (second-growing) PtSi layer thickness vs. time for 250 °C thermal annealing[34]

mined the x^{crit} value for Pt_2Si at which PtSi is expected to grow to 1.3 μm at 250 °C and analogously x^{crit} for Ni_2Si at which NiSi is expected to grow to 0.4 μm at 330 °C. For the Ni_2Si/NiSi transition, Gösele and Tu[8] had estimated about 1 μm. An experimental verification of these values has not yet been possible. A bulk Ni/Si diffusion couple showed the expected NiSi phase, but the interface was irregular and no longer flat and did not allow a quantitative comparison.[36]

In lateral diffusion couples, an example of which is shown in Fig. 12, the sequence of newly appearing phases may conveniently be observed. This has been done for the Pt/Si and Ni/Si systems.[40-42] Since the geometrical constraints and the microstructure differ from those in planar thin-film diffusion couples, no quantitative comparison with predicted x^{crit} values can be made. These experiments nevertheless show that the basic concept of critical thicknesses is the correct one.

2.3.2. Concept of Competing Nucleating Phases (Parallel Model). Recently, Williams *et al.*[43] suggested that at the very beginning of the interdiffusion process new phases may nucleate at the interface of the starting materials but may be kinetically suppressed. They deal with the case indicated in Fig. 13 and assume diffusion-controlled growth. If the starting phases—e.g., $A_\alpha B$ and $A_\gamma B$—are not yet saturated, the condition for sustained growth based on Eq 20 is for x_β, for example,

$$\frac{dx_\beta}{dt} = (j_\beta^A - j_\alpha^A)/(C_{\alpha\beta}^{eq} - C_{\beta\alpha}^{eq}) + (j_\beta^A - j_\gamma^A)/(C_{\beta\gamma}^{eq} - C_{\gamma\beta}^{eq}) > 0 \qquad \text{(Eq 53)}$$

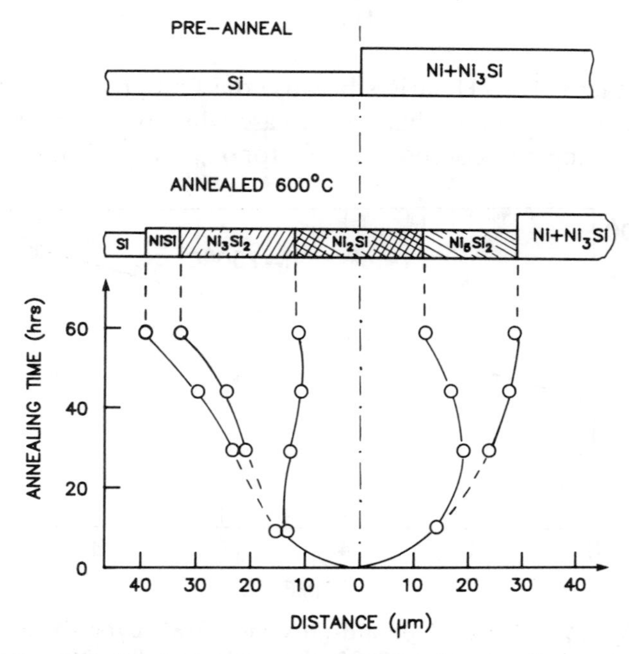

Fig. 12. Phase sequence in lateral growth experiment[40]

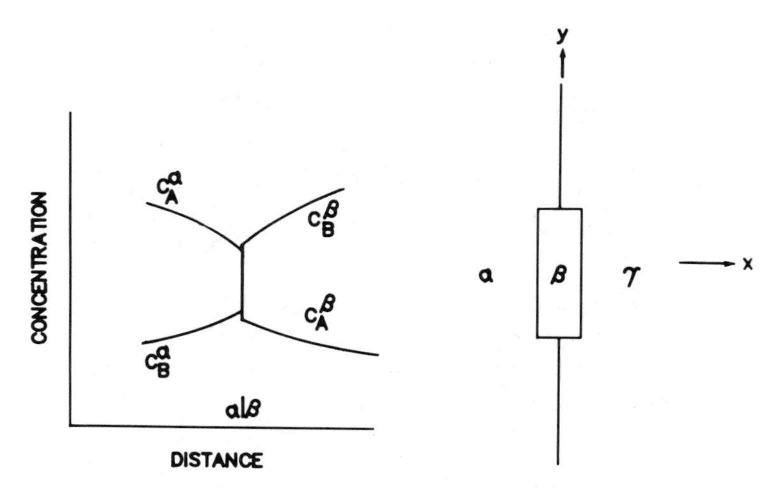

Fig. 13. Schematic diagram of model dealt with by Williams *et al.*,[43] in which a phase β ($A_\beta B$) has just nucleated between phases $A_\alpha B$ and $A_\gamma B$

or, in a formulation equivalent to that given by Williams *et al.*[43] in terms of j_β^A,

$$j_\beta^A > \frac{C_\beta - C_\gamma}{C_\alpha - C_\gamma} j_\alpha^A + \frac{C_\alpha - C_\beta}{C_\alpha - C_\gamma} j_\gamma^A \qquad \text{(Eq 54)}$$

In Eq 54 the C's refer to the average concentrations of A in compounds of narrow composition ranges.

Even if Eq 54 is not fulfilled it cannot directly lead to shrinkage of the $A_\beta B$ phase if the transport is diffusion-controlled, as discussed in earlier sections. However, Williams *et al.*[43] argue that nevertheless the α/γ interface may separate from the $A_\beta B$ phase so that finally either the α or the γ phase completely surrounds the $A_\beta B$ phase, which according to these authors would automatically lead to the dissolution of $A_\beta B$ phase. That this dissolution process would actually occur was not shown explicitly. Since for $x_\beta \rightarrow 0$, Eq 54 is always fulfilled and since the nucleating phase starts practically with $x_\beta \rightarrow 0$, it appears that the mechanism discussed by Williams *et al.*[43] is not consistent with the concept of purely diffusion-controlled growth, unless one takes into account that because of the atomistic structure x_β cannot realistically be smaller than about one lattice constant in the $A_\beta B$ structure, which again leads to a flux limitation and to a possible growth instability similar to that discussed in Section 2.2.1.

As an interesting extension of the Williams *et al.*[43] treatment we consider the parallel nucleation of *two* compound phases—e.g., $A_\beta B$ and $A_\gamma B$—at the $A_\alpha B/A_\delta B$ interface, as indicated in Fig. 14. If both $A_\beta B$ and $A_\gamma B$ grow in a diffusion-controlled manner between saturated phases $A_\alpha B$ and $A_\delta B$, then their growth will be governed by

$$x_\beta = (2\bar{H}_\beta \tilde{D}_\beta \Delta C_\beta^{eq} t)^{1/2} \qquad \text{(Eq 55)}$$

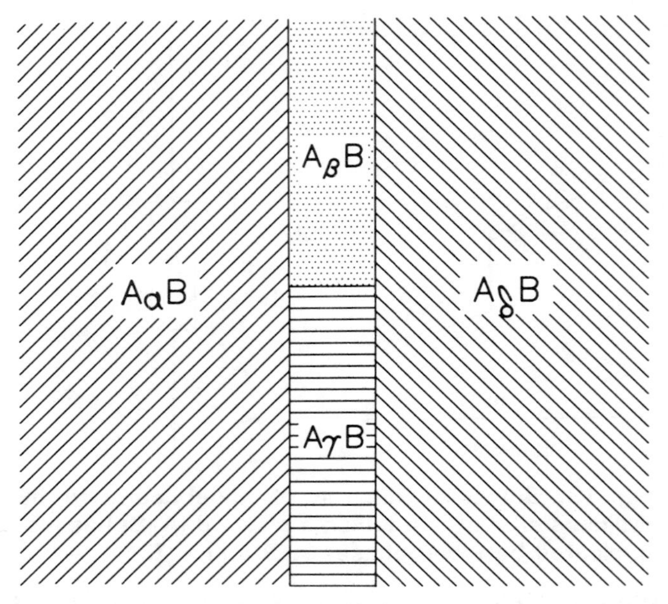

Fig. 14. Extension of the parallel-growth model of Williams *et al.*[43] to two competing phases $A_\beta B$ and $A_\gamma B$ growing between two saturated compound phases $A_\alpha B$ and $A_\delta B$ (schematic)

$$x_\gamma = (2\bar{H}_\gamma \tilde{D}_\gamma \Delta C_\gamma^{eq} t)^{1/2} \quad \text{(Eq 56)}$$

$$\bar{H}_\beta = \Omega_0(1+\beta)^2\left(\frac{1}{\alpha-\beta}+\frac{1}{\beta-\delta}\right) \quad \text{(Eq 57)}$$

$$\bar{H}_\gamma = \Omega_0(1+\gamma)^2\left(\frac{1}{\alpha-\gamma}+\frac{1}{\gamma-\delta}\right) \quad \text{(Eq 58)}$$

Based on the concept of parallel growth competition, the faster-growing compound would overtake and surround the slower-growing one, which would lead to the dissolution of the slower-growing phase. Therefore, the first-growing compound would be the compound with the largest $\tilde{D}\Delta C^{eq}$ product or, in terms of the approximation given in Eq 28, the one with the largest $\tilde{D}\Delta\mu$ product.

We finally mention again that the growth-suppression mechanism suggested by Williams *et al.*[43] implicitly requires a nucleation or an interface-reaction barrier for the phase formation, since otherwise at the appropriate interfaces the missing phases would continuously start to show up and no single-phase growth could be expected in the first place, and it would be futile to discuss the parallel competition case.

2.4. Models of First-Phase Formation

Various attempts have been made to predict the first-forming phase based on thermodynamic data. The highest gain of Gibbs' free energy per atom would result from the formation of the compound with the highest Gibbs' free energy of formation, but this turns out not to be a reliable selection criterion, as can be seen in the case of the Co/Si diffusion complex in which Co_2Si forms first, and not CoSi as expected from this criterion.[14] Walser and Bené[44,45] suggested that the first-forming phase actually will be an *amorphous* phase that in the course of growth transforms into a crystalline compound, which is closest to the lowest eutectic composition in the phase diagram. This first-phase rule is rather successful in the case of silicides but does not hold in all cases.[46] Experimental indications of an amorphous first-forming phase are limited (see, for example, Ref 47-52), which in principle might be due to the fact that the analysis usually is performed not *in situ* during formation but later on, often after rather lengthy experimental procedures, which could allow the initially amorphous phase to transform into a crystalline phase.

Recently, Bené[53] suggested that the first-forming compound in a diffusion couple will be that compound which leads to the *largest rate of decrease in Gibbs' free energy*. This rate he expressed approximately in terms of

$$\frac{d\Delta G}{dt} \propto \tilde{D}\Delta G \qquad \text{(Eq 59)}$$

and included *metastable* phases, such as *amorphous* phases, as competing phases. Although ΔG for amorphous phases is lower than for crystalline phases this lower ΔG may be overcompensated by the higher (and often much higher) diffusivities found in amorphous phases. Eq 59 basically coincides with the result of Williams *et al.*[43] (Section 2.3.2). These authors implicitly suggest the first-growing compound to be that with the largest $\tilde{D}\Delta\mu$ product, and $\Delta\mu$ and ΔG are directly related—e.g., via Eq 30. The only essential difference is the consideration of *metastable* phases by Bené.[53]

In Section 2.3.1 we concluded that a requirement for the first-growing compound is that its initial interface-reaction-controlled rate be higher than the corresponding one of the second-growing phase. We can combine this conclusion with Bené's suggestion that the rate of $d\Delta G/dt$ should be largest. Then we arrive at the more likely selection rule that the first-growing phase is that phase with the maximal

$$\frac{d\Delta G}{dt} \propto \kappa^{\text{eff}}\Delta G \qquad \text{(Eq 60)}$$

or, equivalently,

$$\frac{d\Delta G}{dt} \propto \kappa^{\text{eff}}\Delta \mu \qquad \text{(Eq 61)}$$

It can be expected that, in many cases, although not necessarily, κ^{eff} is smaller for amorphous phases since less atomic rearrangement might be involved.

In conclusion, at the present time the question of the selection rule for the first-growing compound in thin-film binary diffusion couples has still not found an answer which is generally agreed upon.

2.5. Kinetic Phase Suppression and Amorphous Thin-Film Formation

In the last few years it has been observed that *amorphous* compounds may be formed by annealing multilayer thin-film binary diffusion couples starting from crystalline materials.[54-58] Examples of the starting materials are La/Au, Ni/Zr, Co/Zr, Ni/Hf, Cu/Zr, Ti/Ni, Fe/Zr, Rh/Si, Sn/Co, and Y/Au. An overview on this subject has been given by Johnson.[59] It is tempting to consider formation of the amorphous phase in terms of the concept of kinetic suppression of compound phases dealt with in Section 2.3.1. The growth of metastable phases introduces additional complications in this concept. For example, the second-growing phase (supposed to be an equilibrium phase) may nucleate somewhere within the metastable phase and not necessarily at the interfaces. In addition, if the crystalline phase has a composition close to that of the amorphous phase, then the amorphous phase will start to shrink away instead of proceeding to grow as in the case of equilibrium compound phases. If $A_\beta B$ denotes the crystalline phase and $A_\gamma B$ the amorphous phase *with γ close to but not exactly equal to β*, then Eq 35 and 36 reduce to

$$dx_\beta/dt = H(j_\beta^A - j_\gamma^A) \quad \text{(Eq 62)}$$

and

$$dx_\gamma/dt = H(j_\gamma^A - j_\beta^A) \quad \text{(Eq 63)}$$

In Eq 62 and 63 it is assumed that the supply phases are pure A and pure B. Then H is given by

$$H \simeq \Omega_0(1 + \beta)^2/(\beta - \gamma) \quad \text{(Eq 64)}$$

From Eq 62 and 63 it is clear that only the amorphous or only the crystalline phase can grow, and not both at the same time. As long as $\beta \neq \gamma$, a critical thickness for the first-growing (say, amorphous) phase may be defined similarly as in Section 2.3.1 as

$$x_\gamma^{crit} \approx \frac{\Delta C_\gamma^{eq} \tilde{D}_\gamma}{\Delta C_\beta^{eq} \kappa_\beta^{eff}} \quad \text{(Eq 65)}$$

Whether a crystalline phase or an amorphous phase forms first will certainly depend critically on the interfacial-reaction barriers which in turn are influenced by the microstructures of the materials involved.[60] As soon as the compositions of the crystalline and amorphous phases are exactly identical ($\beta = \gamma$), the concept of a kinetic phase suppression, in the mathematical form presented in this paper, breaks down and is no longer applicable.

3. COMPOUND FORMATION WITH VOLUME CHANGES

3.1. General Remarks

The formation of compounds is generally associated with a change in the average volume per atom. For the case of oxide scales on metals for which only oxygen diffuses through the oxide scale and forms oxide at the oxide/metal interface, Pilling and Bedworth[61] pointed out in 1923 that the metal lattice would have to expand or contract to accommodate the newly formed oxide. If the volume of oxide was larger than the volume of metal consumed, then the oxide would be in compression, and in the opposite case the oxide would be in tension. The ratio:

$$R_{PB} = \frac{\text{Density of metal atoms in the metal}}{\text{Density of metal atoms in the oxide}} \qquad \text{(Eq 66)}$$

describing this volume change is called the Pilling-Bedworth ratio. Appropriately modified, it may be applied to other thin-film diffusion couples. Volume changes for silicide reactions have been compiled by Nicolet and Lau.[62]

The stresses developed may be relieved by a number of mechanisms including plastic deformation, viscous flow, or point-defect injection into the substrate.[63] If both constituents of the compound can diffuse independently of each other (e.g., as described by Eq 25 and 26), then the developing stress may induce a *coupling of the diffusion fluxes* leading to an almost stress-free compound-layer growth. Flux coupling will occur only if it is thermodynamically and kinetically more favorable than other stress-relief mechanisms. In the general case of an $A_\alpha B/A_\beta B/A_\gamma B$ diffusion couple in which volume changes may occur at both interfaces $A_\alpha B/A_\beta B$ and $A_\beta B/A_\gamma B$, stress cannot be expected to be prevented by flux coupling, since the coupling requirements at the two interfaces will usually be different. In the following subsection we will deal with the case of a transport layer in which flux coupling is likely to occur, and then we will turn to the effect of point-defect injection due to volume changes during compound formation.

3.2. Pd_2Si Transport Layer: An Example of Flux Coupling

Amorphous silicon (α-Si) deposited on Pd_2Si formed on top of crystalline silicon will shrink away, whereas the crystalline silicon substrate will grow[64,65] (see Fig. 15). The thermodynamic driving force for this recrystallization process is the lower Gibbs' free energy of the crystalline compared with the amorphous silicon. The thickness of the Pd_2Si layer does not change. It acts as a transport layer which

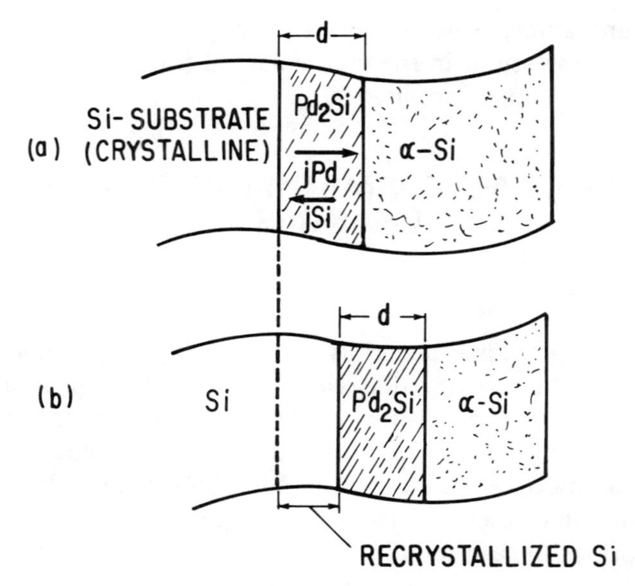

(a) Starting situation, t = 0. (b) Positions of Pd_2Si and recrystallized silicon after a certain annealing time. The fluxes of palladium (j_{Pd}) and of silicon (j_{Si}) are indicated by arrows.

Fig. 15. Schematic illustration of experiment[64] on crystallization of amorphous silicon (α-Si) via a transport layer (Pd_2Si)

shifts its position toward the surface. The growth of the crystalline substrate may occur by (1) diffusion of silicon atoms from the α-Si to the crystalline silicon and/or (2) dissociation of Pd_2Si at the crystalline Si/Pd_2Si interface, regrowth of crystalline silicon, diffusion of palladium to the Pd_2Si/Si interface, and reformation of Pd_2Si at this interface. Since the modified Pilling-Bedworth Ratio R_{PR} (density of silicon atoms in silicon to density of those in Pd_2Si) is close to 2, large stresses would develop, which can be minimized by a coupling of the palladium diffusion flux j_{Pd} to the silicon diffusion flux j_{Si} according to

$$2|j_{Si}| = (R_{PB} - 1)|j_{Pd}| \qquad \text{(Eq 67)}$$

or more generally, for a transport layer $A_\beta Si$,

$$\beta|j_{Si}| = (R_{PB} - 1)|j_A| \qquad \text{(Eq 68)}$$

Under the conditions of flux coupling, the growth rate of crystalline silicon *equals* that of the regrown $A_\beta Si$ at the $A_\beta Si/\alpha$-Si interface. This one-to-one relationship can be seen in the experimental results for the transport layer Pd_2Si, as shown in Fig. 16 for two different thicknesses, d, of the transport layer.[64] After an initial *fast* linear growth of the crystalline silicon a *slow* linear growth follows. The same fast and slow linear growth rates are observed for both Pd_2Si thicknesses, which demonstrates that the transport is an interface-controlled process. The thickness

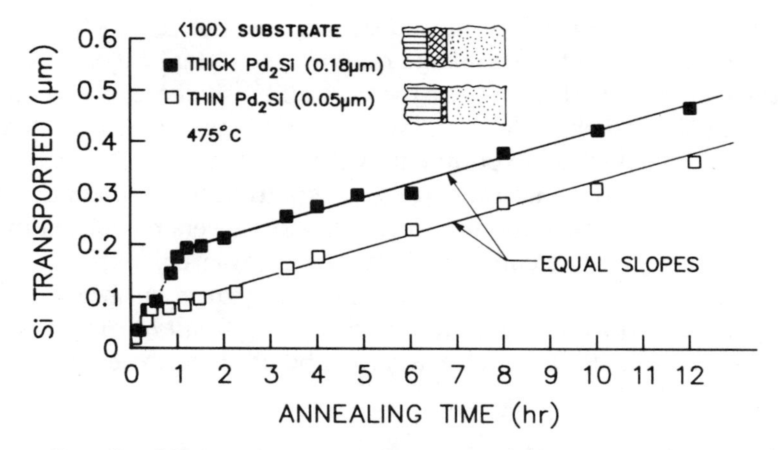

Note the different changeover thickness from the fast initial to the later slower growth rate.

Fig. 16. Equal slopes of transport curves for two different Pd_2Si thicknesses[64]

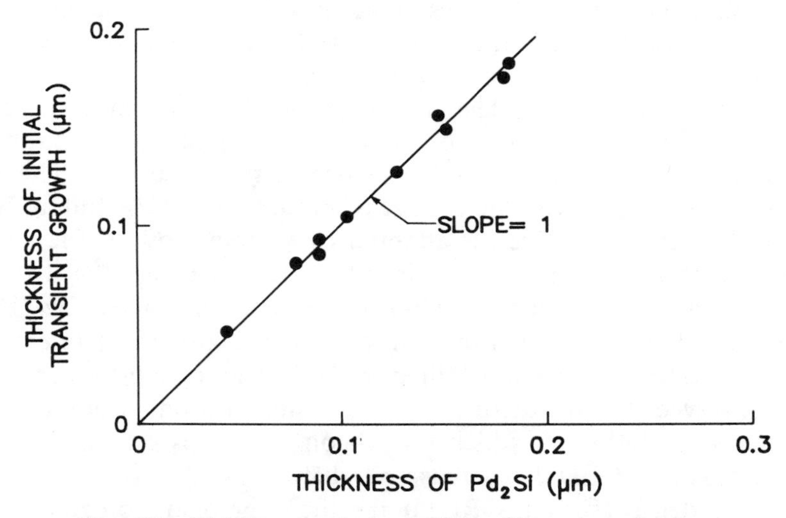

Fig. 17. Thickness of crystalline silicon grown during initial fast transient growth as a function of the thickness of the Pd_2Si transport layer[65]

of newly grown crystalline silicon at which the changeover from *fast* to *slow* linear growth occurs is *equal* to the thickness of the Pd_2Si transport layer (Fig. 17).[65] This result may be explained in terms of flux coupling: The growth rate of crystalline silicon is determined by the interface reaction (dissolution of Pd_2Si) at the crystalline Si/Pd_2Si interface. This dissolution reaction will depend on the microstructure of the Pd_2Si, which is likely to be different for the Pd_2Si formed during the initial reaction between palladium and silicon and for the Pd_2Si regrown at the Pd_2Si/α-Si interface. As soon as the crystalline Si/Pd_2Si interface encounters the regrown Pd_2Si, the growth rate is expected to change. For flux coupling, the growth rate of crystalline silicon and the regrowth rate of Pd_2Si (which is at the same time the shrinkage rate of the original Pd_2Si) are *equal*, and therefore the changeover thickness of the crystalline silicon should coincide with the thickness of the transport layer. The validity of the concept of flux coupling could be further checked by experiments with inert markers,[66-68] which would allow measurement of the flux ratio j_{Si}/j_{Pd}. Equation 67 predicts this flux ratio to be around 0.5. Unfortunately, no other transport layers have been investigated as thoroughly as Pd_2Si. The concept of flux coupling predicts for all other analogous transport layers (e.g., NiSi[69]) that the *changeover thickness* of crystalline silicon from fast to slow linear growth *will always be equal to the thickness of the transport layer*, a prediction which has yet to be checked experimentally.

3.3. Point-Defect Generation

3.3.1. Oxidation-Induced Point Defects in Silicon. Thermal oxidation of crystalline silicon is one of the most relevant processes in the fabrication of integrated circuits. Thermally grown SiO_2 on crystalline silicon is used as the insulating layer in MOS (Metal-Oxide-Silicon) transistors as well as for masking purposes. Thermal oxidation occurs by diffusion of oxygen across the SiO_2 layer and SiO_2 formation at the SiO_2/Si interface. The growth of the SiO_2 layer may approximately be described by a linear/parabolic growth law[10] analogous to Eq 12, although at small thicknesses ($\lesssim$500 Å) deviations have been established.[70,71] The SiO_2 formation is associated with a volume change of more than 100% (Pilling-Bedworth ratio $R_{PB} \approx 2.2$), most of which is adapted by viscous flow of the amorphous SiO_2. The detailed growth mechanism is still under discussion.[72,73] Part of the stress involved may also be relieved by injecting silicon self-interstitials (I) into the silicon substrate (Fig. 18), which in turn leads to the growth of *interstitial-type* stacking faults and to a change in the diffusion properties of substitutional dopants. It has been observed that the diffusion of most substitutional doping elements is *enhanced* below a window in which the silicon surface is oxidized (Oxidation-Enhanced Diffusion, OED), whereas the diffusion of antimony is *retarded* (Oxidation-Retarded Diffusion, ORD) under the same oxidation conditions (Fig. 18).[74] This observation has been rationalized in the following way. In silicon, both vacancies (V) and self-interstitials coexist under thermal equilibrium conditions with concentrations C_V^{eq} and C_I^{eq}, respectively. Surface oxidation leads to the injection of silicon self-interstitials into the substrate, which will react with the vacancies according to

$$I + V \rightleftarrows 0 \qquad \text{(Eq 69)}$$

where 0 denotes the undisturbed lattice. After a short time, local dynamic equilibrium:

$$C_I C_V = C_I^{eq} C_V^{eq} \qquad \text{(Eq 70)}$$

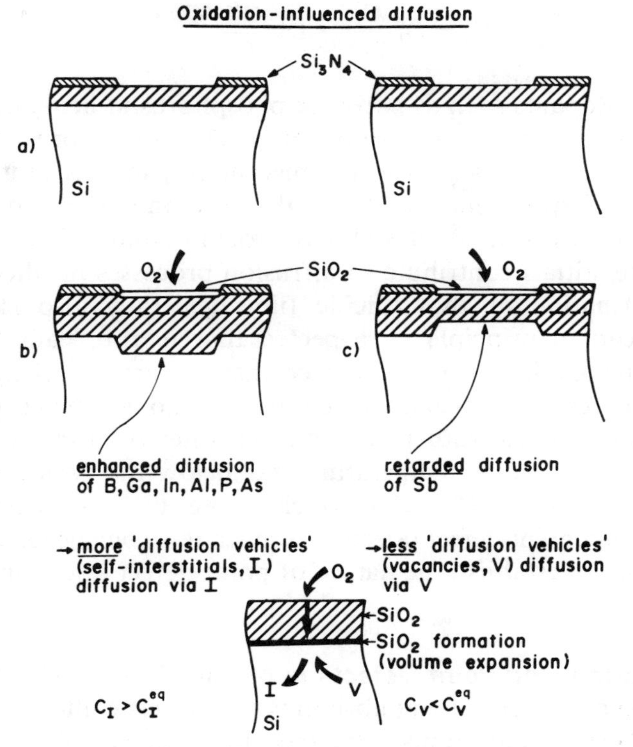

(a) Cross section of a silicon wafer doped near the surface with B, Ga, In, P, Al, As (left-hand side) or Sb (right-hand side) before oxidation. (b and c) Same cross section after surface oxidation, indicating (b) enhanced diffusion or (c) retarded diffusion. At bottom, the oxidation mechanism and the resulting point-defect injection are indicated schematically.

Fig. 18. Influence of surface oxidation on substitutional dopant diffusion in silicon[75]

will be established, where C_I and C_V are the respective actual point-defect concentrations.[74-76] Under thermal equilibrium conditions the diffusivity of substitutional dopants may be attributed partly to a component D_I^s involving self-interstitials and partly to a component D_V^s via the usual vacancy mechanism, so that[74-77]

$$D^s = D_I^s + D_V^s \qquad \text{(Eq 71)}$$

When the point-defect concentrations are perturbed due to the surface oxidation to $C_I > C_I^{eq}$ and $C_V < C_V^{eq}$, D^s changes to

$$D_{ox}^s = D_I^s \frac{C_I}{C_I^{eq}} + D_V^s \frac{C_V}{C_V^{eq}} \qquad \text{(Eq 72)}$$

or, when Eq 70 holds, to

$$D_{ox}^{s} = D_{I}^{s} \frac{C_I}{C_I^{eq}} + D_{V}^{s} \frac{C_I^{eq}}{C_I} \quad \text{(Eq 73)}$$

Based on Eq 63, dopants such as boron or phosphorus utilizing mainly self-interstitials as "diffusion vehicles" will show OED ($D_{ox}^{s} > D^{s}$) for $C_I > C_I^{eq}$, whereas dopants migrating mainly via the vacancy mechanism, such as antimony, will show ORD ($D_{ox}^{s} < D^{s}$). Experiments on OED/ORD and on the related effects of nitridation-influenced diffusion[78] played a key role in establishing that both vacancies and self-interstitials contribute to diffusion processes in silicon.

Since the formation of thin silicide films generally also involves volume changes,[62] it can in principle be expected that it also leads to point-defect injection into silicon. It has been reported that the growth of a Pd_2Si layer on crystalline silicon leads to vacancy injection,[79] although the corresponding Pilling-Bedworth ratio would rather predict a self-interstitial injection. Since most silicide reactions proceed at temperatures far below the typical oxidation temperatures of about 800 to 1100 °C at which dopant diffusion occurs readily, the evidence concerning point-defect injection is much less convincing and more experiments are needed to establish the nature of point-defect injection during growth of silicide films.

3.3.2. Oxidation-Induced Point Defects in Metals. It is commonly assumed that *vacancies* are the only thermal-equilibrium point defects contributing to diffusion processes of substitutional atoms in metals. In 1978, Schilling[80] discussed the possibility that because of their high mobility, *self-interstitials* might contribute to self-diffusion in metals. Further indications of such a self-interstitial contribution have been discussed by Gösele.[81] Since the effects of point-defect injection by surface oxidation played a major role in establishing the coexistence of vacancies and self-interstitials in silicon, it appears reasonable to look into a similar approach for metals. In principle, small substitutionally dissolved impurities would be candidates for showing diffusion effects related to self-interstitials, whereas diffusion of larger impurities is expected to be dominated by the vacancy mechanism.

Unfortunately, experiments analogous to the OED/ORD experiments in silicon are much less straightforward in metals. The OED/ORD effects in silicon are a direct reflection of oxidation-induced super- or undersaturations of intrinsic point defects, which require a sufficiently low density of point-defect sinks. The required low sink density constitutes no problem for *dislocation-free* silicon but severely limits attainable supersaturations in metals, which generally contain dislocation densities above about 10^6 cm^{-2}. The metals for which the most indications of a self-interstitial contribution to diffusion processes are available—gold, platinum, and aluminum[81]—either do not form oxides or show no sustained oxide growth, so that no corresponding experiments can be performed. The only metals for which point-defect injection has been reported are cadmium, zinc, magnesium, and nickel.[82,83] Oxidation of these metals leads to growth of vacancy-type dislocation loops, which is compatible with an oxidation-induced vacancy supersaturation.[82,83] No attempt has been made to investigate the effect of this vacancy supersaturation on diffusion of various substitutionally dissolved impurities.

Finally we mention that other chemically driven processes leading to surface layers associated with volume changes, such as corrosion, may also lead to point-defect injection. Corrosion-induced vacancy injection has been suggested for explaining corrosion-enhanced creep and fatigue at room temperature.[84-86] Since at room temperature the diffusivities of vacancies in the corresponding fcc metals are too low to account for the long-range effect of the surface corrosion, the formation of faster-diffusing divacancies has been assumed.[84] An alternative could possibly be found in terms of corrosion-induced self-interstitials, which can readily move at room temperature.

ACKNOWLEDGMENTS

The author wishes to express his appreciation for financial support by the Westinghouse Educational Foundation and the Duke Endowment during the preparation of this article and for collaboration by K.N. Tu in the area of kinetic suppression of phases and by T.Y. Tan in the area of oxidation-induced point defects in silicon.

REFERENCES

1. J.M. Poate, K.N. Tu, and J.W. Mayer (eds.), *Thin Films-Interdiffusion and Reactions*, John Wiley, New York, 1978
2. P.S. Ho and K.N. Tu (eds.), *Thin Films and Interfaces*, Vol 10, Materials Res. Soc., Pittsburgh, 1982
3. J.E.E. Baglin, D.R. Campbell, and W.K. Chu (eds.), *Thin Films and Interfaces II*, Vol 25, Materials Res. Soc., Pittsburgh, 1984
4. C.R. Aita and K.S. Sree Harsha (eds.), *Thin Films: The Relationship of Structure to Properties*, Vol 47, Materials Res. Soc., Pittsburgh, 1985
5. R.H. Nemanich, P.S. Ho, and S.S. Lau (eds.), *Thin Films-Interfaces and Phenomena*, Vol 54, Materials Res. Soc., Pittsburgh, 1986
6. M. Gibson, G.C. Osbourn, and R.M. Tromp (eds.), *Layered Structures and Epitaxy*, Vol 56, Materials Res. Soc., Pittsburgh, 1986
7. For articles, see the journal *Superlattices and Heterostructures.*
8. U. Gösele and K.N. Tu, *J. Appl. Phys.*, Vol 53, 1982, p. 3252
9. G.V. Kidson, *J. Nuclear Mater.*, Vol 3, 1961, p. 21
10. B.E. Deal and Grove, *J. Appl. Phys.*, Vol 36, 1965, p. 3770
11. H.H. Farrell, G.H. Gilmer, and M. Suenaga, *Thin Solid Films*, Vol 25, 1975, p. 253
12. V.I. Dybkov, *J. Mater. Sci.*, Vol 21, 1986, p. 3078
13. K.N. Tu and J.W. Mayer, in Ref 1, p. 1
14. G. Ottaviani, *Thin Solid Films*, Vol 140, 1986, p. 3
15. H. Schmalzried, *Solid State Reactions*, Academic Press, New York, 1974
16. K.N. Tu, *Ann. Rev. Mater. Sci.*, Vol 15, 1985, p. 147
17. A.L. Geer, *Scripta Met.*, Vol 20, 1986, p. 4571

18. C. Wagner, *Acta Metall.*, Vol 17, 1969, p. 99
19. A.J. Hickl and R.W. Heckel, *Metall. Trans.*, Vol 6A, 1975, p. 431
20. S.R. Shatynski, J.P. Hirth, and R.A. Rapp, *Acta Metall.*, Vol 24, 1976, p. 1071
21. A.T. Fromhold, Jr., *J. Chem. Phys.*, Vol 76, 1982, p. 4260
22. H.S. Hsu, *Oxidation of Metals*, Vol 26, 1986, p. 315
23. R.S Claassen, *Phys. Today*, November 1976, p. 23 (data from W.F. Chambers and E.L. Burgess, unpublished work, 1975)
24. G. Majni, M. Costato, F. Panini, and G. Celotti, *J. Phys. Chem. Solids*, Vol 46, 1985, p. 631
25. G. Majni, G. Ottaviani, and E. Galli, *J. Cryst. Growth*, Vol 47, 1979, p. 583
26. G. Ottaviani, *J. Vac. Sci. Technol.*, Vol 16, 1979, p. 1112
27. G. Ottaviani and M. Costato, *J. Cryst. Growth*, Vol 45, 1978, p. 365
28. G. Ottaviani, G. Majni, and C. Canali, *J. Appl. Phys.*, Vol 18, 1979, p. 285
29. Ya. Ye Geguzin, Yu.S. Kaganovskiy, L.M. Paritskaya, and V.I. Solunskiy, *Phys. Met. Metall.*, Vol 47, 1980, p. 127
30. M. Costato, *Lettere al Nuovo Cimento*, Vol 32, 1981, p. 219
31. F.M. d'Heurle and P. Gas, *J. Mater. Res.*, Vol 1, 1986, p. 205
32. V.I. Dybkov, *J. Mater. Sci.*, Vol 21, 1986, p. 3085
33. V.I. Dybkov, *J. Phys. Chem. Solids*, Vol 47, 1986, p. 735
34. F. Panini, M. Costato, and G. Majni, *Il Nuovo Cimento*, Vol 7, 1986, p. 241
35. F. Nava, S. Valeri, G. Majni, A. Cembili, G. Pignatel, and G. Queirolo, *J. Appl. Phys.*, Vol 52, 1981, p. 6641
36. K.N. Tu, G. Ottaviani, R.D. Thompson, and J.W. Mayer, *J. Appl. Phys.*, Vol 53, 1982, p. 4406
37. C.D. Lien, M.-A. Nicolet, C.S. Pai, and S.S. Lau, *Appl. Phys.*, Vol 36, 1985, p. 153
38. S. Peterson, R. Anderson, J.E.E. Baglin, J. Dempsey, W. Hammer, F.M. d'Heurle, and S. LaPlaca, *J. Appl. Phys.*, Vol 51, 1980, p. 373
39. K.N. Tu, G. Ottaviani, U. Gösele, and H. Föll, *J. Appl. Phys.*, Vol 54, 1983, p. 758
40. L.R. Zheng, E.C. Zingu, and J.W. Mayer, *J. Vac. Sci. Technol.*, Vol A1, 1985, p. 758
41. L.R. Zheng, L.S. Hung, J.W. Mayer, G. Majni, and G. Ottaviani, *Appl. Phys. Lett.*, Vol 41, 1982, p. 646
42. G. Majni, F. Panini, G. Sodo, and P. Cantoni, *Thin Solid Films*, Vol 125, 1985, p. 313
43. D.S. Williams, R.A. Rapp, and J.P. Hirth, *Thin Solid Films*, Vol 142, 1986, p. 47
44. R.M. Walser and R.W. Bené, *Appl. Phys. Lett.*, Vol 28, 1976, p. 624
45. R.W. Bené and R.M. Walser, *J. Vac. Sci. Technol.*, Vol 14, 1977, p. 925
46. R.D. Thompson, B.Y. Tsaur, and K.N. Tu, *Appl. Phys. Lett.*, Vol 38, 1981, p. 535
47. J.W. Allen, A.C. Wright, and G.A.N. Connell, *J. Non-Cryst. Solids*, Vol 42, 1980, p. 509
48. R.W. Bené and H.Y. Yang, *J. Electron. Mater.*, Vol 12, 1983, p. 1
49. R. Beyers and R. Sinclair, *J. Appl. Phys.*, Vol 57, 1985, p. 5240
50. M. Natan, *Appl. Phys. Lett.*, Vol 49, 1986, p. 257
51. M. Natan, *J. Vac. Sci. Technol.*, Vol B4, 1986, p. 1404
52. M.O. Aboelfotoh, A. Alessandrini, and F.M. d'Heurle, *Appl. Phys. Lett.*, Vol 49, 1986, p. 1242
53. R.W. Bené, *J. Appl. Phys.*, Vol 61, 1987, p. 1826
54. R.B. Schwarz and W.L. Johnson, *Phys. Rev. Lett.*, Vol 51, 1983, p. 415
55. B.M. Clemens, W.L. Johnson, and R.B. Schwarz, *J. Non-Cryst. Solids*, Vol 61-62, 1984, p. 817
56. M. Atzmon, J.D. Verhoeven, E.D. Gibson, and W.L. Johnson, *Appl. Phys. Lett.*, Vol 45, 1984, p. 1052
57. H. Schroder, K. Samwer, and U. Koster, *Phys. Rev. Lett.*, Vol 54, 1985, p. 197
58. B.C. Clemens, *Phys. Rev. B*, Vol 33, 1986, p. 7615

59. W.L. Johnson, *Progr. Mater. Sci.*, Vol 30, 1986, p. 81
60. A.M. Vredenberg, J.F.M. Westendorp, F.W. Saris, N.M. van der Pers, and Th.H. de Keijser, *J. Mater. Res.*, Vol 1, 1986, p. 774
61. N.B. Pilling and R.E. Bedworth, *J. Inst. Met.*, Vol 29, 1923, p. 529
62. M.-A. Nicolet and S.S. Lau, in *VLSI Electronics Microstructure Science*, Vol 6, edited by N.G. Einspruch and G.B. Larrabee, Adademic Press, New York, 1983, pp. 457 and 458
63. For an overview, see S.R.J. Saunders, *Sci. Progr. Oxf.*, Vol 63, 1976, p. 163
64. S.S. Lau and W.F. van der Weg, in Ref 1, p. 433
65. Z.L. Liau, S.C. Campisano, C. Canali, S.S. Lau, and J.W. Mayer, *J. Electrochem. Soc.*, Vol 122, 1975, p. 1696
66. U. Gösele, K.N. Tu, and R.D. Thompson, *J. Appl. Phys.*, Vol 53, 1982, p. 8759
67. K.N. Tu and J.W. Mayer, in Ref 1, p. 359
68. E.G. Colgan and J.W. Mayer, *J. Mater. Res.*, Vol 1, 1986, p. 786
69. Z.L. Liau, S.S. Lau, M.-A. Nicolet, and J.W. Mayer, in *Proceedings of the 1st ERDA Semiannual Solar Photovoltaic Conversion Program Conference*, UCLA, Los Angeles, July 22-25, 1975
70. H.Z. Massoud, J.D. Plummer, and E.A. Irene, *J. Electrochem. Soc.*, Vol 132, 1985, p. 1745
71. H.Z. Massoud, J.D. Plummer, and E.A. Irene, *J. Electrochem. Soc.*, Vol 132, 1985, pp. 2685 and 2693
72. N.F. Mott, *Phil. Mag.*, Vol B55, 1987, p. 117
73. B. Leroy, *Phil. Mag.*, Vol B55, 1987, p. 159
74. For an overview and references, see T.Y. Tan and U. Gösele, *Appl. Phys. A*, Vol 37, 1985, p. 1
75. U. Gösele, in *Advances in Solid State Physics*, Vol XXVI, 1986, p. 89
76. D.A. Antoniadis and I. Moskowitz, *J. Appl. Phys.*, Vol 53, 1982, p. 6988
77. S.M. Hu, *J. Appl. Phys.*, Vol 45, 1974, p. 1567
78. P. Fahey, G. Barbuscia, M. Moslehi, and R.W. Dutton, *Appl. Phys. Lett.*, Vol 46, 1985, p. 784
79. I. Ohdomari, K. Konuma, M. Takano, T. Chikyow, H. Kawarada, J. Nakanishi, and T. Ueno, in Ref 5, p. 63
80. W. Schilling, *J. Nat. Mat.*, Vol 69-70, 1978, p. 3
81. U. Gösele, *Scripta Met.*, Vol 19, 1985, p. 93
82. For an overview, see J.E. Harris, *Acta Met.*, Vol 26, 1978, p. 1033
83. C.G. Deacon, M.H. Loretto,and R.F. Smallman, in *Proc. Int. Conf. Point Defects and Defect Interactions in Metals*, edited by J. Takamura, M. Doyama, and M. Kiritani, Univ. Tokyo Press, Tokyo, 1982, p. 253
84. H. Pickering and C. Wagner, *J. Electrochem. Soc.*, Vol 114, 1967, p. 698
85. H. Pickering, *J. Electrochem. Soc.*, Vol 117, 1970, p. 8
86. H. Uhlig, *J. Electrochem. Soc.*, Vol 123, 1976, p. 1699

APPENDIX
General Reference List

Alloying Behavior and Effects in Concentrated Solutions, edited by T.B. Massalski, Metallurgical Society of AIME, Gordon and Breach, New York, 1965, 445 pages
The Theory of Order-Disorder in Alloys, edited by M.A. Krivoglaz and A.A. Smirnov, American Elsevier Pub. Co., New York, 1965, 427 pages
Band Structure Spectroscopy of Metals and Alloys, edited by D.J. Fabian and L.M. Watson, Academic Press, London, 1973, 753 pages
A Configurational Model of Matter, edited by G.V. Samsonov, I.F. Prkiladko, and L.F. Prkiladko, Consultants Bureau, New York, 1973, 289 pages
Theory of Alloy Phases, American Society for Metals, Cleveland, 1956, 378 pages
Energy Bands in Metals and Alloys, edited by L.H. Bennett and J.T. Waber, Metallurgical Society of AIME, Gordon and Breach, New York, 1968, 185 pages
Ordered Alloys: Structural Applications and Physical Metallurgy, edited by B.H. Kear, J.H. Westbrook, and C.T. Sims, Metallurgical Society of AIME, Claitor Publications, Baton Rouge, LA, 1970, 580 pages
The Superalloys, edited by C.T. Sims and W.C. Hagel, Wiley-Interscience, New York, 1972, 614 pages
Calculation of the Thermal Effect Due to the Addition of Alloying Elements to a Steel Bath, by J.-M. Steiler, *Cent. Doc. Sider. Cir. Inf. Tech.*, Vol 36, No. 3, 1979, pp. 385-407 (in French)
The Determination of the Partial Enthalpies of Mixing of Aluminum-Rich Alloy Metals by Solution Calorimetry, by J.-J. Lee and F. Sommer, *Z. Metallkd.*, Vol 76, No. 11, November 1985, pp. 750-754 (in German)
Titanium and Titanium Alloys: Scientific and Technological Aspects, Vol 3, edited by A.I. Khorev, Moscow, 18-21 May 1976, Plenum Press, New York, 1982
Physical Chemistry of Liquid Metal Solutions, edited by J.F. Elliott, Metallurgical Treatises, Beijing, 13-22 Nov 1981, TMS-AIME, 1981
On the Interpretation of Alloying Tendencies and Impurity Heats of Solutions, by C.H. Hodges, *Philos. Mag. B*, Vol 38, No. 3, September 1978, pp. 205-220
Linear Augmented Slater-Type-Orbital Study of Au-5d-Transition-Metal Alloying, by R.E. Watson, J.W. Davenport, and M. Weinert, *Phys. Rev. B, Condens. Matter*, Vol 35, No. 2, 15 Jan 1987, pp. 508-518
Generalization of the Coherent-Potential Approximation to Compositionally Modulated Alloys, by A. Gonis and N.K. Flevaris, *Phys. Rev. B, Condens. Matter*, Vol 25, No. 12, 15 June 1982, pp. 7544-7557
On the Interpretation of Miedema's Alloying Rules, by C.H. Hodges, in *Theory of Alloy Phase Formation*, New Orleans, 19-20 Feb 1979, TMS-AIME, Warrendale, PA, 1980
What's Special About Transition Metals in Alloy Phase Formation?, by R.E. Watson and L.H. Bennett, in *Theory of Alloy Phase Formation*, New Orleans, 19-20 Feb 1979, TMS-AIME, Warrendale, PA, 1980
L.S. Darken's Contributions to the Theory of Alloy Formation and Where We Are Today,

by K.A. Gschneider, Jr., in *Theory of Alloy Phase Formation*, New Orleans, 19-20 Feb 1979, TMS-AIME, Warrendale, PA, 1980

Electronegativity Scale for Metals, by J.A. Alonso and L.A. Girifalco, *Phys. Rev. B, Condens. Matter*, Vol 19, No. 8, 15 Apr 1979, pp. 3889-3985

Two-Particle Green's Function of Electrons in Binary Metal Alloys, by S.L. Ginzburg, *Sov. Phys. Solid State*, Vol 16, No. 1, July 1974, pp. 5-10

Solution Kinetics of Alloying Additions, by P.M. Shurygin and V.D. Shantarin, paper from *Teoriya i Prakt Intens Prots v Konvert i Martenov Pech Metallurgiya*, Moscow, 1965, pp. 64-71

Theory of Critical Alloying, by A.P. Gulyaev, *Metalloved Term Obrabotka Metall*, No. 8, 1965, pp. 20-25

Development Trends in High Strength, Weldable Structural Steels Alloyed with Nb, V, N, Ti, Zr, and Other Elements, by F. Stadler, *Neue Hutte*, Vol 11, No. 10, October 1966, pp. 606-608

Theory of the Alloying Limit, by A.P. Gulyaev, *Metal Sci. Heat Treat.*, No. 7-9, July-August 1965, pp. 501-505

Alloying of Structural Steel, by A.Ya. Dontsova, *Metalloved Term Obrabotka Metall*, No. 9, 1965, pp. 14-17

Electron Cell Model of Alloys, by P. Bolsaitis and L. Skolnick, *Trans. Met. Soc. AIME*, Vol 242, No. 2, February 1968, pp. 215-225

Alloying Behavior of Ni/Au-Ge Films on GaAs, by M. Ogawa, *J. Appl. Phys.*, Vol 51, No. 1, January 1980, pp. 406-412

Resistometric Investigation of Interdiffusion in Thin Couples of Cu and Ni, by A. Wagendristel, H. Bangert, and K. Esmailpour-Kazerouni, in *Proceedings of the Seventh International Vacuum Congress and the Third International Conference on Solid Surfaces Vol III*, Osterreichische Studiengesellschaft fur Atomenergie Ges m b H Vienna, Austria, 1977, pp. 2103-2105

Alloying During Milling of Mixed Metal Powders, by A. Wagendristel, K. Persy, and E. Tschegg, *Planseeber Pulvermet*, Vol 21, No. 4, December 1973, pp. 241-245 (in German)

Kinetics of Alloy Formation in a Metal-Ge System, by A.M. Orlov, P.M. Shurygin, and N.I. Shadeev, *Izvest Akad Nauk SSSR Neorg Materialy*, Vol 9, No. 9, September 1973, pp. 1477-1480 (in Russian)

Electron Concentration as a Guide to Alloying Behaviour: Engel-Brewer and the Platinum Metals, by A.S. Darling, *Platinum Metals Rev.*, Vol 13, No. 2, April 1969, pp. 53-56

Index

NOTE: The symbol (F) or (T) following a page number indicates that information is presented in a figure or a table, respectively.